THE OCEANS: PHYSICAL-CHEMICAL DYNAMICS AND HUMAN IMPACT

The Pennsylvania Academy of Science Publications

Books and Journal

Editor: Shyamal K. Majumdar
Professor of Biology, Lafayette College
Easton, Pennsylvania 18042

1. *Energy, Environment, and the Economy,* 1981. ISBN: 0-9606670-0-8. Editor: S.K. Majumdar.
2. *Pennsylvania Coal: Resources, Technology and Utilization,* 1983. ISBN: 0-9606670-1-6. Editors: S.K. Majumdar and E.W. Miller.
3. *Hazardous and Toxic Wastes: Technology, Management and Health Effects,* 1984. ISBN: 0-9606670-2-4. Editors: S.K. Majumdar and E. W. Miller.
4. *Solid and Liquid Wastes: Management, Methods and Socioeconomic Considerations,* 1984. ISBN: 0-9606670-3-2. Editors: S.K. Majumdar and E.W. Miller.
5. *Management of Radioactive Materials and Wastes: Issues and Progress,* 1985. ISBN: 0-9606670-4-0. Editors: S.K. Majumdar and E.W. Miller.
6. *Endangered and Threatened Species Programs in Pennsylvania and Other States: Causes, Issues and Management,* 1986. ISBN: 0-9606670-5-9. Editors: S.K. Majumdar, F.J. Brenner, and A.F. Rhoads.
7. *Environmental Consequences of Energy Production: Problems and Prospects,* 1987. ISBN: 0-9606670-6-7. Editors: S.K. Majumdar, F.J. Brenner and E.W. Miller.
8. *Contaminant Problems and Management of Living Chesapeake Bay Resources,* 1987. ISBN: 0-9606670-7-5. Editors: S.K. Majumdar, L.W. Hall, Jr. and H.M. Austin.
9. *Ecology and Restoration of The Delaware River Basin,* 1988. ISBN: 0-9606670-8-3. Editors: S.K. Majumdar, E.W. Miller and L.E. Sage.
10. *Management of Hazardous Materials and Wastes: Treatment, Minimization and Environmental Impacts,* 1989. ISBN: 0-9606670-9-1. Editors: S.K. Majumdar, E.W. Miller and R.F. Schmalz.
11. *Wetlands Ecology and Conservation: Emphasis in Pennsylvania,* 1989. ISBN: 0-945809-01-8. Editors: S.K. Majumdar, R.P. Brooks, F.J. Brenner and R.W. Tiner, Jr.
12. *Water Resources in Pennsylvania: Availability, Quality and Management,* 1990. ISBN: 0-945809-02-6. Editors: S.K. Majumdar, E.W. Miller and R.R. Parizek.
13. *Environmental Radon: Occurrence, Control and Health Hazards,* 1990. ISBN: 0-945809-03-4. Editors: S.K. Majumdar, R.F. Schmalz and E.W. Miller.
14. *Science Education in the United States: Issues, Crises, and Priorities,* 1991. ISBN: 0945809-04-2. Editors: S.K. Majumdar, L.M. Rosenfeld, P.A. Rubba, E.W. Miller and R.F. Schmalz.
15. *Air Pollution: Environmental Issues and Health Effects,* 1991; ISBN: 0-945809-05-0. Editors: S.K. Majumdar, E.W. Miller, and J.J. Cahir.
16. *Natural and Technological Disasters: Causes, Effects and Preventive Measures*, 1992; ISBN: O-945809-06-9. Editors: S.K. Majumdar, G.S. Forbes, E.W. Miller, and R.F. Schmalz.
17. *Global Climate Change: Implications, Challenges and Mitigation Measures*, 1992; ISBN: 0-945809-07-7. Editors: S.K. Majumdar, L.S. Kalkstein, B. Yarnal, E.W. Miller, and L.M. Rosenfeld.
18. *Conservation and Resource Management,* 1993; ISBN: 0-945809-08-5. Editors: S.K. Majumdar, E.W. Miller, D.E. Baker, E.K. Brown, J.R. Pratt, and R.F. Schmalz.
19. *Biological Diversity: Problems and Challenges,* 1994; ISBN: 0945809-09-3. Editors: S.K. Majumdar, F.J. Brenner, J.E. Lovich, J.F. Schalles, and E.W. Miller.
20. *The Oceans: Physical-Chemical Dynamics and Human Impact,* 1994; ISBN: 0945809-10-7. Editors: S.K. Majumdar, E.W. Miller, G.S. Forbes, R.F. Schmalz, and A.A. Panah.

THE OCEANS: PHYSICAL-CHEMICAL DYNAMICS AND HUMAN IMPACT

EDITED BY

S.K. MAJUMDAR, Professor and Head
Department of Biology
Lafayette College
Easton, PA 18042

E.W. MILLER, Professor and Associate Dean (Emeritus)
The Pennsylvania State University
University Park, PA 16802

G.S. FORBES, Associate Professor of Meteorology
The Pennsylvania State University
University Park, PA 16802

R.F. SCHMALZ, Professor of Geoscience (Emeritus)
The Pennsylvania State University
University Park, PA 16802

A.A. PANAH, Professor of Geology and Environmental Science
University of Pittsburgh at Bradford
Bradford, PA 16701

Founded on April 18, 1924

A Publication of
The Pennsylvania Academy of Science

Library of Congress Cataloging in Publication Data

Bibliography
Index
Majumdar, Shyamal K. 1938-, ed.

Library of Congress Catalog Card No.: 94-67523

ISBN-0-945809-10-7

A Publication of
The Pennsylvania Academy of Science, Easton, PA 18042

Printed in the United States of America by

Typehouse of Easton
Easton, PA 18042

PREFACE

Earth is a unique planet in our solar system because of its interconnected bodies of salt water, which cover over 71 percent of its surface—the hydrosphere. With an average depth of four kilometers, these oceans comprise 99 percent of the life supporting space on our planet. Ocean basins hold nearly limitless water with high kinetic energy having a potential of producing the most important source of food, mineral deposits, and energy resources—some yet to be discovered.

It is now recognized that the oceans can be altered by man's activities. These activities will affect not only the oceans, but also the land, as the oceans are a major determinant of the climate throughout the world. A major objective of this book is not only to describe the physical aspects of the oceans, but also to provide some interpretations and analyses as to how man has impacted these vast water bodies.

Topics in this book were chosen to reach a balance between physical dynamics, climatic effects, and human relevance of ocean environments. The most important dynamic processes continuously at work in the marine environment have been examined in the light of modern physical-chemical dynamics and modeling techniques. Oceans interact with surrounding environments, such as the biosphere and atmosphere, while oceanic processes relate to astronomical forces. Human impact on the oceanic environment has been specifically covered in terms of physical-chemical and biological contamination and modification.

This book is divided into three parts. Part one consists of eleven chapters considering such aspects as the thermodynamics and physical dynamics of the oceans, modeling of ocean currents, Gulf Stream processes and oceanic remote sensing. The second part treats a number of aspects of climate including climatic control, heat and moisture transfer and the El Niño effects. These eight chapters show how the variation of the ocean surface temperature affects climatic patterns of the earth. The last section examines the human impact on the ocean environment. The pollution problems are investigated, including oil pollution, waste dumping, chemical and metal pollution and the management of radioactive wastes.

This book is written for those who come in contact with the marine environment and for students and professionals concerned with the physical environment of the earth's surface. The information presented is controlled by size and page limitations, but each section has complete references by topic which interested readers may pursue.

The contributors of this book hope that their insights will be helpful in better understanding the physical, chemical, and biological dynamics of the ocean and will enhance assessments toward protection of the ocean environment. The authors are grateful to Lafayette College, the Pennsylvania State University and the University of Pittsburgh at Bradford for providing facilities for their work. They also extend heartfelt thanks to their wives for their encouragement, help, and patience in the preparation of this book.

S.K. Majumdar
E.W. Miller
G.S. Forbes
R.F. Schmalz
A.A. Panah
Editors
July, 1994

ACKNOWLEDGEMENTS

The publication of this book was aided by contributions from The Pennsylvania Power and Light Company, Allentown, Pennsylvania and E. Willard and Ruby S. Miller book publication fund.

The Oceans: Physical-Chemical Dynamics and Human Impact

Table of Contents

Part Three: Human Impact in Ocean Environment

The Oceans: Physical-Chemical Dynamics and Human Impact. Edited by S.K. Majumdar, E.W. Miller, G.S. Forbes, R.F. Schmalz and Assad A. Panah.

Chapter One

THERMODYNAMIC QUANTITIES IN OCEANOGRAPHY

RAINER FEISTEL and **EBERHARD HAGEN**
Institut für Ostseeforschung
Seestr, 15
D18119 Warnemünde
Germany

INTRODUCTION

During the last decades oceanography faced a rapid progress in measurement techniques along with enhanced numerical possibilities of data treatment and dynamical simulation, both especially on the background of urgent questions about climatology and global change. On the other hand there is an obvious calmness in thermodynamics of seawater, concerning basic quantitative knowledge of seawater properties as well as the incorporation of a wider thermodynamic view into the interpretation of field data. Here we shall briefly summarize the international standard by UNESCO formulas, its inconsistencies and possibilities of improvement. We discuss problems related to the classical concepts of potential temperature and density within the general thermodynamic framework. In conviction that such a review must look beyond textbook knowledge we finally try to outline few aspects of extended use of thermodynamics to interpret observed oceanic properties.

EOS 80 AND UNESCO 83 FORMULAS

Sea-going oceanography measures oceanic water properties usually by CTD-probes, where in-situ values of pressure (p) or depth, temperature (t) and electrical conductivity (and sometimes sound speed) are recorded. Assuming a standard chemical composition of seasalt a salinity figure S is computed. This way the classical triple (S,t,p) of independent thermodynamic variables is obtained, which in turn serves as starting point for the calculation of various other quantities. The practical present-day standard for these computations is given by UNESCO formulas (Fofonoff and Millard, 1983) independently for the Equation Of State (EOS 80), Heat Capacity at Constant Pressure (CP 83) and Sound Velocity (SVEL 83), which we intend to discuss briefly now.

First of all, these formulas do not provide any of the thermal fundamental quantities like Internal Energy, Entropy or Enthalpy, which are essential for a deeper physical understanding of energetic and irreversible properties of the ocean. Secondly, even the so-called chemical properties like chemical potential, osmotic and activity coefficients, which play an important role especially for molecular mixing phenomena, remain unknown.

Finally, the Pressure-Volume-Temperature (PVT) properties evaluated by the UNESCO formulas exhibit inconsistencies that, although almost negligible for most practical purposes, may become crucial for certain applications as computing vertical stability or determining salinity from sound speed. To understand the origin of this problem one has to look in more detail onto the way the formulas had been obtained. There are three independent data sources, density of seawater at one atmosphere at different temperatures and salinities, its heat capacity at atmospheric pressure depending on temperature and salinity, and sound speed. The latter then had been used to extend the first two quantities to their high-pressure values as given by EOS 80 and CP 83. If one now takes these two formulas in order to recompute sound speed, the result differs by about 0.3 m/s from the sound speed values originaly used (Fofonoff and Millard, 1983), which is rather poor compared to the precision of sound speed measurements. As we shall see later, this is actually not a thermodynamic problem but results merely from the algorithm used to compute EOS 80.

In addition to these mechanical PVT properties two chemical quantities were published (Millero, Hansen and Hoff 1973, Millero and Leung 1976, Millero 1983), dilution heat on one hand and osmotic coefficient (based on freezing point depression measurements by Doherty and Kester 1974) of seawater at one atmosphere on the other hand. Combining them properly with the formulas mentioned above leads to a rather complete description of seawater properties at local thermodynamic equilibrium.

GIBBS THERMODYNAMIC POTENTIAL OF SEAWATER

Already 30 years ago Fofonoff proposed to compute the thermodynamic potential of seawater as a compact and well-defined starting point for the derivation of all other quantities by thermodynamic rules (Fofonoff, 1962, Mamayev, 1987). This way a full internal mathematical consistency can be achieved. Moreover, also numerically one single procedure can be used to obtain most of the relevant seawater properties. With the independent (natural) variables (S,t,p) this potential is Free Enthalpy (Gibbs Free Energy) G(S,t,p). To adjust it quantitatively to all the different formulas mentioned above one first needs a suitable mathematical structure of this potential together with a reasonable number of free parameters. For a fast and precise computation of integrals and partial derivatives a polynomial is the simplest choice. Only at small salinities it is required for accuracy to take terms like $\sqrt{S}$ and Sln(S) in order to stay in accordance with the theory of dilute electrolytes. We introduce the auxiliary variables (x,y,z) by

$$S = x^2 \cdot 40 \text{ psu}, \; t = y \cdot 40°C, \text{ and } p = z \cdot 1000 \text{ Bar},$$

which vary between 0 and 1 in the range of most oceanographic interest. With a dimensionless polynomial

$$g(x,y,z) = (g_0 + g_1 \cdot y) \cdot x^2 \ln(x) + \Sigma\, g_{ijk} \cdot x^i \cdot y^j \cdot z^k$$

we can now write Free Enthalpy as

$$G(S,t,p) = 1 J/kg \cdot g(x,y,z).$$

Determination of the unknown coefficients of g(x,y,z) can be carried out in successive steps using least square fits to the corresponding most reliable data. The details of the fitting procedure are outlined elsewhere in detail (Feistel, 1993).

A layout of calculated coefficients g_{ijk} together with their respective origins is given Appendix A. Dots stand for coefficients impossible or unnecessary to determine on present data background. The resulting coefficients are tabulated in Appendix B.

The precision of the new thermodynamic potential can be summarized as follows. Sound speed after G(S,t,p) coincides with SVEL 83 formula within 10 ppm (1.5 cm/s), this is about 20 times better than the sound speed computed from EOS 80. The reason is a better numerical convergence of the nonlinear regression used compared with the iteration scheme proposed by Chen and Millero (1978). This better agreement implies improved high-pressure values of density and heat capacity. At 1 atm, specific volume as given by EOS 80 is experimentally known within 4 ppm tolerance (Millero and Poisson, 1981), while specific volume from G(S,t,p) agrees with EOS 80 within 0.2 ppm, i.e. here both formulas are fully equivalent. At higher pressures they deviate by typically 10 ppm (0.01 kg/m^3), which is about the reliability of direct measurements of seawater density (Chen and Millero 1976, 1978).

Heat capacity at atmospheric pressure is reliable experimentally up to 0.5 J/kgK (Fofonoff and Millard, 1983). The corresponding values from G(S,t,p) and CP 83 (Millero, Perron and Desnoyers, 1973) deviate from each other only by 0.01 J/kgK, i.e. there is no practical difference. At high pressure however the deviation grows up to a few J/kgK as a result of a better reproduction of sound speed figures.

Experimental high-pressure specific heat data are not known to the authors. Osmotic coefficients at 1 atm and 0°C after G(S,t,p) differ from values of Millero and Leung (1976) by 0.001, this is inside their error range of about 0.002. Finally, dilution heat from G(S,t,p) at 25 and 30°C deviates from measurements by Millero, Hansen and Hoff (1973) by 7 J/mol in average.

Even if G(S,t,p) reproduces well-known seawater properties with very good accuracy, a future improvement of the potential function by use of an improved data basis must be aspired to. We briefly discuss some weaknesses of the present state. First of all, it is not yet quite clear if e.g. Del Grosso - Mader sound speed formula (Del Grosso and Mader, 1972) - estimated error is a few cm/s (Wille, 1986) - is better than the UNESCO 83 equation. Its use should further improve high-pressure density and specific heat if this formula or a similar one will be released as a new oceanographic standard. Further, EOS 80 has systematic problems at low temperatures as precise maximum density temperature measurements of Caldwell (1978) indicate (they differ by about 0.2°C from those obtained by EOS 80 or G(S,t,p), 5 times worse than Caldwells error range). Such uncertainties may have significant implications for e.g. modelling the penetration of surface water into the deep ocean at high latitudes. Thirdly, dilution heat data used for G(S,t,p) exhibit relatively big standard deviations, and this error propagates into a number of important chemical and thermal quantities derived from G. A better experimental background is very desirable. Finally, in CP 83 as well as in G only very few coefficients appear to describe the salinity dependence of heat capacity. This obvious simplicity invokes some doubt concerning its precision.

QUANTITIES DERIVED FROM THERMODYNAMIC POTENTIAL

Here is a summary of the most important expressions derived from G(S,t,p) by partial derivatives. Note that instead of salinity S, Celsius temperature t or gauge pressure p we also use seasalt concentration $c = 1.00488\ \mathrm{g/kg} \cdot S$, absolute temperature $T = 273.15\ \mathrm{K} + t$, and absolute pressure $P = 101.325\ \mathrm{kPa} + p$.

Specific Volume (Reciprocal Density)	$V(S,t,p) = \partial G/\partial p = 1/\rho$
Entropy (Specific Entropy)	$\sigma(S,t,p) = -\partial G/\partial t$
Relative Chemical Potential	$\mu(S,t,p) = \partial G/\partial S$
Enthalpy	$H(S,t,p) = G + T\cdot\sigma$
Free Energy	$F(S,t,p) = G - P\cdot V$
Internal Energy	$E(S,t,p) = G + T\cdot\sigma - P\cdot V$
Chemical Potential of Water	$\mu_w(S,t,p) = G(S,t,p) - \mu(S,t,p)\cdot S$
Chemical Potential of Seasalt	$\mu_s(S,t,p) = \mu(S,t,p) + \mu_w(S,t,p)$

Let $G_{tt} = \partial^2 G/\partial t^2$, $G_{tp} = \partial^2 G/\partial t \partial p$, $G_{pp} = \partial^2 G/\partial p^2$, $G_{Sp} = \partial^2 G/\partial S \partial p$.

Heat Capacity (at constant pressure)	$C_P = -T\cdot G_{tt}$
Haline Contraction Coefficient	$\beta = -\rho\cdot G_{Sp}$
Thermal Expansion Coefficient	$\alpha = \rho\cdot G_{tp}$
Isothermal Compressibility	$K = -\rho\cdot G_{pp}$
Sound Speed U	$U^2 = V^2\cdot G_{tt}/(G_{tp}^2 - G_{tt}\cdot G_{pp})$

Note that V/U^2 is adiabatic compressibility.

Adiabatic Lapse Rate	$\Gamma = -G_{tp}/G_{tt}$
Vertical Stability	$N_E^2 = -g^2\rho^2\cdot G_{tp}^2/G_{tt}$

with g the local effective gravity acceleration.

(This definition of N_E^2 is a thermodynamic one after Eckart, 1958. See next page for some comments on its relation to the dynamical definition of Brunt-Väisälä Frequency N.)

Relative Apparent Molal Enthalphy: $\Phi_L(S) = 62.488\text{kg/mol}\cdot(H(S) - H(0) - S\cdot\partial H(0)/\partial S)/S$

Differences in Φ_L between two different dilution states are dilution enthalpies (Millero, Hansen and Hoff 1973).

Osmotic Coefficient Φ: $-2mRT\cdot\Phi = (1+c^2)\cdot G(S) - G(0) - (1-c^2)\cdot S\cdot\partial G(S)/\partial S$ with the gas constant $R = 8.3143$ J/mol°C, molality $m = I/1.245$, and ion strength $I = 19.9201\cdot S/(1000 + 1.00488\cdot S)$ (Millero and Leung, 1976).

POTENTIAL TEMPERATURE AND DENSITY

Potential Temperature $\theta(S,t,p,p_0)$ was introduced into oceanography by Helland-Hensen (1912) and is defined as the temperature of a water parcel with in-situ values S,t,p if it is brought to a reference pressure p_0 without exchange of heat and salt. In other words θ is the temperature $\theta = t(S,\sigma,p_0)$ of water with salinity S and entropy $\sigma = \sigma(S,t,p)$. To compute it from G(S,t,p) one needs the inverse formula $t = t(\sigma)$ of $\sigma = \sigma(t)$. Inversion can be computed by Newton iteration, but because entropy is an essential quantity for seawater thermodynamics, an explicit formula has been derived by polynomial inversion (Feistel, 1993):

$$t(S,\sigma,p) = 40°C\cdot\Sigma\, t_{ijk}\cdot x^i\cdot s^j\cdot z^k$$

with x, z as before and a dimensionless entropy variable

$$s = (\sigma(S,t,p) - \sigma(S,0,p))/625\ \text{J/kgK}$$

The coeficients t_{ijk} are tabulated in appendix C. The error of this inversion polynomial is less than 0.02°C for t = 0 to 40°C, S = 0 to 40 psu, p,p_0 = 0 to 1000 Bar. For improved precision it can still be corrected by re-inversion and becomes reliable within 0.05 mK. (Note that this value refers only to the numerical properties of the polynomial, not to the significance of physical measurements behind it). By means of the inversion formula potential temperature is given as

$$\theta(S,t,p,p_0) = t(S,\sigma(S,t,p),p_0)$$

and in the recorrected high-precision form as

$$\theta(S,t,p,p_0) = 2\cdot t(S,\sigma(S,t,p),p_0) - t(S,\sigma(S,t(S,\sigma(S,t,p),p_0),p_0),p_0).$$

Potential density ρ_θ is defined as the density of a water parcel with in-situ values S,t,p if it is brought to a reference pressure p_0 without exchange of heat and salt, in other words, as the density of water with in-situ values of salinity and entropy, but under reference pressure:

$$\rho_\theta(S,t,p,p_0) = \rho(S,\theta(S,t,p,p_0),p_0)$$

Note that the difference between density and potential density can be expressed as a pressure integral over $1/U^2$, if sound speed is expressed in terms of entropy instead of temperature (Kamenkovich and Monin, 1978).

Potential temperature and density have proved to be very useful tools in practical oceanography because they allow to trace water "with the same properties" moving through different pressure levels, where often heat and salinity losses are negligible. Their main disadvantage is that they are no proper thermodynamic state quantities because they depend on an arbitrarily chosen reference state. For considering the history of a single water parcel or for a stationary streamline of parcels, each passing the same state, these potential quantities are fully justified means to compare water at different states. However, if water masses of unknown different origins are compared in, say, an oceanographic section, these quantities have to be interpreted with great care.

For potential temperature the perfect substitute is entropy. It is the quantity that is actually almost conserved during advective motion, and it is a genuine thermodynamic state variable. Indeed Temperature-Salinity diagrams should better read as Entropy-Salinity diagrams, because the latter couple represents the two typical conservative properties of oceanic water masses. This is no new insight but is usually ignored because temperature is easily available and very precise, but by now entropy has not been available at all, and still it is coined by a certain lack of accuracy. Examples are shown in Figure 1. Sometimes the temperature is replaced by the potential temperature. Surfaces of constant density "Sigma-t" ($p = 0$) are usually involved.

With potential density the situation is more complicated. We return to this point in a later section.

THERMODYNAMIC VERTICAL STABILITY

An infinitesimal vertical displacement of a water mass element in a stratified ocean causes an adiabatic change of its temperature and hence of its density, too. This density slightly differs from the density of the surrounding, such that backdriving buoyancy forces lead to harmonic oscillations with the so-called Brunt-Väisälä frequency N:

$$N^2 = g^2(d\rho/dp - (\partial\rho/\partial p)_{S,\sigma}) = g^2(d\rho/dp - 1/U^2)$$

While sound speed U as a thermodynamic quantity is easily computed by in-situ values S,t,p, the vertical density gradient must be taken from the actual stratification of water. A first difficulty therefore is a merely technical one, namely calculating a gradient from usually noisy profile data. The second point is related to the fact that both pressure derivatives of density in the difference N^2 are of same order of magnitude and almost cancel each other. If on top of that different and inconsistent formulas for ρ and U are applied like in the case of EOS 80 and SVEL 83, reliable values for N^2 can hardly be expected (Wang, 1974, Millero, 1982, Siedler and Peters, 1986). This situation is significantly relaxed if both ρ and U have a common origin since then the big but cancelling terms can be removed from the formula analytically. We write the exact differential of density as

$$d\rho = (\partial\rho/\partial p)_{s,t}\, dp + (\partial\rho/\partial t)_{S,p}\, dt + (\partial\rho/\partial S)_{t,p}\, dS$$

and find

$$N^2 = N_E^2 + g^2\rho\cdot(\beta\cdot(dS/dp) - \alpha\cdot(dt/dp))$$

or, assuming the validity of hydrostatics $dp = \rho \cdot g \cdot dz$,

$$N^2 = N_E^2 + g \cdot (\beta \cdot (dS/dz) - \alpha \cdot (dt/dz))$$

(Gill, 1982) with depth z and N_E, α, β defined in the beginning. Evidently in some oceanic situations the vertical gradients of S and t provide only small corrections to N_E^2 and can be dropped completely (for the standard ocean α is some 1.E-4/°C, β is about 8.E-4/psu and N_E^2 is 1.E-8..1.E-6/s²). Under these circumstances vertical stability can simply be calculated as a thermodynamic in-situ state variable independent of actual stratification properties. From these considerations it is clear

T-S-DIAGRAMS: Section: EBC2zs: 21.09.-26.09.1991: r/v "A.v.Humboldt":IOW

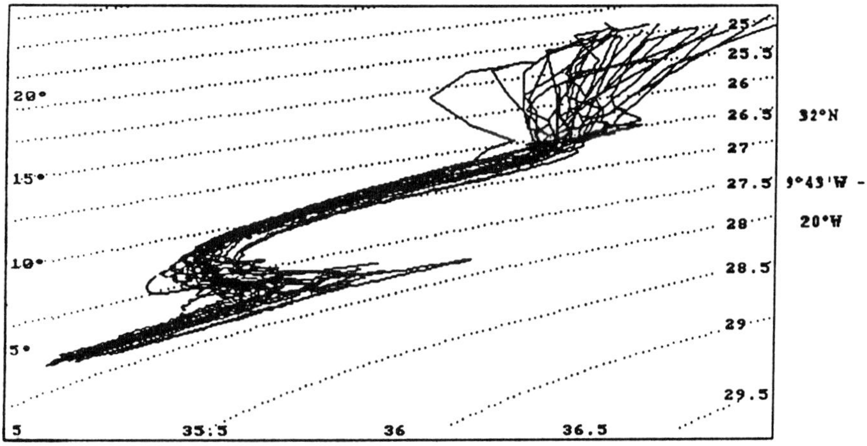

FIGURE 1A.

σ-S-DIAGRAMS; Section; EBC2zs; 21.9.-26.9.1991; r/v "A.v.Humboldt":IOW

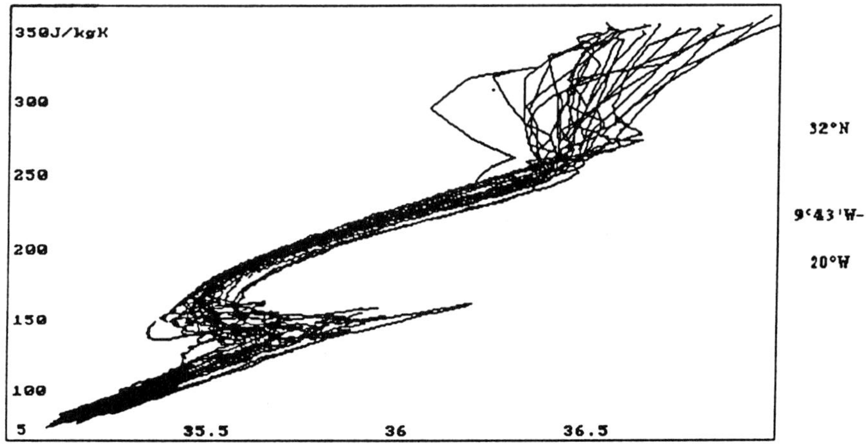

FIGURE 1B. Comparison between a usual temperature-salinity diagram (above) with entropy-salinity (below) resulting from a zonal section in the north-east Atlantic Ocean.

that physically N_E is just the Brunt-Väisälä frequency of an ocean homogeneous in temperature and salinity (see e.g. Landau and Lifschitz, 1974). It can approximately be applied to weakly inhomogeneous situations too, as long as the vertical in-situ density profile is mainly due to pressure and only little to salinity and temperature, which even tend to compensate each other. In concrete situations one has to estimate the error

$$\epsilon = |(N^2/N_E^2)\text{-}1| = C_P \cdot |\beta \cdot (dS/dz) - \alpha \cdot (dt/dz)| / g\alpha^2 T$$

to be much less than 1 or not.

NEUTRAL BUOYANCY SURFACES

Potential density was introduced as a concept to determine surfaces in the ocean where water parcels may move by random or weak forces without feeling stabilizing backdriving buoyancy. From the definition of potential density it is clear that these surfaces are not uniquely defined but depend on an arbitrary reference level. The question arises if a proper thermodynamic state function exists which is constant on such surfaces and thus may substitute the classical approach. Properties of "neutral surfaces" were investigated by McDougall (1987).

On the background of a given spatial distribution of (S,t,p) we consider the restoring buoyancy force vector field. A water parcel at position **r** has the density $\rho(\mathbf{r})$ under pressure p(**r**). Displaced adiabatically to **r** + d**r** under new pressure p + dp its density becomes $\rho(\mathbf{r}) + (1/U^2)dp$, while the surrounding density is $\rho(\mathbf{r}) + \nabla\rho\,d\mathbf{r}$. This difference in specific volume δV causes an excess buoyancy force $\delta\mathbf{F} = -\delta V \cdot \nabla p = V^2 \cdot (\mathbf{a}(\mathbf{r}) \cdot d\mathbf{r}) \cdot \nabla p$ with the vector field

$$\mathbf{a}(\mathbf{r}) = (\nabla\rho - (1/U^2)\nabla p) = \rho \cdot (\beta' \nabla S - \alpha' \nabla\sigma),$$

which may be called not very precisely the "gradient of potential density," but is not evidently a gradient of any scalar function. The second form is obtained using the equation of state with entropy variable as

$$\nabla\rho(S,\sigma,p) = \rho\beta'\nabla S - \rho\alpha'\nabla\sigma + (1/U^2)\nabla p,$$

where α' and β' are the adiabatic coefficients of thermal expansion and haline contraction. Note that the vertical component $a_z = \rho \cdot N^2/g$, which is commonly used as vertical derivative of potential density, is proportional to the Brunt-Väisälä frequency (see Gill, 1982). A displacement from position **r** to **r** + d**r** causes a potential density excess $\delta\rho_\theta = \mathbf{a} \cdot d\mathbf{r}$. The question is now if we can find a thermodynamic quantity X(S,σ,p) in a way that no deviation $\delta\rho_\theta = 0$ appears if d**r** is taken inside a surface X(**r**) = const, in other words, such that the vector **a** is always orthogonal to surfaces of constant X, on which displacements then are "neutral". This in turn states the problem if the Pfaffian form $\delta\rho_\theta = \mathbf{a}(\mathbf{r}) \cdot d\mathbf{r}$ possesses an integrating factor $\lambda(\mathbf{r})$ such that $dX = \lambda(\mathbf{r}) \cdot \delta\rho_\theta$ is an exact differential, or equivalently that $\lambda(\mathbf{r}) \cdot \mathbf{a}(\mathbf{r}) = \nabla X$ is a gradient. The corresponding necessary and sufficient mathematical condition for **a**(**r**) is (see e.g. Margenau and Murphy, 1964)

$$\mathbf{a}(\mathbf{r}) \cdot \mathbf{curl}\ \mathbf{a}(\mathbf{r}) = 0.$$

If the condition is satisfied, neutral buoyancy surfaces exist and we have to look for a way to find them (i.e. to solve the Pfaffian differential equation). If the condition is violated, such surfaces do not at all exist and the solution of the Pfaffian form is a complex manifold of lines, along which neutral displacements occur, but which do not form any compact two-dimentional object. Even more strictly, in the non-integrable case *every* point in the neighborhood of a given one can be reached along a path with $\delta\rho_\theta = 0$, while in the integrable case infinitely many points cannot (Caratheodory's principle). Only in the latter case a unique definition of a neutral surface in our sense is possible. Applying the integrability condition to the definition of **a**(**r**) we get eventually

$$\mathbf{J}(\nabla S, \nabla t, \nabla p) = 0.$$

The notion $\mathbf{J(a,b,c)} = \mathbf{abc} = \mathbf{a}(\mathbf{b} \times \mathbf{c})$ means the spat product. $\mathbf{J}(\nabla S, \nabla t, \nabla p)$ is therefore the Jacobian of the scalar functions S(**r**), t(**r**), p(**r**), which vanishes identically if and only if the three functions are dependent on each other. Seawater thermodynamics has three independent variables, hence in general the integrability condition is *not* fulfilled in the ocean. For important special cases however it can be true, namely if only two independent variables suffice for its description, like in fresh water (without S), or if an additional constraint function C(S,t,p) = 0 holds everywhere in the volume, like in the barotropic ocean, where density is a function of pressure alone, or in central water masses, where entropy (potential temperature) is a function of salinity alone (using any of these restricted cases in the definition equation of **a**(**r**) shows immediately that then **a**(**r**) reduces to a gradient times a scalar function). We propose to call these cases "thermodynamic 2D states" of the ocean. If one takes various CTD profiles of a certain region and the resulting point cloud of samples (S,t,p) in S-t-p space is distributed only on a surface C(S,t,p) = 0 instead of being scattered over a volume, then the ocean region is in a 2D state.

In general 3D state, mixing by neutral displacements exhibits evidently diapycnal contributions because the neutral lines must penetrate density surfaces (otherwise they would form surfaces themselves). It seems plausible, that this is a fast and effective mixing mechanism which irreversibly transforms young (3D) water into mature (2D) water where the mechanism then is switched off by Caratheodory's principle and only slower processes continue the water aging process. Such a hypothesis can possibly help to understand the observed striking conservativity of T-S diagrams of the central water masses.

ENERGY, ENTROPY, ENTHALPY, AND CHEMICAL POTENTIAL

With the new thermodynamic potential G(S,t,p) a number of additional seawater properties have become quantitatively available. Their role and importance for interpreting oceanic CTD sections is not yet fully understood.

Internal energy as the "true energy" is subject to the First Law of thermodynamics. In the ocean the internal energy change of a given volume is always balanced by the enthalpy current through its boundaries (Landau and Lifschitz, 1974). As a function of density and entropy, $E(V,\sigma,S)$, its change during an isopycnal ($dV = 0$) mixing process without salt exchange of the mixing volume with its surrounding is just the mixing heat $dE = -pdV + Td\sigma + \mu dS = Td\sigma = \delta Q$.

For reversible processes, entropy can mostly be considered as synonymous to potential temperature, as we discussed in the beginning. Indeed there is no significant difference between the shapes of T-S and σ-S diagrams of CTD profiles. For irreversible processes entropy is, due to the Second Law of thermodynamics, a proper measure of "water age" if one compares water volumes with the same internal energy.

If water parcels are moving, their energy is changed by work of compression pdV and buoyancy Vdp. Therefore enthalpy $H = E + pV$ describes these changes by $dH = dE + pdV + Vdp$. Adding kinetic and potential energy of the water parcel, the Bernoulli function $B(\mathbf{r}) = H + \mathbf{v}^2/2 + \mathbf{gr}$ is conserved through the parcels history (along the streamlines), if the pressure field is stationary and no heat and salt are exchanged (Gill, 1982).

The relative chemical potential $\mu(S,t,p)$ is defined as the change of free enthalpy $G = \mu_S c + \mu_w(1-c)$, $dG = Vdp - \sigma dT + \mu\, dS$, if at constant pressure and temperature a certain mass fraction dc of salt is replaced by water, therefore accounting for dilution energy of salt. But probably its role in nonequilibrium thermodynamics is much more interesting, which we like to stress briefly now.

If a system is not at equilibrium, there are so-called thermodynamic forces in the system causing thermodynamic fluxes, which permanently produce entropy and tend to drive the system back into the equilibrium state (Fofonoff, 1962, Kamenkovich, 1973, Glansdorff and Prigogine, 1971, Landau and Lifschitz, 1974, Ebeling and Feistel, 1981). Besides viscose friction there are two irreversible processes in seawater - thermal and Fickian molecular diffusion. They are driven by forces which are the gradients of temperature and chemical potential. Turbulent mixing consists of two steps, first tiny water filaments are built by the irregular flow, which are then rapidly smoothed out by molecular diffusion. Note that only the second step produces mixing heat and causes density to change. On the other hand molecular diffusion can effectively trigger thermohaline instabilities like double diffusion. Thus gradients of temperature and chemical potential are important aspects not only for molecular but for mixing in general. Figure 2 shows, for instance, such gradients of the chemical potential at surfaces of constant potential density along a zonal section crossing the coastal upwelling area off Morocco. Note the depth-independent vertical structures, especially in the zone of the continental slope strictly separating the properties of near coastal waters from those of offshore waters. In general, we can expect that irreversible processes in the ocean have the tendency to make the gradients of t and μ smaller and smoother. From the inspection of many CTD profiles we found an impressively linear behavior of chemical potential with depth which was only disturbed in layers with intense mixing. Ocean sections with $\mu(S,t,p)$ clearly emphasize such areas by bended isolines. After first experiences we believe that chemical potential may become a new key property for interpreting CTD profiles. Among other things, we suggest to use the chemical potential as natural vertical axis in hydrographic section plots in order to identify "potential mixing regions".

CONCLUSIONS

Now, the equation of state of sea water is available in a consistent form not only for sea-going oceanographers but also for modellers. Its usage is enabled by the progress in computer power and by the development of precise CTD sensors.

It was the main aim to recommend the necessity to strictly use physical properties of thermodynamics, especially in the practical analysis of hydrographic data, in order to open the way for a deeper understanding of mixing processes detected in different spatial and temporal scales. Related climatic questions concern the formation, maintenance, and spreading of deep and central water masses in the same way as the seasonally generated mode waters by the winter convection. What kind of combination in hydrographic parameters could play a key role in this concert? Many efforts were done to find an answer during the oceanographic history. However, it seems that two parameters could be very important for the physical interpretation of data observed. The first one is given by the Väisälä frequency, which can be obtained for some cases alone from thermodynamic equations or in complete manner. In frequency space, for example, N is the upperst frequency for internal waves. On the other hand N essentially determines the horizontal scale of baroclinic motions by the radius of deformation and is also an ingredient of potential vorticity, which should be conserved along streamlines within auto-barotropic layers.

Section: EBC2zs: 21.09.-26.09. 1991: r/v "A.v. Humboldt"; IOW

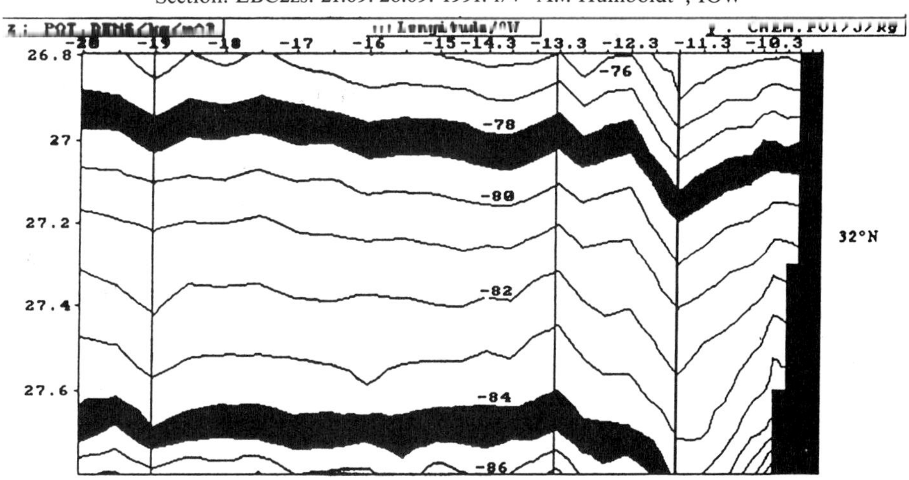

FIGURE 2. Plotted isolines of the chemical potential at surfaces of potential density relative to the reference level at sea surface from the same data as in Figure 1.

The second one could be the practical usage of the chemical potential. "Somehow," this quantity describes the "stand by" for molecular mixing conditions locally. In this way we may interpret regions with drastic lateral (isopycnal) gradients as trigger areas for processes like double diffusion, or more in general for those of diapycnal mixing. Until now it seems that the chemical potential could be also practically used as vertical coordinate in order to inspect small-scale mixing conditions while the choice of Bernoulli surfaces could be suitable with respect to more large scale investigations. We would like to note that "neutral surfaces," which are discussed in the literature extensively, can only exist for our 2D-case (central water masses, etc.) otherwise diapycnal mixing processes are permanently present in the ocean.

REFERENCES

1. Fofonoff, N.P. and R.C. Millard, Jr. 1983. Algorithms for computation of fundamental properties of seawater. *Unesco technical papers in marine science* 44. Unesco.
2. Millero, F.J., L.D. Hansen and E.V. Hoff. 1973. The Enthalpy of Seawater from 0 to 30°C and from 0 to 40 Salinity. *J. Marine Sci.* 31:21-39.
3. Millero, F.J. and W.H. Leung. 1976. The Thermodynamics of Seawater at One Atmosphere. *Am. J. Sci.* 276:1035-1077.
4. Millero, F.J. 1983. The Thermodynamics of Seawater. Part II. Thermochemical properties. *Ocean Sci. Eng.* 8:1-40.
5. Doherty, B.T. and D.R. Kester. 1974. Freezing Point of Seawater. *J. Marine Res.* 32:285-300.
6. Fofonoff, N.P. 1962. Physical Properties of Sea-Water. In: M.N. Hill (Ed.): *The Sea.* Wiley New York, pp. 3-30.
7. Mamayev, O.I. 1987. *Thermohaline Analysis of World Ocean Waters* (in Russ.) Gidrometeoizdat Leningrad.
8. Feistel, R. 1993. Equilibrium Thermodynamics of Seawater Revisited. *Progr. Ocean.* 31:101-179
9. Chen, C.-T. and F.J. Millero. 1978. The equation of state of seawater determined from sound speeds. *J. Marine Res.* 36:657-691.
10. Millero, F.J. and A. Poisson. 1980. Summary of data treatment for the international one atmosphere equation of state for seawater. In: *Background papers and supporting data on the International Equation of State of Seawater,* pp. 19-85 Unesco technical papers in marine science 38. Unesco 1981.
11. Chen, C.-T. and F.J. Millero. 1976. The specific volume of seawater from 0 to 40 salinity, 0 to 40°C, and 0 to 1000 bars. *Deep-Sea Res.* 23:595-612.
12. Millero, F.J., G. Perron and J.F. Desnoyers. 1973. Heat capacity of seawater solutions from 5°C to 35°C and .05 to 22 chlorinity. *J. Geophys. Res.* 78:4499-4506.
13. Del Grosso, V.A. and C.W. Mader. 1972. Speed of Sound in Seawater Samples. *J. Acoust. Soc. Am.* 52:961-974.

14. Wille, P. 1986. Acoustical Properties of the Ocean. In: J. Sündermann (Ed.): *Oceanography,* pp. 265-382. Landolt-Börnstein N.S. V/3a, Springer.
15. Caldwell, D.R. 1978. The maximum density points of pure and saline water. *Deep-Sea Res.* 25:175-181.
16. Eckart, C. 1958. Properties of Water. *Am. J. Sci.* 256:225-240.
17. Helland-Hensen, B. 1912. The Ocean Waters. *Intern. Rev. Hydrobiol. Hydrog.,* Hydrog. Supp. Ser. 1 No. 2. Leipzig.
18. Kamenkovich, V.M. and A.S. Monin. 1978. *Physics of the Ocean.* Vol. 1 (in Russ.). Izdatelstvo Nauka, Moscow.
19. Wang, D.-P. 1974. Computation of oceanic stability (Brunt-Väisälä frequency) with an improved density equation. *Earth Planet. Sci. Letters* 24:87-90.
20. F.J. Millero. 1982. The Thermodynamics of Seawater. Part I. The PVT properties. *Ocean Sci. Eng.* 7:403-460.
21. Siedler, G. and H. Peters. 1986. Properties of Sea Water. In: J. Sündermann (Ed.): *Oceanography,* p. 233-264. Landolt-Börnstein N.S. V/3a, Springer.
22. Gill, A. 1982. *Ocean Atmosphere Dynamics.* Academic Press San Diego.
23. Landau, L.D., E.M. Lifschitz. 1974. *Hydrodynamik.* Lehrbuch der Theoretischen Physik, Bd. VI. Akademie-Verlag Berlin.
24. McDougall, T.J. 1987. Neutral Surfaces. *J. Phys. Ocean.* 36:1950-1964.
25. Margenau, H. G.M. Murphy. 1964. *Die Mathematik für Physik und Chemie,* Bd.I. B.G. Teubner Leipzig.
26. Kamenkovich, V.M. 1973. *Fundamentals of Ocean Dynamics* (in Russ.). Gidrometeoizdat Leningrad.
27. Glansdorff, P., I. Prigogine. 1971. *Thermodynamic Theory of Structure, Stability and Fluctuations.* Wiley Interscience New York.
28. Ebeling, W., R. Feistel. 1981. *Physik der Evolutionsprozesse.* Akademie-Verlag Berlin.

APPENDIX A.

Data Origin of Coefficients gijk of Gibbs Potential

1. index (left to right) - order in salinity root, 2. index (downward inside blocks) - order in temperature, 3. index (downward block by block) - order in pressure, Logarithm terms g0,g1 are from Ideal Solution theory, further

			Ideal Sol.	Limiting Law				
Null	g000	0	g200	g300	g400	g500	g600	Freezing Point
Null	g010	0	g210	g310	g410	g510	g610	Dilution Heat
CP(0,t,0)	g020	0	g220	g320	.	.	.	CP(S,t,0)
Heat	g030	0	g230	g330	.	.	.	Heat
Capacity	g040	0	g240	g340	.	.	.	Capacity
of Water	g050	0	g250	.	.	.	.	of Seawater
at 1 atm	g060	0	.	.	.	.	.	at 1 atm
V(0,t,0)	g001	0	g201	g301	g401	.	.	V(S,t,0)
Specific	g011	0	g211	g311	g411	.	.	Specific
Volume	g021	0	g221	g321	.	.	.	Volume
of Water	g031	0	g231	.	.	.	.	of Seawater
at 1 atm	g041	0	g241	.	.	.	.	at 1 atm
	g051	0	.	.	.	.	.	
	g061	0	.	.	.	.	.	
U(0,t,p)	g002	0	g202	g302	g402	.	.	U(S,t,p)
Sound	g012	0	g212	g312	g412	.	.	Sound
Speed	g022	0	g222	g322	.	.	.	Speed
in Water	g032	0	g232	.	.	.	.	in Seawater
	g042	0	g242	.	.	.	.	
	g052	0	.	.	.	.	.	
	g003	0	g203	g303	g403	.	.	
	g013	0	g213	g313	.	.	.	
	g023	0	g223	.	.	.	.	
	g033	0	g233	g333	.	.	.	
	g043	0	g243	.	.	.	.	
	g004	0	g204	g304	.	.	.	
	g014	0	.	.	.	.	.	
	g024	0	.	.	.	.	.	
	g034	0	g234	.	.	.	.	
	g005	0	g205	.	.	.	.	
	g015	0	g215	.	.	.	.	
	.	0	g225	.	.	.	.	

APPENDIX B.

Coefficients g_{ijk} of Gibbs Potential

($S = x^2 \cdot 40$, $t = y \cdot 40°C$, $p = z \cdot 1000$ bar)

Logarithm terms: $(5814.9808 + 851.5440 * y) * x^2 \ln(x)$

	x^0	x^2	x^3	x^4	x^5	x^6
z^0:						
y^0	0.0000	-4204.5207	-2439.3585	2152.9589	-1261.1070	376.6353
y^1	0.0000	-615.7087	-492.0479	740.6876	-513.5010	169.6353
y^2	-12351.7946	895.0019	-130.7028	0.0000	0.0000	0.0000
y^3	747.2003	-209.2323	44.7489	0.0000	0.0000	0.0000
y^4	-161.4944	55.5650	-11.0394	0.0000	0.0000	0.0000
y^5	58.2142	-3.1539	0.0000	0.0000	0.0000	0.0000
y^6	-11.8302	0.0000	0.0000	0.0000	0.0000	0.0000
z^1:						
y^0	100015.7425	-3300.8417	154.7687	15.6356	0.0000	0.0000
y^1	-271.8113	673.3336	-126.7460	-11.6853	0.0000	0.0000
y^2	1455.9928	-572.5697	87.2761	0.0000	0.0000	0.0000
y^3	-646.5726	238.1182	0.0000	0.0000	0.0000	0.0000
y^4	303.7874	-63.2100	0.0000	0.0000	0.0000	0.0000
y^5	-74.7501	0.0000	0.0000	0.0000	0.0000	0.0000
y^6	1.6712	0.0000	0.0000	0.0000	0.0000	0.0000
z^2:						
y^0	-2544.4522	376.4561	-33.9597	-13.5358	0.0000	0.0000
y^1	768.5030	-319.3157	51.1912	9.6466	0.0000	0.0000
y^2	-706.9668	294.2595	-40.5880	0.0000	0.0000	0.0000
y^3	373.3520	-142.2859	0.0000	0.0000	0.0000	0.0000
y^4	-146.7801	37.7994	0.0000	0.0000	0.0000	0.0000
y^5	27.2894	0.0000	0.0000	0.0000	0.0000	0.0000
z^3:						
y^0	278.0569	-63.8458	21.4511	-6.6615	0.0000	0.0000
y^1	-161.6587	65.6957	-9.6994	0.0000	0.0000	0.0000
y^2	178.7986	-77.3472	0.0000	0.0000	0.0000	0.0000
y^3	-90.2555	54.4240	5.6105	0.0000	0.0000	0.0000
y^4	21.9190	-23.8441	0.0000	0.0000	0.0000	0.0000
z^4:						
y^0	-24.5772	2.5891	-1.0780	0.0000	0.0000	0.0000
y^1	4.7977	0.0000	0.0000	0.0000	0.0000	0.0000
y^2	-14.6082	0.0000	0.0000	0.0000	0.0000	0.0000
y^3	4.5851	6.5341	0.0000	0.0000	0.0000	0.0000
z^5:						
y^0	0.4264	0.8411	0.0000	0.0000	0.0000	0.0000
y^1	2.8998	1.3398	0.0000	0.0000	0.0000	0.0000
y^2	0.0000	-4.0707	0.0000	0.0000	0.0000	0.0000

APPENDIX C.

Coefficients t_{ijk} of Temperature Polynomial

$t = t(S,\sigma,p) = 40°C * \Sigma\, t_{ijk} \cdot x^i \cdot s^j \cdot z^k$
salinity variable x $= \sqrt{(S/40)}$
entropy variable s $= (\sigma - \sigma(S,0,p))/(625\ J/kgK)$
pressure variable z $= p/(1000\ bar)$

	x^0	x^2	x^3	x^4	x^5
z^0:					
s^1	1.01199869	0.07332868	-.01070866	0.00531334	-.00155188
s^2	0.09293048	-.00582152	0.00261540	-.00272922	0.00118086
s^3	-.01003438	-.00190527	0.00043579	0.00049918	-.00023472
s^4	0.00383290	0.00485965	-.00104592	0.00136429	-.00050200
s^5	0.00214479	0.00054381	-.00011778	-.00028516	0.00013452
s^6	-.00097839	-.00084418	0.00014720	-.00023408	0.00007600
z^1:					
s^1	0.11929139	-.02962385	0.00462604	-.00491937	0.00148027
s^2	-.04755215	-.00046622	0.00509983	0.00283581	0.00004813
s^3	0.01872400	0.01219247	-.00215827	0.00178989	-.00092546
s^4	0.00985116	-.00341945	-.00018639	-.00274734	0.00078037
s^5	-.00674854	-.00502027	0.00090928	-.00066550	0.00034686
s^6	-.00211085	0.00109833	-.00021132	0.00107281	0.0
z^2:					
s^1	-.04386100	0.00771215	-.00086018	0.00279497	-.00080045
s^2	0.00978776	0.00684013	-.00218845	-.00067542	-.00055276
s^3	0.01294435	-.00604777	-.00031026	-.00268915	0.00089014
s^4	-.01125311	-.00442632	0.00163436	0.00120457	0.00017836
s^5	-.00193648	0.00481976	-.00054126	0.00216013	0.0
s^6	0.00402401	0.00132724	-.00063444	0.0	0.0
z^3:					
s^1	0.00265124	0.00239614	-.00125115	-.00066086	0.00005142
s^2	0.01033309	-.00293981	0.00050630	-.00052604	0.00058906
s^3	-.01156548	-.00363062	0.00144641	0.00089540	-.00001460
s^4	-.00119867	0.00462333	-.00069145	0.00093403	0.0
s^5	0.00675107	0.00039290	-.00075485	0.0	0.0
s^6	-.00131314	-.00371846	0.0	0.0	0.0
z^4:					
s^1	0.00335288	-.00234637	0.00010876	-.00022722	0.00010797
s^2	-.00434832	0.00178441	0.00073153	0.00098149	-.00019404
s^3	-.00172649	0.00236119	-.00047955	0.00005173	0.0
s^4	0.00603565	-.00069933	-.00055366	0.0	0.0
s^5	-.00180586	-.00342089	0.0	0.0	0.0
s^6	-.00275865	0.0	0.0	0.0	0.0
z^5:					
s^1	-.00053251	-.00045676	0.00008712	0.00011761	-.00004279
s^2	-.00096563	0.00140584	-.00019741	-.00018523	0.0
s^3	0.00296275	-.00090421	-.00024329	0.0	0.0
s^4	-.00109989	-.00146753	0.0	0.0	0.0
s^5	-.00245146	0.0	0.0	0.0	0.0
s^6	0.00225980	0.0	0.0	0.0	0.0

The Oceans: Physical-Chemical Dynamics and Human Impact. Edited by S.K. Majumdar, E.W. Miller, G.S. Forbes, R.F. Schmalz and Assad A. Panah.

Chapter Two

SIMULATING THE OCEANIC CIRCULATION WITH ISOPYCNIC-COORDINATE MODELS

RAINER BLECK and ERIC P. CHASSIGNET

Division of Meterology and Physical Oceanography
University of Miami
Miami, FL 33149

INTRODUCTION

Numerical modeling of geophysical fluid systems using electronic computers began approximately four decades ago. Until the mid- to late 1960s, ocean model development lagged significantly behind that of atmospheric models. A number of reasons (apart from societal needs) can be cited for this lag, among them the inherently greater complexity of circulation systems confined to closed basins, a highly nonlinear equation of state for seawater, and the lack of three-dimensional synoptic observations for ocean model initialization and verification. Furthermore, the computing power required to resolve the relevant hydrodynamic instability processes is far greater in the ocean than in the atmosphere because such processes occur on much smaller scales in the ocean. The above factors, taken together, imply that a major commitment of institutional resources was required in the past to mount a credible oceanic modeling effort. Thus, oceanic modeling never became a "cottage industry" like numerical weather prediction. Fiscal constraints often prohibited the luxury of a "duplication of efforts," and thus steered the oceanographic community away from model diversity. Only now, with supercomputers being within easy reach of university researchers, can such a diversity be achieved and should, in fact, be encouraged.

Before a system of differential equations can be solved on a computer, it must be converted into an algebraic system, a process which entails numerous approximations. Models have become "better" over the years not only because our knowledge of physics has improved, but also because modern computers permit a more faithful representation of the differential equations by their algebraic analogs. Specifically, the discretization or "truncation" error introduced when approximating differentials by finite differences decreases as advances in computing speed and memory size allow smaller and smaller mesh sizes.

Rather than counting on future computers that ultimately will allow algebraic and differential equations to converge, numerical modelers try to reduce the truncation error in their model on existing machines by using sophisticated ("higher-order") approximations or distributing a fixed number of grid points in physical space such that they optimally define the spatial structure of the modeled system. For example, representation of flow in the x, y plane near a ragged coastline can be improved by placing grid points at the intersections of curvilinear coordinate lines hugging the coast. Map projections such as the Mercator projection can be used as a tool to reduce the mesh size toward higher latitudes to account for the general poleward decrease in eddy size.

Another method of reducing the truncation error in geophysical fluid modeling entails replacing the independent variable z in the governing equations[1] by another variable. For example, numerical weather prediction models often use the ratio of *in-situ* and surface pressure ("σ") as vertical coordinate to improve the smoothness of the numerical solution near topographic obstacles.[2] Ocean models based on σ coordinates are gaining popularity,[3,4] ironically at a time when the atmospheric modeling community is beginning to seek alternatives to the σ system because of problems with computing the horizontal pressure force on coordinate surfaces following the bottom relief.

Mixing in stratified fluids where buoyancy effects play a role takes place predominantly along isopycnic or constant-density surfaces (*potential* density surfaces, to be precise).[5] If the conservation equations for salt and temperature are discretized in x, y, z space, that is, if the three-dimentional vectorial transport of these quantities is numerically evaluated as the sum of scalar transports in x, y, z direction, experience shows that it is virtually impossible to avoid diffusing the transported variables in those same directions. Thus, regardless of how the actual mixing term in the conservation equations is formulated, numerically induced mixing is likely to have a cross-isopycnal ("diapycnal") component which may well overshadow the diapycnal processes occurring in nature. This is known to have serious consequences for long-term climate simulations.

Isopycnic-coordinate modeling seeks to eliminate this type of truncation error by transforming the dynamic equations from x, y, z to x, y, ρ coordinates and discretizing them on an x, y, ρ model grid. Transport in x, y direction then takes place in the model on isopycnic surfaces, and the associated numerical diffusion has no diapycnal component. Transport in z direction translates into transport along the ρ axis and can be entirely suppressed if so desired; that is, it also has no unwanted diapycnal component. As a result, spurious heat exchange between warm surface waters and cold abyssal waters does not occur in isopycnic models, nor does horizontal heat exchange across sloping isopycnals such as those marking the Gulf Stream front.

A second advantage of isopycnic coordinates is their tendency to concentrate grid points vertically in regions where grid resolution is needed most. The near-geostrophic balance characterizing large-scale oceanic motion implies that horizontal jet-like currents must be supported by inclined and tightly packed density surfaces. By their very nature, grid points attached to isopycnals congregate in such so-called frontal zones and provide optimal resolution for the current and the density contrast supporting it.

There is a price to be paid for the advantages outlined above. The ocean, when viewed in x, y, ρ space, has a ragged, sloping upper boundary because isopycnals intersect the sea surface. Thus, coordinate layers sandwiched between near-surface isopycnals and the sea surface form wedges in physical space whose sharp edge marks the outcrop line of the respective isopycnal. This also happens where a subsurface coordinate layer intersects a topographic obstacle. The requirement to inflate and deflate these wedges in an orderly fashion while solving the mass conservation equation (also called the layer thickness tendency equation) is a complicated and not yet completely solved numerical problem. Other shortcomings of the isopycnic coordinate system worth mentioning are the lack of vertical grid resolution in neutrally stable water columns and the nonexistence of a reference pressure-independent definition of the potential density of seawater.

It is fair to say that there is no "best" choice of vertical coordinate in a geophysical fluid model because the solutions of the discretized finite-difference equations should all converge toward the solutions of the differential equations as mesh size goes to zero. Each coordinate system is afflicted with its own set of truncation errors whose implications sometimes become known only after extended experimentation.

ISOPYCNIC SURFACES AND MODELING

Perhaps the most elegant way to construct a 3-dimensional density-coordinate numerical model is to form a stack of 2-dimensional shallow-water models, arranged so that lighter isopycnic layers rest on top of denser layers. The resulting discrete-layer structure is a valid finite-difference approximation to the continuous equations transformed to density coordinates. The fact that a density-coordinate *layer* model has an exact physical analog (at least as far as vertical discretization is concerned) makes this a very attractive concept in model-building. Each layer is characterized by a constant value of density and is governed by dynamic equations resembling the shallow-water equations. The layers interact even in the inviscid limit through hydrostatically transmitted pressure forces.

As mentioned earlier, the advantages of isopycnic-coordinate representation are offset by numerical difficulties encountered where interior density surfaces come into contact with either the ocean surface or the ocean bottom, both of which are treated for practical reasons as additional coordinate surfaces. A hybrid, quasi-isopycnic model can avoid such coordinate intersections by locally reverting to a Cartesian coordinate representation wherever minimum isopycnic grid-point separation in the vertical cannot be maintained.[6] By contrast, a pure isopycnic model depicts outcrop lines as boundaries between regions of zero and nonzero isopycnic layer thickness and solves the mass continuity equation in such a way that the collapse of a layer to zero thickness does not cause numerical instability.[7] Two algorithms, the Flux Corrected Transport (FCT) scheme[8] and the MPDATA scheme,[9] have been found to be well suited for this purpose.[10]

Diabatic forcing of models whose coordinate layers maintain their density requires translating density tendencies $d\rho/dt$ into vertical mass fluxes across layer interfaces. In order to accommodate numerically a variety of physical processes occurring in the ocean's surface mixed layer, the uppermost coordinate layer in an isopycnic-coordinate oceanic general circulation model (OGCM) is usually defined as a *non*-isopycnic layer in which termodynamic variables are allowed to vary continuously in x, y, and t. The depth of the mixed layer is typically determined by assuming that sources and sinks of turbulence kinetic energy, such as wind stirring, buoyancy changes at the surface, and entrainment of denser water from below, balance at all times. Since ocean-atmosphere heat transfer has a large effect on the rate and sign of buoyancy production, the mixed-layer depth varies markedly in the course of a year. Consequently, the mixed layer must be allowed to exchange mass with underlying isopycnic layers. This process is referred to as mixed layer entrainment/detrainment.

The response of the ocean to changes in atmospheric conditions, both man-made and natural, is one of the central topics of climate research today. Processes that determine the thermal structure of the thermocline, its motion field, and its buffering capacity are among the topics being studied. Many of these investigations have been inspired by the ventilated thermocline theory.[11] Analytic solutions can be derived for the flow in isopycnic layers forced by density and potential vorticity boundary conditions at the bottom of the Ekman layer in regions of downward Ekman pumping (subduction). Isopycnic modeling opens the door for a generalization of these results, because it allows a coupling of the subduction mechanism with the important process of mixed-layer entrainment and detrainment.

The specific numerical problem posed in the isopycnic framework by mixed-layer entrainment/detrainment is that only a finite number of constant-density layers are available to accept the detrained water (which, as long as it resides in the mixed layer, is allowed to have a density varying continuously in space and time). Some errors in the timing and the rate of detrainment, and consequently in the potential vorticity of the detrained water masses, are therefore inevitable if the model is run in very low vertical resolution mode. How significant these errors are compared to the truncation errors encountered in traditional z-coordinate models using the same number of grid points in the vertical is difficult to assess. There is reason to believe, though, that a Cartesian model with a comparable number of grid points in the vertical would be unable to simulate the annual cycle of mass exchange between the thermocline and the mixed layer with the same degree of realism.

A MODEL OVERVIEW

In this section, we describe the salient features of an isopycnic-coordinate model developed at the University of Miami. Similar efforts have been undertaken at the Florida State University,[12,13] the Naval Research Laboratory,[14,15] Nova University,[16] Oregon State University[17] and the Hamburg Max-Planck Institute,[18] to name a few. While we have attempted to make the description as self-contained as possible, the reader is referred for technical details to previous publications.[19,20,21]

Contrary to some wind-driven isopycnic models which gain computational efficiency by keeping the lowermost coordinate layer at rest, thereby eliminating barotropic gravity waves, the present model permits motion in all layers and time-integrates the barotropic flow component in "split-explicit" mode. With motion being allowed in the abyssal layers, much of the development effort has been directed at stabilizing the numerical solution in places where the flow interacts with topography. Another challenge was the development of transport algorithms that maintain their conservative properties, including positive-definiteness, in the presence of significant spatial and temporal variations in layer thickness.

Governing Equations

The Miami model is a primitive-equation model containing 4 prognostic equations - one for the horizontal velocity vector, a mass continuity or layer thickness tendency equation, and two conservation equations for salt and heat. The equations, written here for a generalized vertical coordinate "s" to make them formally applicable to both the isopycnic domain and the surface mixed layer,[22] are

$$\frac{\partial \mathrm{v}}{\partial t_s} + \nabla_s \frac{\mathrm{v}^2}{2} + (\zeta + f)\mathrm{k} \times \mathrm{v} + \left(\dot{s}\frac{\partial p}{\partial s}\right)\frac{\partial \mathrm{v}}{\partial p} + \nabla_\alpha M = -g\frac{\partial \tau}{\partial p} + \left(\frac{\partial p}{\partial s}\right)^{-1} \nabla_s \cdot \left(\nu \frac{\partial p}{\partial s} \nabla_s \mathrm{v}\right) \quad (1)$$

$$\frac{\partial}{\partial t_s}\left(\frac{\partial p}{\partial s}\right) + \nabla_s \cdot \left(\mathrm{v}\frac{\partial p}{\partial s}\right) + \frac{\partial}{\partial s}\left(\dot{s}\frac{\partial p}{\partial s}\right) = 0 \quad (2)$$

$$\frac{\partial}{\partial t_s}\left(\frac{\partial p}{\partial s}T\right) + \nabla_s \cdot \left(\mathrm{v}\frac{\partial p}{\partial s}T\right) + \frac{\partial}{\partial s}\left(\dot{s}\frac{\partial p}{\partial s}T\right) = \nabla_s \cdot \left(\nu\frac{\partial p}{\partial s}\nabla_s T\right) + \mathcal{H}_T \quad (3)$$

Here, $\mathrm{v} = (u, v)$ is the horizontal velocity vector, p is pressure, T represents any one of the model's thermodynamic variables (temperature, salinity, buoyancy), $\alpha = \rho^{-1}$ is the specific volume, $\zeta = \partial v/\partial x_s - \partial u/\partial y_s$ is the relative vorticity, $M \equiv gz + p\alpha$ is the Montgomery potential, gz is the geopotential, f is the Coriolis parameter, k is the vertical unit vector, ν is a variable eddy viscosity coefficient, and τ is the wind- and/or bottom drag induced shear stress vector. H_T represents the sum of the diabatic source terms acting on T. Subscripts indicate which variable is held constant during partial differentiation. Distances in x, y direction, as well as their time derivatives $\dot{x} \equiv u$ and $\dot{y} \equiv v$, are measured in the projection onto a horizontal plane, a convention which eliminates metric terms related to the slope of the s surface.[22]

Metric terms created when terms involving $(\nabla\cdot)$ or $(\nabla\times)$ are evaluated in a non-Cartesian grid (e.g., in spherical coordinates) can be formally eliminated by evaluating ζ and horizontal flux divergences in (1) – (3) as line integrals around individual grid boxes. Note that applying ∇ to a scalar, such as $v^2/2$ in (1), does not give rise to metric terms.

After vertical integration over a coordinate layer bounded by two surfaces s_{top} s_{bot} (the ocean surface, the bottom of the mixed layer, the sea floor, and all isopycnic layer interfaces are s surfaces in this context), the continuity equation (2) becomes a prognostic equation for the layer weight per unit area, $\Delta p = p_{bot} - p_{top}$:

$$\frac{\partial}{\partial t_s}\Delta p + \nabla_s \cdot (\mathrm{v}\Delta p) + \left(\dot{s}\frac{\partial p}{\partial s}\right)_{bot} - \left(\dot{s}\frac{\partial p}{\partial s}\right)_{top} = 0. \tag{4}$$

The expression $(\dot{s}\partial p/\partial s)$ represents the vertical mass flux – taken to be positive if in $+p$ or downward direction – across an s surface.

Multiplication of (1) by $\partial p/\partial s$ and integration over the interval $(s_{top}\ s_{bot})$, followed by division by $\Delta p/\Delta s$, changes the shear stress term in that equation into

$$\frac{g}{\Delta p}(\tau_{top} - \tau_{bot})$$

while the lateral momentum mixing term integrates to

$$(\Delta p)^{-1}\nabla_s \cdot (\nu \Delta p \nabla_s \mathrm{v}). \tag{5}$$

All other terms in (1) retain their formal appearance.

Wind-induced stress is assumed to be zero at the bottom of the mixed layer, which therefore must not be allowed to become shallower than the Ekman layer. Bottom stress is prorated among coordinate layers in accordance with the assumption that it linearly decreases from $-\rho c_D|\mathrm{v}|\mathrm{v}$ to zero over the lowest $10m$ of the water column. v in this formula is the velocity vector averaged over the lowest $10m$.

The layer-integrated form of (3) is

$$\frac{\partial}{\partial t}T\Delta p + \nabla_s \cdot (\mathrm{v}T\Delta p) + \left(\dot{s}\frac{\partial p}{\partial s}T\right)_{bot} - \left(\dot{s}\frac{\partial p}{\partial s}T\right)_{top} = \nabla_s \cdot (\nu \Delta p \nabla_s T) + \mathcal{H}_T. \tag{6}$$

The above prognostic equations are complemented by diagnostic equations including the hydrostatic equation

$$\frac{\partial M}{\partial \alpha} = p, \tag{7}$$

an equation of state linking T, S to α, and a turbulence kinetic energy balance equation[19] yielding the depth of the mixed layer.

Horizontal and Vertical Grid

Isotropy in horizontal grid resolution is considered an important aspect of model numerics; therefore, the equations are solved on a regular mesh superimposed on a conformal (presently: Mercator) projection of the earth's surface. Thus, for a mesh size of n degrees longitude the mesh size in meridional direction at latitude φ is ($n \cos \varphi$) degrees. The map-scale factor is a function of φ, but not of direction: grid squares on the map project onto (nearly) square areas on the sphere.

Velocity and mass field variables are staggered horizontally such that $u(v)$ grid points are positioned halfway between mass grid points in x (y) direction. This staggering arrangement, known as the Arakawa C grid, is not considered optimal for coarse-mesh (noneddy-resolving) models,[23] but was chosen with eddy-resolving applications in mind.

Thermodynamic variables and the velocity vector v are treated as "layer" variables which are vertically constant within layers but change discontinuously across layer interfaces. p, z and $\dot{s}\partial p/\partial s$, on the other hand, are "level" variables defined on interfaces. Since M only changes according to (7) where α changes vertically, M is a layer variable.

Time Differencing Scheme

The time step permissible when solving hyperbolic differential equations such as those governing fluid motion is limited by the requirement that no signal may travel faster than one grid interval per time step. By far the fastest signals in the ocean are those associated with barotropic gravity waves; these travel nearly 20 times faster than any other signal. Since barotropic motion in a weakly stratified medium is nearly independent of depth, barotropic wave propagation is essentially a 2-dimensional process. As mentioned earlier, the Miami model uses a split-explicit scheme[20] to circumvent the limitation posed by barotropic gravity waves on the permissible time step for advancing the 3-dimensional model fields. The barotropic flow field in the model is separated from the rest of the solution and advanced in time in a separate 2-dimensional "sub"-model using a very small time step.

Explicit integration of the barotropic mode was not only found to be faster than the traditional "rigid-lid" scheme,[1] which eliminates barotropic gravity waves altogether but requires the solution of a Poisson equation for the barotropic stream function at each time step, but also has the advantage of producing a sea-surface height field allowing direct comparison with radar altimetry data.

Liberal use is being made in the model of the time-splitting concept. For example (6), is actually decomposed into 4 separate equations which in turn advect the T field horizontally and vertically, diffuse it, and force it diabatically. Vertical advection, being caused in part by cabbeling (defined below) and by mixed-layer entrainment/detrainment, is itself a 2-step process.

Isopycnal Outcropping

Outcropping of isopycnic coordinate layers is the Achilles' heel of isopycnic modeling. The modeling philosophy employed in the Miami model is to allow each isopycnic layer to exist at every (x, y) point in the model domain, but also to allow it to become massless or dynamically "invisible".[7] Inflation or deflation of each layer, and thus the geographic location where the layer outcrops at a given time, is entirely controlled by the interplay between the continuity and momentum equations. Special algorithms referred to earlier[10] are required to make this scheme work; these are somewhat more expensive than conventional finite-difference methods for solving the mass continuity equation but do not require a special bookkeeping effort[18] to keep track of layer outcrop lines or time-dependent lateral boundaries in individual layers. Note that the presence in the model of a nonisopycnic surface layer does not alleviate the outcropping problem: isopycnic layers in the present model outcrop into the mixed layer.

Variable bottom topography requires that the layers also be allowed to intersect the ocean bottom. The bottom intersection problem is numerically similar to the surface outcrop problem.

Surface Forcing Functions

The model ocean can be driven by a variety of forces emulating those found in nature. These forces are (*i*) large-scale wind stress; (*ii*) small-scale mechanical stirring; (*iii*) two imposed (*i.e.*, model-state independent) fluxes representing solar radiation and precipitation; and (*iv*) two relaxation boundary conditions representing sensible heat flux and evaporation. Salinity varies according to the balance between evaporation and imposed rainfall. This type of salinity forcing, which differs from the commonly employed Newtonian relaxation toward observed surface salinity, is prefered for two reasons: (*a*) a salinity relaxation boundary condition is nonphysical because evaporation is virtually independent of surface salinity and precipitation is totally so; (*b*) imposed boundary fluxes are more likely to generate unrealistic salinity features, thus the model's reaction to such features provides valuable insight into its ability to simulate climates different from the present one.

Mixed Layer

The modes of interaction between the thermodynamically active mixed layer and the interior isopycnic coordinate layers are thoroughly documented elsewhere.[19,21,24] The main problem is that water to be discharged from the mixed layer into underlying layers during periods of surface heating generally does not match the density of an interior layer. Thus, the model mixed layer is too deep most of the time, *i.e.*, it contains a certain amount of "fossil" mixed-layer water. This water is being heated in the model at the same rate as the "true" mixed layer and thus remains indistinguishable from the latter. For obvious physical reasons, the heat required to warm the fossil mixed layer cannot be taken from the atmosphere; rather, it is taken from the bottom of the fossil mixed-layer column. The heat loss there creates water dense enough to match an interior coordinate layer, allowing it to be detrained into that layer. The unknown in this problem is the relative thickness of the heated and cooled portions of the fossil mixed layer.

The model is able to dispose of the fossil mixed layer each time the mixed-layer density during springtime warming matches the density of an interior coordinate layer. This gives rise to sudden detrainment events which are numerically undesirable. They represent a special truncation error in the sense that they can be suppressed by reducing the mesh size in ρ space, *i.e.*, by adding more layers.

Equation of State

The equation of state for seawater currently used in the Miami model approximates density by a polynomial linear in salinity and of 3rd-degree in temperature.[25] The mixed-layer detrainment algorithm requires that this equation be inverted to yield T as a function of S and ρ. Finding the roots of a 3rd-degree polynomial involves repeated calls to trigonometric functions to avoid complex-valued arithmetic and is therefore rather time-consuming. Of the three roots of the polynomial, the one guaranteed to be real lies near 180°C and is therefore nonphysical. The other two roots, if they exist, form a pair straddling a temperature value in the range -3° to -5°C where ρ for seawater reaches its maximum; the root above this value is the physical one and is used in computing $T(\rho, S)$.

In the present model implementation, no distinction is made between density and potential density, and the effect of pressure on these quantities is ignored. These simplifications can (and, in fact, do) lead to erroneous conclusions about the static stability of abyssal water masses marked by vertical salinity gradients. One possible solution to this problem is to reference potential density to 2000m depth rather than the surface.

The nonlinearity of the equation of state causes the mixture of two water parcels of equal density but different temperature and salinity to be denser than the original parcels. The German verb *kabbeln*, which is used to describe this phenomenon, suggests that such mixing usually produces a "choppy" or fine-grain vertical velocity field. Cabbeling in a coarse-mesh ocean model is therefore a severely aliased process.

In an isopycnic coordinate model, cabbeling generates a positive vertical velocity component $d\rho/dt$, i.e., a gradual transfer of mass from each coordinate layer to the one beneath it. In the Miami model, the cabbeling process is treated as part of the more general problem of maintaining the reference density in coordinate layers where the T, S fields evolve independently of each other.

Thermodynamic Equations

As stated earlier, the model accomodates two independent thermodynamic variables. The prognostic equations for these variables (buoyancy b and salinity S in the mixed layer, temperature T and S in the interior) are all of the form (6). Horizontal advection (note that "horizontal" translates into isopycnal everywhere in the model except in the mixed layer) is handled by the conservative, monotony-preserving, version of the Smolarkiewicz transport scheme.[26] The diabatic source term H_T in (3) symbolizes three different processes - cabbeling, diabatic diffusion, and mixed-layer entrainment/detrainment. Due to the need to keep density constant in isopycnic layers, all diabatic processes must be converted into vertical mass fluxes $\dot{s}\partial p/\partial s$. Thus, in the absence of diabatic processes, no transport of thermodynamic variables takes place across layer interfaces, either explicitly or implicitly.

Continuity Equation

The split-explicit time integration scheme requires that changes in layer thickness caused by the barotropic velocity component $\bar{\mathrm{v}}$ be evaluated separately from those caused by the baroclinic component v'. The barotropic velocity is approximated in the model by the vertically averaged velocity $\bar{\mathrm{v}} = (\Sigma_k \mathrm{v}_k \Delta p_k) / (\Sigma_k \Delta p_k)$ where k is the coordinate layer index and the summation extends over all layers. The terms "barotropic" and "depth-averaged" will be used here interchangeably. The deviation of the actual velocity in a layer from the depth-averaged velocity will be termed "baroclinic".

Depth-independent motion stretches or shrinks the water column uniformly; that is, it changes the thickness Δp of a layer in proportion to Δp itself. This suggests that the individual effects of $\bar{\mathrm{v}}$ and v' on Δp may be isolated by setting $\Delta p = (1 + \eta)\Delta p'$ where temporal changes in $(1 + \eta)$ and $\Delta p'$ are caused, respectively, by mass flux divergences associated with $\bar{\mathrm{v}}$ and v'. Hence, η is independent of k while $p_b' \equiv \Sigma_k \Delta p_k'$ is time-independent.

The layer-integrated continuity equation (8), with vertical mass fluxes omitted for brevity, can now be written as

$$\frac{\partial}{\partial t_s}[(1+\eta)\Delta p_k'] + \nabla_s \cdot [(\bar{\mathrm{v}} + \mathrm{v}_k')(1+\eta)\Delta p_k'] = 0 \tag{8}$$

Summation over k and linearization yields the barotropic continuity equation

$$\frac{\partial}{\partial t}(p_b'\eta) + \nabla \cdot [\bar{\mathrm{v}}\Delta p_b'] = 0 \tag{9}$$

Subtracting (9) from (8) and approximating $(1 + \eta)$ by 1 yields the baroclinic continuity equation

$$\frac{\partial}{\partial t_s}\Delta p_k' + \nabla_s \cdot (\mathrm{v}_k \Delta p_k') = \frac{\Delta p_k'}{p_b'}\nabla \cdot (\bar{\mathrm{v}} p_b') \tag{10}$$

where $\mathrm{v}_k = \bar{\mathrm{v}} + \mathrm{v}_k'$. Note that replacing $(1 + \eta)$ by 1 in both (9) and (10) preserves the condition $\partial(\Sigma_k \Delta p_k') / \partial t = 0$.

The corresponding split of the momentum equations into their barotropic and baroclinic parts will be discussed in the following section.

Momentum Equations

Even though the nonlinear terms in (1) play no significant role in coarse-mesh basin-scale simulations, the equations are written in their fully nonlinear form to accomodate fine-mesh applications. The Coriolis and nonlinear terms conserve potential vorticity and potential enstrophy except in the vicinity of zero layer thickness zones.[20] The mixing terms are written in conservative form and use an eddy viscosity that reverts from a constant value to one proportional to the total deformation in regions of large horizontal shear.[27] Specifically, the eddy viscosity used in isopycnal mixing of momentum is defined as

$$\nu = max\left(u_d\Delta x, \eta[(u_x - v_y)^2 + (v_x + u_y)^2]^{1/2}\Delta x^2\right) \tag{11}$$

where Δx is the mesh size, subscripts denote partial derivatives, u_d is a background diffusive velocity, and η is a dimensionless factor.

Near steep bottom topography, the horizontal pressure force $\nabla_\rho M$ is computed in spatially non-centered form to avoid extrapolating pressure to underground grid points. Details of the scheme, which basically infers a layer-depth weighted average of $\nabla_\rho M$ from points surrounding the grid point in question, are provided elsewhere.[20]

Lateral sidewalls are usually chosen to coincide with grid points of the *normal* velocity component to allow enforcement of the kinematic boundary condition. In the C grid, this implies that the tangential velocity grid point nearest a sidewall is displaced from the wall by half a grid distance. A "mirror image" of this grid point on the opposite side of the wall is needed to solve the momentum, equation at the inside point. Depending on how one defines the velocity at the mirror image point, one can run a C grid model with free-slip or no-slip lateral boundary conditions.

The convention of extending coordinate layers intersecting the ocean bottom as massless layers all the way to the coast is followed. This convention, in conjunction with a layer depth-weighted lateral mixing formulation, makes it nearly impossible to communicate lateral sidewall drag to, say, grid points in the abyssal ocean next to the continental rise. A "submerged sidewall" boundary condition has been developed to remedy this problem.[20]

A barotropic momentum equation is needed to complement (9) in the 2-dimensional sub-model for the barotropic gravity wave field. Due to the small amplitude of these waves, the equation for $\overline{\mathrm{v}}$ is stated in the simple linearized form

$$\frac{\partial\overline{\mathrm{v}}}{\partial t} + f\mathrm{k}\times\overline{\mathrm{v}} + \alpha_0\nabla(p_b'\eta) = \frac{\partial\overline{\mathrm{v}}^*}{\partial t} \tag{12}$$

The term on the right-hand side is a source term for barotropic momentum. It arises during the solution of the v' equation which one obtains by subtracting (12) from (1):

$$\frac{\partial \mathrm{v}_k'}{\partial t_s} + \nabla_s\frac{\mathrm{v}_k^2}{2} + (\zeta + f)\mathrm{k}\times\mathrm{v}_k - f\mathrm{k}\times\overline{\mathrm{v}} + \left(\dot{s}\frac{\partial p}{\partial s}\right)\frac{\partial \mathrm{v}_k}{\partial p} + \nabla_\alpha(M - \alpha_0 p_b'\eta) =$$
$$-g\frac{\partial\tau}{\partial p} + \left(\frac{\partial p}{\partial s}\right)^{-1}\nabla_s\cdot\left(\nu\frac{\partial p}{\partial s}\nabla_s\mathrm{v}_k\right) - \frac{\partial\overline{\mathrm{v}}^*}{\partial t} \tag{13}$$

The term in question is added here to assure that $\Sigma_k \mathrm{v}_k'\ \Delta p_k'$ remains zero at all times. It is evaluated as a residual after (13) has been solved in all model layers.

Subgrid-Scale Mixing

Let Z denote one of the prognostic variables in the model. Since the finite-difference analog of the product $\nu\partial Z/\partial x$ is numerically equivalent to the finite-difference analog of $u_d\partial Z$ (where ∂Z is the Z difference between neighboring grid points, u_d is a "diffusive flux velocity" of magnitude $\nu/\partial x$, and ∂x is the grid size), the practice of specifying u_d rather than ν as the parameter controlling isopycnal mixing was adopted. Setting u_d constant throughout the model domain has the advantage of removing any geographic bias from the relative weight of diffusive and advective processes in shaping the character of the numerical solution. The u_d value also provides a convenient measure of the degree to which diffusion overshadows advection.

Subgrid-scale turbulent flux terms in the prognostic differential equations formally result from Reynolds averaging of the various nonlinear flux terms. Since the term $\nabla\cdot(\mathrm{v}\Delta p)$ in (4) is nonlinear, it would seem inconsistent not to add a "thickness diffusion" term to (4). Following the parameterization of subgrid-scale fluxes in the other model equations, the turbulent thickness diffusion term is expressed as $-\nabla\cdot(\nu\nabla\Delta p)$. In order to avoid injecting mass into areas where Δp vanishes due to layer outcropping or intersection with bottom topography, and to prevent thickness smoothing from changing the total weight of a column, thickness smoothing is actually implemented in the model in the form of "interface smoothing". Special precautions are followed to avoid smoothing interfaces coinciding with the sea floor.

Diapycnal Mixing

The diapycnic mixing algorithm in the Miami model is a simplified version of the algorithm developed by Hu.[28] Let θ be the variable that serves as vertical coordinate in the model (potential density), and let $F = K_z\partial\theta/\partial z$ be a parameterization of the vertical turbulent buoyancy flux $\overline{w'\theta'}$. Diapycnal gradients, i.e., gradients normal to θ surfaces, are approximated in the model by gradients in z direction.

To solve the diffusion equation

$$\left(\frac{\partial\theta}{\partial t}\right)_z = \frac{\partial}{\partial z}\left(K_z\frac{\partial\theta}{\partial z}\right) \tag{14}$$

in an (x,y,θ) reference frame, the roles of θ and z as dependent and independent variables must be reversed. This is done through the identity

$$\left(\frac{\partial a}{\partial b}\right)_c\left(\frac{\partial b}{\partial c}\right)_a\left(\frac{\partial c}{\partial a}\right)_b = -1.$$

which changes (14) into

$$\left(\frac{\partial z}{\partial t}\right)_\theta = -\frac{\partial F}{\partial\theta}\,; \qquad F = K_z\frac{\partial\theta}{\partial z}. \tag{15}$$

The above equation is solved in individual grid columns subject to the boundary conditions $F=0$ at top and bottom, expressing conservation of total θ in each column.

Since (15) requires knowledge of vertical F derivatives on layer interfaces, the fluxes F themselves must be *layer* variables. Physical intuition dictates that these fluxes, to first order, should be inversely proportional to isopycnal layer thickness. However, the possible presence of massless and near-massless layers in the water column requires special measures to insure boundedness of the finite-difference estimates of $\partial\theta/\partial z$. We achieve boundedness by formulating $\partial\theta/\partial z$ in terms of θ values representing integrals over a finite depth interval Δz.

SPECIFIC APPLICATIONS

Outcropping

Layered models of stratified flow have been favored by theoreticians because of their conceptual simplicity, but have only seen limited use until recently in numerical modeling due to difficulties related to layer outcropping (vanishing layer depth). Two mass transport schemes capable of handling outcropping, the Flux Corrected Transport (FCT) and the MPDATA algorithms, were tested against analytic solutions describing wind- and buoyancy-forced flows.[10]

The essential aspects of the circulation[29,30,31] in a two-layer, two-gyre wind-forced ocean basin (one aspect being the size of the outcrop region) are controlled by a single nondimensional number $\lambda = \tau_m L/g' H_1^2$ where τ_m is the maximum wind stress; L the basin width; $g' = g\Delta\rho/\rho$ the buoyancy difference between the two layers, and H_1 the mean upper-layer thickness. In order to illustrate the model's ability to handle zero layer thicknesses, one of the analytic solutions[29] for a finite-depth lower layer (Figure 1 a,b) has been reproduced numerically. The parameters τ_m, L and g' were chosen such that $\lambda = 0.30$ yields an outcrop region which spans virtually the entire cyclonically-forced northern half of the domain region and extends into the anticyclonically-forced southern half (Figure 1 a,b). After approximately 5 years of integration starting from rest, the model reached a steady state characterized by upper- and lower-layer mass transport streamfunctions shown in Figure 1 c,d. These model results were obtained with the Flux Corrected Transport algorithm. As in the analytic solution (Figure 1 a,b), the outcrop area in both figures is observed to extend south of the latitude of maximum wind stress (the zero wind stress curl line, ZWCL). The upper- and lower-layer flow combine to reduce the total mass flux across the ZWCL to zero, satisfying the Sverdrup balance. This result combined with others[10] alleviates longstanding reservations about the severity of the outcrop problem in layer models.

Bottom Topography

Another aspect which also retarded the development of layered models is the difficulty associated with isopycnal intersecting steep topography. Not only does one need to accomodate regions of zero isopycnal layer thickness in the presence of topography, but also the drag exerted laterally by steep slopes must be taken into account whenever bottom topography affects layer thickness.[20] The model's ability to handle this aspect was well illustrated in a series of experiments performed in a 4000 meter deep periodic channel with a tall isolated seamount at the center and with 15 constant-density layers in the vertical.[32] The Coriolis parameter, fluid stratification, coefficient of lateral friction and length scale of the topography were held constant, while the mean downstream flow speed and height of the obstacle varied up to an extreme case of a 20 cm s^{-1} current impinging on a 3500 m seamount.

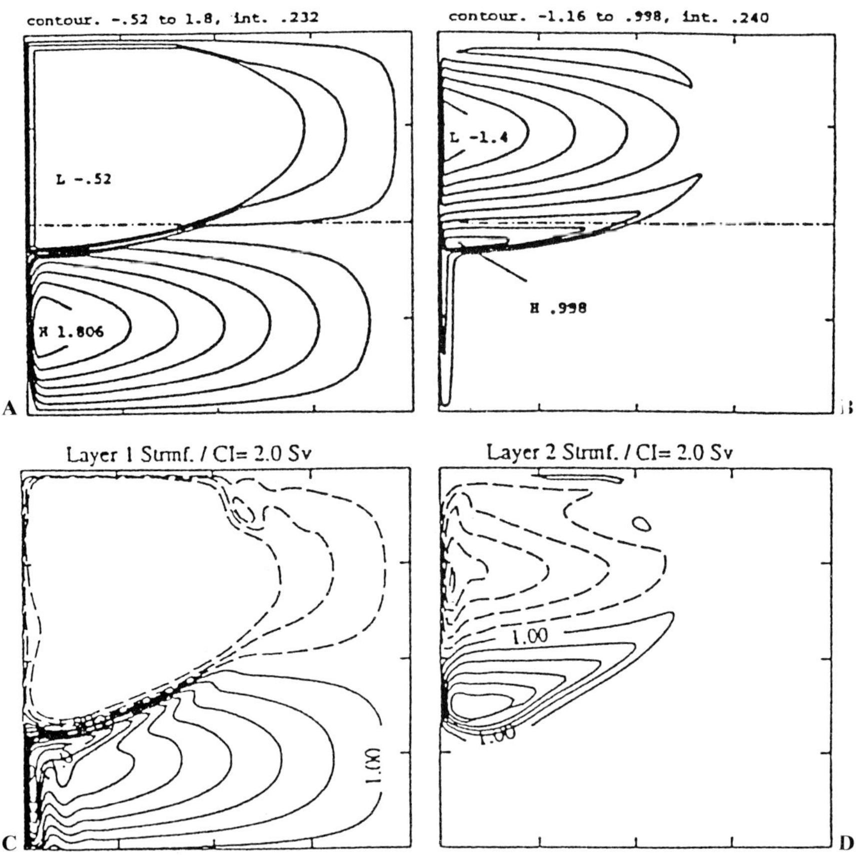

FIGURE 1. Mass transport stream function in upper layer (left) and lower layer (right). (a,b) analytic solution (nondimensional units correspond approximately to 12 × $10^6 m^3 s^{-1}$); (c,d) numerical solution using the Flux Corrected Transport algorithm.

The cross-stream vertical section through flow near the tall seamount in Figure 2 illustrates the model's behavior in the presence of isopycnals intersecting steep topography. The behavior of the fluid is found to differ little from simulations with smaller seamounts, as far as the formation and interaction of relative vorticity anomalies over and in the lee of the obstacle and the downstream transport of a cyclonic vortex are concerned.[32,33]

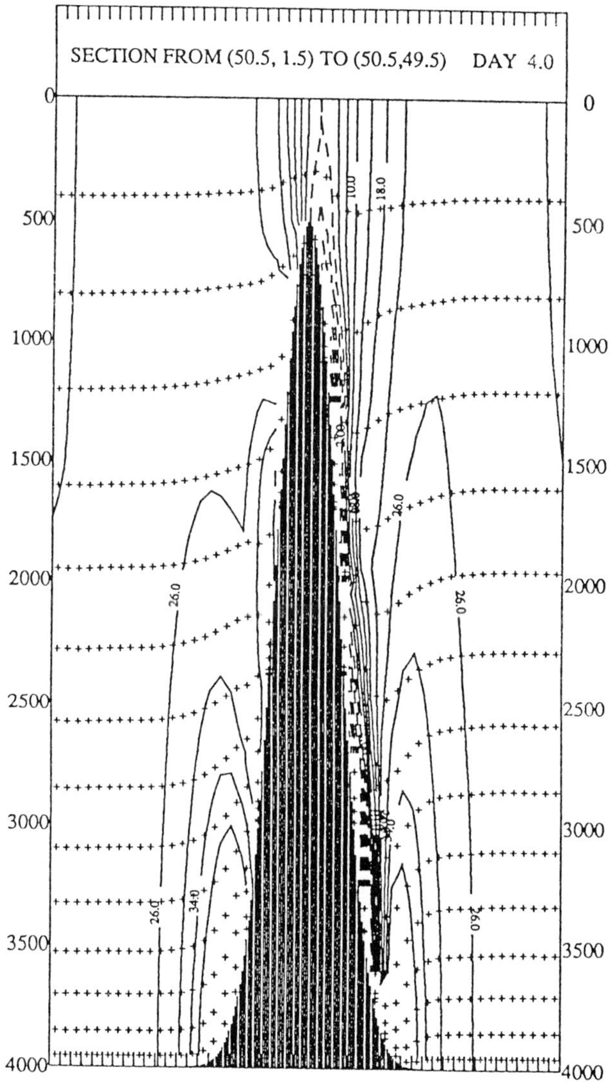

FIGURE 2. Cross-stream vertical section through center of seamount. Plus signs denote layer interfaces. Solid isotachs denote flow away from viewer; contour interval 4 *cm s*$^{-1}$.

A SIMULATION OF THE ATLANTIC OCEAN

As part of a multi-year, multi-institutional series of model applications and analyses, with the goal of establishing a sequence of eddy-resolving baseline calculations of the wind- and thermohaline-driven large scale ocean circulation, both the University of Miami and the Rennell Centre for Ocean Circulation in Southampton, England, are presently performing high resolution simulations of the circulation in the Northern and Equatorial Atlantic on decadal time scales using the above isopycnic-coordinate model. A similar experiment was completed a few years ago at the National Center for Atmospheric Research (NCAR) with the z-coordinate primitive-equation model developed at the Geophysical Fluid Dynamics Laboratory (GFDL).[34] Until recently, the GFDL model was the only "general-purpose" numerical model available to the physical oceanography community. The main purpose of the Miami and Southampton simulations is to answer questions concerning the degree to which the numerical representation of physical processes affects the realism of the numerical solution.

The simulation described in the remainder of this section covers the Atlantic between approximately 80°N (the Fram strait) and 15°S. The domain is spanned by a 1° mesh consisting of 126 by 103 grid points.[39] No artificial smoothing has been applied to the bathymetry, so that the features are relatively steep and complex. The model comprises a total of 20 layers: a Kraus-Turner-type mixed layer (vertically homogeneous but of density varying in the horizontal) and 19 isopycnic layers with potential density values of 25.65-27.75$kg\ m^{-3}$ in increments of 0.15, and from 27.75-28.15$kg\ m^{-3}$ in increments of 0.10 (the densities are specified with the two leading digits omitted following standard oceanographic practice). The model, started from a state of rest and spun up for a period of 30 years, was initialized from the September temperature, salinity and interface depth values of the Levitus dataset.[35] The base of the mixed layer was initially defined to coincide with the 25.575 density surface, or 50m, whichever was deeper.

The forcing functions driving the circulation are based on interpolations onto the model grid of the Hellerman and Rosenstein winds,[36] the Esbensen and Kushnir heat fluxes,[37] and the Jaeger precipitation[38] minus the Esbensen and Kushnir evaporation fields. Thermodynamic fluxes applied at the sea surface are based on the above plus a correction term which provides a moderate and partial relaxation (with approximately a 4-month timescale for a 100m thick water column) of the mixed layer temperature and salinity to Levitus climatology. The ocean interior, on the other hand, is not relaxed to climatological conditions in any way.

The northern and southern boundaries are represented by solid walls and no thermodynamic restoration was employed, so that no lateral fluxes of either heat or salt were allowed in or out of the model domain. At all lateral boundaries of the model domain, including coastlines and islands, a no slip condition was used for the velocity fields. Finally, a simple sea-ice model was incorporated into the code for the present integration.

Figure 3 shows the mixed layer velocity patterns in March after 30 years of integration. A prominent feature is of course the Gulf Stream, which flows up the coast of Florida and separates from the US shelf at about 40°N, near Cape Cod. This is somewhat too far north (satellite imagery indicating separation at Cape Hatteras, about 34°N) but is in line with results from other models.[34,40,41] The transport in the Gulf Stream is about 33 Sv at 26°N, increases steadily to about 45 Sv at 32°N, also in general agreement with the results of other studies. After separation from the coast, the current is associated with a marked temperature front extending across much of the central Atlantic and separating the cyclonic subpolar gyre to the north from the anticyclonic subtropical gyre to the south. Other features worth noting are the equatorial current system producing a strong North Brazil current, and upwelling resulting from divergence of the surface currents, at both the equator and off the coast of Africa, both being known phenomena.

The model mixed layer depth fields after 30 years of integration are compared with observations in Figure 4. The observations[42] show the deepest mixing in an area occupied by water with densities in excess of 27.75. This region covers most of the Labrador basin, but also extends in a thin tongue around Cape Farewell into the Irminger basin. This near correspondence between mixing in excess of 1000m and densities above 27.75 is also seen in the model results, even though the area over which this occurs in the model is somewhat smaller in the Labrador basin and more pronounced in the Irminger basin. Overall, the agreement between the observations and the model output in the mixed layer is quite good.

The model's representation of the annual cycle of mixed layer deepening and retreat is shown schematically in Figure 5. Layer interfaces are shown in these sections, which cut through the top 2400m of the model domain, as solid

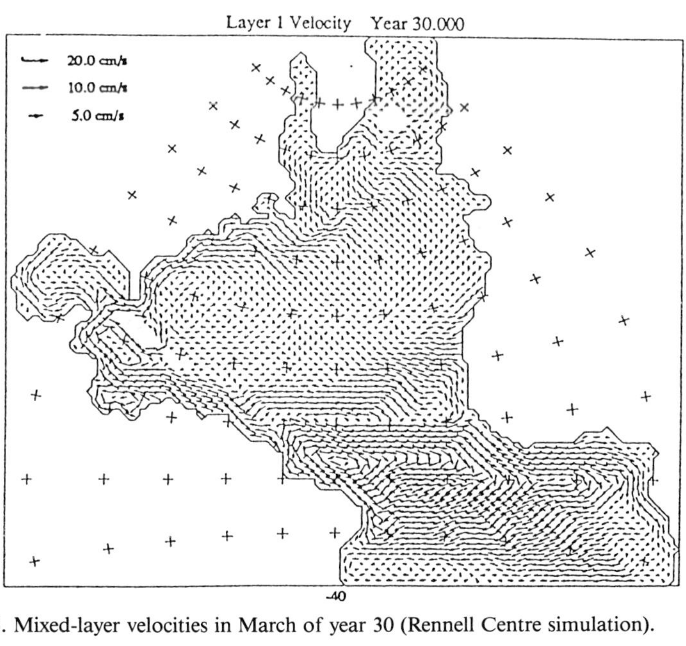

FIGURE 3. Mixed-layer velocities in March of year 30 (Rennell Centre simulation).

lines; isotherms are dashed. The stairsteps built into the isotherms are intentional: they illustrate the fact that T is a layer variable in the model and thus piecewise constant in the vertical. Stairsteps have been rounded off to improve legibility.

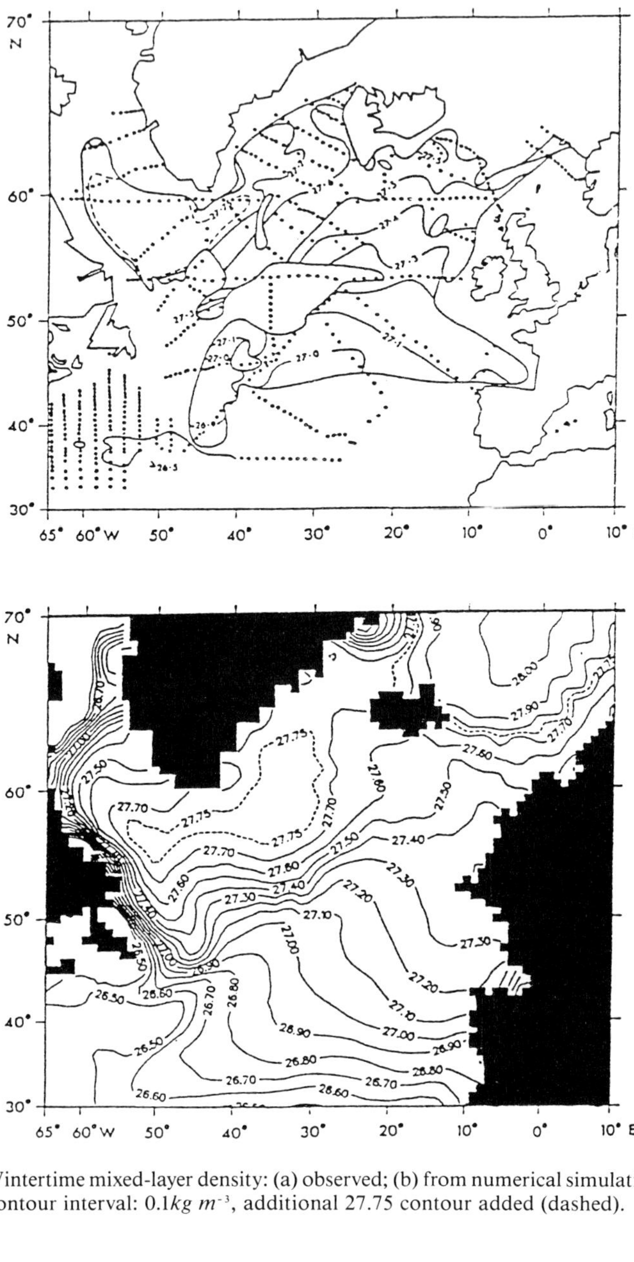

FIGURE 4. Wintertime mixed-layer density: (a) observed; (b) from numerical simulation in March of year 30. Contour interval: 0.1$kg\ m^{-3}$, additional 27.75 contour added (dashed).

In late winter (Figure 5a), the interface marking the mixed layer bottom is sharply inclined polewards of 50°N, allowing the mixed layer to reach depths in excess of 1000m. In summer (Figure 5b), almost the entire volume occupied by the subpolar wintertime mixed layer is seen to be filled with isopycnal layers; a seasonal thermocline has formed. Layer interfaces there rise sharply to the surface in a manner reminiscent of the wintertime mixed-layer isotherms.

The "bowl" formed by the upper isopycnic layers between 15°N and 50°N in Figure 5b represents the (primarily wind-driven) subtropical gyre. Figure 5a shows that this bowl is truncated along its northern edge by the wintertime mixed layer;

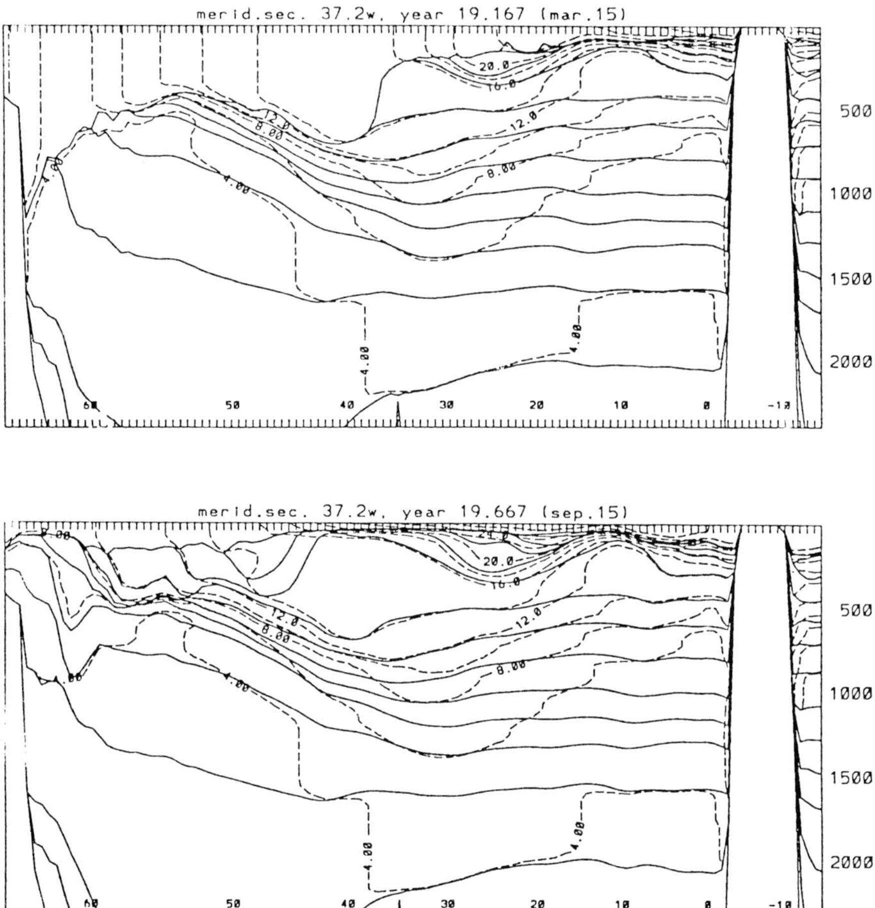

FIGURE 5. Vertical-meridional section through model grid along 34.5°W (a) in March and (b) in September of year 19. Solid lines: layer interfaces; dashed lines: isotherms. The sharp rise south of the equator corresponds to the eastern tip of the South American continent.

that this bowl is truncated along its northern edge by the wintertime mixed layer; in other words, the bowl resides partly in the seasonal and partly in the permanent thermocline. The fact that the surface mixed layer reaches down into the subtropical gyre on an annual basis means that the subtropical gyre is a "ventilated" body of water.

OUTLOOK

Recognition of the quasi-adiabatic character of oceanic motion on fairly long time scales has encouraged the use of layer models in studying dynamic interaction modes, but a variety of technical difficulties has impeded their development into full-fledged OGCMs. Most of these difficulties have been overcome by now. The intrinsic capacity of layer models to dynamically adapt their spatial vertical resolution to the occurrence of baroclinic zones, as well as the prospect of being able to handle long-term climate-oriented OGCM integrations in a framework where diabatic exchange processes are totally under the experimenter's control, has encouraged several modeling groups to pursue the development of practical isopycnic OGCM codes to study the oceanic circulation on time scales ranging from days to centuries.

At the time of this writing, both the Miami and the Rennell group are gearing up for basin-scale, eddy-resolving simulations. The Miami group intends to do its simulations on a massively parellel processor (MPP). During the past year, the Los Alamos National Laboratory has provided generous programmer assistance to port the Miami model to one particular MPP, the CM-5 built by Thinking Machines Inc. Array sizes on a Connection Machine must be padded to the next power of 2 for good performance, implying that an ocean basin fitting snugly into an array of size 2^m x 2^n minimizes the number of unused processors. (Many processors remain unused as it is because they end up on dry land.)

A 0.9-degree version of the model (128 x 128 horizontal grid points) runs on 64 nodes of the CM-5 at 1.3 times the speed of a single-processor Cray Y-MP, or 3.5 hrs. per model year. Optimal machine utilization will be reached with a further doubling of the horizontal grid resolution. To achieve satisfactory eddy resolution, the number of horizontal grid points will have to be quadrupled (0.225-degree mesh).

ACKNOWLEDGEMENTS

Many co-workers contributed toward the development of the Miami model in the past ten years as well as to the results presented in this chapter. To name a few, we would like to acknowledge the late D. Boudra, A. Bennett, S. Dean, H. Hanson, D. Hu, E. Kraus, K. Maillet, A. New, C. Rooth, L. Smith and S. Sun. Support was provided by the National Science Foundation, the Office of Naval Research and the Department of Energy.

REFERENCES

1. Bryan K. 1969. A numerical method for the study of the circulation of the world ocean. *J. Comput. Phys.,* 4,347-376.
2. Phillips, N.A. 1957. A coordinate system having some special advantages for numerical forecasting. *J. Meteor.,* 14,184-185.
3. Blumberg, A.F. and G.L. Mellor. 1987. A description of a three-dimensional coastal ocean circulation model. *Three-Dimensional Coastal Ocean Models,* 4, edited by N. Heaps. American Geophysical Union, 208 pp.
4. Haidvogel, D.B., J. Wilkin and R. Young. 1991. A semi-spectral primitive equation ocean circulation model using vertical sigma and orthogonal curvilinear horizontal coordinates. *J. Comp. Phys.,* 94,151-185.
5. Montgomery, R.B. 1937. A suggested method for representing gradient flow in isentropic surfaces. *Bull. Amer. Meteor. Soc.,* 18,210-212.
6. Bleck, R. and D.B. Boudra. 1981. Initial testing of a numerical ocean circulation model using a hybrid (quasi-isopycnic) vertical coordinate. *J. Phys. Oeanogr.,* 11,755-770.
7. Bleck, R. and D.B. Boudra. 1986. Wind-driven spin-up in eddy-resolving ocean models formulated in isopycnic and isobaric coordinates. *J. Geophys. Res.,* 91C,7611-7621.
8. Zalesak, S. 1979. Fully multidimensional flux-corrected transport algorithms for fluids. *J. Comp. Phys.,* 31,335-362.
9. Smolarkiewicz, P. 1984. A fully multidimensional positive definite advection transport algorithm with small implicit diffusion. *J. Comput. Phys.,* 54,325-362.
10. Sun, S., R. Bleck and E.P. Chassignet. 1993. Layer outcropping in numerical models of stratified flows. *J. Phys. Oceanogr.,* 23, 1877-1884.
11. Luyten, J.R., J. Pedlosky and H. Stommel. 1983. The ventilated thermocline. *J. Phys. Oceanogr.,* 13,292-309.
12. Luther, M.E. and J.J. O'Brien. 1985. A model of the seasonal circulation in the Arabian Sea forced by observed winds. *Prog. Oceanogr.,* 14,353-385.
13. Jensen, T.G. 1991. Modeling the seasonal undercurrents in the Somali Current system. *J. Geophys. Res.,* 96,22151-22167.
14. Hurlburt, H.E. and J.D. Thompson, 1980. A numerical study of loop intrusions and eddy shedding. *J. Phys. Oceanogr.,* 10,1611-1651.
15. Thompson, J.D. and W.J. Schmitz. 1989. A limited-area model of the Gulf Stream: Design, initial experiments, and model-data intercomparison. *J. Phys. Oceanogr.,* 19,791-814.
16. McCreary, J.P. and P.K. Kundu. 1988. A numerical investigation of the Somali current during the Southwest Monoon. *J. Mar. Res,* 46,25-58.
17. Bennett, A. 1992. Personal communication.
18. Oberhuber, J. 1993. Simulation of the Atlantic circulation with a coupled sea ice - mixed layer - isopycnal general circulation model. Part I and II. *J. Phys. Oceanogr.,* 23,808-845.
19. Bleck, R. H.P. Hanson, D. Hu and E.B. Kraus. 1989. Mixed layer-thermocline interaction in a three-dimensional isopycnic coordinate model, *J. Phys.*

Oceanogr., 19,1417-1439.
20. Bleck, R. and L. Smith. 1990. A wind-driven isopycnic coordinate model of the north and equatorial Atlantic Ocean. 1. Model development and supporting experiments. *J. Geophys. Res.,* 95C, 3273-3285.
21. Bleck, R., C. Rooth, D. Hu and L. Smith. 1992. Salinity-driven thermocline transients in a wind- and thermohaline-forced isopycnic coordinate model of the North Atlantic. *J. Phys. Oceanogr.,* 22,1486-1505.
22. Bleck, R. 1978. Finite difference equations in generalized vertical coordinates. Part I: Total energy conservation. *Contrib. Atm. Phys.,* 51, 360-372.
23. Arakawa, A. and V.R. Lamb. 1977. Computational design of the basic processes of the UCLA general circulation model, *Methods Comput. Phys.,* 17,174-265.
24. Kraus, E.B. and J.S. Turner. 1967. A one-dimensional model of the seasonal thermocline: II. The general theory and its consequences. *Tellus,* 19,98-106.
25. Friedrich, H. and S. Levitus. 1972. An approximation to the equation of state for sea water, suitable for numercial ocean models. *J. Phys. Oceanogr.,* 2,514-517.
26. Smolarkiewicz, P. and W. Grabowski. 1990. The multidimensional positive definite advection transport algorithm: Nonoscillatory option. *J. Comput. Phys.,* 86,355-375.
27. Smagorinsky, J.S. 1963. General circulation experiments with the primitive equations. I: The basic experiment. *Mon. Wea. Rev.,* 91,99-164.
28. Hu, D. 1991. A joint mixed-layer/isopycnic coordinate numerical model of wind- and thermohaline-driven ocean general circulation with model sensitivity study. Ph. D. dissertation. U. of Miami. 220 pp.
29. Huang, R.X. 1984. The thermocline and current structure in subtropical/subpolar basins. Ph.D. thesis, WHOI-84-42, 218 pp.
30. Huang, R.X. 1986. Numerical simulation of wind-driven circulation in a subtropical/subpolar basin, *J. Phys. Oceanogr.,* 16,1636-1650.
31. Huang, R.X. and G. Flierl. 1987. Two-layer models for the thermocline and current structure in subtropical/subpolar gyres. *J. Phys. Oceanogr.,* 17,872-884.
32. Smith, L.T. 1992. Numerical simulations of stratified rotating over finite amplitude topography. *J. Phys. Oceanogr.,*22,686-696.
33. Huppert, H.E. and K. Bryan. 1976. Topographically generated eddies. *Deep-Sea Res.,* 23,655-679.
34. Bryan, F.O. and W.R. Holland. 1989. A high resolution simulation of the wind- and thermohaline-driven circulation in the North Atlantic Ocean. In *Parameterization of Small Scale Processes.* Proceedings Aha Huliko'a Hawaiian Winter Workshop. U. of Hawaii. January 17-20, 1989. P. Muller and D. Anderson, Eds.
35. Levitus, S. 1982. *Climatological Atlas of the World Ocean.* NOAA Professional Paper 13, 173 pp.
36. Hellerman, S. and M. Rosenstein. 1983. Normal monthly wind stress over the World Ocean with error estimates. *J. Phys. Oceanogr.,* 13,1093-1104.
37. Esbensen, S.K. and Y. Kushnir. 1981. Heat budget of the global ocean: estimates from surface marine observations. Rep. No. 29, Climatic Research Institute, Oregon State University, Corvallis, OR, 27 pp.

38. Jaeger, L. 1976. Monatskarten des Niederschlags fur die ganze Erde. Berichte des Deutschen Wetterdienstes, 18, No. 139, Ofenbach, FRG.
39. New, A., R. Bleck, Y. Jia, R. Marsh, M. Huddleston and S. Barnard. 1994. An isopycnic model study of the North Atlantic. Part 1: Model experiment and water mass formation. Part 2: Interdecadal variability of the subtropical gyre. *J. Phys. Oceanogr.,* submitted.
40. Semtner, A.J. and R.M. Chervin. 1988. A simulation of the global ocean circulation with resolved eddies. *J. Geophys. Res.,* 93,15502-15522.
41. Semtner, A.J. and R.M. Chervin. 1992. Ocean general circulation from a global eddy-resolving model. *J. Geophys. Res.,* 97,5493-5550.
42. McCartney, M.S. and L.D. Talley, 1982. The subpolar mode water of the North Atlantic Ocean. *J. Phys. Oceanogr.,* 12,1169-1188.

The Oceans: Physical-Chemical Dynamics and Human Impact. Edited by S.K. Majumdar, E.W. Miller, G.S. Forbes, R.F. Schmalz and Assad A. Panah. © 1994, The Pennsylvania Academy of Science.

Chapter Three

NONLINEAR OCEAN DYNAMICS

A.D. KIRWAN, JR., B.L. LIPPHARDT, JR., and K.L. GREGORY

Center for Coastal Physical Oceanography
Old Dominion University
Norfolk, VA 23529-0276

ABSTRACT

Kinematic aspects of the nonlinear terms in the hydrodynamic equations are explored through the use of a three-layer model on an *f* plane with the middle layer restricted to a finite size. The effects of horizontal divergence, spin (vorticity/2) and deformation are illustrated with special emphasis given to a comparison of the nonlinear quasigeostrophic and fully nonlinear solutions. The latter are characterized by an elevated level of superinertial scale variability.

INTRODUCTION

One of the latest buzz words in oceanography is "nonlinear." It is commonly used in conjunction with

- weakly nonlinear wave phenomena such as solitons. See, for example, Rizzoli [1982], Osborne et al. [1986], Osborne and Boffetta [1989] and Sakuma and Ghil [1991];
- nonlinear quasigeostrophic (QG) phenomena such as modons. A recent review and extension to a rotating frame is given in Mied et al. [1992]; and
- chaos. A substantial number of studies have been reported on a wide variety of topics. These include Samelson and Allen's [1987] analysis of flow over variable topography, analysis of trajectory data reported by Brown and Smith [1990, 1991]

and the important theoretical analyses of Osborne and Caponio [1990] and Osborne et al. [1991].

Nonlinear hydrodynamic processes, in fact, encompass much more than the above. The goal here is to illustrate some of the general kinematic properties of the nonlinear terms in the hydrodynamic equations rather than provide a synopsis of some specific nonlinear phenomena. It is hoped that this approach will unify some views on nonlinear dynamics as they may pertain to submesocale oceanographic processes.

Two recent developments that do not appear to fall into the above three categories have motivated the approach taken here. The first is the discovery of small intense eddies in nearly every part of the world ocean. As these flows have nearly zero potential vorticity they may be referred to as near zero potential vorticity vortices, or NZPVs. Newton et al. [1974], Perkins [1976], Dugan et al. [1982], Kunze and Sanford [1984], Howe [1984], Houghton et al. [1986], Elliot and Sanford [1986a, b], Kunze and Lueck [1986], D'Asaro [1988a], Armi et al. [1989] and Kennelly et al. [1989a-c] have provided useful descriptions. The abundance of such observations suggests that NZPVs are produced at a greater rate than typical mesoscale rings and eddies spawned by western boundary currents. Because of their large numbers and ubiquitous nature NZPVs may surpass the larger mesoscale eddies in their ability to transport energy and momentum in the ocean.

Several characteristics of NZPVs are relevant to the present study. Their horizontal extent is of the order of 10 km while their thickness is typically 300 m. Such small structure means that they are hard to detect by hydrographic surveys and remote sensing. As to their shape, the very detailed measurements recorded by Kennelly et al. [1989a-c] indicate significant ellipticity whereas the analysis of McWilliams [1985] is based on nearly circular structure. Ellipticity is consistent with the notion that these vortices are advected by large scale flow, in which case environmental deformations likely will distort the vortex shape. Estimates of potential vorticity for these flows range from slightly positive [Kunze, 1986; Houghton et al., 1986; Elliot and Sanford, 1986a] to zero [Newton et al., 1974; D'Asaro, 1988a] to negative [Howe, 1984; Houghton et al., 1986]. It is commonly believed that negative potential vorticity (superinertial) vortices are "inertially unstable" [Fjørtoft, 1946]; however, Ripa [1987] has shown that this is not necessarily true. Finally, analyses of observations by Elliot and Sanford [1986b] and D'Asaro [1988b] indicate a nonlinear momentum balance within these vortices.

The second development is the realization that standard Garrett Munk internal wave theory cannot account for all superinertial scale variability seen in the internal wave data. Analyses of internal wave observations in the late 1970s by Briscoe [1977] and Müller et al. [1978] showed that some high frequency and high wavenumber spectra were not consistent with the standard internal wave model. Recent studies substantiate this. Pinkel [1985] suggests the possibility of "two-dimensional geostrophic turbulence" as a source of this variability while Müller et al. [1988] suggests that "vortical or potential vorticity-carrying motion exists at scales traditionally associated with internal waves." These and other relevant results have been reviewed by Gregg [1987].

It is not yet possible to say whether the NZPVs described above are responsible for the superinertial fluctuations observed by Pinkel [1985] and Müller et al. [1988]. However, it should be clear from the phenomenological scales involved that a complete understanding of their dynamics requires an approach which takes into account the full nonlinearity in the hydrodynamic equations. Pursuant to this, the goal here is to characterize some relevant kinematic properties of the nonlinear terms in the hydrodynamic equations. Specific focus is directed to a comparison of nonlinear QG solutions with fully nonlinear solutions.

A three-layer model in which the middle layer is confined to a finite compact region is used for the analysis. As the model is fully nonlinear, it admits both QG and nonquasigeostrophic solutions. It was first used for geophysical fluid dynamics (GFD) flows by Cushman-Roisin et al. [1985]. Since then there have been a plethora of papers dealing with theoretical aspects as well as applications. Kawano and Nakamoto [1992] have even extended the approach to cyclonic eddies.

The section FORMULATION AND REVIEW OF PREVIOUS WORK describes the model and presents a novel scaling for the nonlinear terms. The section ISOLATED LENS SOLUTIONS gives examples of solutions when the vortex is isolated from motions of its environment. Here, nonlinear quasigeostrophic solutions are compared with those from a fully nonlinear, nonquasigeostrophic case. Environmentally forced solutions are considered in the section ENVIRONMENTALLY FORCED SOLUTIONS. The robust nature of nonlinear solutions is demonstrated by considering a case of resonant forcing of the middle layer. It is seen that stable solutions exist even when the frequencies in the environmental flow are resonant with those of the unforced solution for that layer. The final section, DISCUSSION / CONCLUSION summarizes the findings and discusses the possible significance of the results.

FORMULATION AND REVIEW OF PREVIOUS WORK

The hydrodynamic equations for a three-layered fluid on a rotation frame are well known to be

$$\frac{D}{Dt}U_i^{(1)} + f\sigma_{ij}U_j^{(1)} + g\frac{\partial}{\partial x_i}(h^{(1)} + h^{(2)} + h^{(3)}) = 0 \tag{1A}$$

$$\frac{D}{Dt}h^{(1)} + h^{(1)}\frac{\partial U_j^{(1)}}{\partial x_j} = 0 \tag{1B}$$

$$\frac{D}{Dt}U_i^{(2)} + f\sigma_{ij}U_j^{(2)} + g\frac{\partial}{\partial x_i}[(1 - \epsilon_{21})h^{(1)} + h^{(2)} + h^{(3)}] = 0 \tag{2A}$$

$$\frac{D}{Dt}h^{(2)} + h^{(2)}\frac{\partial U_j^{(2)}}{\partial x_j} = 0 \tag{2B}$$

$$\frac{D}{Dt}U_i^{(3)} + f\sigma_{ij}U_j^{(3)} + g\frac{\partial}{\partial x_i}[(1-\epsilon_{31})h^{(1)} + (1-\epsilon_{32})h^{(2)} + h^{(3)}] = 0 \tag{3A}$$

$$\frac{D}{Dt}h^{(3)} + h^{(3)}\frac{\partial U_j^{(3)}}{\partial x_j} = 0. \tag{3B}$$

Here, U_i^α and h^α are respectively the ith component of the velocity and the thickness of the α layer. Also, f and g are respectively the Coriolis parameter and local gravity. Finally, $D/Dt = \partial/\partial t + U_j\partial/\partial x_j$ is the material derivative, $\epsilon_{pq} = (\rho_p - \rho_q)/\rho_p$ is the density contrast between layers p and q, and $\sigma_{ij} = \begin{bmatrix} 0 & -1 \\ 1 & 0 \end{bmatrix}$.

The middle layer is prescribed as a patch of fluid having compact support, while the other layers are unbounded. Figure 1 is an illustration of the geometry for this situation.

Attention here is focused on the dynamics of the middle layer when the exterior flow, i.e., layers 1 and 3, is prescribed. Thus, the method employed by Ruddick [1987] is appropriate. In this procedure, (1A) and (3A) are inverted to express the gradients of $h^{(1)}$ and $h^{(3)}$ in terms of the accelerations of $U_i^{(1)}$and $U_i^{(3)}$and the gradient of $h^{(2)}$. These are then substituted into (2A) to yield (dropping superscripts for layer 2)

$$\frac{D}{Dt}U_i + f\sigma_{ij}U_j + g_*\frac{\partial h}{\partial x_i} = W^{(1)}\mathcal{L}_i^{(1)} + W^{(3)}\mathcal{L}_i^{(3)} \tag{4}$$

where

- $g_* = g\epsilon_{32}\epsilon_{21}/\epsilon_{31}$ is the "reduced" gravity for the middle layer;
- $W^{(1)} = (\epsilon_{31} - \epsilon_{21})/\epsilon_{31}$ is equal to 1 - $W^{(3)}$; and
- $L_i^{(\alpha)} = \frac{D}{Dt} U_i^{(\alpha)} + f\sigma_{ij}U_j^{(\alpha)}$ is the inertial and Coriolis acceleration of the α layer.

Note that when the upper layer is absent, ($\rho_1 = 0$), $W^{(1)} = 0$ and (4) reduces to the two-layer case. The barotropic shallow water or reduces gravity equations follow accordingly.

Solutions to equations (2B) and (4) are the focus here. There are a number of approaches that can be taken, none of which is without some caveat. As indicated above, we have elected to employ a polynomial expansion, due to Goldsbrough [1930], that has been widely used in many previous studies of this problem. See, for example, Thacker [1981], Ball [1963, 1965], Cushman-Roisin et al. [1985], Young [1986], Cushman-Roisin [1987], Kirwan and Liu [1991], Rogers [1989], Kirwan et al. [1990], Brickman and Ruddick [1990] and Kawano and Nakamoto [1992].

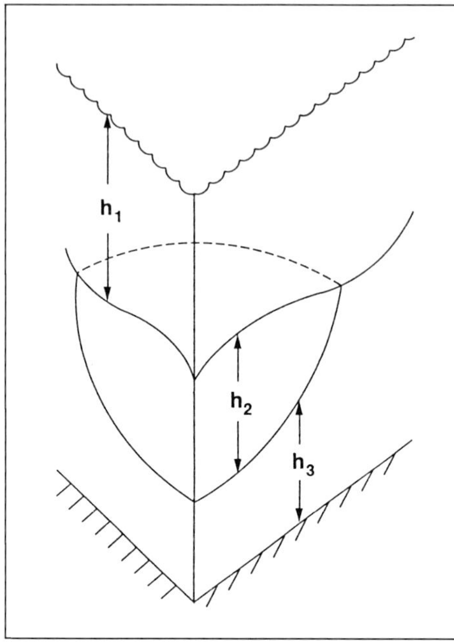

FIGURE 1. Illustration of the geometry of the three layer model.

In this expansion the lens swirl velocity and thickness are expressed respectively as linear and quadratic functions of spatial coordinates relative to the center of mass of the middle layer. Clearly this oversimplifies the horizontal velocity profile, constrains the lens geometry to only one asymmetric mode and precludes viscous and mixing processes. On the other hand, this procedure preserves exactly the nonlinearity in the layer version of the hydrodynamic equations without limitations or complications imposed by asymptotic expansions.

In conjunction with the expansion it will also be convenient to nondimensionalize (2B) and (4). For this purpose, the horizontal scale is specified as L, the vertical scale as H, the swirl velocity scale as U and the evolutionary time scale as T. The Goldsbrough expansion then gives

$$\tilde{U}_i = \tilde{G}_{ij}(\tilde{t})\tilde{x}_j = Uu_i = L[\alpha\delta_{ij}G(t)/2 + \beta G^d_{ij}(t) + \gamma G^r_{ij}(t)]x_j \quad \text{(5A)}$$

and

$$\tilde{h} = \tilde{h}_0(\tilde{t}) + \tilde{B}_{ij}(\tilde{t})\tilde{x}_i\tilde{x}_j = Hh = Hh_0(t) + L^2[(1/L_1)\delta_{ij}B(t)/2 + (1/L_2)B^d_{ij}(t)]x_ix_j \quad \text{(5B)}$$

where $\tilde{x}_i = Lx_i$ and $\tilde{t} = Tt$ and the ~ variables are dimensional. Also, the tensors G_{ij} and B_{ij} have been decomposed into their "invariant" forms (see Sedov [1966]) as

$$G = G_{kk}$$
$$G_{ij}^{d} = (G_{ij} + G_{ji})/2 - \delta_{ij}G/2$$
$$G_{ij}^{r} = (G_{ij} - G_{ji})/2$$
$$B = B_{kk}$$
$$B_{ij}^{d} = (B_{ij} + B_{ji})/2 - \delta_{ij}B/2. \tag{6}$$

Here, G, G_{ij}^{d} and G_{ij}^{r} are recognized respectively as the nondimensional horizontal divergence, deformation tensor and spin (vorticity/2) tensor. The kinematic interpretations of these three components are

- G — horizontal cross-sectional area changes preserving shape and orientation;
- G_{ij}^{d} — shape changes preserving area and orientation; and
- G_{ij}^{r} — orientation changes preserving area and shape of a generic parcel.

See Dutton [1986] or Sedov [1966] for further discussion. The reciprocal time scales associated with these three processes are α, β and γ. Young [1986] has shown that for this model G is associated with inertial time scales and G_{ij}^{d} are associated with superinertial time scales. Time scales associated with G_{ij}^{r} may be either super- or subinertial.

The tensor B_{ij} also has geometric significance. For example, for a circular middle layer, $B_{ij}^{d} = 0$. Then, L_1 is a scale appropriate for the lens radius. For elliptical lenses, L_1 and L_2 can be related to scales for the major and minor axes.

The introduction of the α, β, γ scales has the advantage of permitting a scaling associated with different kinematic modes of the velocity field. As will be seen, this leads to a Rossby number scaling by $(\alpha/f, \beta/f, \gamma/f)$ rather than the conventional U/fL. The presence of three Rossby numbers deserves some comment. The spin component of the velocity gradient, γ/f, has been used previously to specify the Rossby number in studies of quasigeostrophic lenses by McWilliams [1985], D'Asaro [1988a] and Hebert et al. [1990]. However, for nonquasigeostrophic flows, deformation and divergence are important ingredients in the dynamics. Hence, scaling by typical magnitudes of these components seems equally appropriate. Of course, the classical Rossby number is recovered by setting $\alpha = \beta = \gamma$.

Applying (5) to (2B) and (4), using (6) and extensive algebra, yields

$$r\,\dot{h}_0 + \epsilon_\alpha h_0 G = 0 \tag{7A}$$

$$r\,\dot{B} + 2[\epsilon_\alpha BG + 2\epsilon_\beta(L_1/L_2)(B_N G_N + B_S G_S)] = 0 \tag{7B}$$

$$r\,\dot{B}_S + 2\epsilon_\alpha B_S G + \epsilon_\beta(L_2/L_1)G_S B - 2\epsilon_\gamma(L_2/L_1)B_N G_R = 0 \tag{7C}$$

$$r\,\dot{B}_N + 2\epsilon_\alpha B_N G + \epsilon_\beta(L_2/L_1)G_N B + 2\epsilon_\gamma(L_2/L_1)B_S G_R = 0 \tag{7D}$$

$$\alpha r\,\dot{G} + \alpha\epsilon_\alpha G^2/2 + 2\beta\epsilon_\beta(G_N^2 + G_S^2) - 2\epsilon_\gamma(\gamma G_R^2 + G_R) + 2(\alpha/\epsilon_\alpha)(L_i/L_1^*)^2 B = W^{(1)}\mathcal{L}_G^1 + W^{(3)}\mathcal{L}_G^3 \tag{7E}$$

$$\gamma r\,\dot{G}_R + \gamma\epsilon_\alpha G G_R + \epsilon_\alpha G/2 = 0 \tag{7F}$$

$$\beta r\,\dot{G}_N + \beta\epsilon_\alpha G G_N - \epsilon_\beta G_S + 2(\beta/\epsilon_\beta)(L_i/L_2^*)^2 B_N = W^{(1)}\mathcal{L}_N^i + W^{(3)}\mathcal{L}_N^3 \tag{7G}$$

$$\beta r\,\dot{G}_S + (\beta\epsilon_\alpha) G G_S + E_\beta G_N + 2(\beta/\epsilon_\beta)(L_i/L_2^*)^2 B_S = W^{(1)}\mathcal{L}_S^1 + W^{(3)}\mathcal{L}_S^3 \tag{7H}$$

where

- $r = 1/fT$, the ratio of the pendulum day to evolutionary time, or Kibel number;
- $(\epsilon_\alpha, \epsilon_\beta, \epsilon_\gamma) = (\alpha/f, \beta/f, \gamma/f$, the Rossby numbers;
- $L_i/L_p^* = \sqrt{g_*H/f)}/\sqrt{HL_p}$, $p = 1, 2$. These are the ratios of the internal deformation radius to the geometric scales;
- $(\dot{\cdot}) = d/dt$;

and

- $\mathcal{L}_G^{(\alpha)} = \dot{G}^{(\alpha)} + (G^{(\alpha)})^2/2 + 2[(G_N^{(\alpha)})^2 + (G_S^{(\alpha)})^2 - (G_R^{(\alpha)})^2 - G_R^{(\alpha)}];$
- $\mathcal{L}_N^{(\alpha)} = \dot{G}_N^{(\alpha)} + G^{(\alpha)}G_N^{(\alpha)} - G_S^{(\alpha)};$
- $\mathcal{L}_S^{(\alpha)} = \dot{G}_S^{(\alpha)} + G^{(\alpha)}G_S^{(\alpha)} + G_N^{(\alpha)}.$

Equations equivalent to a homogenous version of (7) were first derived by Ball [1963, 1965] who also showed that there are five properties that are conserved when integrated over the middle layer volume. Four of these have succinct physical meaning: volume, potential vorticity, total energy and total angular momentum. These are general results depending only upon (2) and plausible flux divergence conditions at the lens edge, and not on specific solution forms such as (5).

ISOLATED LENS SOLUTIONS

A number of special solutions to (7) have been reported when the lens is isolated from the environment, i.e., all L_i are zero. These can be categorized by selection of the scaling parameters α, β and γ. Figure 2 summarizes the results. The simplest solutions are the three associated with any one particular kinematic mode: horizontal divergence, deformation or spin. These are obtained by setting any two of α, β and γ equal to zero while requiring the third to be nonzero. As indicated in the figure, these solutions are manifested respectively as cross-sectional area changes, shape changes and orientation changes of the middle layer. Note that the $\alpha \neq 0$

case can exist only for $f = 0$ since an f plane will induce vorticity with radial motion. The $\gamma \neq 0$ case, on the other hand, requires $f \neq 0$ if the lens is not to evolve into an infinitesimally thin pancake with an infinite radius. The $\beta \neq 0$ case, reported by Liu [1992], exists for both $f = 0$ and $f \neq 0$.

The next level of complication involves solutions that mix two of the three kinematic modes. Here, one of three scales is set to zero. As indicated in the figure, there are three cases of these solutions. Wave-like solutions are obtained by requiring that $\gamma = 0$. These can exist only if $f = 0$ for the reason cited above. Such solutions can be obtained from a velocity potential and have some superficial similarity to stationary gravity solitons. Quasigeostrophic solutions are obtained by requiring that $\alpha = 0$. These solutions can be developed from streamfunctions. Such solutions have been thoroughly dissected in the GFD literature. One of these is the *rodon,* which is analytic. It arises from a special initial dynamical balance. Liu [1992] has a summary of recent applications of this class to observations. The third class of solutions ($\beta = 0$) are sometimes called *pulsons* since the lens pulsates at the intertial frequency. These solutions are also analytic, however, they are unlike most other nonlinear solutions as the oscillation frequency is always exactly inertial and independent of velocity amplitudes or vertical displacement of the layer oscillations.

The most general solutions permit all three kinematic modes. These modes are nonlinearly superposed so that there can be several frequencies interacting nonlinearly with each other. Cushman-Roisin [1987] and Rogers [1989] have discovered special analytic solutions that include all three kinematic modes. Numerical investigations of the general case have been reported by Kirwan et al. [1992] and Kirwan and Lipphardt [1993]. Their results suggest behavior that is substantially different from that predicted by QG theory.

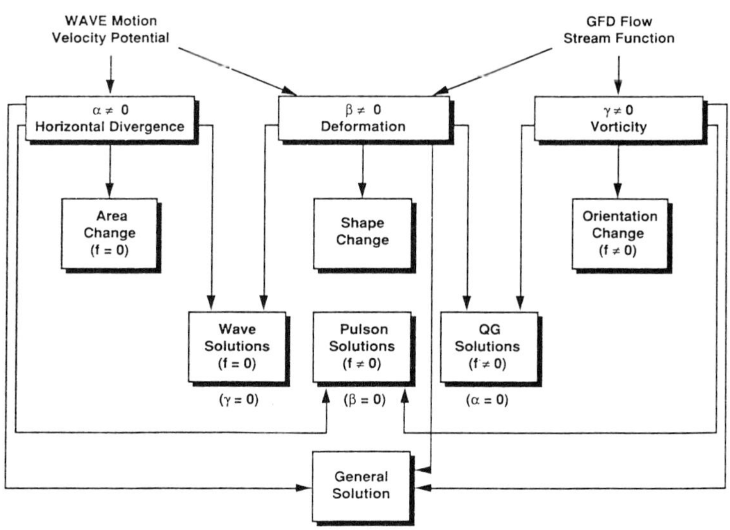

FIGURE 2. Schematic relating the classes of solutions to the invariant components of the velocity gradient, i.e., the divergence, spin and deformation.

TABLE 1

Initial conditions for numerical experiments described in text.

	h_0	B	B_N	B_S	G	G_R	G_N	G_S
Circular vortex	1.0	-0.10	0.0	0.000	0.00	-0.450	0.5	0.0000
Subinertial rodon	1.0	-0.25	0.0	-0.039	0.00	-0.404	0.0	-0.0964

To illustrate this we compare the evolution of a lens predicted by QG dynamics with that obtained from consideration of all three kinematic modes. The initial conditions for all simulations are given in Table 1. For the QG case, the solution was obtained numerically (see Kirwan and Lipphardt [1993] for details) by solving (7) with L and α equal to zero. This was repeated for the general case except that α was flagged to 1.

Figure 3 compares the solutions. The dotted curves in this figure give the QG solution ($\alpha = 0$). The deformation field of the QG solution ($G_D = \sqrt{G_N^2 + G_S^2}$) oscillates at a single superinertial frequency with a cuspy amplitude. The lens velocity in this case, calculated at point (1.712, 0) and (0.919, 0) for the $\alpha = 0$ and $\alpha = 1$ cases respectively, shows that both super- and subinertial frequencies are present. The other three fields are stationary.

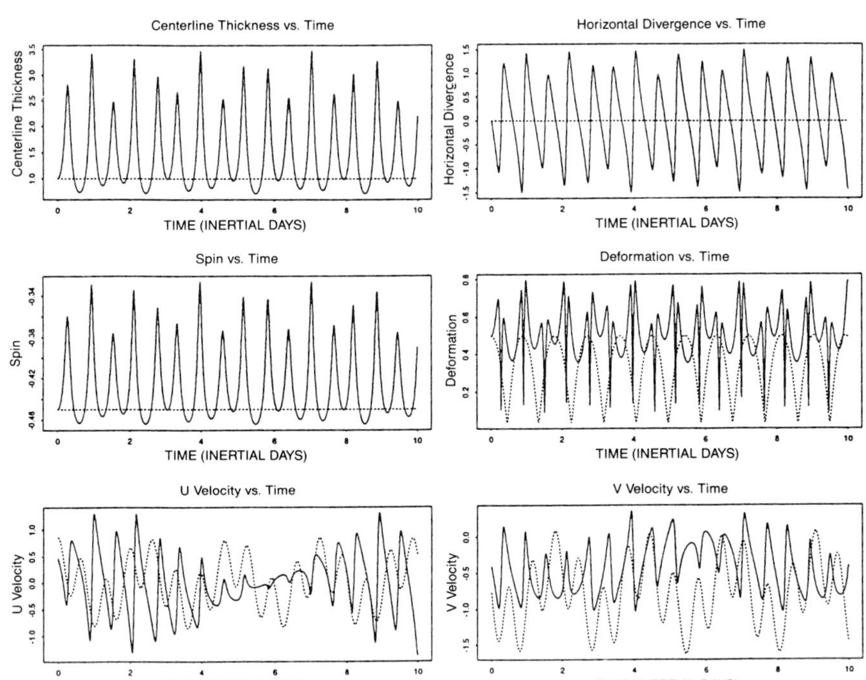

FIGURE 3. Solutions for the unforced case for centerline thickness (h_o), horizontal divergence (G), spin (G_R), deformation ($\sqrt{G_N^2 + G_S^2}$) U velocity and V velocity.

The general solutions ($\alpha = 1$) are given by the solid curves. All the fields show superinertial fluctuations. It is noteworthy that the oscillations of the thickness (h_o) and spin (G_R) fields show significant asymmetry; for example, during any one cycle the thickness field spends more time in the shallow configuration with only a short spurt into the deep configuration. The deformation and velocity fields are particularly skewed and erratic when compared with the QG solutions.

Most of these characteristics have been noted previously by other investigators. They clearly show that QG-based models, although nonlinear, cannot account for a significant amount of the variability contained in the general solutions. Much of this occurs at superinertial time scales. The superinertial scale fluctuations are associated with deformation in the velocity field. Horizontal divergence seems to play the role of catalyst. Its presence in the general case enhances those fluctuations. The presence of such fluctuations also suggests that these NZPVs could play a significant role in both mesoscale and small scale process.

ENVIRONMENTALLY FORCED SOLUTIONS

Here we compare the responses of the QG and general nonlinear balances in the upper layer when it is forced by prescribed flow in the lower layer. To do this, the $W^{(\alpha)}$ in (7) are set to 0.5 and the flow fields in these two layers are prescribed as described below. This approach was first used by Ruddick [1987].

It is not well appreciated that nonlinearity may actually enhance the stability of flows. In order to illustrate this, a somewhat artificial case of resonant forcing is considered. This is achieved by using the unforced solution for the middle layer as the prescribed flow for the upper and lower layers. If the dynamics of the intermediate layer were linear then clearly this prescription would produce secular growth in the solutions, since the forcing would include natural frequencies of the middle layer. As shown below, such behavior does not occur when the middle layer can respond nonlinearly.

Initial conditions for all three layers are given in the second line of Table 1. For an isolated layer these initial conditions would produce a rodon solution. This is an analytic QG solution in which the intermediate layer has a constant elliptical shape that rotates anticyclonically at a constant rate.

Time series of the solutions are shown in Figure 4. The dotted curves represent the QG response while the solid curves depict the general solution. It is seen that the forcing produces subinertial fluctuations in the thickness, divergence and spin fields in the general case. Unlike the unforced case, the thickness field tends to a configuration substantially thinner than the QG solution. However, the basic asymmetry of the solutions, in which the lens spends more time during one cycle in the shallow configuration, is preserved. As with the unforced case, the QG assumption prevents evolution of these three fields in this model.

It was surprising to find such large differences in the evolution of the deformation fields of the two cases. The QG solution showed a rather large subinertial oscillalation of this field but with a smaller superposed superinertial component also present.

Other experiments have shown that as the magnitude of the forcing increases the subinertial fluctuations increase in magnitude and shift to lower frequencies. The response of the deformation for the general case was considerably subdued.

The velocity fields, calculated at (1.316, 0) for the $\alpha = 0$ case and at (2.393, 0) for the $\alpha = 1$ case, also showed considerable difference in their responses. The QG solutions contained an ultralow frequency component (approximately 70 days) along with lesser sub- and superintertial scale fluctuations. Subinertial scale fluctuations were also present in the general case solutions but the most striking feature was the distortion in the fluctuations.

Figure 5 shows Poincaré maps for four selected combinations of variables. These were obtained by sampling the time series once per inertial period for 1000 inertial periods. In this figure the aspect ratio is the ratio of the minor to major ellipse axis. The x's are for the QG solution while the dots are for the general case. There is a great deal more scatter and less structure in the general case. Curiously, the general case maps also differ qualitatively from the maps given in Kirwan and Lipphardt [1993] where the forcing was at the inertial frequency. It appears from this that Poincaré maps may be useful for characterizing the solutions.

It is important to note that resonant forcing does not necessarily induce an instability in the solutions. It is also clear that there is a significant difference in the

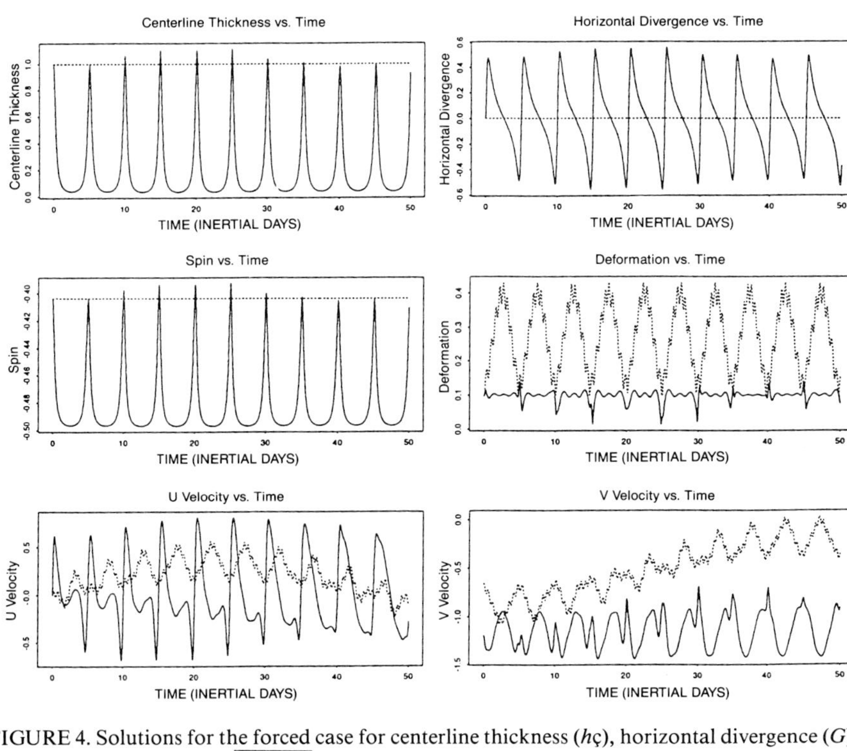

FIGURE 4. Solutions for the forced case for centerline thickness ($h_ç$), horizontal divergence (G), spin (G_R), deformation $(\sqrt{G_N^2 + G_S^2})$, U velocity and V velocity.

responses of the QG and general cases. Detailed stability studies of these two cases could be enlightening.

DISCUSSION / CONCLUSION

The model developed here appears to be applicable to submesoscale vortices, which are common in the ocean. Both observations and scaling arguments suggest that there is a nonlinear dynamical balance in these structures. The calculations presented here support the notion that the general nonlinear balance is more appropriate than the QG balance.

The general solutions for the isolated case exhibited considerably more superinertial scale variability than did the nonlinear QG solutions. The velocity deformation is responsible for this variability, but it is also necessary for horizontal divergence to be present for its effects to be fully felt. There was also considerable difference in the response of the QG and general cases to environmental forcing. Additional study, particularly in regard to the stability of the solutions, seems warranted.

There are two broader issues that deserve mention. First, spectral analysis is a standard time series tool used in oceanography. The time series generated by the general model often showed substantial distortion, even when the basic signal was

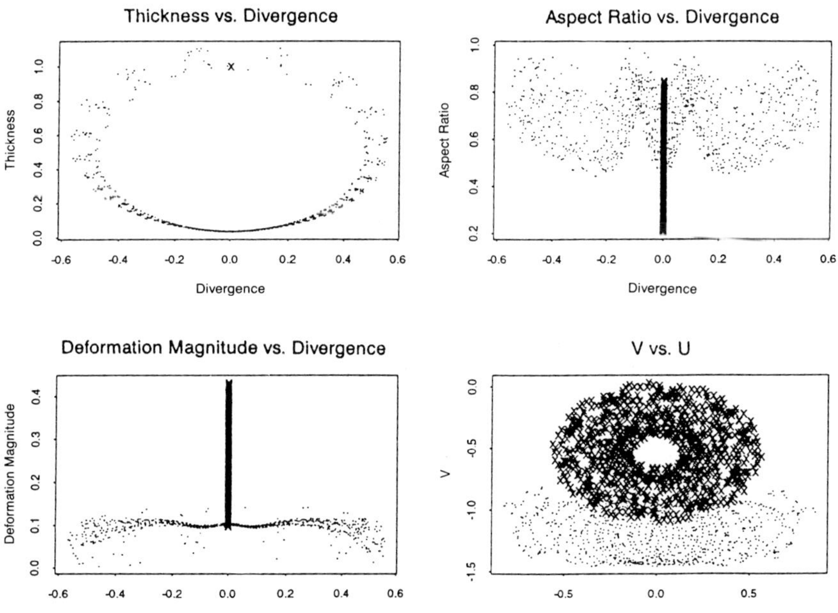

FIGURE 5. Poincaré maps for h_0 versus G, aspect ratio (defined in text) versus G, G_D versus G, V versus U. The *x's* are for the unforced case; the dots are for the resonantly forced case. These were obtained from a time series of 1000 inertial days.

nearly periodic. The distortion has obvious potential to alias the Fourier spectra but to date no effort has been made to quantify this effect. Second, we were struck by the power of the Poincaré maps to characterize the solutions. This tool should find more use in both experiment and theory.

Finally, we return to the question of the applicability of the approach and its results to both NZPVs and superinertial variability. An experimental test of this hypothesis will require better collaboration between the mesoscale and superinertial scale oceanographers than has existed in the past. The analysis here, however, is consistent with the notion that complete nonlinearity in the hydrodynamic equations is an important mechanism for "transinertial" variability.

ACKNOWLEDGEMENTS

This effort was sponsored in part by the Office of Naval Research and the National Science Foundation under contracts N00014-91-J-1560 and NSF-OCE-9019960 to Old Dominion University. A.D. Kirwan, Jr. acknowledges the Samuel L. and Fay M. Slover endowment to Old Dominion University. B.L. Lipphardt, Jr. acknowledges the Department of Defense National Defense and Engineering Graduate Fellowship Program.

REFERENCES

1. Rizzoli, P. Malanotte. 1982. Planetary Solitary Waves in Geophysical Flows, pp. 147-224. In: E. Saltzman (Ed.) *Advances in Geophysics*. Vol. 24. Academic Press, NY, pp. 328.
2. Osborne, A.L., A.D. Kirwan, Jr., A. Provenzale and L. Bergamasco. 1986. The Korteweg-deVries Equation in Lagrangian Coordinates. *Phys. Fluids.* 29: 656-660.
3. Osborne, A.R. and G. Boffetta. 1989. The Shallow-Water Nonlinear Schrödinger Equation in Lagrangian Coordinates. *Phys. Fluids. A.* 1: 1200-1210.
4. Sakuma, H. and M. Ghil. 1991. Stability of Propagating Modons for Small-Amplitude Perturbations. *Phys. Fluids. A.* 3: 408-414.
5. Mied, Richard P., A.D. Kirwan, Jr. and G.J. Lindemann. 1992. Rotating Modons over Isolated Topographic Features. *J. Phys. Oceanogr.* 22: 1570-1582.
6. Samelson, Roger M. and J.S. Allen. 1987. Quasi-Geostrophic Topographically Generated Mean Flow over the Continental Margin. *J. Phys. Oceanogr.* 17: 2043-2064.
7. Brown, Michael G. and Kevin B. Smith. 1990. Are SOFAR Float Trajectories Chaotic? *J. Phys. Oceanogr.* 20: 139-149.
8. Brown, Michael G. and Kevin B. Smith. 1991. Ocean Stirring and Chaotic Low-Order Dynamics. *Phys. Fluids A.* 3: 1186-1192.
9. Osborne, A.R. and R. Caponio. 1990. Fractal Trajectories and Anomalous Diffusion for Chaotic Particle Motions in 2D Turbulence. *Phys. Rev. Lett.* 64: 1733-1736.

10. Osborne, A.R., R. Caponio and A. Pastorello. 1991. Hamiltonian Chaos and Multi-fractal Particle Dynamics in 2D Turbulence. *Mod. Phys. Lett. A.* 6: 83-91.
11. Newton, J.L., K. Aagaard and L.K. Coachman. 1974. Baroclinic Eddies in the Arctic Ocean. *Deep Sea Res.* 21: 707-719.
12. Perkins, H. 1976. Observed Effect of an Eddy on Inertial Oscillations. *Deep Sea Res.* 23: 1037-1042.
13. Dugan, J.P., R.P. Mied, P.C. Mignerey and A.F. Schuetz. 1982. Compact Intrathermocline Eddies in the Sargasso Sea. *J. Geophys. Res.* 87: 385-393.
14. Kunze, E. and T.B. Sanford. 1984. The Mean and Fluctuating Velocity Structure in a Warm Core Ring. *EOS Trans.* 65: 913.
15. Howe, M.R. 1984. Current and Hydrographic Measurements in the Mediterranean Undercurrent near Cape St. Vincent. *Oceanol. Acta.* 7: 163-168.
16. Houghton, R.W., D.B. Olson and P.J. Celone. 1986. Observation of an Anticyclonic Eddy Near the Continental Shelf Break South of New England. *J. Phys. Oceanogr.* 16: 60-71.
17. Elliot, B.A. and T.B. Sanford. 1986a. The Subthermocline Lens D1, Part I: Description of Water Properties and Velocity Profiles. *J. Phys. Oceanogr.* 16: 532-548.
18. Elliot, B.A. and T.B. Sanford. 1986b. The Subthermocline Lens D1, Part II: Kinematics and Dynamics. *J. Phys. Oceanogr.* 16: 549-561.
19. Kunze, E. and R. Lueck. 1986. Velocity Profiles in a Warm Core Ring. *J. Phys. Oceanogr.* 16: 991-995.
20. D'Asaro, E.A. 1988a. Observations of Small Eddies in the Beaufort Sea. *J. Geophys. Res.* 93: 6669-6684.
21. Armi, L., D. Hebert, N. Oakey, J.F. Price, P.L. Richardson, H.T. Rossby and B. Ruddick. 1989. Two Years in the Life of a Mediterranean Salt Lens. *J. Phys. Oceanogr.* 19: 354-370.
22. Kennelly, M.A., T.B. Sanford and T.W. Lehman. 1989a. CTD Data from the Gulf of Cadiz Expedition: *R/V Oceanus* Cruise 202, Gulf of Cadiz Expedition Contribution Number 2. *Technical Report 8917.* Applied Physics Laboratory, University of Washington.
23. Kennelly, M.A., M.D. Prater and T.B. Sanford. 1989b. XBT and XSV Data from the Gulf of Cadiz Expedition: *R/V Oceanus* Cruise 202, Gulf of Cadiz Expedition Contribution Number 3. *Technical Report 8920.* Applied Physics Laboratory, University of Washington.
24. Kennelly, M.A., M.D. Prater, J.H. Dunlap, E.L. Kunze and T.B. Sanford. 1989c. XCP Data from the Gulf of Cadiz Expedition: *R/V Oceanus* Cruise 202, Gulf of Cadiz Expedition Contribution Number 4. *Technical Report 8925.* Applied Physics Laboratory, University of Washington.
25. McWilliams, J. 1985. Submesoscale Coherent Vortices in the Ocean. *Rev. of Geophys.* 23: 165-182.
26. Fjϕrtoft, R. 1946. On the Frontogenesis and Cyclogenesis in the Atmosphere, Part I: On the Stability of the Stationary Circular Vortex. *Geofys. Pub.* 16: pp. 28.
27. Ripa, P. 1987. On the Stability of Elliptical Vortex Solutions of the Shallow

Water Equations. *J. Fluid Mech.* 183: 343-363.
28. D'Asaro, E.A. 1988b. Generation of Submesoscale Vortices: A New Mechanism. *J. Geophys. Res.* 93: 6685-6693.
29. Briscoe, M.G. 1977. On Current Finestructure and Moored Current Meter Measurements of Internal Waves. *Deep Sea Res.* 24: 1121-1131.
30. Müller, P., D.J. Ohlers and J. Willebrand. 1978. The IWEX Spectrum. *J. Geophys. Res.* 83: 479-500.
31. Pinkel, R. 1985. A Wavenumber-Frequency Spectrum of Upper Ocean Shear. *J. Phys. Oceanogr.* 15: 1453-1469.
32. Müller, P., R.C. Lien and R. Williams. 1988. Estimates of Potential Vorticity at Small Scales in the Ocean. *J. Phys. Oceanogr.* 18: 401-416.
33. Gregg, M.C. 1987. Diapycnal Mixing in the Thermocline: A Review. *J. Geophys. Res.* 92: 5249-5286.
34. Cushman-Roisin, B., W.H. Heil and D. Nof. 1985. Oscillations and Rotations of Elliptical Warm-Core Rings. *J. Geophys. Res.* 90: 11,756-11,764.
35. Kawano, T. and S. Nakamoto. 1992. Rotational Modes for Warm and Cold Eddies. *J. Geophys. Res.* 97: 12,659-12,669.
36. Ruddick, B.R. 1987. Anticyclonic Lenses in Large-Scale Strain and Shear. *J. Phys. Oceanogr.* 17: 741-749.
37. Goldsbrough, G.R. 1930. The Tidal Oscillations in an Elliptic Basin of Variable Depth. *Proc. R. Soc. London.* A30: 157-167.
38. Thacker, W.C. 1981. Some Exact Solutions to the Nonlinear Shallow-Water Wave Equations. *J. Fluid Mech.* 107: 499-508.
39. Ball, F.K. 1963. The Effect of Rotation on the Simpler Modes of Motion of a Liquid in an Elliptic Paraboloid. *J. Fluid Mech.* 17: 240-256.
40. Ball, F.K. 1965. Some General Theorems Concerning the Finite Motion of a Shallow Liquid Lying on a Paraboloid. *J. Fluid Mech.* 22: 529-545.
41. Young, W.R. 1986. Elliptical Vortices in Shallow Water. *J. Fluid Mech.* 171: 101-119.
42. Cushman-Roisin, B. 1987. Exact Analytical Solutions for Elliptical Vortices of the Shallow Water Equations. *Tellus.* Ser. A. 39: 235-244.
43. Kirwan, A.D., Jr. and J. Liu. 1991. Shallow Water Equations on an *F*-Plane, pp. 99-127. In: A.R. Osborne (Ed.) *Nonlinear Topics in Ocean Physics, Course CIX, International School of Physics "Enrico Fermi."* North Holland, Amsterdam, pp. 996.
44. Rogers, C. 1989. Elliptic Warm-Core Theory: The Pulsrodon. *Phys. Lett. A.* 138: 267-273.
45. Kirwan, A.D., Jr., A.W. Indest, J. Liu and N. Clark. 1990. Ring Evolution in General Circulation Models from Path Analysis. *J. Geophys. Res.* 95: 18,057-18,073.
46. Brickman, D. and B. Ruddick. 1990. The Behavior and Stability of a Lens in a Strain Field. *J. Geophys. Res.* 95: 9657-9670.
47. Sedov, L.I. 1966. *Foundations of the Non-Linear Mechanics of Continua.* J.E. Adkins, and A.J.M. Spencer (Ed.) Pergamon, Oxford, pp. 252.
48. Dutton, John A. 1986. *The Ceaseless Wind: An Introduction to the Theory*

of Atmospheric Motion. Second edition. Dover Publications, Inc., NY, pp. 617.
49. Hebert, D., N. Oakey and B. Ruddick. 1990. Evolution of a Mediterranean Salt Lens: Scalar Properties, *J. Phys. Oceanogr.* 20: 1464-1483.
50. Liu, J. May 1992. Some Solutions to a Lens Model with Applications to Warm Core Eddies. Ph.D. thesis. Old Dominion University, Norfolk, Virginia, pp. 141.
51. Kirwan, A.D., Jr., B.L. Lipphardt, Jr. and J. Liu. 1992. Negative Potential Vorticity Lenses. *Int. J. Engng. Sci.* 30(10): 1361-1378.
52. Kirwan, A.D., Jr. and B.L. Lipphardt, Jr. 1993. CoherentKI Flows with Near Zero Potential Vorticity. *J. Mar. Sys.* 4(2-3): 95-115.

The Oceans: Physical-Chemical Dynamics and Human Impact. Edited by S.K. Majumdar, E.W. Miller, G.S. Forbes, R.F. Schmalz and Assad A. Panah.

Chapter Four

PHYSICAL MODELING OF OCEAN CURRENTS

DON L. BOYER and **XIUZHANG ZHANG**

Department of Mechanical and Aerospace Engineering
Arizona State University
Tempe, Arizona 85287-6106

INTRODUCTION

The development of physical models of fluid dynamical systems is standard practice in such fields as aerospace, civil and mechanical engineering. One can cite the modeling of airfoils, wind effects on tall buildings and drag reduction of racing cars, respectively, as but a few of the many examples. If one considers the modeling of a section of a long airfoil, with the objective of determining by experiment the lift and drag per unit length, for example, it is a straightforward matter to show that the respective lift and drag coefficients (i.e., the lift and drag made dimensionless by a suitable scaling) are functions of only the Reynolds number, $Re = UD/\nu$; here U is the air speed approaching the airfoil, D is a characteristic dimension of the wing, say its chord length, and ν is the kinematic viscosity of the air. This simple dependence of the lift and drag on only one dimensionless parameter makes the process of modeling in aerodynamics relatively straightforward. While such effects as the level of turbulence in the free stream and the surface roughness of the airfoil may also be important, these can also be modeled by incorporating such features in the experiment tests. Model testing, employing wind or water tunnels, continues to be an important technique in engineering practice.

The physical modeling of geophysical flow systems, such as the interaction of atmospheric or oceanographic currents with topographic features, is vastly more complicated. The reason is simply that the geometry of geophysical systems and the physical processes that come into play are far more complex. In the present context we focus on the modeling of ocean currents. Consider first the geometry of the oceans. The bottom bathymetry and coastlines are exceedingly complex and include a wide range of characteristic length scales. An additional complicating factor is that the vertical scales of the basins are far smaller than the horizontal ones.

A second issue is that the scales of oceanic motions of interest range from those of the ocean basin (e.g., Gulfstream, Kuroshio and Antarctic Circumpolar Current) to the dissipation scale of turbulent kinetic energy. Such a wide range of length scales is not at issue in the aerodynamics example discussed above.

Ocean currents are forced by a wide range of factors. On the global scale one has tides as driven by the motion of the sun and the moon relative to the earth and differential heating at the ocean surface owing to solar insolation. The earth's atmosphere drives the ocean at the air-sea interface on a range of scales. The pressure and wind shear stress distributions of weather systems are primary driving forces of ocean currents. These systems also lead to evaporation and precipitation processes at the ocean surface which in turn drive oceanic motions. An additional complicating factor is the feedback between the atmosphere and the ocean occurring along the ocean surface; i.e., the atmosphere clearly drives the ocean but there is also an effect of the nature of the oceanic surface in the driving of the atmosphere. In the present discussion it is assumed that the atmosphere is a driver of the ocean but not the reverse; the writers are unaware of any attempts to develop coupled atmosphere-ocean physical models.

The ambient stratification of the ocean is a primary factor in the determination of the nature of ocean currents. It is well known that this "background stratification" is the result of salinity and temperature gradients in the ocean; the resulting ambient density distribution is a nonlinear one. For large scale motions, the earth's rotation is also an important factor in determining the character of ocean current systems. In summary, the nature of the geometry, the ambient fluid and the fluid forcing combine to make oceanic flows exceedingly complex and physical modeling thereof, difficult. One is thus led to making simplifying assumptions concerning the scales of motion of interest, the nature of the driving forces and the geometry.

In the present discussion we focus on ocean current scales for which the earth's rotation is a primary factor in determining the nature of the motion field of interest. This, as will be seen, requires the Rossby number Ro of the flow to be small. By this restriction, we do not mean to imply that other motion scales are unimportant, only that some small piece of the range of possibilities will be considered.

In a historical context it is important to cite the physical model of the wind-driven ocean circulation developed by von Arx (1952).[1] This model was a rotating circular basin having a parabolic bottom to simulate the increase of the importance of the earth's rotation with increasing latitude. Only the Northern Hemisphere was considered, with the model continents being appropriately placed in the basin. The working fluid was water and was taken as homogeneous; i.e., stratification effects were not considered. Three equally-spaced blowers were used to simulate the surface wind stresses of the westerlies on the fluid surface. Some interesting features were demonstrated by the model such as the westward intensification of ocean currents as represented by the Gulf Stream and the Kuroshio.

The general area of geophysical fluid dynamics deals with the mechanics of rotating and/or stratified fluids. Much of the vast literature in this field is related in one way or the other with the dynamics of ocean current systems. The reader is referred to some of the following excellent reviews of topics of interest in the better understanding of the dynamics of a range of problems related to ocean currents: buoyant plumes and thermals, Turner (1969)[2]; multicomponent convection, Turner (1985)[3]; turbulent mixing, Sherman et al. (1978)[4] and Fernando (1991)[5]; wakes in stratified flows, Lin and Pao (1979)[6]; gravity currents in rotating systems, Griffiths (1986)[7]; transport and diffusion in stratified fluids, Snyder (1985)[8]; baroclinic instability, Hart (1979)[9]; and rotating strait and sill flow, Pratt and Lundberg (1991).[10]

Perhaps the review article most closely related to the present discussion is that of experiments in rotating and stratified flows by Maxworthy and Browand (1975).[11] That review concentrated on two problems; i.e., (i) ocean mixing whereby the smallest scales of stratification are developed and maintained and (ii) the dynamics of the wind-generated oceanic circulation. In the latter, emphasis was given to those models for which rotation was important but for which stratification played little role. In contrast, in the present communication we will focus on those systems for which rotation and stratification both are of leading order importance. Furthermore, emphasis will be given to examples for which the modeling of specific oceanic current systems is being addressed.

The plan for the review is to discuss first the general considerations that need be made in modeling large-scale ocean current systems; we also discuss the specific similarity conditions that apply to the case for which a broad, spatially-independent current impinges on an isolated topography. Some brief comments are made on the experimental facilities and techniques used in the modelling, but the reader is in general referred to other communications for details. Examples of specific physical models, including (i) the flow in the vicinity of Fieberling Guyot in the North Pacific, (ii) the effects of ocean floor bathymetry on the path of the Antarctic Circumpolar Current, (iii) the flow in the western Mediterranean, including the transport through the Strait of Gibraltor and (iv) a model of ocean currents in the Barents Sea, are discussed.

SIMULATION OF FLOW PAST ISOLATED TOPOGRAPHY

Recognizing that one will never be able to simulate the details of the motion fields of ocean current systems to the extent possible for say the flow past an airfoil, we consider the question of the simulation of the motion field in the vicinity of an isolated seamount resulting from its interaction with a defined background current. We address those factors that need be considered in the simulation process and discuss the degree to which each can be modeled. In so doing the similarity parameters of the modeling are defined.

Consider the idealized system given schematically in Figure 1. This sketch is to represent both the prototype (i.e., ocean) and the physical model (i.e., laboratory). The following factors are considered of leading order importance in the modeling process:

Topography. The prototype bathymetry is, of course, highly complex and is characterized by a wide range of length scales with the largest being the horizontal dimension D of the feature; other important scales for our purposes are the topographic height h_0 and the fluid depth H. The model topography is "smoothed" to the extent that horizontal and vertical undulations of the bathymetric feature having scales much smaller than D and h_0, respectively, are not reproduced. Furthermore the prototype is generally taken as having a flat bottom and the assumption made that other seamounts are sufficiently far from the feature in question as to not influence the motion field near the seamount. Because we are restricting to motion field scales in which the background rotation of the earth is important (i.e., small Rossby numbers), we are necessarily dealing with a prototype whose vertical scale is much smaller than its horizontal one. Because physical models need to be placed on turntables (i.e., one has a limit on the horizontal scale of the model) and because the ratio of the bottom boundary layer thickness to the fluid depth must also be modeled, strict geometric modeling is not possible. One is thus led to vertical exaggeration in the model. Fortunately this is allowable to the extent that both the model and prototype are in approximate hydrostatic equilibrium. One final note is that in eliminating surface roughness in the model, the small scale motions (e.g., turbulence) generated by the background current impinging on the model are excluded. This boundary turbulence is known to lead to mixing and the possible intrusions of fluid motions away from the seamount. Such intrusions can interact with the mean flow and alter the large-scale motion field near the topographic feature.

Stratification: The background stratification of the ocean owes to both vertical gradients in the salinity and the temperature. The resulting density stratification is nonlinear with depth with the upper mixed layer being roughly homogeneous, a strong gradient in the vicinity of the thermocline and a weaker gradient in the deep ocean. This density distribution is characterized in the modeling process by the Brunt-Väisälä frequency distribution in the laboratory using a variation of the Oster (1965)[12] two-tank method. For the present model, however, we assume a linear density distribution leading to a constant Brunt-Väisälä frequency $N = (g\Delta\rho/\rho_0 H)^{1/2}$, where $\Delta\rho$ is the density difference from bottom to top, ρ_0 is the mean density and g is the acceleration of gravity.

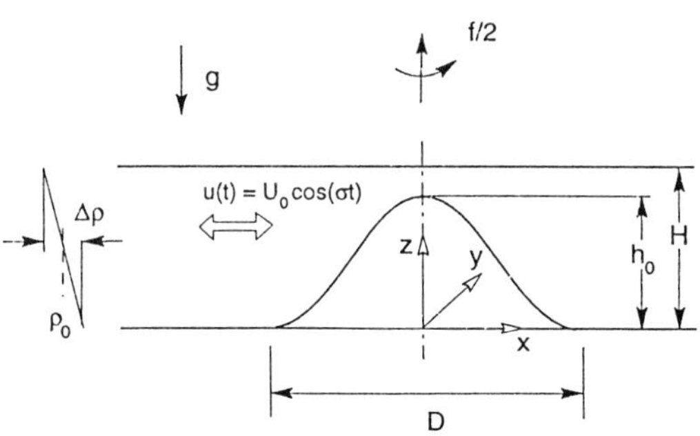

FIGURE 1. Physical system for oscillatory flow past an isolated topography; a model for Fieberling Guyot.

Rotation: The background rotation of the earth is, of course, a crucial modeling factor. This rotation is simulated by placing the experimental facility on a turntable whose rotation rate is Ω or Coriolis parameter $f = 2\Omega$. For situations with large meridional extent, the vertical component of the earth's rotation varies significantly over the region to be modeled. In such cases this variation must be taken into account. For homogeneous models this is rather straightforward in the sense that this variation in rotation rate can be modeled by sloping the lower bounding surface of the model from south to north, with the thinnest fluid layer being toward the north. This so-called beta-plane approximation accounts for lowest order effects of the increase in the importance of rotation with latitude as one goes toward the north. Unfortunately, it has not been possible to establish a simple method of mechanically simulating such beta effects for the case of stratified fluids and this matter remains one of the important unanswered questions related to the physical modeling of geophysical systems. One last remark is that in certain applications (e.g., the Antarctic Circumpolar Current), the effective beta effect resulting from the sloping bottom topography can overwhelm the planetary beta effect so that the latter can be neglected, even for motion fields with rather large north to south extents.

Viscosity: Oceanic motions are turbulent with length scales varying from those of the ocean basin itself to the dissipation scale. Most laboratory models developed to date consider laminar flows, although the study of turbulent motions in rotating and stratified fluid systems is certainly possible. It is not reasonable, however, to assume that it will ever be possible to model occanic turbulence in the laboratory over the range of velocity and length scales found in the ocean. In the laminar models, it has thus become customary to consider the working fluid kinematic viscosity coefficient as the effective viscosity and to relate the model to an ocean with effective mean vertical and horizontal eddy viscosity coefficients, ν_v and ν_H, respectively. While such an assumption is a good engineering approximation, it must be emphasized that it is not backed by first principles.

Background Current: In the present modeling situation we are dealing with an analogy to the testing of the section of an airfoil in a wind tunnel. In this spirit the oncoming flow stream is an important consideration. While this flow characteristic in the aerodynamics case is a rather straightforward one, for the ocean, that quantity is rather poorly and imprecisely defined. Oceanic currents are by their very nature exceedingly complex and thus the case of a uniform, spatially-independent current is never found, even in an approximate way, in practice. The simplest approximation that can be made is that of a broad uniform oncoming current of constant speed U_0. This can rather easily be realized in the laboratory by setting up a rotating water tunnel facility similar in concept to wind tunnels used in aerodynamics; see Vaziri and Boyer (1971).[13] Alternatively, the topography can be towed through a rotating fluid, otherwise at rest with respect to a rotating observer; see, for example, Boyer and Zhang (1990).[14]

Background currents in the ocean can also be time dependent; for example, periodic, in the case of tidal motions. Such time-dependent motions can be modeled in the laboratory with tow-tank facilities by driving the towing carriage at a prescribed function of time through the working fluid. One can show that the resulting motions are dynamically equivalent to those occurring by moving a time-dependent current past a model topography fixed with respect to the turntable. This similarity can only be shown for background currents that are spatially independent; i.e., no horizontal or vertical shears can be simulated.

Background currents in the ocean are also characterized by the presence of vertical and horizontal shear. Furthermore it is well known that the general motion systems in the ocean include the presence of advecting anticyclonic and cyclonic eddy structures having vertical shear. In fact isolated advecting eddies are found in the ocean interior; e.g., the meddies discussed by Armi et al. (1989).[15] Recent laboratory studies have shown the feasibility of developing controlled jets having both horizontal and vertical shear in the laboratory; see Chen and Boyer (1990)[16] and Chen et al. (1992).[17] These models can also include advecting eddy structures resulting from flow instabilities. While laboratory studies are just now beginning on the interaction of jets with isolated topography, this area of research seems a fruitful one for the future.

For the purpose of defining a set of similarity conditions for the physical system depicted in Figure 1, we restrict to the case of (i) a broad current (i.e., no vertical or horizontal shear) consisting of a constant background mean flow U_0 and a tidal current in the same direction $U_1 \sin(\omega t)$, where U_1 and ω are the magnitude and frequency of the tide; (ii) a linear stratification (i.e., a constant N); (iii) an f-plane (i.e., a constant background rotation with no beta effects); (iv) negligible wind shear stress effects; (v) and constant horizontal and vertical eddy viscosity coefficients. Utilizing the Buckingham Pi Theorem from dimensional analysis, one can show that the following set of dimensionless parameters characterize the system under consideration:

$$Ro = \frac{U_0}{fD} \quad \text{, Rossby number}$$

$$S = \frac{N^2H^2}{f^2D^2} \quad \text{, Burger number,}$$

$$E = \frac{\nu}{fH^2} \quad \text{, Ekman number,}$$

$$Ro_t = \frac{\omega}{f} \quad \text{, temporal Rossby number,}$$

$$\frac{U_1}{U_0} \quad \text{, velocity ratio,}$$

$$\frac{h_0}{D} \quad \text{, topography aspect ratio and}$$

$\frac{h_o}{H}$, normalized topography height.

An important parameter, but not independent of the above, is the normalized tidal excursion distance $X = 2U_1/\omega D = 2(U_1/U_0)(Ro/Ro_t)$; in the absence of a mean current U_o, X is simply the distance an undisturbed fluid parcel travels in one-half a tidal cycle.

Dynamical and kinematical modeling of the flow past Fieberling requires that the above dimensionless parameters be matched in the laboratory at their prototype vaules. This is generally possible except for the topographic aspect ratio h_o/D which is substantially larger in the laboratory model. An examination of the equations of motion and boundary conditions, however, shows that under the conditions $Ro << O(1)$, $S \sim O(1)$, $E << O(1)$, $Ro_t \sim O(1)$, $U_1/U_o \sim O(1)$ and $h_0/H \sim O(1)$, the parameter H/D does not enter the problem provided $(H^2/D^2)(Ro_t^2/S^2) = \omega^2/N^2 << 1$. Physically, this condition assures that the laboratory model is in hydrostatic equilibrium, a very good approximation for ocean currents of the scales being considered.

EXPERIMENTAL FACILITIES AND TECHNIQUES

Depending on the type of modelling to be accomplished, there can be a wide variety of experimental facilities that can be employed. To exemplify those facilities we focus on the isolated topography problem. Such experiments were conducted in a rectangular tow-tank of dimensions 3.90 m long, 1.20 m wide and 0.25 m deep. The stratified fluid can be established by employing the two-tank method described by Oster (1965).[12] After the tank is filled, the system is slowly spun up by small increments to the desired Coriolis parameter f; roughly six hours should be allowed for the system to reach a state of solid body rotation (the spin-up time for stratified fluids is substantially longer than that for homogeneous ones). The background flow, composed of a sinusoidal oscillation superimposed on a uniform motion in the same direction, is realized by towing the model along the long axis of the tow tank. Cross-stream or oblique oscillations to the constant free stream are also possible but are not considered here. This method of towing the model through a fluid otherwise at rest with respect to a rotating observer is dynamically equivalent to moving the fluid as an appropriate function of time past the model; see Boyer and Zhang (1990).[14]

Flow visualization is achieved by means of neutrally buoyant particles and dye tracers. The latter can use tracers such as fluoroscein or employ the pH technique discussed by Baker (1966)[18] or the electrolytic precipitation method detailed in Honji et al. (1980).[19] The particles and dye tracers are illuminated by horizontal or vertical light sheets, roughly 0.5 cm in width for the present application. A 35 mm camera and/or video camera are attached to the towing carriage and thus move synchronously with the model topography and are used to obtain data for later analysis. The towing system, cameras and lights are all controlled by a computer. Neutrally buoyant particles are used to determine the approximate instantaneous streamline fields. The magnitude and direction of the illuminated particles can be determined by appropriately chopping the light sheets during the short time exposure used to obtain particle streak photographs. The pH and electrolytic precipitation techniques are used to determine streakline fields and in particular are useful in determining the nature of the mean flow behavior over extended periods. Techniques such as laser Doppler velocimetry, image processing of particle streaks, laser-induced flourescence and traversing conductivity probes are also used extensively to obtain laboratory velocity and density fields. The interested reader is referred to Maxworthy and Browand (1975)[11] for more details on the facilities and instrumentation used to study models incorporating background rotation and stratification.

SOME EXAMPLES

Flow Past Isolated Topography: Fiberling Guyot

The Office of Naval Research has recently supported a combined field observational and numerical and laboratory modeling program to investigate the influence of isolated topography on its biological and physical environment. The site chosen for the study was the region in the vicinity of Fieberling Guyot in the North Pacific. The initial hypothesis was that a broad current consisting of a uniform component superimposed on an oscillatory component to simulate the tides would be a reasonable model for the background flow. Numerical models of a uniform steady flow past finite amplitude topography were thus developed with the focus being on Taylor caps and cones that might be observed above Fieberling; see Chapman and Haidvogel (1993).[20] The physical modeling addressed the problem of incorporating the tidal component as well as a uniform stream as the background flow; Boyer and Zhang (1990).[14,21] This physical model demonstrated the regions of parameter space (e.g., Ro and Ro_t values for fixed E, S and geometry) for which fully-attached flow, standing eddies or vortex shedding might be found in the vicinity of the seamount; i.e., suitable flow regime diagrams were developed.

Field observations taken early in the program, however, indicated that the background flow near Fieberling was far more complex than these simplified numerical and physical models. In fact the mean current in the vicinity of Fieberling was found to be very weak. Roden (1991)[22] found that the flow was characterized by a weak mean flow, strong tidal components, jet-like structures at various depths and advecting eddy fields in the vicinity of the surface. Genin et al. (1989)[23] reported that a mean anticyclonic circulation and a diurnal (subinertial, $Ro_t < 1$) current are trapped around the seamount. Eriksen (1991)[24] reported that the mean and diurnal currents are bottom intensified and that the M2 tide (superinertial, $Ro_t > 1$) is surface intensified.

With the observations as background, new physical modeling experiments were conducted. In the first, a homogeneous model of the pure oscillation (i.e., no mean current) over an isolated seamount was conducted. These experiments were carried out in the 13 m diameter, 1 m deep rotating test facility of the Institut de

Mecanique de Grenoble, France and are reported in Boyer et al.(1991)[25] This study showed that, owing to the lack of stratification, and for rotationally dominated flows, the Taylor-Proudman constraint leads to flows which are approximately depth independent. The general flow field is exemplified by the sketch of a number of observed particle paths on the fluid surface given in Figure 2. The particles exhibit anticyclonic loops for each flow oscillation cycle and these loops are superimposed on a mean anticyclonic current. The study parameterized the mean or rectified flow in terms of the system parameters and in terms of the distribution over the seamount. The experiments were also in good agreement with numerical results and in qualitative agreement with field observations.

Similar experiments were then conducted for linearly stratified fluids and are reported in Zhang and Boyer (1993).[26] That study delineates a series of flow regime diagrams which characterize the flow as a function of the system parameters. It is shown that flow rectification is a ubiquitous feature of the experiments but that its magnitude varies greatly with the system parameters. For subinertial frequencies the

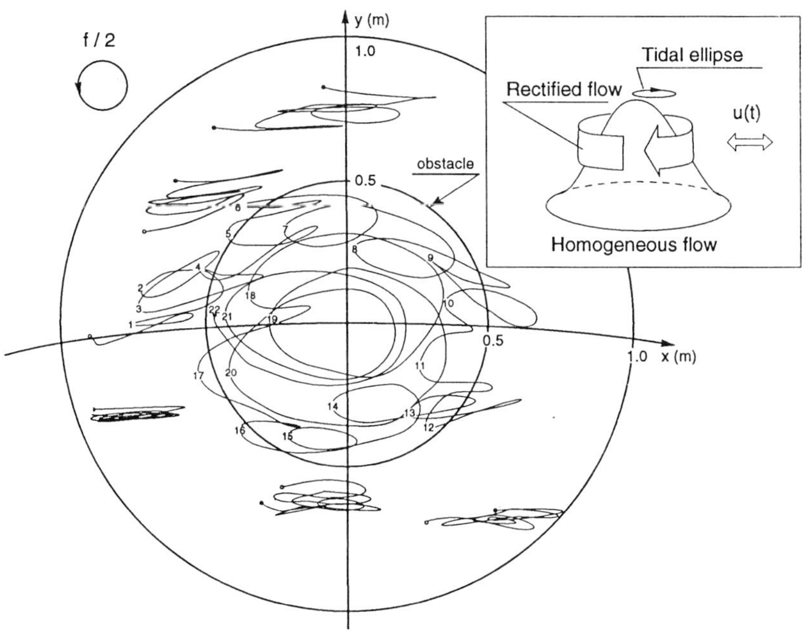

FIGURE 2. Free surface particle paths from laboratory experiments for barotropic oscillatory (right-left-right...) flow in the vicinity of an isolated obstacle; the parameter values are Ro = 0.051, Ro_t = 0.5, E = $4.1(10)^{-5}$ and X = 0.32. The obstacle has a radius of 0.5 m and a height of 0.3 m, while the fluid depth is 0.35 m; the background rotation is anticlockwise. The small numbers of the particle paths represent the cycle number of oscillation, starting at zero. There is a weak wind drift current from left to right. The path starting along the negative x-axis at "0" and ending at "22" depicts a single trajectory and demonstrates the mean anticyclonic motion occurring above the topography. Note that from cycles "18" to "22" the tracer fully passes around the obstacle center during each oscillatory cycle. The insert is a schematic which depicts the depth independent nature of the rectified flow.

rectified current is bottom trapped in the sense that the mean flow is stronger near the surface of the topographic feature; see Figure 3. At superinertial and near-intertial frequencies, the flow is strongly enhanced near the surface as shown in Figure 4.

Recently experiments have been initiated to investigate the effects of a current having both horizontal and vertical shear impinging on an isolated topographic feature. These experiments have been motivated by the observations of Roden (1991)[22] that jet-like motions are a common feature of the motion field in the vicinity of Fieberling Guyot. Jets in rotating and stratified fluids can be established by source-sink methods as described by Chen and Boyer (1990)[16] for two-layer fluids and Chen et al. (1992)[17] for linearly stratified flows. The technique is indicated by the schematic diagram of Figure 5. A homogeneous, two-layer or continuously stratified fluid is introduced to a depth H in a circular test cell. The surface water

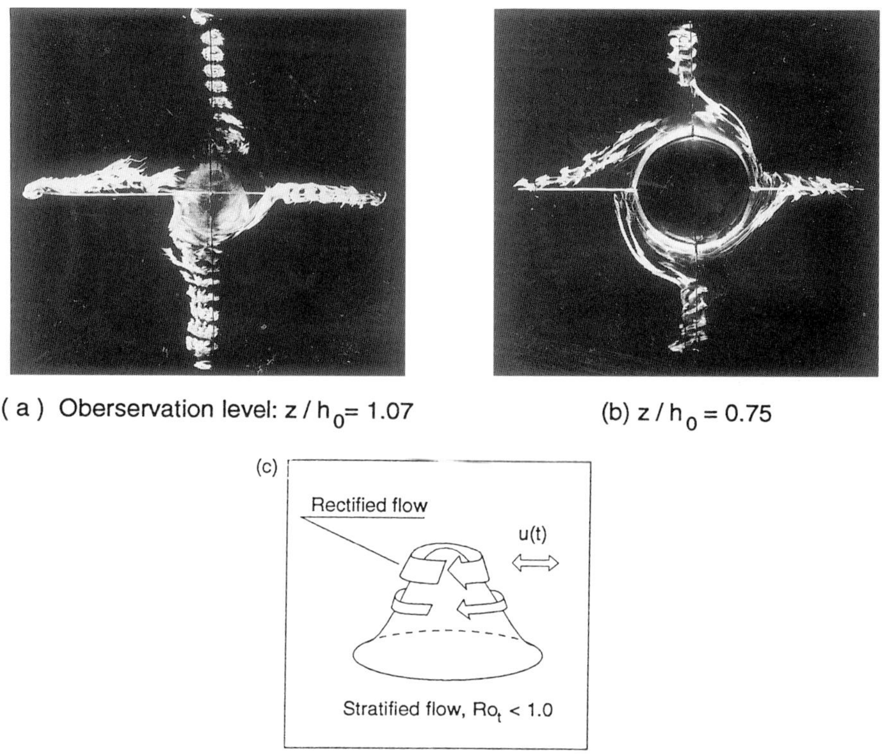

FIGURE 3. Dye tracer photographs (a,b) in a horizontal plane and an interpretive sketch (c) depicting the nature of the rectified flow developed by a pure oscillatory subinertial ($Ro_t < 1.0$) flow in the left-right-left direction past an axially symmetric obstacle for a linearly stratified fluid; the parameter values are $Ro = 0.013$, $Ro_t = 0.5$, $S = 1.0$, $E = 7.8(10)^{-5}$ and $X = 0.052$. The tracer was released for 8 oscillation cycles along a horizontal and a vertical line. The observation level in (a) is $z/h_o = 1.07$ (i.e., above the topography) and in (b) is $z/h_o = 0.75$. In (b), the bright circle depicts the outer boundary of the obstacle at that level. In stratified flows the rectified current is found to be strongest near the top of the topography and quite weak at levels above the obstacle.

is then recirculated by a uniform source circumscribing the outer reaches of the cell and a uniform sink along an inner radius. This surface forcing is equivalent to a surface wind stress that varies inversely with the distance from the test cell center; see Boyer et al. (1993).[27] For the case of homogeneous flow, the resulting jet has horizontal shear with a velocity profile approximately independent of depth; the horizontal scale is that of the radial distance between the source and sink. For the two-layer case, the jet is stronger in the upper layer with the profile being roughly independent of depth in each layer. For the continuously (linearly) stratified case, the vertical shear is continuous with the surface velocities being the strongest. In all cases the flow can be unstable in the sense that wavelike motions occur in the jet which can lead to eddy structures being established in the vicinity of the jet. These

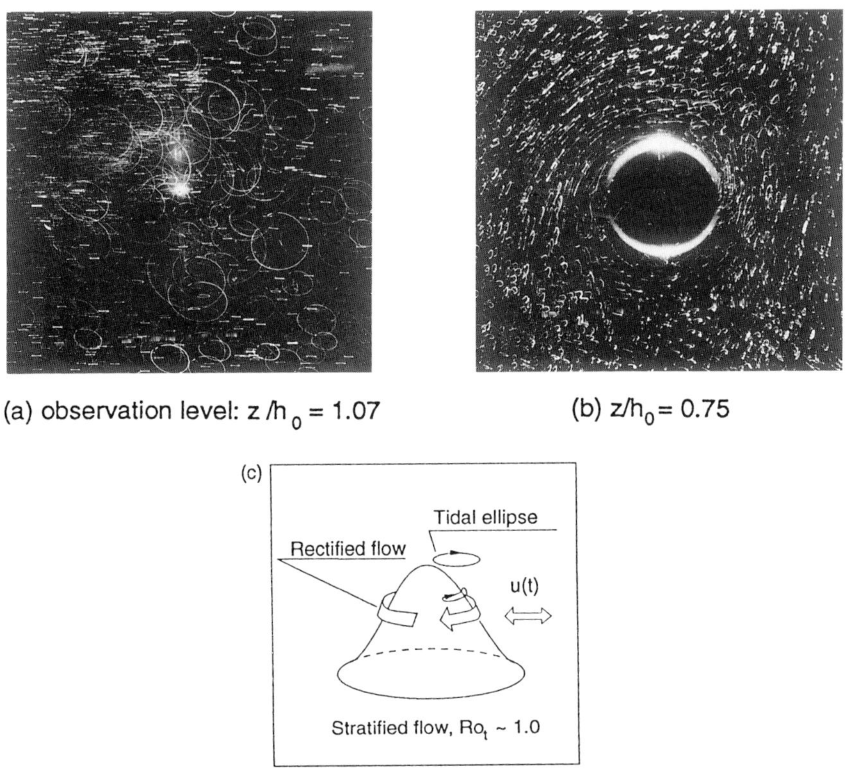

FIGURE 4. Particle streak photographs (a,b) obtained with a camera exposure time of one oscillation cycle and an interpretive sketch (c) of pure oscillatory inertial flow past an axially symmetric obstacle for a linearly stratified fluid; the parameter values are Ro = 0.01, Ro_t = 1.0, S = 1.0, E = $7.8(10)^{-5}$ and X = 0.02. The observation level in (a) is z/h_o = 1.07 while that in (b) is 0.75. The short dashed lines are tracer particles that have adhered to the horizontal upper bounding surface of the model. They thus represent the undisturbed tidal excursion length; note that the tidal oscillation loops are substantially larger, indicating a resonance effect above the obstacle. The observation in (b) shows that strong tidal loops are suppressed below the top of the obstacle, but that some weak anti-cyclonic flow rectification is taking place.

systems seem excellent ones for the study of the interaction of currents and eddies having horizontal and vertical shear with bottom topography.

Antarctic Circumpolar Current

The physical system depicted in Figure 5 and described above for establishing a jet in a rotating and stratified fluid was used to investigate the effect of the bathymetry of the Southern Oceans on the path of the Antarctic Circumpolar Current; see Boyer et al. (1993).[27] A scale model of the bathymetry was developed using a vertical exaggeration of 8.0. This exaggeration again was allowable owing to the hydrostatic nature of both the oceanic and the model flows. Planetary beta effects were not included owing to the fact that topographic beta effects due to the large slopes associated with the mid-ocean ridges were substantially larger than planetary beta effects. Both homogeneous and linearly stratified models were considered.

The experiments were initiated by introducing the desired stratification into the test cell in which the model topography had been placed, bringing the fluid to solid body rotation and finally engaging the pump driving the source-sink flow. The temporal nature of the flow field as it reached a steady state was investigated; here, however, we focus on the flow after it has reached an approximate steady state. Flow field observations were made from both the near surface and at mid-depth by employing streak photography of neutrally buoyant particles. Figure 6 is an interpretive sketch of the surface flow for an experiment with a homogeneous fluid.

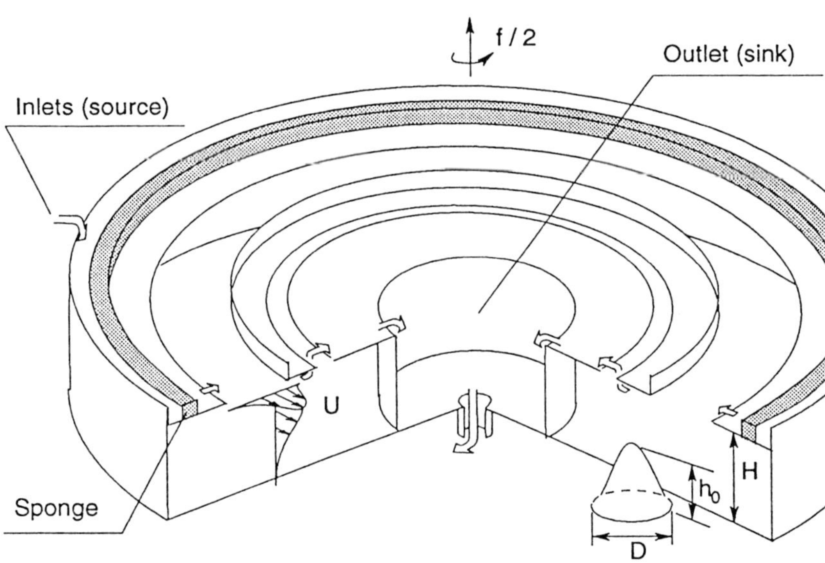

FIGURE 5. Schematic diagram depicting how a cylindrically symmetric source and sink can generate an azimuthal jet in a rotating system. For homogeneous fluids, the jet in the fluid interior has horizontal but no vertical shear. For a continuously stratified fluid, the jet has both horizontal and vertical shear.

The qualitative nature of the streamline patterns in Figure 6 are in good agreement with oceanic observations of Gordon, Molinelli and Baker. (1978).[28] In particular the simulation includes well-defined strong currents along the mid-ocean ridge, strong perturbations in the vicinity of the Macquarie Ridge, Campbell Plateau and Kerguelen Gaussberg Plateau, strong meridional transport to the east of the Drake Passage and anomalies to the south (wave troughs) and to the north (wave ridges) of the main circumpolar current over ocean basins and submarine mountain ridges, respectively. The experiments demonstrated the importance of the Eltanin and Udintsev fracture zones in the vicinity of 135° W longitude in directing the ACC eastward across the Southeast Pacific Basin to the Drake Passage. Experiments without these fractures led to a northwest flow along the mid ocean ridge before being deflected by the outermost boundary of the test cell and then passing through the Drake Passage after traversing along the western coast of South America. The model was also used to estimate the transport through the Drake Passage, and, while this quantity was somewhat greater than obtained by oceanic observations, it was in reasonable agreement with accepted oceanic values.

Experiments with a linearly stratified model were also conducted. Qualitatively these demonstrated similar horizontal surface flow patterns as their homogeneous counterpart with the exception that for the same forcing, the surface currents were

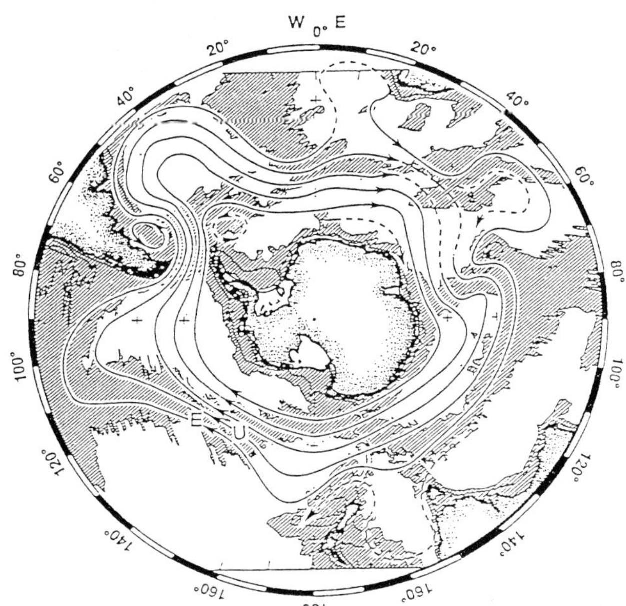

FIGURE 6. Sketch of the surface streamline patterns from neutrally buoyant particle streak photographs obtained from a physical model of the effects of ocean floor bathymetry on the nature of the Antarctic Circumpolar Current; taken from Boyer et al. (1993). The symbols E and U represent the locations of the Eltanin and Udintsev fracture zones through the mid-ocean ridge. These features are critical in simulating a current having characteristics similar to observations west of the Drake Passage.

stronger. The stratified model exhibited vertical shear and predicted a smaller transport through the Drake Passage than the homogeneous model. Overall the modeling demonstrated nicely the feasibility of modeling wind-driven stratified, rotating currents past irregular bottom topography.

Simulation of the Large-Scale Motion in the Alboran Sea

It is well known that the Mediterranean Sea is more salty than the Atlantic Ocean and that this density imbalance at the Strait of Gibraltor creates density-driven currents through the Strait. These flows have profound influence on the motion fields in the Atlantic to the west of Gibraltor (i.e., the Bay of Cadiz) and in the western Mediterranean (i.e., the Alboran Sea).

Whitehead and Miller (1979)[29] have developed a physical model of this region that depicts some of the motion features observed in the Alboran Sea. The model consists of two basins, representing the eastern Atlantic and the western Mediterranean, connected by a narrow channel simulating the Strait of Gibraltor; the shorelines of the basins are taken as vertical and no attempt is made to model the bottom bathymetry. The model is mounted on a rotating table. The connecting channel includes a sliding door which can be used to fully separate the two large basins. Before the initation of the experiment, the door is closed and the model Atlantic and Mediterranean basins are filled with fresh and salt water, respectively, with the density difference being appropriate for the model. The liquids are brought to the required rotation rate and a sufficient time is allowed for the fluids to achieve a state of approximately solid body rotation.

The experiment is then initiated by removing the sliding door. The lighter fluid from the Atlantic rises over the heavier fluid of the Mediterranean and enters the Sea as a surface current; at the same time the heavier Mediterranean fluid flows into the Atlantic as a subsurface current. While the model is limited by the fact that it never can reach a steady state, it nicely depicts the general characteristics of the anticyclonic gyre of the surface flow in the Alboran Sea. The model gyre has roughly the correct form and lateral extent, reaching eastward to about Alboran Island, in reasonable agreement with observations.

Simulation of Ocean Currents in the Barents Sea

The Barents Sea is an open, marginal sea north of Norway: see Figure 7. Here, two oceans interact, forming the Polar Front (PF) of the Barents Seas. There is much interest in this region owing to petroleum exploration, fishing, the desire for pollution control and the importance of this area in considerations related to global climate change. McClimans and Nilsen (1993)[30] have developed a physical model of this region which exhibits many of the ocean current features noted in field observations.

The model is a 5 m diameter rotating basin in which the bathymetry of the region is reproduced with a vertical exaggeration permissible within the constraints of the model being hydrostatic, as discussed previously; the outer edge of the model is that depicted on Figure 7. The model is filled with a homogeneous liquid (water) and brought to solid body rotation.

The experiment is initiated (driven) by appropriate inflows of Atlantic Water (AW), southwest of Tromosϕflaket and Arctic Surface Water (ASW), past Franz Josef Land (FJL); here, the AW source was made appropriately salty and the ASW was fresh. Two other water sources simulating the Norwegian Coastal Current (NCC) (fresh) and the Norwegian Sea Deep Water (NSDW) (salty) beneath the AW, were also introduced. Bottom water formation was simulated during the winter months by introducing Rhodamine dyed brine (very salty) through nozzles at the locations marked by the x's on Figure 7.

The physical model was able to simulate such features as (i) the spreading of the Norwegian Coastal Current over Tromsϕflaket, (ii) a warm core jet along the southeastern slope of the Svalbardbanken, (iii) the anticyclonic circulation around Sentralbanken, (iv) Norwegian Coastal Water flowing north across the Bear Island Channel, (v) deep water outflows north through the Franz-Victoria Trough and west through the Bear Island Channel and (vi) tidally driven circulation on the Svalbard-banken and around Bear Island. This rather complex physical model was thus able to simulate most of the important physical features of the motion fields in the Barents Seas.

FIGURE 7. The geographical extent of the Barents Sea model in a 5 m diameter basin as investigated by McClimans and Nilsen (1993).[30] The surface current features are adapted from Loeng (1991).[31]

CONCLUSIONS

Physical models of ocean current systems have reached a stage of development whereby quite realistic simulations can be achieved. An essential component of the modeling process is a basic understanding of the pertinent physics governing the motion field in question. This leads to a determination of the important parameters defining the problem; a study of the similarity parameters for a particular application follows naturally from the consideration of the governing equations of motion and the appropriate boundary conditions. For large-scale current systems for which the earth's rotation is an essential determining factor (i.e., small Rossby numbers), it is important that the background rotation and stratification, the coastline shape and ocean floor bathymetry and boundary conditions be modeled in accordance with the similarity parameters obtained from the analysis.

Physical models should be considered as compliments to numerical investigations. Laboratory models have considerable advantage over numerical ones in their ability to determine overall flow behavior over wide ranges of system parameter space both quickly and relatively inexpensively; e.g., one can readily determine whether eddies will be shed in the lee of islands and/or seamounts. Laboratory studies, coupled with suitable scaling analyses utilizing the governing equations of motion, can also be used to investigate the physical processes involved in a particular type of flow situation. Physical models can isolate particular physical processes one at a time to study and thus can lead to a deeper understanding of the physics of a particular ocean current system. This is contrasted to the ocean current system itself, which integrates over a wide range of physical processs making it difficult to isolate the important physics.

Future physical modeling development should include:

1. The simulation of realistic geometry including coastlines and sea-floor bathymetry, should be emphasized. Continuing attention should also be given to utilizing realistic inflow and outflow boundary conditions in the spirit of the Barents Sea model above.
2. The use of more realistic Brunt-Väisälä frequency distributions to simulate the upper mixed layer, pycnocline and deep ocean.
3. Continued development of observational techniques to improve the quantitative nature of the data; e.g., it should become a matter of course that vorticity and velocity fields are presented.
4. Laboratory results should be carefully compared with numerical models as tests for the latter and as a means to better understand the physical processes at work in the former.
5. Both the laboratory and the numerical models should be compared increasingly with ocean observations. Such comparisons should prove useful in improving our understanding of both the models and the ocean itself.
6. Field programs should include a physical modeling component to aid in addressing such experimental design issues as the determination of the type of sensors to be employed and their location and to help in the physical interpretation of the observations.

ACKNOWLEDGEMENTS

The authors wish to express their gratitude to the Office of Naval Research and the National Science Foundation for supporting much of the work described herein.

REFERENCES

1. von Arx, W.S. 1952. A laboratory study of the wind-driven ocean. circulation. Tellus, **4,** 311-318.
2. Turner, J.S. 1969. Buoyant plumes and thermals. Annu. Rev. Fluid Mech. **1,** 29-44.
3. Turner, J.S. 1985. Multicomponent convection. Annu. Rev. Fluid Mech., 17, 11-44.
4. Sherman, F.S., J. Imberger and G.M. Corcos. 1978. Turbulence and mixing in stably stratified waters. Annu. Rev. Fluid Mech., **10,** 267-288.
5. Fernando, H.J.S., 1991. Turbulent mixing in stratified flows. Annu. Rev. Fluid Mech., **23,** 455-493.
6. Lin, J.T. and Y.H. Pao. 1979. Wakes in stratified flows. Annu. Rev. Fluid Mech., **11,** 317-338.
7. Griffiths, R.W. 1986. Gravity currents in rotating systems. Annu.. Rev. Fluid Mech., **18,** 59-89.
8. Snyder, W.H. 1985. Fluid modeling of pollutant transport and diffusion in stably stratified flows over complex terrain. Annu. Rev. Fluid Mech., **17,** 239-266.
9. Hart, J.E. 1979. Finite amplitude baroclinic instability. Annu. Rev. Fluid Mech., **11,** 147-172.
10. Pratt, L.J. and P.A. Lundberg. 1991. Hydraulics of rotating strait and sill flow. Annu. Rev. Fluid Mech., **23,** 81-106.
11. Maxworthy, T. and F.K. Browand. 1975. Experiments in rotating and stratified flows: Oceanographic applications. Annu. Rev. Fluid Mech., **7,** 273-305.
12. Oster, G. 1965. Density gradients. Sci. Amer., **213,** 70-76.
13. Vaziri, A. and D.L. Boyer. 1971. Rotating flow over shallow topographies. J. Fluid Mech., **50,** 79-95.
14. Boyer, D.L. and X. Zhang. 1990. The interaction of time-dependent rotating stratified flow with isolated topography. Dyn. Atm. Oceans, **14,** 543-575.
15. Armi, L., D. Hebert, N. Oakey, J.F. Price, P.L. Richardson, H.T. Rossby and B. Ruddick. 1989. Two years in the life of a Mediterranean salt lens. J. Phys. Ocean., **19,** 354-370.
16. Chen, R. and D.L. Boyer. 1990. Jet instability and the stabilizing effect of topography on jets in a two-layer rotating system. Geophys. Astro. Fluid Dyn., **52,** 45-70.
17. Chen, R., D.L. Boyer, and L. Tao. 1992. Stability characteristics of jets in linearly-stratified rotating fluids. Geophys. Astro. Fluid Dyn., **65,** 1-29.
18. Baker, D.J. 1966. A technique for the precise measurement of small fluid velocities. J. Fluid Mech., **26,** 573-575.

19. Honji, H., S. Taneda, and A. Tatsuno. 1980. Some practical details of the electrolytic precipitation method of flow visualization. Res. Inst. Appl. Mech., Kyushu Univ., **28,** 83-89.
20. Chapman, D.C. and D.B. Haidvogel. 1992. Formation of Taylor caps over a tall isolated seamount in a stratified ocean. Geophys. Astrophys. Fluid Dyn., **64,** 31-65.
21. Boyer, D.L. and X. Zhang. 1990. Motion of oscillatory currents past isolated topography. J. Phys. Ocean., **20,** 1425-1448.
22. Roden, G.I. 1991. Mesoscale flow and thermohaline structure around Fieberling seamount. J. Geophys. Res., **96,** 16653-16672.
23. Genin, A., M. Nobel, and P.F. Lonsdale. 1989. Tidal currents and anticyclonic motions on two North Pacific seamounts. Deep-Sea Res., **36,** 1803-1816.
24. Eriksen, C. 1991. Observations of amplitude flows atop a large seamount. J. Geophys. Res., **96,** 15227-15236.
25. Boyer, D.L., G. Chabert d'Hieres, H. Didelle, J. Verron, R. Chen, and L. Tao. 1991. Laboratory simulation of tidal rectification over seamounts: Homogeneous model. J. Phys. Ocean., **21,** 1559-1579.
26. Zhang, X. and D.L. Boyer. 1993. Laboratory study of rotating, stratified, oscillatory flow over a seamount. J. Phys. Ocean., **23,** 1123-1141.
27. Boyer, D.L., R. Chen, L. Tao, and P.A. Davies. 1993. Physical model of bathymetric effects on the Antarctic Circumpolar Current. J. Geophys. Res., **98,** 2587-2608.
28. Gordon. A.J., E. Molinelli, and T. Baker. 1978. Large-scale relative dynamic topography of the Southern Ocean. J. Geophys. Res., **83,** 3023-3032.
29. Whitehead, J.A. and A.R. Miller. 1979. Laboratory simulation of the gyre in the Alboran Sea. J. Geophys. Res., **84,** 3733-3742.
30. McClimans, T.A. and J.N. Nilsen. 1993. Laboratory simulation of the ocean currents in the Barents Sea. Dyn. Atm. Oceans, in press.
31. Loeng, H. 1991. Features of the physical oceanographic conditions of the Barents Sea. Polar Res. **10,** 5-18.

The Oceans: Physical-Chemical Dynamics and Human Impact. Edited by S.K. Majumdar, E.W. Miller, G.S. Forbes, R.F. Schmalz and Assad A. Panah. © 1994, The Pennsylvania Academy of Science.

Chapter Five

TROPICAL ATLANTIC CIRCULATION

JAMES A. CARTON
Department of Meteorology
University of Maryland
Computer and Space Sciences Building
College Park, MD 20742

INTRODUCTION

Kirchner, in his study *Mundus Subterraneous*,[1] correctly identified the dominant pattern of surface circulation in the tropical Atlantic as a broad northwestward flow crossing the equator into the northern hemisphere. This early idea of the ocean's circulation was based on the mistaken impression that tropical currents are driven by the mass fluxes due to rainfall and evaporation. The first detailed analysis of surface circulation was provided by Major James Rennell. In his *Currents of the Atlantic Ocean,*[2] based on ships' log-books, Rennell was able to document the extent of what we now call the South Equatorial and North Equatorial Currents, carrying water across the tropical Atlantic from the southeastern basin toward the Caribbean Sea. He recognized that the tropical Atlantic contains a band of surface water that travels eastward toward the Gulf of Guinea, against the prevailing direction of the trade winds. However, it was not until later in the nineteenth century that the full extent and seasonality of the North Equatorial Countercurrent was recognized.

The first comprehensive hydrographic survey of the tropical Atlantic was carried out during the Meteor expeditions of 1925-1927.[3] Analysis of these observations allowed scientists to identify the major types and sources of water in the tropical Atlantic and provided some of the first indications of the circulation at mid- and deep levels in the ocean. By the 1960s, research on the physical oceanography of the Atlantic began to be influenced by a growing awareness of the coupling between the tropical oceans and the atmosphere. A series of increasingly extensive experimental programs were carried out to investigate major features of the variability of the upper ocean. The first of these was EQUALANT in 1963.[45] This early program was primarily exploratory, but succeeded in defining a number of important characteristics of the circulation, such as the zonal equatorial pressure gradient. This effort was followed by a joint meteorological/oceanographic study, GATE, in 1974.[67] Among the oceanographic achievements of this program was the identification of 20-30 day instability waves. GATE was followed by FGGE in 1979, although this program had limited oceanographic goals in the Atlantic. Interest in the oceanographic characteristics of the strong seasonal cycle culminated in the SEQUAL/FOCAL program of 1983-4 (a special issue of the *Journal of Geophysical Research*, 1987 summarizes much of this work). This program provided a broad range of upper ocean measurements during what turned out to be an unusual two-year period. Soviet scientists made extensive hydrographic surveys as well during the 1980s under the SECTIONS program. Very recently the WESTRAX program has concentrated on the interesting circulation along the western boundary and its exchanges with the equatorial currents. These six programs provide a fairly clear picture of the mean and seasonal variability of the Atlantic.

THE OBSERVED OCEAN

Time-Mean Conditions

Annual averaged sea surface temperature (SST) shows a zonal gradation with a maximum of about 30°C in the west decreasing to 26°C in the east. This zonal change in SST results from both zonal changes in surface heating and in surface winds. SST also varies with latitude, decreasing towards the poles as the incident solar radiation decreases.

The subsurface temperature field is characterized by a sharp thermocline separating the warm surface waters from the cool (<12°C, see Figure 1, top) waters at middepth. Near the equator the middle of the thermocline can be defined by the depth of the 20°C isotherm. Along the equator, this isotherm varies in depth between 150 meters in the west and 20 meters at the Greenwich Meridian. Eastward of that meridian it deepens again.

Surface salinity is maximum in the subtropics of both hemispheres, reaching salinities well above 37 parts per thousand, while in the high rainfall regions of the tropics it remains below 35.5 parts per thousand. The high values of salinity in the subtropics result from the fact that in the tropical Atlantic, unlike the tropical Pacific, evaporation exceeds precipitation. Between 20°S and 20°N the ocean loses 1.1×10^5 m^3/s to the atmosphere.[89] Throughout much of the western tropical Atlantic a subsurface salinity maximum exists at or above the thermocline (Figure 1, bottom). This salinity structure results from equatorward flow of high salinity subtropical water, which is subducted below the lower salinity surface water of the tropics.

The circulation of the upper tropical Atlantic, above the thermocline, is characterized by a complex system of zonal and western boundary currents. The circulation can be thought of as two counterclockwise gyres, one in the northern hemisphere and one in the southern hemisphere, with the *North Equatorial Countercurrent* (NECC) defining their common boundary. To the north and south of the NECC are the westward flowing *North Equatorial Current* (NEC) and *South Equatorial Current* (SEC). The NECC is more strongly variable than the currents on either side of it, reaching its maximum strength in northern summer-fall when surface currents in the west can exceed 1 m/s. These currents are primarily restricted to the thermocline waters and above (however some observations by Richardson and Reverdin[10] indicate the NECC may extend somewhat below the thermocline). The coastal Guinea Current carries some of the water from the NECC further eastward into the Gulf of Guinea. Near the equator a second eastward current, the *Equatorial Undercurrent* (EUC), persists at thermocline depths throughout the year.

SEASONAL VARIABILITY

Atmospheric Changes

The strong seasonal variability observed in the tropical Atlantic Ocean is largely

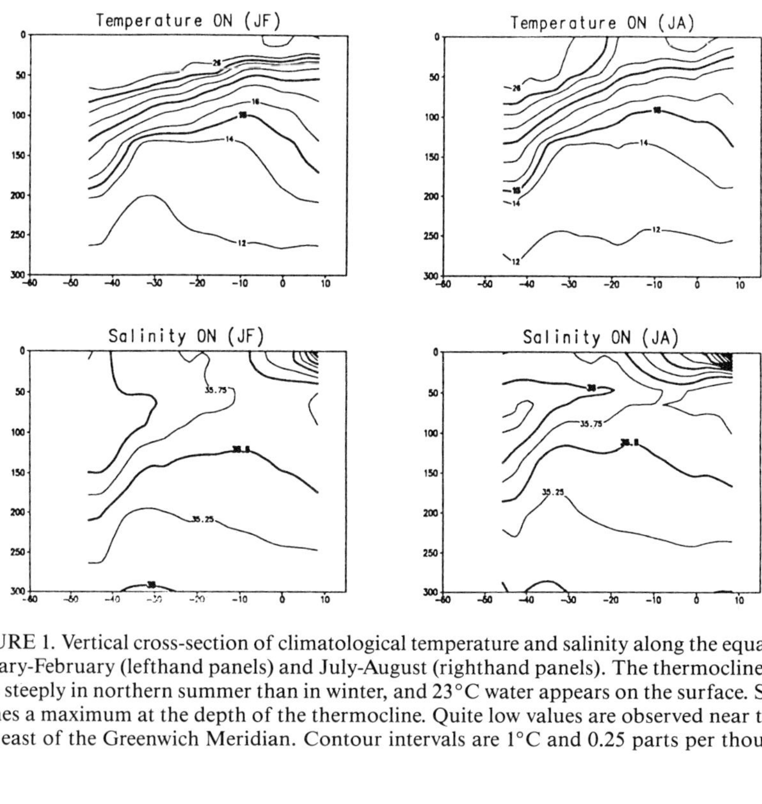

FIGURE 1. Vertical cross-section of climatological temperature and salinity along the equator for January-February (lefthand panels) and July-August (righthand panels). The thermocline slopes more steeply in northern summer than in winter, and 23°C water appears on the surface. Salinity reaches a maximum at the depth of the thermocline. Quite low values are observed near the surface, east of the Greenwich Meridian. Contour intervals are 1°C and 0.25 parts per thousand.

the result of seasonal fluctuations of surface winds, but there is also a contribution due to seasonal heating and freshwater flux. The wind pattern in the tropical Atlantic consists of two systems circulating anticyclonically around high pressure zones located in the subtropics of each hemisphere.

The boundary between these two systems is formed by the Intertropical Convergence Zone (ITCZ). This band of high convection coincides with the northern edge of the southeast trade winds (compare Figures 2, top left and right). Because

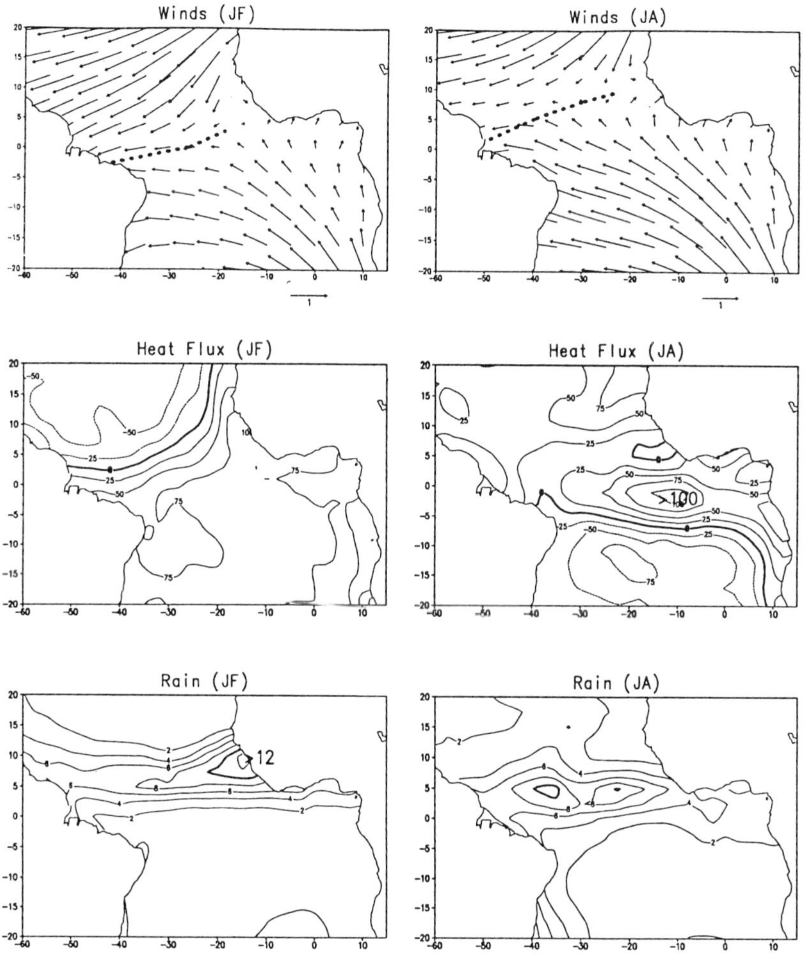

FIGURE 2. Seasonal surface fluxes for January-February (lefthand panels) and July-August (righthand panels) based on an eight-year average (upper panels). Surface wind stress based on a ten-year average of analyses of the European Centre for Medium Range Forecast forecasts. Units are dynes/$cm.^2$ The position of the Intertropical Convergence zone is marked with a dotted line. Surface net heat flux from Oberhuber[43] (center panels). Contour interval is 25 Watts/m^2. Downward fluxes are positive. Rainfall from Yoo and Carton[44] based on satellite measurements (lower panels). Contour interval is 2 mm/day.

atmospheric convergence is high in this region, winds are generally weak. The latitude of the ITCZ varies from near the equator during February-March to >10°N in July-August. At higher latitudes the winds strengthen, but the seasonal variability of the winds weaken. The seasonal shift in the position of the ITCZ causes the trade winds along the equator to weaken late in the year and strengthen abruptly in late spring. This discription applies primarily to the western and central basins. East of the Greenwich Meridian the character of the wind field changes. Here the seasonal patterns of wind are largely controlled by changes in heating over the desert regions of Africa. In northern summer a monsoon wind pattern develops in the eastern Gulf of Guinea so that the predominant wind direction is northeastward rather than northwestward as it is in the west.

The surface freshwater flux is composed of seasonally varying source terms, rainfall and river discharge, and steady losses due to evaporation. Rainfall has the strongest annual and semiannual variations, resulting from seasonal shifts of the ITCZ (Figure 2, bottom). On either side of the mean position of the ITCZ, rainfall rates may change by up to 7 mm/day between the dry and rainy seasons. River discharge is a small part of the net water budget, except in coastal regions.[11] The largest contribution is in May and October.

In the tropical oceans, surface heat flux results primarily from the balance of shortwave and longwave radiative terms and latent heating. The largest of these terms, shortwave radiation, varies seasonally as the latitudinal position of the sun changes. Since the sun crosses the equator twice a year in March and September, this term has a semiannual component on the equator. In addition, it is altered by seasonal changes in cloud distributions. The second largest term, the longwave contribution, is primarily a function of SST except in the eastern basin. Here the presence of low level stratus clouds increases the amount of downward radiation in northern summer months at the same time that it reduces incident solar radiation. Latent heating is also reduced over the cooler eastern waters, as well as in low wind regions such as along the equator in northern wintertime. When all these terms are summed together, net heat flux at the surface varies seasonally between ± 50 Watts/m^2 (Figure 2, center).

Oceanic Changes

The main way in which the oceans communicate with the atmosphere is through changes in SST. In comparison with the Pacific and Indian Oceans, the Atlantic is relatively cool, with limited areas where SST exceeds 28°C. In northern winter when net surface heating is high south of the equator, southern SSTs reach their maximum. Along the equator the warmest temperatures are found in the eastern basin. By northern summer the highest surface temperatures are to be found in the northern hemisphere (Figure 3, top). The eastern equatorial zone is now a region of anomalously cool sub-25°C water, while in the western tropical Atlantic SST exceeds 28°C. Particularly in the equatorial regions changes in SST reflect changes in the thermal field throughout the upper ocean. We can summarize changes to the thermal structure of the subsurface ocean by examining *heat content*. For this discussion we will define heat content to be the average 0/300 meter temperature in units of degrees C (Figure 3, bottom). This layer is deep enough to include the thermocline transition from near-surface water to cooler water below the thermocline. Heat content increases from minima in the eastern basin near 10°S-10°N toward the equator, and westward. Seasonal changes in heat content are associated with seasonal changes in some of the major current systems.

North Equatorial Countercurrent

North of the equator, the largest seasonal changes in heat content occur in the western tropical basin. In summer a minimum in heat content occurs at 9°N, where the vertically averaged temperature decreases by over one degree from its equatorial value. This meridional change in temperature implies a pressure distribution which decreases with increasing latitude. The maximum sea level drop across the NECC has been estimated to be 20-35 cm during late summer-fall.[12] Associated with this southward pressure gradient is an implied eastward geostrophic current, the NECC. The NECC traverses the Atlantic basin zonally in the latitude band between 3°N and 9°N. Maximum surface speeds are attained in the northern summer/early fall, while weak westward flows (i.e., no Countercurrent) prevail in the late winter/early spring. Maximum transport may be 20 to 25 × 10^6 m^3/s in midbasin, with higher

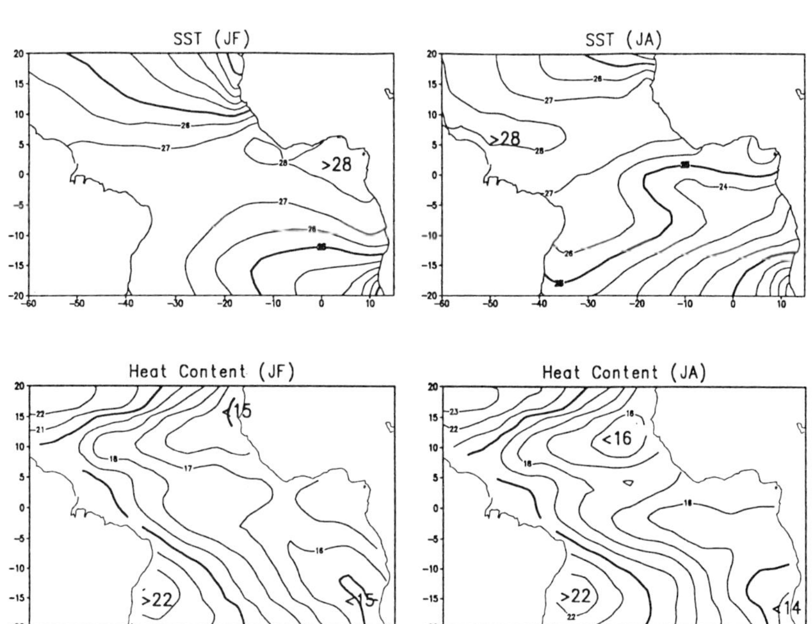

FIGURE 3. Seasonal oceanic variables for January-February (lefthand panels) and July-August (righthand panels). (a) Sea surface temperature from Reynolds,[45] based on an eight year average. Contour interval is 1°C. (b) Upper ocean heat content from Levitus,[46] computed as the average 0-300 m temperature. Contour interval is 2°C.

values expected further west. A schematic diagram of the structure of this and other equatorial currents is shown in Figure 4.

A second feature of the NECC which is immediately striking is its abrupt origin near the coast of Brazil. The source of water for the NECC is apparently the retroflection, or offshore turning of the *North Brazil Current.*[13,14,15] This is discussed in Section *Western Boundary Currents.*

The spatial variation of NECC transport is still uncertain. Since the meridionally-averaged thermocline shallows considerably eastward (from 110 to 50m from 43°W to 19°W, according to Garzoli and Katz;[16] but the NECC does not broaden eastward;[17] the expectation is that the transport will substantially decrease in this direction. This has not been directly measured, though the low transport values reported by Bubnov and Egorikhin[18] and Henin and Hisard[19] at 23°W are suggestive.

For many years it has been assumed that the southward pressure gradient associated with the NECC is the result of local wind curl. During summer the zonal trade winds weaken with increasing latitude between the equator and 7-10°N. This leads to a local deepening of the thermocline resulting from an Ekman-convergence of surface water. However, the strong northward winds which occur along the equator in summer and fall may also contribute to an increase in sea level just north of the equator and thus lead to an eastward geostrophic current. The possible importance of this effect has not yet been explored.

Western Boundary Currents

Surveys of the subsurface temperature field in the western tropical Atlantic have revealed high levels of both quasi-stationary and moving eddies. In the surveys of Bruce, et al.[20] two quasi-stationary anticyclonic eddies were identified. These eddies, which have shown up consistently in other surveys, are located in front of the mouth

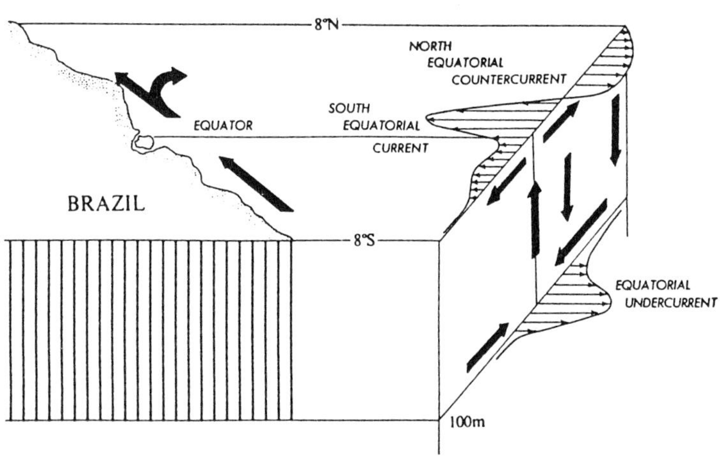

FIGURE 4. Schematic diagram of the circulation in the western tropical Atlantic (from Philander and Pacanowski[47]).

of the Amazon, and a few hundred kilometers further north off the Demerara Rise. They represent locations where the North Brazil Current retroflects into the NECC during northern summer-winter. Further northwest along the coast, the fate of this current is less clearly defined. Evidence from current meter moorings and ocean color suggest that the North Brazil Current is no longer a continuous current, but rather, is made up of a series of eddies.[21] The source of these eddies and their contributions to cross-equatorial mass and heat flux are still unclear.

Equatorial Undercurrent

A striking feature of the equatorial oceans is that the strong eastward flowing EUC persists throughout the year with transports of 13×10^6 m^3/s at 23°W, and maximum speeds of 0.7-0.8 m/s.[22] Zonally, it is confined within a couple of degrees of the equator. In the west the depth of maximum speed is near 100 meters, somewhat above the thermocline, while in the east the speed maximum is below the thermocline. The source of water for the EUC seems to be the surface waters of the southern subtropics resulting in higher salinities and Oxygen than the surrounding water.[23] The EUC mass transport gradually decreases as the current flows eastward, losing mass to the nearsurface currents. By 0°E the current is weak and variable.

The balance of forces giving rise to the EUC also varies with longitude and depth.[24] In the west the undercurrent accelerates in response to the westward pressure gradient set up by the westward trade winds. Further east in the thermocline and above, the EUC maintains a balance between the downward exchange of westward momentum imparted by the trade winds and the eastward-acting pressure gradient.

Thc zonal pressure gradient along the equator is itself closely related to the zonal wind stress. In northern summer the equatorial trade winds accelerate abruptly leading to an increase in sea level in the west by 15 cm. By February the pressure gradient is very weak, in response to the weakening winds.[25] Compensation occurs in the temperature field so that the high summertime sea levels are associated with a deep summertime thermocline.[26]

Tropical Instability Waves

Satellite SST observations reported in Legeckis and Reverdin[27] confirmed the existence of tropical instablility waves in the eastern tropical Atlantic. These wave-like disturbances propagate westward along the surface temperature front between the cool equatorial water and the warmer waters of the NECC. Their meridional amplitudes are typically 500 km, while their zonal wavelengths are approximately 1000 km.[28] These instability waves have vertical structure which penetrates through the thermocline and add 20-30 day variability to the velocity field throughout equatorial zone.[29] The possible role these instability waves play in mixing and cross-equatorial mass and heat transport is still unclear.

INTERANNUAL AND DECADAL VARIABILITY

In comparison with the Pacific, interannual fluctuations of variables such as SST in the Atlantic are weaker and have a broader range of scales. At decadal time scales Hastenrath,[30] Ward and Folland,[31] and Servain[32] have shown that tropical Atlantic SSTs undergo fluctuations that have patterns extending into midlatitudes with low amplitude near the equator. The possible causes and consequences of these variations are discussed later. However, at less than decadal time scales the SST anomalies are more closely confined to the eastern equatorial region. This latter pattern of fluctuation seems most similiar to the ENSO warmings in the eastern Pacific.

As discussed in the previous section, SST in the eastern Atlantic has a strong seasonal variation. It reaches its maximum in northern hemisphere spring, when the equatorial winds are weakest and the thermocline is deepest. During this season the sun is directly overhead, providing maximum incident solar radiation at the equator. Minimum SST occurs two seasons later. Even as the sun returns to the equator in northern fall, increased evaporation and the uplifting thermocline due to the intensification of the trade winds leads to surface cooling along the equator and the southern coast of Africa. Superimposed on this annual variation are fluctuations that frequently exceed 0.5°C.[33,34] Warm events of this magnitude occur with varying strengths every few years. Here we will discuss two of the best-documented events, those of 1963 and 1984.

Observations from the EQUALANT program during the summer of 1963 showed that SST was two degrees above normal in the eastern Atlantic. This warming occurred simultaneously with an anomalous relaxation of the trade winds and the zonal pressure gradient along the equator.[35] At the same time, further south along the southwest African coast of Namibia massive environmental changes were taking place. Warm saline tropical water replaced the normally cool Benguela Current.[36] These physical changes also had substantial affects on fish and bird populations documented in Crawford, et al.[37] Through fall and winter the main SST anomaly shifted southward, the equatorial wind anomalies are primarily southward, and conditions gradually returned to normal along the southern coast.

It is evident from the meteorological observations that the origin of the 1984 warm event extends back into 1983. Early in 1983, the northern extension of the southeast trade winds in the western basin, (which had been somewhat weak during the preceding year) intensified by 0.3 dyn/cm^2 as a result of changes in the winds and convection in the eastern Pacific. The trade wind anomaly during this period was primarily confined to a 10° band surrounding the equator and its narrow confinement led to anomalous heat storage in the west, both on and off the equator. In the west the unusual character of the trade winds was recorded at St. Peter and Paul's Rocks (2°N, 29°W), where the onset of the seasonal trades occurred a month early in 1983, and remained anomalously intense through the summer and fall. Late in 1983 the winds relaxed to near-normal conditions while in the spring of 1984 the reappearance of the trade winds occurred in May, as expected climatologically.[38] Rainfall was enhanced south of the equator in early 1984, indicating that the ITCZ had shifted anomalously southward, actually crossing the equator early in that year.

Decadal Variability

As mentioned on the previous page, a striking feature of decadal variability in the tropical Atlantic is the slow evolution of a dipole pattern of SST which is anticorrelated in the two hemispheres. Time series of this pattern of SST anomaly seems to be related to rainfall changes in both the northern Nordeste region of Brazil and the southern Sahel region of Africa.[39,40,41]

Hastenrath[42] explains the response of the atmosphere to the dipole of SST anomaly in the following way. Surface pressure is reduced over warm water, so that when SST is low in the northern hemisphere and high in the southern hemisphere, there is a southward pressure gradient at the equator. The ITCZ is perturbed southward of its climatological position. In March-April, when the ITCZ is normally closest to the equator, a southward perturbation of even a few hundred kilometers will radically increase rainfall rates dramatically in regions just south of the equator, like the northern Nordeste. Higher than normal SST in the northern hemisphere and lower than normal SST in the southern hemisphere leads to a northward shift of the ITCZ and droughts to the south. A secondary factor is the effect of southern SSTs on the stability of the southeast trade winds. Increasing southern SST will tend to destabilize the boundary layer of the southeast trade winds causing them to convect more easily. The causes of the patterns in SST which drive the changes in winds and rainfall have not yet been identified.

SYNTHESIS

The tropical Atlantic is similar to the eastern tropical Pacific to the extent that both have strong seasonal variability. Along the equator SST reaches a minimum of 22-23°C in northern summer and fall when the equatorial trade winds are intense. The primary cause of this drop in temperature is the shallowing of the thermocline in response to the strengthening trade winds. During these seasons the equatorial thermocline in the west is deep and sea level is high. Seasonal changes also occur off the equator.

Several factors make interannual variability of the tropical Atlantic more complex than that of the tropical Pacific. Perhaps the most important is that the Atlantic appears to support a wider variety of phenomena, including modes of variability that are not confined to the tropics. The presence of continents and the geometry of their coastlines add complexity. However, evidence is accumulating that the tropical Atlantic is subject to a phenomenon similar to ENSO in the Pacific.

The eastern tropical Atlantic is characterized by events with anomalously warm SSTs. These events occur with varying strengths every few years. The timing of the warm events is linked to the seasonal cycle, so that the maximum anomaly occurs during the normally cold northern summer-fall. By winter a broad, weak, warm anomaly appears south of the equator and gradually spreads westward during the first half of the year following the warm event. The warm SST anomalies are accompanied by changes in the surface winds. These changes frequently include a weakening of the northeast trade winds early in the year, and a relaxation of the southeast trade winds along the equator in the spring to the west of the region where the anomalously warm SST develop. The similarities between these changes and the changes associated with El Niño in the Pacific are striking.

Yet an examination of convection in the atmosphere shows that there are significant differences between the Atlantic and Pacific. In contrast to the eastern Pacific, the eastern equatorial Atlantic has strong seasonal convection. Anomalies of convection during warm events are primarily associated with a southward shift and intensification of the Intertropical Convergence Zone, in contrast to the eastward shift with time observed in the Pacific. The likely reason for these differences is the extra geographic control over the position of convective zones in the Atlantic because of the presence of the African and South American continents. Other reasons may be the smaller amplitude and shorter duration of the fluctuations of SST in the Atlantic. A second difference between events in the Atlantic and Pacific has to do with the relative influence of the global circulation. It is accepted that El Niño is the result of unstable interactions between the Pacific Ocean and its overlying atmosphere. It seems that the initial cause of many of the warm events in the Atlantic may be traced back to changes in the large-scale wind field which do not have their origin in the Atlantic.

Like their Pacific counterpart, warm events in the tropical Atlantic have broad climate effects. We have shown that rainfall increases along the northern Gulf of Guinea during the normally dry summer months. On the western side of the basin, along the northeast coast of South America, seasonal rainfall normally occurs in spring. In the spring of the year after a warm event, this rainfall is enhanced, contributing to the floods that plague the Nordeste region of Brazil. The eastward shift of heat along the equatorial wave guide, which is mainly responsible for the elevated SSTs in the simulation, also causes a variety of changes along the southern coast of Africa. These include a reversal of the normally equatorward Benguela Current, and a southward intrusion of warm, saline, tropical water. Observations confirm the appearance of the anomalous physical changes during years with warm events. The resulting effects on the fish and bird populations are profound.

REFERENCES

1. Kirchner, A. 1678. *Mundus Subterraneous.* in XII *Libros digestus,* 3rd edition, Amstelodami.
2. Rennell, James. 1832. *Currents of the Atlantic Ocean and of those which prevail between the Indian Ocean and the Atlantic.* London.
3. Defant, A. 1936. Schichtung and Zirkulation des Atlantischen Ozeans (die Troposphaere). *Wissenschaftliche Ergebnise der Deutschen Atlantischen Expedition "Meteor"* 1925-1927. 6:289-411. Also published as The Troposphere. Vol. 6, Part 1. Translated by W.J. Emergy, 1981, Amerind Publishing Co. 113 pp.
4. Neumann, G. and R. Williams. 1965. Observations of the Equatorial Undercurrent in the Atlantic Ocean at 15W during Equalant I. *J. Geophys. Res.* 70:297-304.

5. Katz, E. and collaborators. 1977. Zonal pressure gradient along the equatorial Atlantic. *J. Mar. Res.* 35:293-307.
6. Philander, S.G.H. 1974. The oceanographic subprogram of GATE. *Bull. of the American Met. Soc.* 55:738-744.
7. Philander, S.G.H. and W. Duing. 1980. The oceanic circulation of the tropical Atlantic, and its variability, as observed during GATE. *Deep-Sea Res. GATE supplement to* 26:1-27.
8. Baumgartner, A. and E. Reichel. 1975. *The world water balance, mean annual global, continental and maritime precipitation, evaporation and run-off.* Elsevier, 179 pp.
9. Yoo, J.-M. and J.A. Carton. 1990. Annual and interannual variation of the freshwater budget in the tropical Atlantic Ocean and the Caribbean Sea. *J. Phys. Oceanogr.*
10. Richardson, P.L. and G. Reverdin. 1987. Seasonal cycle of velocity in the Atlantic North Equatorial Countercurrent as measured by surface drifters, current meters, and ships. *J. Geophys. Res.* 92:3691-3708.
11. Muller-Karger, C.R. McClain and P.L. Richardson. 1988. The dispersal of the Amazon's water. *Nature.* 333:56-59.
12. Katz, E.J. 1981. Dynamic topography of the sea surface in the equatorial Atlantic. *J. Marine Res.* 39:53-63.
13. Cochrane, J.D., F.J. Kelly and C.R. Olling. 1979. Subthermocline countercurrents in the western equatorial Atlantic Ocean. *J. Phys. Oceanogr.* 9:724-738.
14. Bruce, J.G., J.L. Kerling and W.H. Beatty III. 1985. On the north Brazilian eddy field. *Progress in Oceanography.* 14:57-63.
15. Flagg, C.N., R.L. Gordon and S. McDowell. 1986. Hydrographic and Current Observations on the continental slope and shelf of the western equatorial Atlantic. *J. Phys. Oceanogr.* 16:1412-1429.
16. Garzoli, S. and E.J. Katz. 1983. The forced annual reversal of the Atlantic North Equatorial Countercurrent. *J. Phys. Oceanogr.* 13:2082-2090.
17. Merle, J. and S. Arnault. 1985. Seasonal variability of the surface dynamic topography in the tropical Atlantic Ocean. *J. Mar. Res.* 43:267-288.
18. Bubnov, V.A. and V.D. Egorikhin. 1977. Investigations of large-scale current structure in the tropical Atlantic along 23,30'W. *Oceanology.* 17:631-635.
19. Henin, C. and P. Hisard. 1987. The North Equatorial Countercurrent observed during the Programme Francais Ocean et Climat Dans l'Atlantic Equatorial Experiment in the Atlantic Ocean, July, 1982 to August, 1984. *J. Geophys. Res.* 92:3751-3758.
20. Bruce, J.G., J.L. Kerling and W.H. Beatty III. 1985. On the north Brazilian eddy field. *Progress in Oceanography* 14:57-63.
21. Johns, W.E., T.N. Lee, and F.A. Schott. 1991. The North Brazil Current retroflection: seasonal structure and eddy variability. *J. Geophys. Res.* 95: 22103-22120.
22. Hisard, P. and C. Hennin. 1987. Response of the equatorial Atlantic Ocean to the 1983-4 wind from the FOCAL cruise data set. *J. Geophys. Res.* 92: 3759-3768.

23. Metcalf, W.G. and M.C. Stalcup. 1967. Origins of the Atlantic equatorial undercurrent. *J. Geophys. Res.* 72:4959-4975.
24. Wacongne, S. 1988. The dynamics of the Equatorial Undercurrents and its termination. Ph.D. Thesis, Woods Hole Oceanographic Institution, Woods Hole, MA.
25. Carton, J.A. and E.J. Katz. 1990. Estimates of the zonal slope and seasonal transport of the Atlantic North Equatorial Countercurrent. *J. Geophys. Res.* 95:3091-3100.
26. Merle, J. and S. Arnault. 1985. Seasonal variability of the surface dynamic topography in the tropical Atlantic Ocean. *J. Mar. Res.* 43:267-288.
27. Legeckis, R. and G. Reverdin. 1987. Long waves in the equatorial Atlantic during 1983. *J. Geophys. Res.* 92:2835-2842.
28. Steger, J. and J. Carton. 1991. Long waves and eddies in the tropical Atlantic: 1984-1990. *J. Geophys. Res.* 96:15,161-15,172.
29. Weisberg, R.H., J.H. Hickman, T.Y. Tang and T.J. Weingartner. 1987. Velocity and temperature observations during the Seasonal Response of the Equatorial Atlantic Experiment at 0N, 28W. *J. Geophys. Res.* 92:5061-5075.
30. Hastenrath, S. 1978. On modes of tropical circulation and climate anomalies. *J. Atmos. Sci.* 35:2222-2231.
31. Ward, M.N. and C.K. Folland. 1991. Prediction of seasonal rainfall in the North Nordeste of Brazil using eigenvectors of sea-surface temperatures. *Int. J. Climatology.* 11:711-743.
32. Servain, J. 1991. Simple climatic indices for the tropical Atlantic Ocean and some applications. *J. Geophys. Res.* 96:15,137-15,146.
33. Carton, J.A. and B. Huang. 1993. Warm Events in the tropical Atlantic. *J. Phys. Oceanogr.* Submitted.
34. Zebiak, S. 1993. Air-sea interaction in the equatorial Atlantic region. Submitted.
35. Katz, E. and collaborators. 1977. Zonal pressure gradient along the equatorial Atlantic. *J. Mar. Res.* 35:293-307.
36. Shannon, L.V., A.J. Boyd, G.B. Brundrit and J. Taunton-Clark. 1986. On the existence of an El Niño type phenomenon in the Benguela system. *J. Mar. Res.* 44:495-520.
37. Crawford, R.J.M., W.R. Siegfried, L.V. Shannon, C.A. Villacastin-Herroro and L.G. Underhill. 1990. Environmental influences on marine biota off southern Africa. *S. African J. Sci.* 86:330-339.
38. Colin, C. and S.L. Garzoli. 1987. In situ wind measurements and the ocean response in the equatorial Atlantic during the Programme Francais Ocean et Climat Dans l'Atlantic Equatorial and Seasonal Response of the Atlantic Ocean Experiment. *J. Geophys. Res.* 92:3741-3750.
39. Hastenrath, S. and L. Heller. 1977. Dynamics of climate hazards in Northeast Brazil. *Quart. J. Roy. Met. Soc.* 103:77-92.
40. Markham, C.G. and D.R. McLain. 1977. Sea surface temperature related to rain in Ceara, Northeastern Brazil. *Nature.* 265:320-323.
41. Palmer, T.N. 1986. Influence of the Atlantic, Pacific and Indian Oceans on Sahel rainfall. *Nature.* 322:251-253.

42. Hasenrath, S., 1985. *Climate and circulation of the tropics,* D. Reidel Publishing Co. Boston, 455pp.
43. Oberhuber, J. 1988. *An atlas based on the 'COADS' data set.* Max-Planck-Institut fur Meteorologie, Report 15.
44. Yoo, J.-M. and J.A. Carton. 1990. Annual and interannual variation of the freshwater budget in the tropical Atlantic Ocean and the Caribbean Sea. *J. Phys. Oceanogr.* 20:831-845.
45. Reynolds, R.W. 1988. A real-time global sea surface temperature analysis. *J. Climate.* 1:75-86.
46. Levitus, S. 1982. *Climatological Atlas of the World Ocean.* NOAA Professional Paper #13.
47. Philander, S.G.H. and R.C. Pacanowski. 1986. The heat and mass budget in a model of the tropical Atlantic Ocean. *J. Geophys. Res.* 91:14212-14220.

The Oceans: Physical-Chemical Dynamics and Human Impact. Edited by S.K. Majumdar, E.W. Miller, G.S. Forbes, R.F. Schmalz and Assad A. Panah.

Chapter Six

THE GULF STREAM SYSTEM: An Observational Perspective

JOHN M. BANE, Jr.
Marine Sciences Program
University of North Carolina
Chapel Hill, NC 27599-3300

INTRODUCTION

It was Ponce de Leon who, in 1513, provided the first written account of the Gulf Stream. While sailing south along the east coast of Florida from about Cape Canaveral to the Dry Tortugas, his three ships frequently found the Stream too strong to allow them progress, and they would anchor to wait for the current to abate [*Herrera y Tordesillas*, 1601; *Scisco*, 1913; *Stommel*, 1965]. Near the end of that same century, John White noted that he had to stand far out to sea while sailing northward along the Carolinas, because closer to shore there were "eddy currents setting to the south and southwest" [*Kohl*, 1868]. Together these early observations point to the swiftness of the northward flow of the Gulf Stream and its complex variability.

Today we have observed this current system more than any other in the world's oceans, and we still are discovering new aspects of its personality. New observations seem always to provide unexpected insights and new challenges. This paper presents a description of the Gulf Stream system based primarily on physical oceanographic observations of the past twenty years. A companion paper in this book [*Malanotte-Rizzoli*, 1994] discusses dynamical and modeling studies that have attempted to provide physically correct explanations for the observed nature of the Gulf Stream, and which are used today in efforts to predict the future behavior of the Stream.

LARGE-SCALE CIRCULATION IN THE NORTH ATLANTIC OCEAN

The large-scale circulation of the world's major oceans can be divided roughly into two categories according to the nature of the forcing. Circulation systems in the upper ocean arise primarily because the winds that blow across the ocean provide momentum to the ocean's surface layer. The thickness of this surface layer varies, but it typically extends from the ocean surface to somewhere between a few meters and a few hundred meters depth. The currents in this layer, driven directly by the wind, redistribute surface water in a fashion that creates a pattern of "hills" and "valleys" on the ocean surface, thereby inducing an oceanic surface topography that does not correspond to a level surface. The hills and valleys have horizontal extents on the order of the size of the ocean basins, and the geometric relief of this surface topography (*i.e.*, vertical distance from an undisturbed ocean surface to this dynamically deformed surface) is on the order of one meter. This sloping surface topography creates a horizontal pressure gradient force that extends much deeper into the water column than does the directly wind-driven surface layer. Water responds to this horizontal pressure gradient force wherever it is experienced, whether in the surface layer or below. The resulting currents, referred to as the *wind-driven circulation,* form basin-scale gyres in the middle and higher latitudes, and east-west banded currents near the equator. [One notable exception to this general pattern is the Antarctic Circumpolar Current, which flows generally eastward around the globe just north of the Antarctic continent.] *Thermohaline circulation* is induced by differences in ocean water density that are generated by ocean-atmosphere interactions. In various locations at certain times of the year, surface water density will increase as temperature drops or salinity increases. Temperature decreases can be caused by atmospheric cooling and evaporation; salinity increases can result from evaporation or formation of sea ice by freezing. Such processes are typical in high-latitude regions during winter, and the density increases that occur there may be sufficient to cause the surface water to sink. This downward motion introduces water into the deeper portions of the ocean basins, and the result is a lateral deep flow that is primarily thermohaline in origin.

Turning our attention to the North Atlantic Ocean, we see clearly the patterns of wind-driven and thermohaline circulations. The two volume transport cartoons shown in Figure 1, produced recently by *Schmitz and McCartney* [1993], reveal the mean, wind-driven, subtropical gyre of the North Atlantic (Figure 1a) as well as the mean thermohaline flow in the deeper portion of the basin (Figure 1b). A conspicuous feature in each panel is the east-west asymmetry of the current patterns; there are consistently stronger flows on the western side of the basin in each cartoon. This "westward intensification" results when the ocean current flows meridionally along a western ocean boundary on a rotating, spherical earth. The first dynamically satisfactory explanation of this aspect of large-scale ocean circulation was given by Henry Stommel in a classic paper [*Stommel*, 1949; see *Malanotte-Rizzoli*, 1994]. The westward intensification in the wind-driven gyre is responsible for the Gulf Stream, which is that portion of the circulation in Figure 1a depicted as the bold transport line that begins in the Yucatan Straits, flows through the Gulf of

Mexico and the Florida Straits, then follows the east coast of the United States northward and northeastward past Cape Hatteras and into the central North Atlantic.

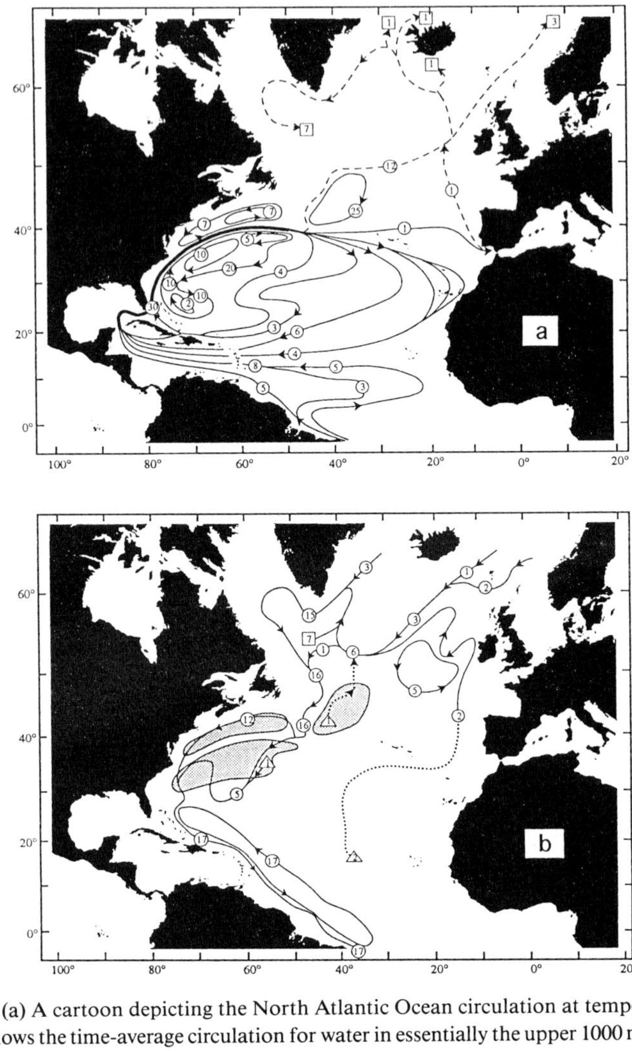

FIGURE 1. (a) A cartoon depicting the North Atlantic Ocean circulation at temperatures above 7°C. This shows the time-average circulation for water in essentially the upper 1000 m of the ocean, which is primarily wind-driven. The subtropical gyre may be seen as the anticyclonic (clockwise) pattern filling most of the central portion of the figure. The Gulf Stream is the bold line beginning in the Yucatan Straits (between Cuba and Mexico) and extending to the center of the North Atlantic. (b) A cartoon showing circulation of the deep waters of the North Atlantic Ocean. The time-average thermohaline flow (lines) plus the deep manifestations of the strong recirculations associated with the Gulf Stream (shaded patches) may be seen. In either panel, numbers indicate volume transport values in Sverdrups (1 Sverdrup = $10^6 m^3/sec$); circles are horizontal transports, squares denote sinking associated with the generation of thermohaline flow, and triangles denote upwelling. [Figures from *Schmitz and McCartney*, 1993].

THE BATHYMETRY OF THE WESTERN NORTH ATLANTIC OCEAN

Before looking more closely at the Gulf Stream, it is instructive to consider the bathymetry of the western North Atlantic Ocean (Figure 2). The continental shelf along the entire east coast of the United States and Canada extends from the shoreline out to about the 100-m isobath. There, at the shelf break, the bottom slope changes to fall off more steeply over the continental slope. North of Cape Hatteras there is a single continental slope that drops down from the shelf break to the deep floor of the North Atlantic, which is at a depth of around 5,000-5,500 m. The continental slope has varying degrees of steepness, but there are basically two regions—a steep region between 100 m and 2000 m formally known as the continental slope, and a gently sloping region between 2000 m and about 4500 m known as the continental rise. Following the continental slope southward from Cape Hatteras, one can see that the continental slope divides into two portions. The western portion has depths ranging from the shelf break depth (~100 m) to about 800 or 900 m. This "western" or "upper" slope runs along the western edge of the Blake Plateau and through the Florida Straits. The eastern portion of the slope extends from the

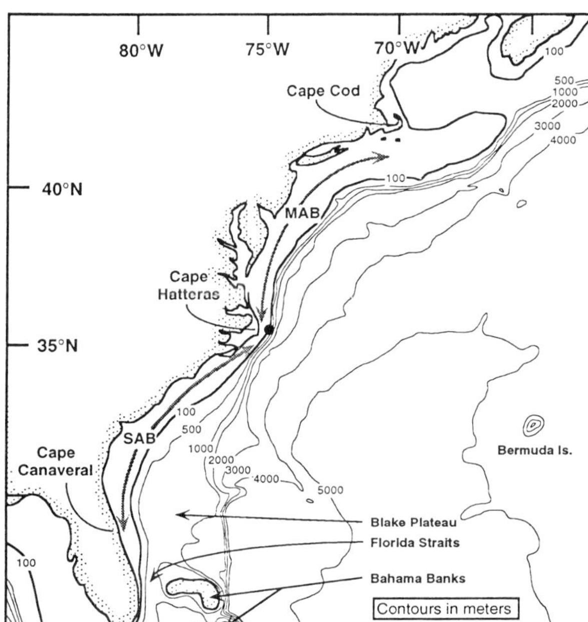

FIGURE 2. Bathymetry of the western North Atlantic ocean. The South Atlantic Bight (SAB, from Cape Canaveral to Cape Hatteras) and the Middle Atlantic Bight (MAB, from Cape Hatteras to Cape Cod) are denoted. The continental shelf is confined between the coast and the 100 m depth contour; the continental slope is between 100 m and 2000 m; the continental rise stretches from the 2000 m contour to about 4500 m depth. Note the change in orientation of the continental slope at the Hatteras Corner (dot), from SW-NE to S-N as one proceeds northward.

outer Blake Plateau to the deep Atlantic floor. The Bahama Banks are carbonate platforms with very shallow banktops (~ 3-10 m), located at the southern end of the Blake Plateau. Their western edges form the eastern side of the Florida Straits, and their eastern edges adjoin the western portion of the "divided" continental slope southeast of the Blake Plateau.

When considering the flow of the Gulf Stream along the east coast of the United States and into the deep Atlantic, an important aspect of the topography is the change in general orientation of the isobaths along the continental slope approaching from the south. Notice how the continenal slope orientation changes from SW-NE to about S-N at the location of the dot placed just northeast of Cape Hatteras. This corner in the slope geometry denotes the change from the South Atlantic Bight (SAB) to the Middle Atlantic Bight (MAB). We will refer to this location as the "Hatteras Corner" in the following discussion.

THE GULF STREAM

One may see in Figure 1a that the Gulf Stream forms the western limb of the North Atlantic's wind-driven, sub-tropical gyre. The gyre flows anticyclonically (clockwise in the northern hemisphere), and it typically extends from the surface to depths of about 1,000 m. The Gulf Stream current system may be traced from its roots in the tropical Atlantic Ocean and the Caribbean Sea through the Yucatan Straits, through the Gulf of Mexico and the Florida Straits, and along the upper (western) portion of the divided continental slope in the SAB to Cape Hatteras. As the Stream's flow passes Cape Hatteras, it separates from the continental slope near the Hatteras Corner and enters deeper water, as indicated in Figures 1a, 3 and 4 [*Stommel*, 1965; *Worthington*, 1976; *Fofonoff*, 1981; *Watts*, 1983].

The Gulf Stream's flow has remarkable coherence along this lengthy path, and it is often considered to be one current. It has different names in various locations, however. Within the Gulf of Mexico it is called the Loop Current, since its path occasionally penetrates northward so as to form an anticyclonic loop that fills much of the eastern Gulf [*Maul*, 1977; *Hurlburt and Thompson*, 1980; *Vukovich*, 1988; *Johnson* et al., 1992]. Through the Florida Straits it is called the Florida Current. Emerging from the Straits at about 27°N, the current is usually called the Gulf Stream, although *Iselin* [1936] proposed that the current be called the Florida Current all the way to Cape Hatteras. Past Cape Hatteras the flow is referred to as the Gulf Stream by essentially all authors. (We will use the name "Gulf Stream" beginning at the northern end of the Florida Straits.) Around 45°W, the Gulf Stream "branches" into two currents, the northward flowing North Atlantic Current and the southeastward flowing Azores Current [*Mann*, 1967; *Clarke* et al., 1980; *Sy*, 1988]. The nature of the branching process is not well documented. In particular, it is not known if the single Gulf Stream jet actually splits into two currents or if the Stream flows into the North Atlantic Current part of the time and into the Azores Current part of the time, thereby giving the average pattern suggested in Figure 1a.

The main flow in the Gulf Stream is jet-like, with its maximum speed at the surface (Figures 5 and 6). The jet has a width at the ocean surface of about 80-150 km and a depth of about 800-1200 m. The maximum current speed is typically observed to be about 200 cm/sec, although during strong events speeds exceeding 250 cm/sec have been seen. Moving away from the surface maximum, the current decreases in strength to either side of the jet or to greater depth within the jet. Thus, the basic jet-like current structure has strong vertical and lateral shears [*Stommel*, 1965; *Richardson* et al., 1969; *Halkin and Rossby*, 1985; *Hall and Bryden*, 1985; *Hall*, 1986; *Joyce* et al., 1986; *Johns and Shay*, 1992]. There is a distinct temperature structure associated with the jet. The geostrophic balance, the primary dynamical force balance in the Gulf Stream, results in an asymmetric pattern of ocean temperature across the Stream. Figure 5 shows how the permanent thermocline slopes upward from right to left through the jet, separating the warm water within the core of the jet

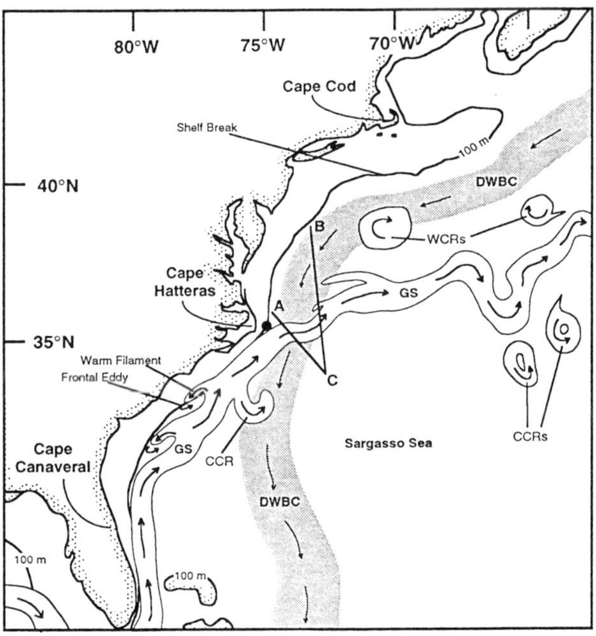

FIGURE 3. Schematic depiction of the major ocean currents in the western North Atlantic Ocean. The Gulf Stream (GS) flows northward from the Florida Straits and along the upper continental slope past the Hatteras Corner (dot), where it usually departs from the slope. Meanders and frontal eddies propagate along the Stream's shoreward edge in the SAB, and some are energetic enough to make it past Hatteras and into deeper water. Several WCRs and CCRs are depicted. The deep western boundary current (DWBC) follows the lower continental slope and continental rise from high latitudes southward, and it crosses beneath the Gulf Stream in the vicinity of the Hatteras Corner, interacting with the Stream to varying degrees as it does so. The features in this schematic diagram correspond roughly to those in the satellite image in Figure 4. (Note that this schematic does not extend as far east and south as does Figure 4.) The A-C and B-C lines show the locations of the sections in Figure 8.

from the cooler water on its left side (looking downstream). This surfacing thermocline is often known as either the cold wall or the north wall of the Gulf Stream [*Stommel*, 1965; *Halkin and Rossby*, 1985].

Since the permanent thermocline (the layer in which temperature rapidly decreases with increasing depth) intersects the sea surface along the left hand edge of the

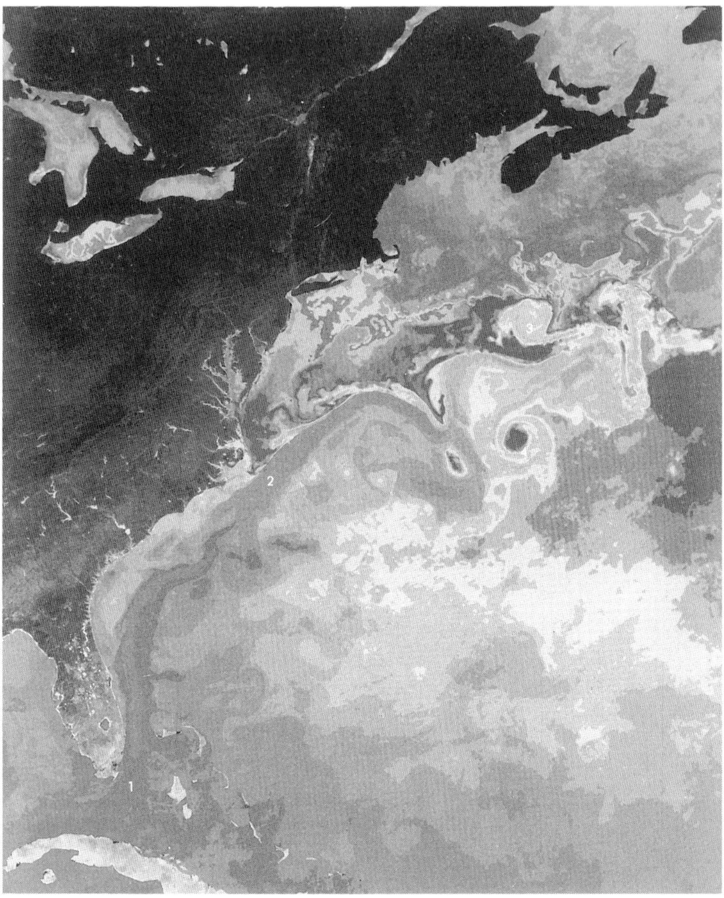

FIGURE 4. A satellite image showing the sea surface temperature field of the western North Atlantic Ocean during June 1984. Temperatures correspond to a gray scale. The warm temperatures of the northward flowing Gulf Stream allow it to be identified along its path from the Florida Straits (1) to Cape Hatteras (2) and eastward into the deeper Atlantic. The surface water in the Stream cools slightly as it proceeds along this course. A warm core ring is identified as (3), and a cold core ring is identified as (4). The Sargasso Sea Water south and east of the Gulf Stream is quite warm, while the Slope Water north of the Stream is cool to cold. One may compare the features in this image with those in the schematic in Figure 3. (Image courtesy of Otis Brown, Robert Evans and Mark Carle, RSMAS, University of Miami.)

Stream during most of the year, satellite infrared sensors can delineate the position of the Stream by locating this strong, sea surface temperature (SST) gradient. The satellite image in Figure 4 is a striking example of the patterns exhibited in the SST field because of oceanic currents. The Gulf Stream is clearly evident in this image from its flow through the Florida Straits to its meandering exit from the area along the eastern edge of the figure. Cold core and warm core Gulf Stream rings are also clearly defined by the SST variations.

The volume transport of the Gulf Stream jet is about 30 Sverdrups (1 Sv = $10^6 m^3/s$) within the Florida Straits and increases in the downstream direction. Off Cape Hatteras, the transport is about 70 Sv, at 68°W it is about 110 Sv, and by the time the current reaches 60°W, the transport is around 150 Sv [*Schmitz and Richardson*, 1968; *Knauss*, 1969; *Richardson* et al., 1969; *Worthington*, 1976; *Hendry*, 1982; *Halkin and Rossby*, 1985; *Hall*, 1986; *Hogg*, 1992; *Johns and Shay*, 1992; *Hall and Fofonoff*, 1993; *Schmitz and McCartney*, 1993]. The downstream increase in Gulf Stream transport is depicted in Figure 1a by the finer transport lines that join and follow the bold transport line from Florida northward. These lines represent inflow into the Gulf Stream, which becomes part of the Stream

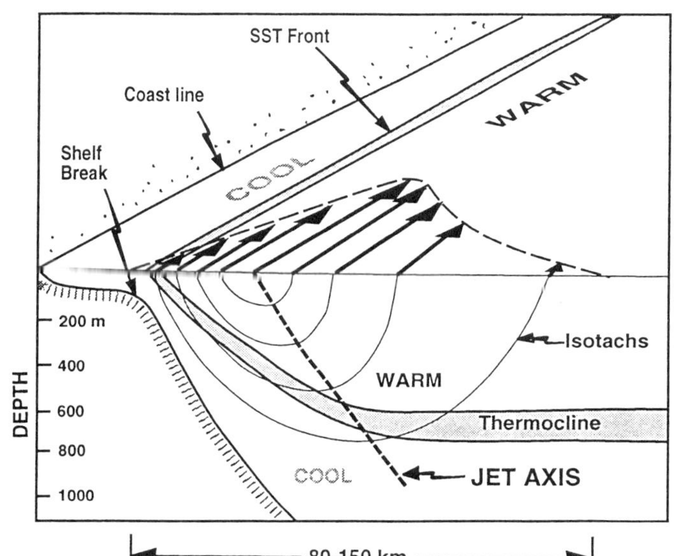

FIGURE 5. A schematic representation of the Gulf Stream jet, showing what the Stream would look like as it flows along the upper continental slope in the absence of any meandering or strong ring/Stream or DWBC/Stream interactions. Note that the jet current is strongest at the surface, and that the jet's axis is positioned further seaward with depth. The basic temperature structure is determined by the permanent thermocline, which separates warm and cold waters, and which shallows in the shoreward direction. The intersection of the thermocline with the sea surface forms a sea surface temperature (SST) front, which can be detected by satellite sensors and used to determine the location of the Gulf Stream (see Figure 4).

and provides the increase in its transport along its path. Note that there is an eastward decrease in transport beginning about 50°W.

Only a portion of the transport increase can be accounted for by the numbers on the fine lines in Figure 1a. The additional increase in transport is due to deep flow provided by recirculation systems on the northern and southern sides of the Gulf Stream. These recirculations reach to the bottom, well below the 7°C isothermal surface bounding the flows in Figure 1a. The shaded patches in Figure 1b indicate the extensions of these recirculations to depths below the 7°C isothermal surface. The two primary recirculations involved in the Gulf Stream's deep flow are indicated by the two shaded patches closest to the North American continent in Figure 1b; their upper level expressions are easily seen in Figure 1a. The southern recirculation has been called the "Worthington Gyre" since its discovery by Val Worthington in the 1970's [*Worthington*, 1976]. More recently, Nelson Hogg, Henry Stommel and collaborators have provided a foundational description of the "Northern Recirculation Gyre" [*Hogg and Stommel*, 1985; *Hogg* et al., 1986; *Hogg*, 1992].

GULF STREAM MEANDERS

The Florida Current essentially fills the northern Florida Straits, which is about 100 km wide (from the peninsula of Florida to the Bahama Banks) and 800 m deep

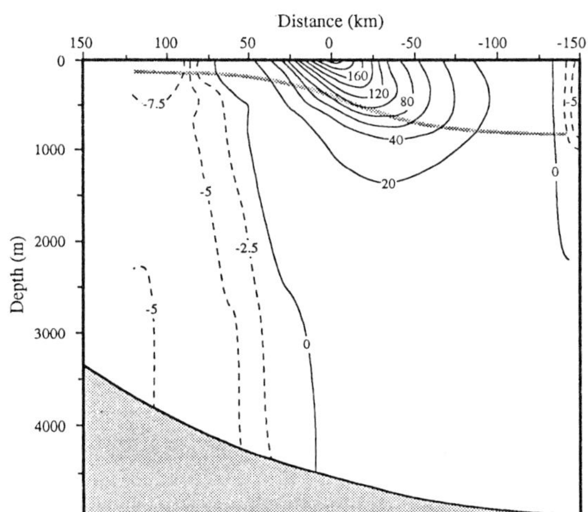

FIGURE 6. A vertical, North-South section through the Gulf Stream showing downstream velocity in cm/sec. This section was constructed with data collected along 68°W longitude (*cf.* Figure 3) during the SYNOP program. Positive values are eastward velocities (into the page) and negative values are westward velocities (out of the page). The location of the center of the thermocline is denoted by the shaded line that deepens from about 150 m on the left side of the figure to about 1000 m on the right side. A portion of the deep western boundary current was observed; it is the -5 cm/sec flow to the north (left) of the Gulf Stream. The observed structure of the Gulf Stream jet shown here may be compared with that of the schematic jet in Figure 5. [This figure was adapted from *Johns and Shay*, 1992.]

(see Figures 2, 3 and 4). As the current emerges from the Florida Straits and becomes known as the Gulf Stream, it no longer has the topographic constraint of the Bahama Banks on its eastern side. It flows along the upper (western) portion of the continental slope towards the Hatteras Corner with the left-hand edge of its surface typically following the 100-m isobath (see Figures 3, 4 and 5). As the Stream passes the Hatteras Corner, it usually departs the continental slope and no longer makes contact with it. Although this is typical, there are periods during which the Stream turns northward at the corner and follows the upper slope for many tens of kilometers before finally separating from it [*Richardson* et al., 1969; *Bane and Brooks*, 1979; *Halliwell and Mooers*, 1979; *Bane*, 1983; *Olson* et al., 1983; *Casagrande*, 1987; *Gangopadhyay* et al., 1992].

Time-dependent, wavelike meanders in the Gulf Stream's path cause variations in the location of the Stream. These meander patterns typically grow as they travel downstream from the Florida Straits (Figures 3 and 4). Gulf Stream meanders within the SAB travel in the downstream direction with speeds ranging from about 25 to over 60 km/day. Wavelengths can range from about 90 to over 250 km, which results in periods of about 2 to 10 days. Downstream from the Hatteras Corner, the meanders change in character. Longer period, longer wavelength meanders grow to become dominant in this region. Wavelengths in the 200 to 1,000 km range and periods of 12 to 60 days are common. Many of the meander motions observed in the SAB and downstream of Cape Hatteras show that the jet simply oscillates laterally with only small changes in the structure of the jet's flow. This results in increasing or decreasing current and temperature at a fixed point of observation. If the point of observation is within the positive lateral shear zone along the Stream's left side, then an increase in temperature is observed concurrently with an increase in downstream velocity as a meander "crest" (the shoreward excursion in a meander pattern) passes the point. As the meander pattern continues its progress to the northeast, the temperature and downstream velocity drop with the approach of the meander "trough" (the seaward excursion in a meander pattern) [*Webster*, 1961; *Brooks and Bane*, 1981, 1983; *Maul* et al., 1978; *Bane and Brooks*, 1979; *Halliwell and Mooers*, 1979, 1983; *Legeckis*, 1979; *Bane* et al., 1981; *Lee* et al., 1981, 1989; *Watts and Johns*, 1982; *Bane*, 1983; *Lee and Atkinson, 1983; Lee and Waddell*, 1983; *Olson* et al., 1983; *Pietrafesa* et al., 1985; *Singer* et al., 1983; *Watts*, 1983; *Bane and Dewar*, 1988; *FRED Group*, 1989].

Cyclonic (counterclockwise), frontal eddy motions are observed to be part of the meander phenomenon (see Figure 3). A frontal eddy is spun up in a meander trough as the meander grows while progressing to the north and northeast. The eddy circulation often causes warm surface water to be drawn out from the main Gulf Stream jet at the meander crest that leads the frontal eddy. This forms a warm filament that trails upstream from the crest and along the shoreward side of the frontal eddy. The filaments in the SAB are typically centered at about the shelf break, thereby bringing warm, Gulf Stream water onto the outer continental shelf (see Figure 4). Their warm temperatures make filaments observable from satellites during most of the year. Oftentimes the warm filament water separates from its parent meander crest and slows to become stranded over the outer continental shelf. It may subsequently mix with the ambient shelf water or may be entrained back into the northward flow of the Gulf Stream [*Bane* et al., 1981; *Lee* et al., 1981; *FRED Group*, 1989].

GULF STREAM RINGS

As Gulf Stream meanders propagate and grow downstream from the Hatteras Corner, one may occasionally become large enough to form a loop to the south with an extent sufficient to bring the jet back upon itself. This loop may close and detach from the jet, forming a ring of Gulf Stream water that encloses slope water in its center. Such a ring will swirl in the cyclonic sense, as shown in Figure 7 [see also Figure 2 in *Malanotte-Rizzoli*, 1994]. Due to its core of cold slope water, the name Cold Core Ring (CCR) is given to such a feature. A typical CCR has a diameter in the 200-300 km range, and its depth of influence may exceed 2,000 m. CCRs are formed east of 70°W, and about 10 CCRs exist at any one time (see examples in Figures 3 and 4). They travel generally westward at about 4 km/day when not in contact with the Gulf Stream, and northeastward to eastward when they are attached to it. Often the rings will break away from the Stream after an eastward "ride" and transit back to the southwest and west, thereby moving in a large, clockwise loop. Apparently, the ultimate fate of all CCRs is complete coalescence with the Gulf Stream jet [*Richardson*, 1983]. This usually happens east of 75°W. CCRs are sometimes observable in satellite SST images, with a warm, circular band surrounding the cooler water of the core. Air-sea interaction and mixing processes can destroy this SST signature, although the swirling ring may still exist beneath an almost constant SST region. Sometimes surface cooling in the fall and winter months will restore the SST contrast, since the warmed layer over the cool core is thin compared to the ring of warm water swirling around the core. Interactions with the Gulf Stream will often highlight a CCR position in a SST image, since such interactions may

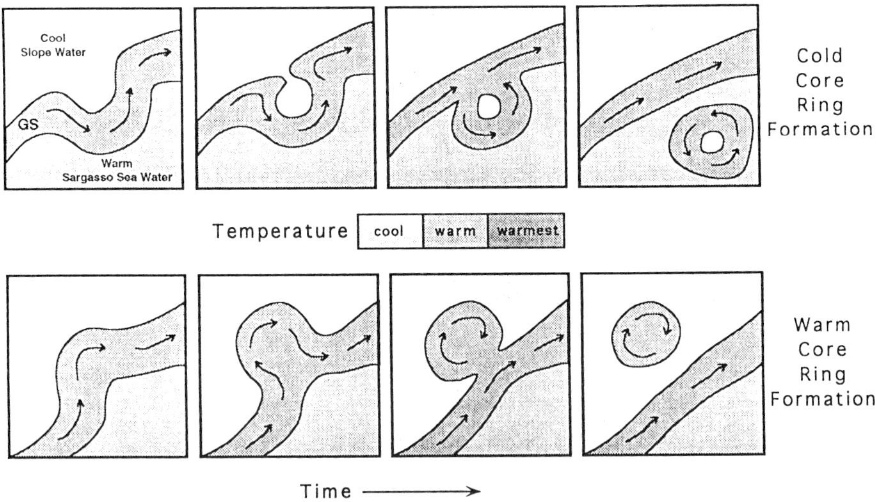

FIGURE 7. Idealized views of the formation of a cold core ring (top) and a warm core ring (bottom). As the Gulf Stream loops well to the south (north), a region of cyclonic (anticyclonic) flow is established, which oftentimes detaches from the jet to become a cold core (warm core) ring.

cause the ring to draw warm surface water out of the main Gulf Stream jet and wrap it around the core [*Lai and Richardson*, 1977; *Richardson*, 1980, 1981, 1983, 1993].

Warm Core Rings (WCRs) are formed several hundred km northeast of Cape Hatteras in a fashion similar to CCRs, although in this case the original Gulf Stream meander that gives birth to a WCR is an extended anticyclonic meander to the north (Figure 7). If the meander closes on itself and detaches, it forms an anticyclonic swirl of warm water that is usually 60-200 km in diameter and extends to several hundred meters depth (see Figures 3 and 4). It has been shown that WCR lifetimes in the 10-year period 1974 through 1983 had a bimodal distribution. Short-lived rings had a mean lifetime of 54 days and long-lived rings had a mean lifetime of 229 days. A few rings lived to between 350 and 400 days. In this same study period, the average number of WCRs existing in any one year was 8 [*Brown* et al., 1986]. During its lifetime, a typical WCR will travel west-southwestward between the Gulf Stream jet and the continental slope. Translation speeds average 5.5 km/day. Short-lived WCR's rarely make it past the apex of the MAB (~ 73°W), whereas the long-lived WCR's usually move into the southern MAB between the New Jersey-to-Virginia slope and the Gulf Stream. The fate of every WCR is apparently complete coalescence with the Gulf Stream jet. This usually occurs between 70°W and Cape Hatteras, as a ring traveling to the southwest must approach the Gulf Stream due to the decreasing distance between the Stream and the shelf break along its trajectory. It is common for a WCR to have one or more interactions with the Gulf Stream between its birth and its demise [*Watts and Olson*, 1978; *Halliwell and Mooers*, 1979; *Richardson*, 1983; *Brown et al.,* 1986].

GULF STREAM PATH VARIABILITY

During interactions with the Gulf Stream, both CCRs and WCRs can cause considerable contortions in the path of the main Gulf Stream jet. The jet path has been observed to move considerably north or south of its typical location during a strong ring/Stream interaction, and to remain displaced for weeks to months afterwards. A distinct change in the nature of the Gulf Stream's meandering is usually caused by such a ring-induced shift in the Stream's path. Studies in the SAB suggest that the general path of the Gulf Stream between about 32°N and Cape Hatteras is biomodal, having a "weakly deflected" state and a "strongly deflected" state. While weakly deflected, the Gulf Stram flows along the upper slope and supports meanders and frontal eddies as described above. Periods of these meanders are in the 4-10 day band, and they can progress past the Hatteras Corner into deeper water. While in a strongly deflected state, the Gulf Stream jet turns directly eastward near 32°N and flows almost 100 km out to sea before turning back towards Cape Hatteras. While in this state the Stream can support larger amplitude meanders that progress more slowly from 32°N to Hatteras, than do those meanders of the weakly deflected state. Meanders in the strongly deflected state appear to strongly contort the path of the jet, and they do not produce warm filaments as often as do the meanders occurring in the weakly deflected state. The transition between these two states can be rapid (e.g. a few days), yet once in a particular state the Stream will tend to stay there for weeks to months [*Bane and Dewar*, 1988; *FRED Group*, 1989; *Lee* et al., 1989; *Savidge* et al., 1992].

A case study during the Frontal Eddy Dynamics Study [*FRED Group*, 1989] has shown that the meanders which arrive at Cape Hatteras from the southwest, may have distinctly different amplitudes at different times, depending on how close the Stream's path is to the shelf break just to the southwest of the Hatteras Corner. In early 1987, the Stream was in a weakly deflected state and relatively close to the shelfbreak there, and the meanders had small lateral amplitudes. Their associated frontal eddies created rather narrow warm filaments over the outer continental shelf. During late spring an interaction began between a CCR and the Gulf Stream near 32°N. The ring proceeded to the northeast, interacting with the Stream as it went. During the several-day period of the interaction, the Stream's path was moved slightly offshore near Cape Hatteras, although the path remained near the upper slope at 32°N; there was not a shift into a strongly deflected state. The meander activity near 32°N did not change significantly as a result of this ring/Stream interaction, although it did near Cape Hatteras. This may be understood by considering what happens to a meander/frontal eddy event as it progresses from 32°N to Hatteras. As the event progresses northeastward, it encounters increasingly steeper topography along the upper slope. This causes energy to be converted from meander energy into mean flow energy, which decreases the amplitude of the meander and its associated frontal eddy. After the ring/Stream interaction in the FRED study, the meanders progressed along a Stream path that was somewhat farther removed from the slope influence, and so did not lose as much energy during their transit from 32°N to Hatteras as did those meanders prior to the ring/Stream interaction. Thus, when the meanders arrived at Cape Hatteras, they were larger than the earlier meanders. The larger meanders produced larger warm filaments as well, each of which swept cyclonically around its frontal eddy and quite close to shore. Interestingly, such larger filaments can cause Gulf Stream water to get closer to shore near Hatteras at a time when the Gulf Stream jet is farther from shore [*Ebbesmeyer and Bane*, 1989; *FRED Group*, 1989].

The Gulf Stream path between the Hatteras Corner and about 68°W undergoes a long-period variation, with a northward, bow-like displacement occurring every few years [*Halliwell and Mooers*, 1979, 1983; *Cornillon*, 1986]. Data collected between 1979 and 1986 show there is also a tendency for a seasonal oscillation in the path in this area, with the most southern position occurring in June and the most northern in August. A secondary southern shift occurs in November. Large variations about this general pattern have been seen, however, with amplitudes sufficient to have extreme positions occur at other than the above times during a given year. Rapid shifts of the jet to either the north or the south can occur when CCRs and WCRs interact strongly with the Stream within the first hundred km or so northeast of Hatteras. Such was the case during 1984 and 1985. A WCR/Gulf Stream interaction that happened just at the end of 1983, turned the Stream to the north at the Hatteras Corner, causing it to follow the shelf edge until it separated at about

37°N, roughly 150 km north of its more usual separation point. Four months later, in late April 1984, a CCR coalesced with the Stream near 73°W, and that interaction resulted in a rapid shift of the Stream's path back to the south. A similar CCR/Gulf Stream coalscence in September 1984 caused a path shift back to the north. While in a northward path configuration, the Gulf Stream meanders had lower frequencies than when in a southward (more typical) path configuration [*Tracey and Watts*, 1986; *Casagrande*, 1987; *Bane* et al., 1989; *Churchill* et al., 1989; *Cronin* et al., 1992].

THE DEEP WESTERN BOUNDARY CURRENT

The general thermohaline flow beneath the permanent thermocline in the western North Atlantic Ocean is southwestward (Figure 1b). This flow has its origins at high latitudes where it is generated primarily by the sinking of surface waters, which experience density increases during periods of strong winter cooling. The flow is intensified along the deep continental slope and rise, and this forms the deep western boundary current (DWBC), a jet-like flow about 100 km wide (Figure 3). The DWBC extends upward from the bottom and has maximum speeds in its core of around 20 cm/sec (Figure 8). DWBC volume transports have been estimated to be between 4 and 50 Sv. The trajectory of the DWBC along the base of the continental slope crosses under the Gulf Stream in the vicinity of the Hatteras Corner, as can be seen in Figure 3 [*Fine and Molinari*, 1988; *Pickart* et al., 1989; *Leaman and Harris*, 1990; *Lee* et al., 1990; *Molinari* et al., 1992; *Pickart*, 1992; *Pickart and Smethie*, 1993].

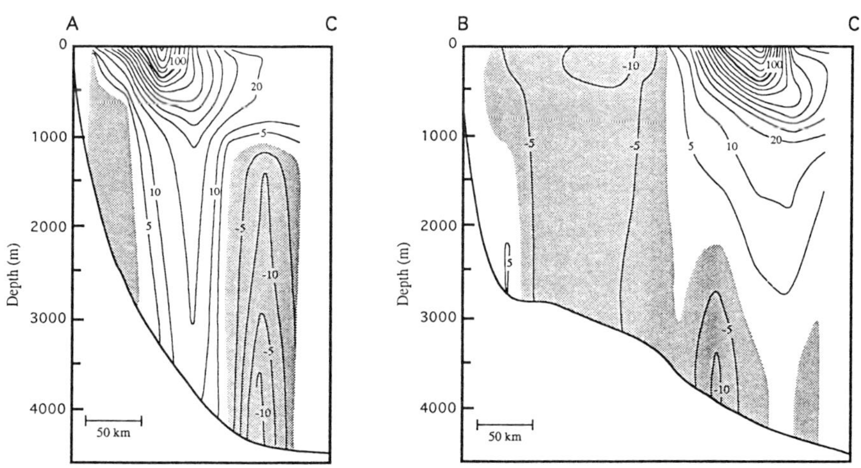

FIGURE 8. Two vertical sections through the Gulf Stream and DWBC taken near the Hatteras Corner. These sections show the structure of the horiontal velocity on either side of the cross-over of the Stream and the DWBC (the locations of these two sections are shown in Figure 3). The shading indicates the DWBC flow in a direction out of the page. Numbers are velocities in cm/sec, with positive values into the page and negative values out of the page. [Figure adapted from *Joyce* et al., 1986.]

Observations from several deployments of current meters show that the subthermocline waters over the continental slope and rise support topographic Rossby wave motions, which cause oscillations in the along-isobath velocity component at periods of tens of days. These oscillations can have amplitudes of 20 cm/sec or more, and so may enhance or reduce the southward DWBC flow at a particular time [*Richardson and Knauss*, 1971; *Richardson*, 1977; *Thompson*, 1977; *Hogg*, 1981, 1983; *Hendry*, 1982; *Hogg and Stommel*, 1985; *Joyce* et al., 1986; *Johns and Watts*, 1985, 1986; *Pickart and Watts*, 1990].

THE FUTURE

During the last few years, the largest observational program ever conducted in the Gulf Stream was undertaken. With sponsorship primarily from the U.S. Office of Naval Research and the National Science Foundation, the SYNOP (*SYN*optic *O*cean *P*rediction) experiment measured the Gulf Stream in detail from near its separation point at the Hatteras Corner to just south of the Grand Banks, near 50°W longitude [see Figure 3 in *Malanotte-Rizzoli*, 1994]. The data collected during SYNOP are beginning to reveal details of the Gulf Stream that were only suggested from earlier studies. It is becoming clear that the Gulf Stream current does not operate in isolation from its neighboring water masses and circulations; it is truly a system. Changes in one portion of the system can have far-reaching impacts on currents and variability in another part. That aspect of the Gulf Stream system poses significant challenges for future observational and modeling studies: Observations must be made simultaneously over vast areas to see the complete system; the interior of the system must be accurately and repeatedly probed in order to discern its three-dimensional, time-varying anatomy; the cause-and-effect relationships that drive the system in its ever-changing patterns must be clearly sorted out; and it is important to look ahead and know the future of the Gulf Stream, in a fashion similar to predicting the atmosphere's weather. These challenges are driving ocean scientists to develop and apply new technologies to the study of the oceans at a pace not seen before. Examples such as satellite altimetery [*Fu et al.*, 1988]; ocean acoustic tomography [*Munk and Worcester*, 1988]; autonomous, ocean-exploring, instrumented, surface and subsurface floats [*Stommel*, 1989]; and sophisticated numerical models which incorporate newly acquired data for analysis and prediction [*Thompson et al.*, 1992] are either in use or under development. Together these powerful tools promise to provide an increasingly accurate and well-focused look into the true nature of the Gulf Stream.

ACKNOWLEDGMENTS

Preparation of this paper was supported by the Office of Naval Research, grant number N00014-92-J-1683, and the National Science Foundation, grant number OCE-9302816. Figures were expertly prepared by Sara Haines. We thank Otis Brown, William Johns, Michael McCartney, William Schmitz and Thomas Shay for use of figures.

REFERENCES

Bane, J.M., Jr. 1983. Initial Observations of the Subsurface Structure and Short-Term Variability of the Seaward Deflection of the Gulf Stream off Charleston, South Carolina. *J. Geophys. Res.* 88:4673-4684.

Bane, J.M., Jr. and D.A. Brooks. 1979. Gulf Stream Meanders along the Continental Margin from the Florida Straits to Cape Hatteras. *Geophys. Res. Letters* 6:280-282.

Bane, J.M., Jr., D.A. Brooks and K.R. Lorenson. 1981. Synoptic Observations of the Three-Dimensional Structure and Progagation of Gulf Stream Meanders along the Carolina Continental Margin. *J. Geophys. Res.* 86:6411-6425.

Bane, J.M., Jr. and W.K. Dewar. 1988. Gulf Stream Bimodality and Variability Downstream of the Charleston Bump. *J. Geophys. Res.* 93:6695-6710.

Bane, J.M., Jr., L.M. O'Keefe and D.R. Watts. 1989. Mesoscale Eddies and Submesoscale, Coherent Vortices: Their Existence Near and Interations with the Gulf Stream, pp. 501-518. In: J.C.J. Nihoul and B.M. Jamart (Eds.) *Mesoscale/Synoptic Coherent Structures in Geophysical Turbulence.* Elsevier Science Publishers, Amsterdam, pp. 841.

Brooks, D.A. and J.M. Bane, Jr. 1981. Gulf Stream Fluctuations and Meanders over the Onslow Bay Upper Continental Slope. *J. Phys. Oceanogr.* 11:247-256.

Brooks, D.A. and J.M. Bane, Jr. 1983. Gulf Stream Meanders off North Carolina During Winter and Summer 1979. *J. Geophys. Res.* 88:4633-4650.

Brown, O.B., P.C. Cornillon, S.R. Emmerson and H.M. Carle. 1986. Gulf Stream Warm Rings: A Statistical Study of Their Behavior. *Deep-Sea Res.* 33:1459-1473.

Casagrande, C. 1987. Study of Physical Processes on the U.S. Mid-Atlantic Continental Slope and Rise. OCS Study MMS-87-0024, U.S. Department of Interior, Minerals Management Service, Atlantic Outer Continental Shelf Office, Vienna, VA.

Clarke, R.A., H.W. Hill, R.F. Reiniger and B.A. Warren. 1980. Current System South and East of the Grand Banks of Newfoundland. *J. Phys. Oceanogr.* 10:25-65.

Cornillon, P. 1986. The Effect of the New England Seamounts on Gulf Stream Meandering as Observed from Satellite IR Imagery. *J. Phys. Oceanogr.* 16:386-389.

Cronin, M., E. Carter and D.R. Watts. 1992. Prediction of the Gulf Stream Path from Upstream Paramaters. *J. Geophys. Res.* 97:7257-7269.

Ebbesmeyer, C.C. and J.M. Bane, Jr. 1989. Relationship between Gulf Stream Path and Frontal Eddy Variability off Raleigh Bay, North Carolina. *Trans. Amer. Geophys. Union.* 71:1403.

Fine, R.A. and R.L. Molinari. 1988. A Continuous Deep Western Boundary Current between Abaco (26.5°N) and Barbados (13°N). *Deep-Sea Res.* 35:1441-1450.

Fofonoff, N.P. 1981. The Gulf Stream System, pp. 112-139. In: B. Warren and C. Wunsch (Eds.) *Evolution of Physical Oceanography.* MIT Press, Cambridge, MA, pp. 623.

FRED Group. 1989. Frontal Eddy Dynamics (FRED) Experiment off North Carolina. C.C. Ebbesmeyer (Ed.) Prepared by Evans-Hamilton, Inc. OCS Study MMS 89-0028 Vol 2 Technical Report, March 1989. U.S. Department of the Interior, Minerals Management Service, Herndon, VA, pp. 325.

Fu, L-L., D.B. Chelton and V. Zlotnicki. 1988. Satellite Altimetery: Observing Ocean Variability from Space. *Oceanography* 2:4-11.

Gangopadhyay, A., P. Cornillon and D.R. Watts. 1992. A Test of the Parsons-Veronis Hypothesis on the Separation of the Gulf Stream. *J. Phys. Oceanogr.* 22:1286-1301.

Halkin, D. and H.T. Rossby. 1985. The Structure and Transport of the Gulf Stream at 73°W. *J. Phys. Oceanogr.* 15:1439-1452.

Hall, M.M. 1986. Horizontal and Vertical Structure of the Gulf Stream Velocity at 68°W. *J. Phys. Oceanogr.* 16:1814-1818.

Hall, M.M. and H.L. Bryden. 1985. Profiling the Gulf Stream with a Current Meter Mooring. *Geophys. Res. Letters* 12:203-206.

Hall, M.M. and N.P. Fofonoff. 1993. Downstream Development of the Gulf Stream from 68°W to 55°W. *J. Phys. Oceanogr.* 23:225-249.

Halliwell, G.R., Jr. and C.N.K. Mooers. 1979. The Space-Time Structure and Variability of the Shelf Water-Slope Water and Gulf Stream Surface Temperature Fronts and Associated Warm Core Eddies. *J. Geophys. Res.* 84:7707-7725.

Halliwell, G.R., Jr. and C.N.K. Mooers. 1983. Meanders of the Gulf Stream Downstream from Cape Hatteras 1975-1978. *J. Phys. Oceanogr.* 13:1275-1292.

Hendry, R.M. 1982. On the Structure of the Deep Gulf Stream. *J. Mar. Res.* 40:119-142.

Herrera y Tordesillas, Antonio de. 1601. Historia General de los Hechos de los Castellanos en las Islas i Tierra Firme del Mar Oceano, 5 vols., Emprenta Real, por Iuan Flamenco, Madrid.

Hogg, N.G. 1981. Topographic Waves along 70°W on the Continental Rise. *J. Mar. Res.* 39:627-649.

Hogg, N.G. 1983. A Note on the Deep Circulation of the Western North Atlantic: Its Nature and Causes. *Deep-Sea Res.* 30:945-961.

Hogg, 1992. On the transport of the Gulf Stream between Cape Hatteras and the Grand Banks. *Deep-Sea Res.* 39:1231-1246.

Hogg, N.G., R.S. Pickart, R.M. Hendry and W.S. Smethie. 1986. The Northern Recirculation Gyre of the Gulf Stream. *Deep-Sea Res.* 33:1139-1165.

Hogg, N.G. and H. Stommel. 1985. On the Relationship between the Deep Circulation and the Gulf Stream. *Deep-Sea Res.* 32:1181-1193.

Hurlburt, H.E. and J.D. Thompson. 1980. A Numerical Study of Loop Current Intrusions and Eddy Shedding. *J. Phys. Oceanogr.* 10:1611-1651.

Iselin, C. O'D. 1936. A Study of the Circulation of the Western North Atlantic, PPOM 4, No. 4 MIT-Woods Hole Oceanographic Institution, pp. 101.

Johns, W.E. and T.J. Shay. 1992. Stream-Averaged Structure and Transport of the Gulf Stream at 68°W. *The SYNOPtician* 5:4-6 and 24-27.

Johns, W.E. and D.R. Watts. 1985. Gulf Stream Meanders: Observations on the Deep Currents. *J. Geophys. Res.* 90:4819-4832.

Johns, W.E. and D.R. Watts. 1986. Time Scales and Structure of Topographic Rossby Waves and Meanders in the Deep Gulf Stream. *J. Mar. Res.* 44:267-290.

Johnson, D.R., J.D. Thompson and J.D. Hawkins. 1992. Circulation in the Gulf of Mexico from Geosat Altimetry during 1985-1986. *J. Geophys. Res.* 97:2201-2214.

Joyce, T.M., S.D. Pierce and C. Wunsch. 1986. Synoptic Gulf Stream Velocity Profiles through Simultaneous Inversion of Hydrographic and Acoustic Doppler Data. *J. Geophys. Res.* 91:7573-7586.

Knauss, J.A. 1969. A Note on the Transport of the Gulf Stream. Frederick C. Fuglister Sixtieth Anniversary Volume. *Deep-Sea Res.* 16:117-123.

Kohl, J.G. 1868. Geschichte des Golfstroms and seiner Erforschung. Bremen: C.E. Muller.

Lai, D.Y. and P.L. Richardson. 1977. Distribution and Movement of Gulf Stream Rings. *J. Phys. Oceanogr.* 7:670-683.

Leaman, K.D. and J.E. Harris. 1990. A Note on the Average Absolute Transport of the Deep Western Boundary Current East of Abaco Island, the Bahamas. *J. Phys. Oceanogr.* 20:467-475.

Lee, T.N. and L.P. Atkinson. 1983. Low Frequency Current and Temperature Variability from Gulf Stream Frontal Eddies and Atmospheric Forcing along the Southeastern US Outer Continental Shelf. *J. Geophys. Res.* 88:4541-4568.

Lee, T.N., L.P. Atkinson and R. Legeckis. 1981. Observations of a Gulf Stream Frontal Eddy on the Georgia Continental Shelf, April 1977. *Deep-Sea Res.* 28:347-348.

Lee, T.N., W. Johns, F. Schott and R. Zantopp. 1990. Western Boundary Current Structure and Variability East of Abaco, Bahamas at 26.5°N. *J. Phys. Oceanogr.* 20:446-466.

Lee, T.N. and E. Waddell. 1983. On Gulf Stream Variability and Meanders over the Blake Plateau at 30°N. *J. Geophys. Res.* 88:4617-4632.

Lee, T.N., E. Williams, J. Wang, R. Evans and L.P. Atkinson. 1989. Response of South Carolina Continental Shelf Waters to Wind and Gulf Stream Forcing During Winter of 1986. *J. Geophys. Res.* 94:10,715-10,754.

Legeckis, R. 1979. Satellite Observations of the Influences of Bottom Topography on the Seaward Deflection of the Gulf Stream off Charleston, South Carolina. *J. Phys. Oceanogr.* 9:483-497.

Malanotte-Rizzoli. 1994. The Gulf Stream System: A Modeling Perspective. This Volume.

Mann, C.R. 1967. The Termination of the Gulf Stream and the Beginning of the North Atlantic Current. *Deep-Sea Res.* 14:337-359.

Maul, G.A. 1977. The Annual Cycle of the Loop Current, I, Observations during a One-Year Time Series. *J. Marine Res.* 35:29-47.

Maul, G., P.W. DeWitt, A. Yanaway and S.R. Baig. 1978. Geostationary Satellite Observations of Gulf Stream Meanders: Infrared Measurements and Time Series Analysis. *J. Geophys. Res.* 83:6123-6135.

Molinari, R.L., R.A. Fine and E. Johns. 1992. The Deep Western Boundary Current in the Tropical North Atlantic Ocean. *Deep-Sea Res.* 39:1967-1984.

Munk, W.H. and P.F. Worcester. 1988. Ocean Acoustic Tomography. *Oceanography* 1:8-10.

Olson, D.B., O.B. Brown and S.R. Emmerson. 1983. Gulf Stream Frontal Statistics from the Florida Straits to Cape Hatteras Derived from Satellite and Historical Data. *J. Geophys. Res.* 88:3227-3236.

Pietrafesa, L.J., G.S. Janowitz and P.A. Wittman. 1985. Physical Oceanographic Processes in the Carolina Capes. pp. 23-32. In: L.P. Atkinson, D.W. Menzel and K.A. Bush (Eds.) *Oceanography of the Southeastern U.S. Continental Shelf.* American Geophysical Union, Washington, DC, pp. 156.

Pickart, R.S. 1992. Space-Time Variability of the Deep Western Boundary Current Oxygen Core. *J. Phys. Oceanogr.* 22:1047-1061.

Pickart, R.S., N. Hogg and W.S. Smethie. 1989. Determining the Strength of the Deep Western Boundary Current Using the Chlorofluoromethane Ratio. *J. Phys. Oceanogr.* 19:940-951.

Pickart, R.S. and W.S. Smethie. 1993. How Does the Deep Western Boundary Current Cross the Gulf Stream? *J. Phys. Oceanogr.* 23:In Press.

Pickart, R.S. and D.R. Watts. 1990. Deep Western Boundary Current Variability at Cape Hatteras. *J. Mar. Res.* 48:765-791.

Richardson, P.L. 1977. On the Crossover between the Gulf Stream and the Western Boundary Undercurrent. *Deep-Sea Res.* 24:139-159.

Richardson, P.L. 1980. Gulf Stream Ring Trajectories. *J. Phys. Oceanogr.* 10:90-104.

Richardson, P.L. 1981. Gulf Stream Trajectories Measured with Free-Drifting Buoys. *J. Phys. Oceanogr.* 11:999-1010.

Richardson, P.L. 1983. Gulf Stream Rings, pp. 19-45. In: A.R. Robinson (Ed.) *Eddies in Marine Science.* Springer-Verlag, NY, pp. 609.

Richardson, P.L. 1993. A Census of Eddies Observed in North Atlantic SOFAR Float Data. *Prog. Oceanogr.* 31:1-50.

Richardson, P.L. and J. Knauss. 1971. Gulf Stream and Western Boundary Undercurrent Observations at Cape Hatteras. *Deep-Sea Res.* 18:1089-1109.

Richardson, W.S., W.J. Schmitz and P.P. Niiler. 1969. The Velocity Structure of the Florida Current from the Straits of Florida to Cape Fear. *Deep-Sea Res.* 16:225-231.

Savidge, D.K., J.O. Blanton, T.N. Lee and R.H. Evans. 1992. Influence of an Offshore Shift in the Gulf Stream on Waters of the South Carolina Continental Shelf. *J. Phys. Oceanogr.* 22:1085-1094.

Schmitz, W.J., Jr. and M.S. McCartney. 1993. On the North Atlantic Circulation. *Rev. Geophys.* 31:29-49.

Schmitz, W.J., Jr. and W.S. Richardson. 1968. On the Transport of the Florida Current. *Deep-Sea Res.* 15:679-693.

Scisco, L.D. 1913. The Track of Ponce de Leon in 1513. *Bull. Am. Geogr. Soc.* XLV:721-735.

Singer, J.J., L.P. Atkinson, J.O. Blanton and J.A. Yoder. 1983. Cape Romain and the Charleston Bump: Historical and Recent Hydrographic Observations. *J. Geophys. Res.* 88:4685-4698.

Stommel. 1948. The Westward Intensification of Wind-Driven Ocean Currents. *Trans. Amer. Geophys. Union*, 29:202-206.

Stommel, H. 1965. *The Gulf Stream.* University of California Press, Berkeley, CA, pp. 248.

Stommel, H. 1989. The Slocum Mission. *Oceanography* 2:22-25.

Sy, A. 1988. Investigation of Large-Scale Circulation Patterns in the Central North Atlantic: The North Atlantic Current, the Azores Current, and the Mediterranean Water Plume in the Area of the Mid-Atlantic Ridge. *Deep-Sea Res.* 35:383-413.

Thompson, J.D., T.L. Townsend, A. Wallcraft and W.J. Schmitz, Jr. 1992. Ocean Prediction and the Atlantic Basin: Scientific Issues and Technical Challenges. *Oceanography* 5:36-41.

Thompson, R. 1977. Observations of Rossby Waves near Site D. *Prog. in Oceanogr.* 7:135-162.

Tracey, K.L. and D.R. Watts. 1986. On Gulf Stream Meander Characteristics near Cape Hatteras. *J. Geophys. Res.* 91:7587-7602.

Vukovich, F.M. 1988. Loop Current Boundary Variations. *J. Geophys. Res.* 93:15,585-15,591.

Watts, D.R. and D.B. Olson. 1978. Gulf Stream Ring Coalescence with the Gulf Stream off Cape Hatteras. *Science* 202:971-972.

Watts, D.R. and W. Johns. 1982. Gulf Stream Meanders: Observations on Progagation and Growth. *J. Geophys. Res.* 87:9467-9476.

Watts, D.R. 1983. Gulf Stream Variability, pp. 114-144. In: A.R. Robinson (Ed.) *Eddies in Marine Science*. Springer-Verlag, NY, pp. 609.

Webster, F. 1961. A Description of Gulf Stream Meanders Off Onslow Bay. *Deep-Sea Res.* 8:130-143.

Worthington, L.V. 1976. *On the North Atlantic Circulation.* Johns Hopkins Press, Baltimore, MD, pp. 110.

The Oceans: Physical-Chemical Dynamics and Human Impact. Edited by S.K. Majumdar, E.W. Miller, G.S. Forbes, R.F. Schmalz and Assad A. Panah. © 1994, The Pennsylvania Academy of Science.

Chapter Seven

THE GULF STREAM SYSTEM: A Modeling Perspective

PAOLA MALANOTTE-RIZZOLI
Center for Meteorology and Physical Oceanography
Department of Earth, Atmospheric and Planetary Sciences
Massachusetts Institute of Technology
Cambridge, MA 02139

INTRODUCTION

Among the most interesting and most complex systems of the world ocean are the so-called western boundary currents, that is jet-like, intense currents localized on the western sides of the oceanic gyres and characterized by highly nonlinear dynamics. The Gulf Stream system is the prototype of such energetic boundary currents.

Why is the Gulf Stream important? First, for climate and global ocean circulation studies. Due to its geographical configuration in the Northwestern Atlantic, the Gulf Stream is the boundary current that most profoundly affects the global thermohaline circulation. As the most crucial ingredient of the subtropical gyre wind-driven circulation, the Gulf Stream plays a major role in the poleward heat transport by the ocean in the Northern Atlantic, thus directly affecting the polar convection cells that drive the world "conveyor belt". Understanding the dynamics of the Gulf Stream jet is therefore crucial to the assessment of its role in the global ocean and major dynamical issues remain that are still poorly understood and modeled. Examples are the issue of jet separation at Cape Hatteras; the penetration length into the gyre interior; the interaction with the deep western boundary current.

Equally important is the understanding of the Stream dynamics for regional studies. The discovery of the mesoscale eddy field in the 70's has shown the western boundary currents systems to be the regions of highest eddy energy and variability. Thanks to instability processes leading to meander growth and decay and consequent eddy-mean flow interactions, the Gulf Stream is now believed to be the major source of the mesoscale eddy energy radiating into the ocean interior.

The jet instabilities themselves, barotropic and baroclinic, that lead to meander growth, steepening and bending, are responsible for the formation of cold/warm core rings that pinch-off and propagate away from the Stream. Successive and multiple ring-Stream interactions are possible, with ring merger events, ring rebirths after merger and downstream advection. Such rings are the most efficient mechanism for mixing properties across the Stream as they transport in their core heat, salt, nutrients, biota characteristic of their original water mass (Sargasso Sea water for the warm core; Slope water for the cold core rings).

This richness of highly nonlinear dynamical processes, and the complexity of interactions between the Stream jet and the overall gyre circulation including the northern and southern tight inertial recirculations, make the Gulf Stream system one of the most fascinating and important parts of the world ocean.

In these last decades, since the first publication of Stommel's monograph (1965[1]), an impressive amount of observational evidence and related modeling work has been accumulated that make the Stream system maybe the most intensively studied oceanic region. It is therefore impossible to provide in a single review a comprehensive discussion of the observational, dynamical and modeling aspects. In this review I will focus on the dynamical and modeling issues linked to the Stream, with only a brief summary of the observations that motivate these studies. Also, given the necessity for choice, clearly the selection of dynamical problems reflects my own preferences and judgment of their importance, and does not pretend to be exhaustive. My goal is to focus the reader's attention and interest on some of the high priority questions that have emerged in our rapidly growing knowledge of the Gulf Stream System.

Historical Review

The early theories of the ocean wind-driven circulation modeled the Gulf Stream as a western boundary current in a steady, homogeneous ocean that closes the interior Sverdrup circulation. These early models were basically of two types, frictional (Stommel, 1948[2]; Munk, 1950[3]) an inertial (Fofonoff, 1954[4]; Charney, 1955[5]; Morgan, 1956[6]). Subsequent models included both frictional and inertial processes, and focused upon the numerical study of the nonlinear Stommel problem, with a smooth transition from the frictional Stommel's solution to the inertial Fofonoff's gyre (Veronis, 1966a,b[7,8]) and on the nonlinear Munk's problem (Bryan, 1963[9]). These models evidenced a very important problem not yet fully understood, that is the mechanism of separation of the Gulf Stream from the continental slope at Cape Hatteras. The first model explicitly including isopycnal surfacing as a crucial ingredient for the separation mechanism is the one by Parsons (1969[10]), later reanalyzed by Veronis (1973[11]) in his two-layer model of the World Ocean wind-driven circulation. For a complete review of these early theories see Veronis (1981[12]).

Successive modeling efforts focussed upon determining the shape of the Gulf Stream path, the so-called trajectory models, the first of which proposed by Warren (1963[13]) was based on bulk or integrated properties of the Stream. The thin-jet approach to analyze the finite amplitude behavior of Gulf Stream meanders was subsequently elaborated by Niiler and Robinson (1967[14]) and extended to the time-dependent case by Luyten and Robinson (1974[15]), Robinson et al. (1975[16]), Flierl and Robinson (1984[17]). Rhines (1977[18]) used a two-layer model to study the instability of a thin zonal jet confined to the upper layer. His model however had no bottom topography and lacked an explicit recirculation mechanism.

The first numerical models reproducing the recirculation gyres and allowing for baroclinic instabilities of the western boundary current were developed by Holland and Lin (1975a,b[19,20]) who carried out an extensive series of simulations of the ocean wind-driven general circulation. Subsequently Semtner and Mintz (1977[21]) developed a primitive equation model including the shelf topography and surface heat exchange. Owens and Bretherton (1978[22]) further carried out a numerical study of mid-ocean mesoscale eddies. High-resolution numerical studies exploring the energetics of eddy-mean flow and eddy-eddy interactions were also carried out by Robinson et al. (1977[23]); Semtner and Holland (1978[24]) and Holland (1978[25]). In this latter and important work, Holland showed that a simpler, eddy-resolving quasi-geostrophic two-layer model, forced by a double-gyre wind stress field, produced a mid-ocean jet with a more realistic behavior, closer to the actual Gulf Stream. Strong recirculation gyres are induced in the lower layer by eddy momentum fluxes due to vortex stretching (Holland and Rhines, 1980[26]).

Further important modeling studies were prompted by the discovery of warm and cold core rings and their formation from finite amplitude Gulf Stream meanders. After leaving Cape Hatteras, the Stream develops meanders that, after growing to large amplitude, steepen, and bend over themselves, leading to the formation of quasi-circular coherent structures, the rings, that pinch-off at the breaking of the meander neck. The cold rings detach to the south, the warm ones to the north and subsequently migrate south (north) westward. Successive multiple and vigorous ring-Stream interactions are possible, as described first by Richardson (1980[27]). First models for the cold/warm core rings were developed by Flierl (1977[28]; 1979a,b[29,30]).

These later studies, based upon the observational evidence accumulated in the decade of the 70's, revealed the Gulf Stream System to be much more complex than the one implied by the early theories of the wind-driven ocean circulation. For a review of the observational and phenomenological knowledge of the Stream System up to the 80's, see Fofonoff (1981[31]).

The present picture of the Gulf Stream system shows an intense jet structure coherent with depth that detaches at Cape Hatteras, meandering into the ocean interior. Multiple meanders can grow to finite amplitude, both stationary and propagating downstream (Halliwell and Mooers, 1983[32]). The meanders steepening and bending can give birth to warm and cold core rings, coherent structures endowed with long lifetimes that propagate upstream and can undergo multiple Stream-ring interactions, with mergers and coalescence or rebirth. The meanders' sudden growths and decays, and in general their stochastic behavior, are now believed to be responsible for vigorous radiation of transient eddies into the ocean interior and for the high levels of mesoscale energy and variability in the North Western Atlantic. The Stream mean path undergoes more vigorous oscillations, leading to large broadening of the mean path envelope after crossing over the New England Sea Mountain Chain. Embedded in the overall subtropical gyre wind-driven circulation are two strong closed inertial recirculations closely tight to the Stream southern and northern flanks. The southern recirculation, the so-called Worthington gyre, first proposed by Worthington (1976[33]) is thought to be responsible for the almost order-of-magnitude increase of Stream transport after leaving Cape Hatteras. The northern recirculation gyre is mostly confined to the deep layer and limited in extension by the northern continental slope (Hogg et al., 1986[34]; Hogg, 1992[35]). All these multiple scales simultaneously present, the richness of mesoscale and coherent features, the complexity of multiple nonlinear interactions are schematized in Figure 1.

Jet Instabilities and Ring Processes

A very great number of studies have been produced to investigate the instabilities of Gulf Stream type jets, starting from the classic paper by Orlanski (1969[36]), who

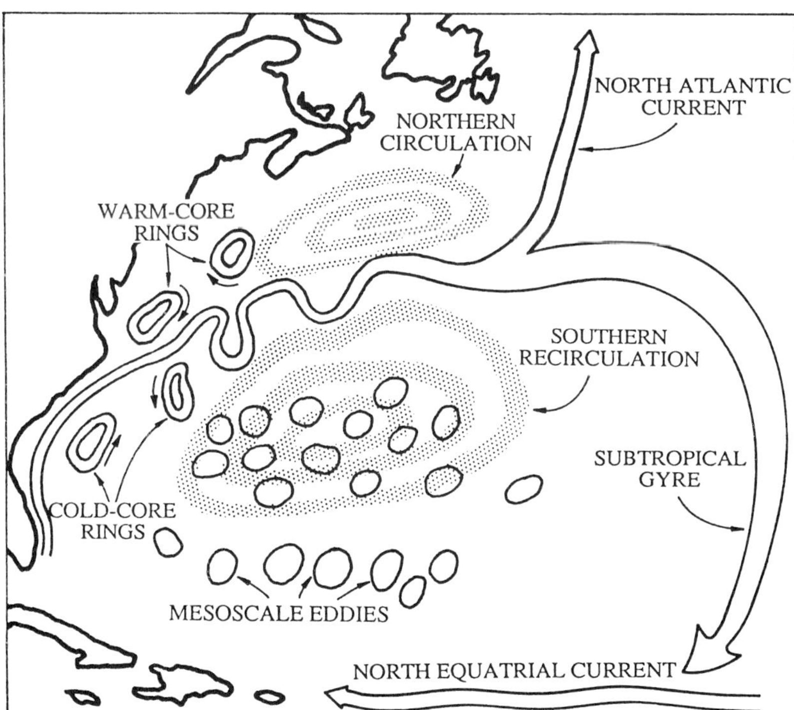

FIGURE 1. Schematic of the Gulf Stream System

used a two layer model with bottom topography. By solving the linearized perturbation equations numerically, he found that bottom topography has two competing effects on the stability of the Gulf Stream jet, the stabilizing factor of bottom slope and the destabilizing one due to the topography amplitude. Since then, studies of jet instabilities, in more or less idealized configurations, have multiplied. Far from giving an exhaustive review of past literature, I quote only some among the most significant works, namely the works by Sela and Jacobs (1971[37]). Hart (1974[38]); DeSzoeke (1975[39]); Ikeda (1983[40]); Killworth (1983[41]); Killworth et al. (1984[42]); Paldor (1983a,b[43,44]); Flierl et al. (1987[45]).

Among the most recent works is the one by Johns (1988[46]) who used a linear instability model for one-dimensional baroclinically unstable waves on the Gulf Stream potential vorticity gradient near Cape Hatteras. This continously stratified model was successful in predicting the time and length scales of growing meanders, but significantly underestimated the growth rate. All these studies consider only temporally unstable modes. Hogg (1976[47]) found that spatial and temporal waves in a baroclinic model are separated in wavenumber, with short waves being spatially unstable and long waves being temporally unstable. Many of the above mentioned authors agree that spatial instability may be more relevant than the temporal one for the Gulf Stream jet (Johns, 1988[46]), however spatially unstable models are much more difficult because of important nonlinearities that cannot be neglected. It is well known that Gulf Stream instabilities producing meander growth, steepening and bending may lead to ring birth and ring structures pinching off the Gulf Stream. For a review of the phenomenology of cold and warm core rings see Richardson (1983[48]). For a review of theoretical models of isolated eddy and vortices see Flierl (1987[49]). Figure 2, from Richardson (1983[48]), visualizes the process leading to ring formation and birth.

More recently, a very interesting body of work has developed following the early trajectory models previously discussed. In these studies, the method of contour dynamics is used, reviewed in the papers by Zabusky et al. (1979[50]) and Zabusky and Overman (1981[51]). Among the pioneering papers are those by Stern and Pratt (1985[52]) and Pratt and Stern (1986[53]) who employ contour dynamics to determine the conditions under which a ring is likely to separate from the Gulf Stream. Later investigations on the interaction of vortex structures with shear flow include those by Stern and Flierl (1987[54]); Pratt (1988[55]), Polvani (1988[56]). Polvani et al. (1989[57]); Pratt et al. (1990[58]); Meacham (1991[59]); Bell and Pratt (1992[60]).

It is emphasized again that, for obvious reasons of conciseness, the above review is not meant to be exhaustive but rather to provide to the interested reader a sample of significant papers on jet instabilities and ring formation processes and an overall picture of different methodological approaches.

The Mesoscale Eddy Radiation Field

The observational evidence of this last decade shows that the eddy kinetic and potential energy densities increase by orders of magnitude in the Northwestern Atlantic when proceeding from the basin interior towards the Gulf Stream (Wyrtki et al., 1976[61]; Dantzler, 1977[62]; Richardson, 1983[63]; Schmitz et al., 1983[64]). Local wind forcing and topographic features do not seem to provide sufficient forcing for the observed high energy levels. Thus the Stream is the most plausible source for this enhanced eddy activity (Wunsch, 1983[65]; Schmitz et al., 1983[64]). Also, low frequency motions on the lower continental slope are dominated by energetic-topographic Rossby waves. Again, the Stream has been implicated as the most likely source (Hogg, 1981[66], Welsh et al., 1991[67]; Bower and Hogg, 1993[68]).

The Stream is an eastward propagating meandering current and theoretical studies show that steady, eastward moving meanders cannot radiate energy into the far field. Wave radiation remains trapped in the near field both in linear models (Pedlosky, 1977[69]; Talley, 1983a,b[70,71]) and in nonlinear ones (Malanotte-Rizzoli, 1984[72]) with an eastward phase speed matching that of the forcing. However, numerical eddy-resolving experiments both for idealized jets (Ikeda, 1981[73]; Ikeda and Apel, 1981[74]; Flierl et al., 1987[45]) and in more realistic configurations (Holland et al., 1983[75]) show mesoscale radiation to be excited and radiate away from an eastward moving jet. This apparent contradiction was solved when the focus shifted from the equilibrium forced component of the Rossby radiation field, the object of the previous theoretical studies and the only one excited by a steadily propagating meander. The meander transient motions, such as sudden growth, decay and

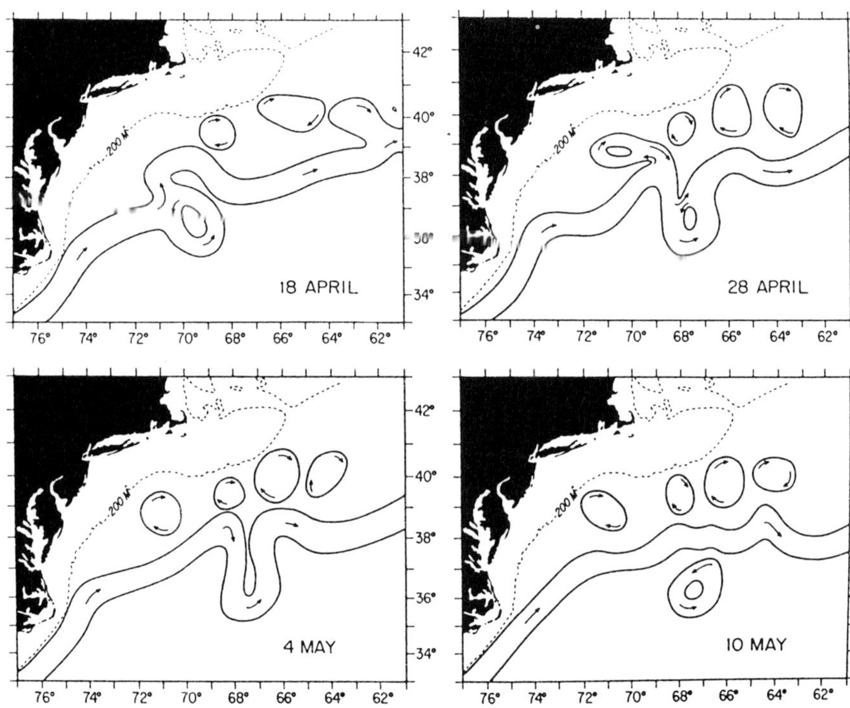

FIGURE 2. Schematic representation of ring-Stream interaction. 18 April, 28 April, 4 May; cold ring advected downstream; 10 May: cold ring birth; 28 April: warm ring birth (from Richardson, 1983[48]).

amplitude pulsations, and the stochastic nature of the life cycles of an ensemble of meanders can excite pulses of Rossby waves. The transient component of the response to meander forcing radiates energy away from the source into the basin interior (Malanotte-Rizzoli, et al., 1987[76]; Hogg, 1988[77]; Malanotte-Rizzoli et al., 1994[78]). In this latter work the simple quasi-geostrophic barotropic vorticity equation on a β-plane with and without relief is used to investigate numerically the eddy radiation from a southern boundary forcing in the fully nonlinear case. The boundary forcing idealizes the Stream motion and consists of a stochastic forcing succession of eastward moving meanders. Eddy statistics are computed from the model and from moored current meter data collected in the Gulf Stream region between the Grand Banks and Cape Hatteras. Considering the model's simplicity, the model predicted kinetic energies and covariances of the radiation field compare rather well with the observed ones, both in intensity and in the meridional distribution.

A very important question related to the mesoscale eddy field is whether the divergence of eddy vorticity fluxes can induce mean flows south or north of the Stream such as the observed recirculation regions (Figure 1). Deep recirculation cells driven by eddy vortex stretching fluxes are found in eddy-resolving simulations of the general gyre circulation (Holland and Rhines, 1980[26]). McEwan et al. (1980[79]) addressed the problem of flow rectification by quasi-geostrophic eddies, but their induced Eulerian circulation is zonally averaged. Charney and Flierl (1981[80]) studied the problem of mean flow induced by a steady boundary forcing. In their model, their propagating meander is not growing in time, like in the case analyzed by Rhines (1977[18]), but the model has friction. Haidvogel and Rhines (1983[81]) show that a spatially local, oscillatory forcing can generate a rectified mean flow with closed circulation regions whose structure depends critically on the shape of the mean quasi-geostrophic contours. Malanotte-Rizzoli et al. (1994[78]) also address the question whether the deep recirculation gyre north of the Stream may be driven by the divergence of the (barotropic) eddy vorticity fluxes. They consider an along-channel, zonal variation of the forcing mechanism which is essential to obtain a meridional component in the induced mean flow. However, in the case in which the potential vorticity contours do not close near the southern boundary, the eddy vorticity fluxes induce a rectified mean flow composed by a series of alternating anti-cyclonic/cyclonic cells. Only when allowing for a variable thickness of the deep layer, following the work of Hogg and Stommel (1985[82]), a mean cyclonic circulation is induced that follows the closed potential vorticity contours and compares well with the observed northern recirculation gyre of Figure 1.

Numerical Modeling of the Gulf Stream System

The first comparison between the eddy kinetic energy distribution of the Gulf Stream evaluated in a numerical model and in the real ocean was carried out by Schmitz and Holland (1982[83]) and later by Hall (1986[84]). In their pioneering study, Schmitz and Holland[83] use a two-layer, eddy resolving quasi-geostrophic model. Even though the temporal and meridional scales of the model were realistic, the zonal penetration scale of the model jet was several times shorter than observed. The problem of the zonal penetration of midlatitude jets was further investigated by Holland and Schmitz (1985[85]). They showed that increased vertical resolution can allow for much better zonal penetration. In an extensive series of numerical calculations Schmitz and Holland (1986[86]) show that greater penetration scale can be achieved by reducing the jet baroclinic instability, thus favoring the inertial character of the flow. See also Evans et al. (1987[87]) for a review of numerical modeling of the ocean circulation up to 1986.

Eddy-resolving quasi-geostrophic models have been used to study the dynamics of Gulf Stream meander and ring interaction processes (Robinson et al., 1988[88]). Spall and Robinson (1990[89]) demonstrate some of the deficiencies of quasi-geostrophic dynamics by comparing a Quasi-Geostrophic (QG) model to a Primitive Equation (PE) model of the Stream. In a study of isolated vortices such as Gulf Stream rings on the β-plane, McWilliams et al. (1986[90]) analyze the inadequacy of quasi-geostrophic dynamics by comparing QG model results with Balance Equation (BE) and PE model results. While the differences are slight between BE and PE simulations, QG results show important discrepancies. In a recent work, Chassignet (1992[91]) analyzes and compares the statistics of warm and cold core rings in four different models, QG in physical coordinates; linear BE in physical coordinates (McWilliams et al., 1990[92]); PE in isopycnic coordinates (Bleck and Boudra, 1986[93]) and PE in sigma coordinates (Haidvogel et al., 1991[94]). He derives ring locus, lifetime, propagation speed, radius and interface displacement statistics from the numerical simulations and compares them to the 10-year statistical analysis of Gulf Stream rings based on time series of satellite infrared imagery carried out by Brown et al. (1986[95]). The statistics differ significantly in the four models, showing the sensitivity of ring formation processes to the physical and numerical assumptions made.

The problem of the separation of a mid-latitude jet from a western boundary is an important issue not yet fully understood and resolved. In fact, in most realistic eddy-resolving models of the North Atlantic circulation the Gulf Stream overshoots north of Cape Hatteras. Thompson and Schmitz (1989[96]) give an overview of the problem. Using a two-layer PE regional model they demonstrate the sensitivity of the separation process to the bottom topography and to the presence of a Deep Western Boundary Current. Bryan and Holland (1989[97]) simulate the North Atlantic wind-driven and thermohaline circulation with the GFDL (Geophysical Fluid Dynamics Laboratory) PE model developed by Cox (1984[98]), with 30 vertical levels, fully realistic bathymetry and coastlines, wind stress forcing and surface heat and salt fluxes. Even in this most realistic simulation, the Stream separation point is farther north than observed.

In a recent study Chassiguet and Gent (1991[99]) investigate the influence of boundary conditions on midlatitude jet separation. They compare the four dynamical models mentioned above, i.e. QG and BE in physical coordinates; PE in isopycnic and sigma coordinates. A systematic difference they find is that in level models the separation occurs poleward of the zero-wind-stress-curl line while in layer models it occurs equatorward. Finally, Mellor and Ezer (1991[100]) and Ezer and Mellor (1992[101]) study the importance of including both wind stress and surface heat flux in the model to achieve a more realistic separation of the Gulf Stream. In particular, they stress the importance for the separation problem of maintaining the density structure and associated geostrophic flow in the northern recirculation gyre. Heat transfer, wind stress and slope water inflow are all essential ingredients to keep the gyre structure. However, Ezer and Mellor (1992[101]) use a regional model rather limited in zonal extent (80°W to 60°W) and it is possible that the open boundary condition at the eastern border crucially affects the location of the separation and successive meandering of the Stream.

CONCLUSIONS

The above review, even though far from exhaustive, shows the impressive body of literature developed during the last twenty years to investigate the richness of processes and of structures characterizing the Gulf Stream System. What are at present the trends of novel research and methodologies that promise to have the greatest potential for the understanding of such processes? One of such new bodies of research is the rapidly growing field of data assimilation. Data assimilation has been very well developed for more than 30 years in meteorology and prompted in the oceanographic community by the forthcoming availability of synoptic datasets with global coverage such as the TOPEX altimetric dataset. As far as the Gulf Stream System is concerned, a major observational project has been recently carried out under the auspices of ONR (Office of Naval Research); the SYNOP (Synoptic Ocean Prediction) experiment. Figure 3 shows the field configuration of SYNOP, with

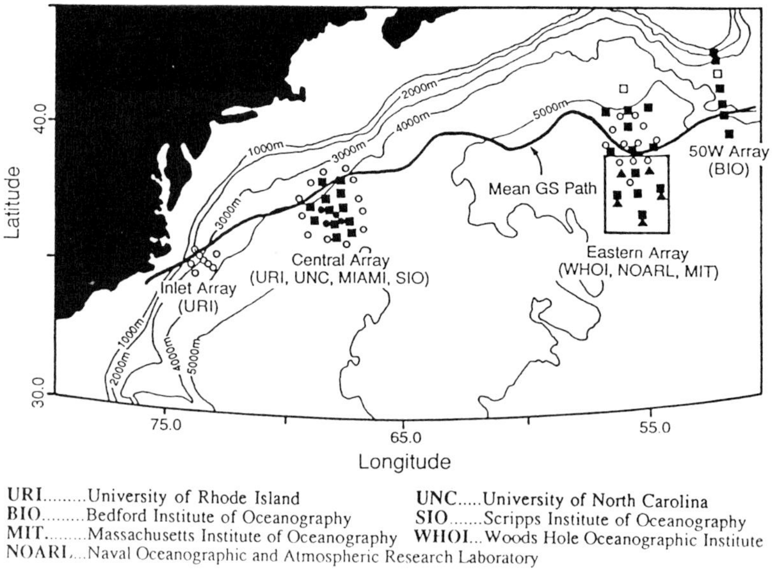

URI.........University of Rhode Island
UNC.....University of North Carolina
BIO.........Bedford Institute of Oceanography
SIO.......Scripps Institute of Oceanography
MIT.........Massachusetts Institute of Oceanography
WHOI...Woods Hole Oceanographic Institute
NOARL...Naval Oceanographic and Atmospheric Research Laboratory

FIGURE 3. Configuration of the SYNOP experiment.

the three major arrays located on the Gulf Stream mean path at different dynamical locations of the jet. Two major scientific objectives motivate the assimilation of the SYNOP datasets in numerical models of the Gulf Stream system. First, and foremost, to use these observations to constrain the models to obey a realistic behavior in the localized region where the data are available. Thus, process studies can be carried out all over the model domain, and especially where data are not available, to quantify and understand the instabilities, energy conversions, eddy-mean flow interactions and the richness of Gulf Stream events in the four dimensional (space plus time) evolution of a numerical ocean consistent both with the data and the model dynamics.

Second, the prediction of the Gulf Stream front itself and related frontal instabilities is of major interest for the Navy. Recent and significant papers belonging to both categories are those by Robinson et al. (1989[102]); Malanotte-Rizzoli et al. (1989[103]); Leese et al. (1992[105]); Malanotte-Rizzoli and Young (1992[106]). Thus the Gulf Stream System, even though one of the most studied regions of the World Ocean both from the observational and from the theoretical point of view, still offers richness of dynamical and physical processes poorly understood, measured and modeled. New growing bodies of research, new evidence and new methodologies open a suite of exciting and original research issues that will lead to further perspectives and increased understanding of this complex system.

ACKNOWLEDGEMENTS

This work was carried out with the support of the Office of Naval Research, Grant No. N00014-90-J-1481.

REFERENCES

1. Stommel, H. 1965. *The Gulf Stream: A Physical and Dynamical Description.* 2nd ed., University of California Press, Berkeley, Ca, pp. 248.
2. Stommel, H., 1948. The westward intensification of wind-driven ocean currents. *Trans. American Geophys. Union.* 29:202-206.
3. Munk, W.H. 1950. On the wind-driven ocean circulation. *J. Meteorol.* 7:79-93.
4. Fofonoff, N.P., 1954. Steady flow in a frictionless homogeneous ocean. *J. Mar. Res.* 13:254-262.
5. Charney, J.G., 1955. The Gulf Stream as an inertial boundary layer. *Proc. Nat. Ac. Sciences, U.S.A.* 41:731-740.
6. Morgan, G.W., 1956. On the wind-driven ocean circulation, *Tellus,* 8:301-320.
7. Veronis, G., 1966a. Wind-driven ocean circulation - Part I. Linear theory and perturbation analysis. *Deep-Sea Res.,* 13:17-29.
8. Veronis, G., 1966b. Wind-driven ocean circulation - Part 2. Numerical solutions of the non-linear problem. *Deep-Sea Res.,* 13:30-35.
9. Bryan, R., 1963. A numerical investigation of a nonlinear model of a wind-driven ocean, *J. Atmos. Sci.,* 20:594-606.

10. Parsons, A.T., 1969. A two-layer model of Gulf Stream separation. *J. Fluid Mech.,* 39:511-528.
11. Veronis, G., 1973. Model of world ocean circulation: I. Wind-driven, two layer. *J. Mar. Res.,* 31:228-288.
12. Veronis, G., 1981. Dynamics of large-scale ocean circulation, pp. 140-183. In: B.A. Warren and C. Wunsch (Ed.) *Evolution of Physical Oceanography.* The MIT Press, Cambridge, MA, pp. 620.
13. Warren, B.A., 1963. Topographic influences on the path of the Gulf Stream, *Tellus,* 15:167-183.
14. Niiler, P.P., and A.R. Robinson, 1967. The theory of free inertial jets, II. A numerical experiment for the path of the Gulf Stream, *Tellus,* 19:601-619.
15. Luyten, J.R. and A.R. Robinson, 1974. Transient Gulf Stream meandering. Part II: Analysis via a quasi-geostrophic time-dependent model. *J. Phys. Oceanogr.:* 4, 256-269.
16. Robinson, A.R., J.R. Luyten and G.R. Flierl, 1975. On the theory of thin rotating jets, a quasi-geostrophic time dependent model. *Geophys. Fluid Dyn.* 6:211-244.
17. Flierl, G.R. and A.R. Robinson, 1984. On the time-dependent meandering of a thin jet. *J. Phys. Oceanogr.,* 14:412-423.
18. Rhines, P.B., 1977. The dynamics of unsteady currents, pp. 189-318. In: E.D. Goldberg, I.N. McCave, J.J. O'Brien, and J.H. Steele (Eds.), *The Sea: Ideas and Observations on Progress in the Study of the Seas. 6: Marine Modeling,* pp. 712.
19. Holland, W.R. and L.B. Lin, 1975a. On the generation of mesoscale eddies and their contribution to the oceanic general circulation. I. A preliminary numerical experiment, *J. Phys. Oceanogr.,* 5:642-657.
20. Holland, W.R. and L.B. Lin, 1975b. On the generation of mesoscale eddies and their contribution to the oceanic general circulation. II. A parameter study. *J. Phys. Oceanogr.,* 5:658-669.
21. Semtner, A.F. and Y. Mintz, 1977. Numerical simulation of the Gulf Stream and mid-ocean eddies. *J. Phys. Oceanogr.,* 7:208-230.
22. Owens, W.B. and F.P. Bretherton, 1978. A numerical study of mid-ocean mesoscale eddies. *Deep-Sea Res.,* 25:1-14.
23. Robinson, A.R., D.E. Harrison, Y. Mintz and A.J. Semtner, 1977. Eddies and the general circulation of an idealized oceanic gyre: a wind and thermally driven primitive equation numerical experiment, *J. Phys. Oceanogr.,* 7:182-207.
24. Semtner, A.J. and W.R. Holland, 1978. Intercomparison of quasi-geostrophic simulations of the western North Atlantic circulation with primitive equation results. *J. Phys. Oceanogr.,* 8:735-754.
25. Holland, W.R., 1978. The role of mesoscale eddies in the general circulation of the ocean. Numerical experiments using a wind-driven quasi-geostrophic model. *J. Phys. Oceanogr.,* 8:363-392.
26. Holland, W.R. and P.B. Rhines, 1980. An example of eddy-induced ocean circulation. *J. Phys. Oceanogr.,* 10:1010-1031.
27. Richardson, P.L., 1980. Gulf Stream Ring trajectories. *J. Phys. Oceanogr.,* 10:90-104.

28. Flierl, G.R., 1977. The application of linear quasi-geostrophic dynamics to Gulf Stream rings., *J. Phys. Oceanogr.,* 7:365-379.
29. Flierl, G.R., 1979a. A simple model for the structure of warm and cold core rings. *J. Geophys. Res.,* 84:781-785.
30. Flierl, G.R., 1979b. Baroclinic solitary waves with radial symmetry. *Dyn. Atmos. Oceans,* 3:15-38.
31. Fofonoff, N.P., 1981. The Gulf Stream system. pp. 112-137. In: B.A. Warren and C. Wunsch (Eds.), *Evolution of Physical Oceanography,* The MIT Press, Cambridge, MA pp. 620.
32. Halliwell, G.R., Jr., and C.N.R. Mooers, 1983. Meanders of the Gulf Stream downstream from Cape Hatteras. *J. Phys. Oceanogr.,* 13:1275-1292.
33. Worthington, L.V., 1976. On the North Atlantic circulation, *The J. Hopkins Oceanogr. Studies,* 6:pp. 110.
34. Hogg, N.G., R.B. Pickart, R.M. Hendry and W. Smethie, 1986. On the northern recirculation gyre of the Gulf Stream, *Deep-Sea Res,* 33:1139-1166.
35. Hogg, N.G., 1992. On the transport of the Gulf Stream between Cape Hatteras and the Grand Banks. *Deep-Sea Res.,* 39, 1231-1246.
36. Orlanski, I., 1969. The influence of bottom topography on the stability of jets in a baroclinic fluid, *J. Atmos. Sci.,* 26:1216-1232.
37. Sela, J. and S.J. Jacobs, 1971. Ageostrophic effects on Gulf Stream instability. *J. Atmos Sci.,* 28:962-967.
38. Hart, J.E., 1974. On the mixed stability problem for quasi-geostrophic ocean currents. *J. Phys. Oceanogr.,* 4:349-356.
39. De Szoeke, R.A., 1975. Some effects of bottom topography on baroclinic stability. *J. Mar. Res.,* 33:93-122.
40. Ikeda, M. 1983. Linear instability of a current flowing along a bottom slope using a three-layer model. *J. Phys. Oceanogr.,* 13:208-223.
41. Killworth, P.D. 1983. Long-wave instability of an isolated front. *Geophys. Astrophys. Fluid Dyn.,* 24:235-258.
42. Killworth, P.D., N. Paldor and M.E. Stern. 1984. Wave propagation and growth on a surface front in two-layer geostrophic current. *J. Mar. Res.,* 42:761-785.
43. Paldor, N. 1983a. Stability and stable modes of coastal fronts. *Geophys. Astrophys. Fluid Dyn.,* 27:217-229.
44. Paldor, N. 1983b. Linear stability and stable modes of geostrophic fronts. *Geophys. Astrophys. Fluid Dyn.,* 24:299-326.
45. Flierl, G.R., P. Malanotte-Rizzoli and N.F. Zabusky. 1987. Nonlinear waves and coherent vortex structures in barotropic β-plane jets. *J. Phys. Oceanogr.,* 17:1408-1438.
46. Johns, W.E. 1988. One-dimensional baroclinically unstable waves in the Gulf Stream potential vorticity gradient near Cape Hatteras. *Dyn. Atmos. Oceans.* 11:323-350.
47. Hogg, N.G. 1976. On spatially growing baroclinic waves in the ocean. *J. Fluid Mech.* 78:217-238.
48. Richardson, P.L. 1983. Gulf Stream rings, pp. 19-45. In: A.R. Robinson (ed.) *Eddies in Marine Science,* Springer-Verlag, pp. 609.

49. Flierl, G.R. 1987. Isolated eddy models in Geophysics, *Am. Rev. Fluid. Mech.,* 19:493-530.
50. Zabusky, W.J., M. Hughes and K.V. Roberts. 1979. Contour dynamics for the Euler equations in two dimensions. *J. Comput. Phys.,* 30:96-106.
51. Zabusky, N.J. and E.A. Overman. 1981. Regularization of contour dynamical algorithms. *J. Comput. Phys.,* 52:357-368.
52. Stern, M.E. and L.J. Pratt. 1985. Dynamics of vorticity fronts. *J. Fluid Mech.,* 161:513-532.
53. Pratt, L.J. and M.E. Stern. 1986. Dynamics of potential vorticity fronts and eddy detachment. *J. Phys. Oceanogr.,* 16-1099-1118.
54. Stern, M.E. and G.R. Flierl, 1987. On the interaction of a vortex with a shear flow. *J. Geophys. Res.,* 92:10,733-10,744.
55. Pratt, L.J., 1988. Meandering and eddy detachment according to a simple (looking) path equation. *J. Phys. Oceanogr.,* 18:1627-1640.
56. Polvani, L.M., 1988. Geostrophic vortex dynamics, PhD thesis, MIT/WHOI Joint Program, pp. 221.
57. Polvani, L.M., G.R. Flierl and N.J. Zabusky, 1989. Filamentation of unstable vortex structures via separatrix crossing: a quantitative estimate of onset time. *Phys. Fluids A,* 1:181-184.
58. Pratt, L.J., J. Earles, P. Cornillon and J.F. Cayhe, 1990. The nonlinear behavior of varicose disturbance in a simple model of the Gulf Stream. *Deep-Sea-Res.,* 30:5591-5622.
59. Meacham, S.P., 1991. Meander evolution on piecewise uniform, quasi-geostrophic jets. *J. Phys. Oceanogr.,* 21:1139-1170.
60. Bell, G.J. and L.J. Pratt, 1992. The interaction of an eddy with an unstable jet. *J. Phys. Oceanogr.,* 22:1229-1244.
61. Wiztky, K., L. Magaard and J. Hager, 1976. Eddy energy in the oceans. *J. Geophys. Res.,* 81:2641-2646.
62. Dantzler, H.L., 1977. Potential energy maxima in the tropical and subtropical North Atlantic. *J. Phys. Oceanogr.,* 7:512-519.
63. Richardson, P.L., 1983. Eddy kinetic energy in the North Atlantic from surface drifters. *J. Geophys. Res.,* 88:4355-4367.
64. Schmitz, W.J., Jr. W.R. Holland and J.F. Price, 1983. Mid-latitude mesoscale variability. *Rev. Geophys. Space Phys.,* 21:1109-1119.
65. Wunsch, C.I., 1983. Western North Atlantic interior, pp. 46-65. In: A.R. Robinson (Ed.) *Eddies in Marine Science,* Springer-Verlag, pp. 609.
66. Hogg, N.G., 1981. Topographic waves along 70°W on the continental rise. *J. Mar. Res.,* 39:627-649.
67. Welsh, E.G., N.G. Hogg and R.M. Hendry, 1991. The relationship of low frequency deep variability near the HEBBLE site to Gulf Stream fluctuations. *Mar. Geology,* 99:303-317.
68. Bower, A.S. and N.G. Hogg, 1993. Evidence for barotropic wave radiation from the Gulf Stream. *J. Phys. Oceanogr.,* in press.
69. Pedlosky, J., 1977. On the radiation of mesoscale energy in the mid-ocean. *Deep-Sea Res.,* 24:591-600.

70. Talley, L.D., 1983a. Radiating barotropic instability. *J. Phys. Oceanogr.,* 13:972-987.
71. Talley, L.D., 1983b. Radiating instabilities of thin baroclinic jets. *J. Phys. Oceanogr.,* 13:2161-2181.
72. Malanotte-Rizzoli, P., 1984. Boundary-forced nonlinear planetary radiation. *J. Phys. Oceanogr.,* 14:1032-1046.
73. Ikeda, M., 1981. Meanders and detached eddies of a strong eastward-flowing jet using a two-layer quasi-geostrophic model. *J. Phys. Oceanogr.,* 11:526-540.
74. Ikeda, M. and J.R. Apel, 1981. Mesoscale eddies detached from spatially growing meanders in an eastward-flowing oceanic jet using a two-layer quasi-geostrophic model. *J. Phys. Oceanogr.,* 11:1638-1661.
75. Holland, W.R., D.E. Harrison and A.J. Semtner, 1983. Eddy-resolving numerical models of large-scale ocean circulation, pp. 379-403. In: A.R. Robinson (Ed.) *Eddies in Marine Science,* pp. 609.
76. Malanotte-Rizzoli, P., D.B. Haidvogel and R.E. Young, 1987. Numerical solution of transient boundary-forced radiation. Part I: The linear regime. *J. Phys. Oceanogr.,* 17:1439-1457.
77. Hogg, N.G. 1988. Stochastic wave radiation by the Gulf Stream. *J. Phys. Oceanogr.,* 18:1687-1701.
78. Malanotte-Rizzoli, P., N.G. Hogg, and R.E. Young, 1994. Stochastic wave radiation by the Gulf Stream: Numerical experiments. *Deep-Sea Res.,* submitted.
79. McEwan, A.D., R.O.R.Y. Thompson and R.A. Plumb, 1980. Mean flows driven by weak eddies in rotating systems. *J. Fluid Mech.,* 99:655-672.
80. Charney, J.G. and G.R. Flierl, 1981. Oceanic analogues of large-scale atmospheric motions, pp. 504-548. In: B.A. Warren and C. Wunsch (Eds.) *Evolution of Physical Oceanography,* The MIT Press, Cambridge, MA, pp. 620.
81. Haidvogel, D.B. and P.B. Rhines, 1983. Waves and circulation driven by oscillatory winds in an idealized ocean basin. *Geophys. Astrophys. Fluid Dyn.,* 25:1-63.
82. Hogg, N.G. and H. Stommel, 1985. On the relationship between the deep circulation and the Gulf Stream, *Deep-Sea Res.,* 32:1181-1193.
83. Schmitz, W.J., Jr., and W.R. Holland, 1982. A preliminary comparison of selected numerical eddy-resolving general circulation experiments with observations, *J. Mar. Res.,* 40:75-117.
84. Hall, M.M., 1986. A diagnostic investigation of kinetic energy budgets in a numerical model. *J. Geophys. Res.,* 91:2555-2568.
85. Holland, W.R. and W.J. Schmitz, Jr., 1985. Zonal penetration scale of model mid-latitude jets. *J. Phys. Oceanogr.,* 15:1859-1875.
86. Schmitz, W.J., Jr. and W.R. Holland, 1986. Observed and modeled mesoscale variability near the Gulf Stream and Kuroshio extension. *J. Geophys. Res.,* 91:9624-9638.
87. Evans, J.C., D.B. Haidvogel and W.R. Holland, 1987. A review of numerical ocean modeling (1983-1986): Midlatitude mesoscale and gyre-scale. *Rev. Geophys.,* 25:235-244.

88. Robinson, A.R., M.A. Spall and N. Pinardi, 1988. Gulf Stream simulations and the dynamics of ring and meander processes. *J. Phys. Oceanogr.,* 18:1320-1353.
89. Spall, M.A. and A.R. Robinson, 1990. Regional primitive equation studies of the Gulf Stream meander and ring formation region. *J. Phys. Oceanogr.,* 20:985-1016.
90. McWilliams, J.C., P.R. Gent and N.J. Norton, 1986. The evolution of balanced low-mode vortices on the β-plane. *J. Phys. Oceanogr.,* 16:838-855.
91. Chassignet, E.P., 1992. Rings in numerical models of ocean general circulation: a statistical study. *J. Geophys. Res.,* 97:9479-9492.
92. McWilliams, J.C., N.J. Norton, P.R. Gent and D.B. Haidvogel, 1990. A linear balance model of wind-driven, mid-latitude ocean circulation. *J. Phys. Oceanogr.,* 20:1349-1378.
93. Bleck, R. and D.B. Boudra, 1986. Wind-driven spin-up in eddy-resolving ocean models formulated in isopycnic and isobaric coordinates. *J. Geophys. Res.*, 91:7611-7621.
94. Haidvogel, D.B., J.L. Wilkin and R.E. Young, 1991. A semi-spectral primitive equation ocean circulation model using vertical sigma and orthogonal curvilinear horizontal coordinates. *J. Comput. Phys.,* 94: 151-185.
95. Brown, O.B., P.C. Cornillon, S.R. Emmerson and H.M. Carle, 1986. Gulf Stream warm rings: a statistical study of their behavior. *Deep-Sea Res.,* 33:1459-1473.
96. Thompson, J.D. and W.J. Schmitz, Jr., 1989. A limited-area model of the Gulf Stream: design, initial experiments and model-data intercomparison. *J. Phys. Oceanogr.,* 19:791-814.
97. Bryan, F.G. and W.R. Holland, 1989. A high-resolution simulation of the wind- and thermohaline-driven circulation in the North Atlantic Ocean, pp. 99-115. In: P. Muller and D. Anderson (Eds.) *Proceedings 'Aha Hulitkoa' Hawaiian Winter Workshop,* pp. 212.
98. Cox, M., 1984. A primitive equation three-dimensional model of the ocean. *Tech Rep. I.* NOAA Geophys. Fluid Dyn. Lab., Princeton Univ., Princeton, N.J. pp. 53.
99. Chassignet, E.P. and P.R. Gent, 1991. The influence of boundary conditions on midlatitude jet separation in ocean numerical models. *J. Phys. Oceanogr.,* 21:1290-1299.
100. Mellor, G.L. and T. Ezer, 1991. A Gulf Stream model and an altimetry assimilation scheme. *J. Geophys. Res.,* 96:8779-8795.
101. Ezer, T. and G.L. Mellor, 1992. A numerical study of the variability and the separation of the Gulf Stream induced by surface atmospheric forcing and lateral boundary flows. *J. Phys. Oceanogr.,* 22:660-682.
102. Robinson, A.R., M.A. Spall, L.J. Walstad, and W.G. Leslie, 1989. Data assimilation and dynamical interpolation in Gulfcast experiments. *Dyn. Atmos. Oceans,* 13:301-316.
103. Malanotte-Rizzoli, P., R.E. Young and D.B. Haidvogel, 1989. Initialization

and data assimilation experiments with a primitive equation model. *Dyn. Atmos. Oceans,* 13:349-378.

104. Leese, J.A., R.C. Willems and L.A. Yeske, 1992. Evaluation capability for the navy ocean modeling and prediction program. *Oceanography,* 5:55-59.
105. Robinson, A.R., 1992. Shipboard prediction with a regional forecast model. *Oceanography,* 5:42-48.
106. Malanotte-Rizzoli, P. and R.E. Young, 1992. How useful are localized clusters of traditional oceanographic measurements for data assimilation? *Dyn. Atmos. Oceans,* 17:23-61.

The Oceans: Physical-Chemical Dynamics and Human Impact. Edited by S.K. Majumdar, E.W. Miller, G.S. Forbes, R.F. Schmalz and Assad A. Panah. © 1994, The Pennsylvania Academy of Science.

Chapter Eight

OCEANIC RESPONSE TO TROPICAL CYCLONES

LYNN K. SHAY
MPO/RSMAS
University of Miami
4600 Rickenbacker Causeway
Miami, FL 33149

INTRODUCTION

The passage of a tropical cyclone over a warm ocean represents one of the more extreme cases of air-sea interactions as the ocean exchanges heat with the atmosphere. The most apparent effect of tropical cyclone passage is the marked cooling of the sea surface temperatures (SSTs) of 1-5°C, which occurs to the right of the storm track by 1-2 radii of maximum winds (R_{max}). Although warm SSTs ($>$ 26°C) are required to maintain the tropical cyclone,[1] the *rightward bias* of the maximum SST decreases and mixed layer depth increases of 20-40 m are primarily due to vertical mixing ($\overline{\omega' T_h'}$) of the cooler thermocline water with the warmer mixed layer water. This entrainment mixing accounts for about 80% of the observed SST decreases,[2] which is due either to vertical shear of the horizontal currents[3] or surface-generated turbulence within the mixed layer.[4] Both of these mechanisms play important roles in mixed layer thermodynamics.

The ocean's momentum response is classified into two regimes: the directly-forced or near-field, and the evolving 3- dimensional wake or far-field. In the near-field, the cyclonically-rotating wind field of a tropical cyclone causes the oceanic mixed layer currents to diverge from the storm track starting within one-quarter of an inertial wavelength behind the eye, where the inertial wavelength (Λ) is defined as the product of the storm translation speed U_h and the inertial period (IP).[5] This divergence causes the upwelling of cooler water underneath the storm track that decreases the mixed layer depth. Over the next one-half inertial cycle ($\frac{3\Lambda}{4}$), mixed layer currents converge toward the storm track, causing an increase in the mixed layer depth as warmer water is downwelled into the thermocline. This alternating cycle of upwelling and downwelling of the isotherms (and isopycnals) occurs over distances of Λ and establishes horizontal pressure gradients that couple the mixed layer to the thermocline as part of a spreading 3-dimensional wake.

The strong atmospheric forcing and the vertical displacements of the isopycnals induce the horizontal pressure gradients that excite baroclinic near-inertial motions in the thermocline. As these forced, near-inertial waves propagate away from the storm track in the oceanic wake, the ocean's mass and current fields adjust to form a spreading *baroclinic* ridge along the periphery of the upwelling and downwelling regimes that remain in the wake for a period of time following passage. In the wake of hurricane Ella, elevated isopycnals were observed for approximately three weeks,[6] whereas the near-inertial current response to hurricanes Frederic and Gloria persisted for similar time scales in the thermocline.[7, 8]

An interesting feature in recent ocean current observations has been the rapid penetration of the near-inertial response through the thermocline and into the bottom layers.[6,7,11] While forced near-inertial motions are characterized by oscillations of large vertical wavelengths associated with the low-order baroclinic modes,[9,10] the immediate pulse and the excitation of near-inertial oscillations induced by hurricane Frederic throughout the water column to 457 m (Figure 1), and the deep ocean currents at 700 m excited by hurricane Allen suggests another mechanism. The rapid penetration of the response during Frederic was due to the sea surface depression (*barotropic* trough) forced by the wind stress and mean mass divergence.[5,12,13] Although the accompanying geostrophic and near-inertial currents are weak compared to the baroclinic response (8-10%), the barotropic response becomes increasingly important for tropical cyclones moving into shallow water. It should also be noted that this trough differs from the sea surface elevation due to the storm surge that results from the combined effect of the inverted barometric effect (1 cm sea level rise per 1 mb decrease in central pressure), the winds driving the water towards the coast on the right side of the track during landfall, and to a lesser extent the phase of the astronomical tides. The distinction is that the barotropic trough remains in the sea surface wake for a longer time scale than the storm surge, which moves with the storm center and dissipates at landfall. This latter effect is beyond the scope of this paper (see [14] for a detailed description).

The objective here is to document the salient features of the ocean's thermal and momentum response to tropical cyclones based on a blend of observational, analytical and numerical investigations. Of particular interest are recent findings and the emerging technologies that may address scientific issues related to the oceanic response in both the directly-forced and wake regimes. Recent field experiments and observations are summarized in RECENT EXPERIMENTS. AIR-SEA PARAMETERS describes air-sea parameters and nondimensional numbers governing the initial response to tropical cyclones where high-resolution measurement were obtained. The thermal response to tropical cyclones is given in THERMAL RESPONSE, and within the framework of near-inertial dynamics, the momentum response is described in MOMENTUM RESPONSE, followed by a summary with concluding remarks for future research in SUMMARY.

RECENT EXPERIMENTS

Hurricane Planetary Boundary Layer Experiments

During 1984 and 1985, the Hurricane Research Division (HRD) of NOAA and a consortium of oil companies sponsored the Ocean Response To A Hurricane (ORTAH) Joint Industry Program as part of the Hurricane PBL (HPBL) experiments in tropical cyclones Norbert, Josephine and Gloria from the NOAA WP-3Ds.

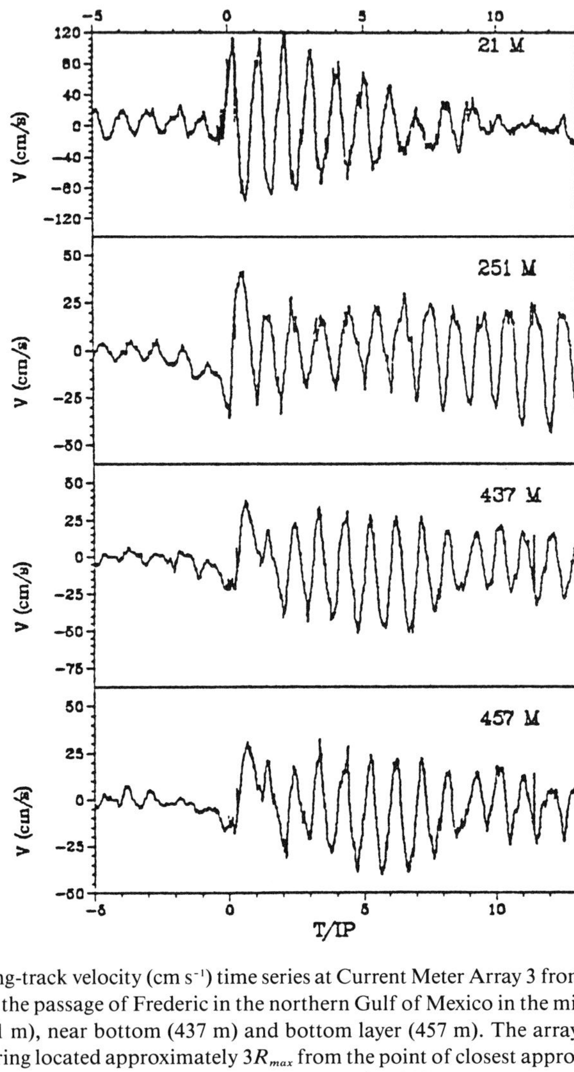

FIGURE 1. Along-track velocity (cm s^{-1}) time series at Current Meter Array 3 from 7-25 Sept. 1979 observed during the passage of Frederic in the northern Gulf of Mexico in the mixed layer (21 m), thermocline (251 m), near bottom (437 m) and bottom layer (457 m). The array was a taut-wire subsurface mooring located approximately $3R_{max}$ from the point of closest approach of hurricane Frederic at 21 UTC 12 Sept. 1979 (t = 0) (from Shay et al., 1990).

For the first time, Airborne eXpendable Current Profilers (AXCPs) were deployed in the directly-forced region of these storms providing a snapshot of the vertical structure of the ocean currents and temperatures.[15,16] Not since the hurricane Hilda experiments,[17] has there been such an organized effort to understand the oceanic response to tropical cyclones. The AXCP operates on the principle of motionally-induced voltage differences between two electrodes.[18] By descending at a rate of 4.5 m s^{-1}, the baroclinic currents are accurately sampled at intervals of about 2-3 m with *rms* errors of 1-2 cm s^{-1} relative to an unknown, but constant depth-independent flow. In addition, drifting buoys were deployed in the ocean 150 km ahead of Josephine from an Air Force WC-130 aircraft,[19] and from a NOAA WP-3D in Gloria.[20] The buoy trajectories indicated the presence of mesoscale variability in the ocean currents and temperatures in the vicinity of the North Atlantic Subtropical Front.[21]

Hurricane Gilbert Experiment

The evolving 3-dimensional ocean response to the passage of hurricane Gilbert in the western Gulf of Mexico was observed by deploying AXCPs and Airborne eXpendable BathyThermograph (AXBTs) from 14-19 September 1988.[22] This experimental effort improved upon previous efforts in measuring the prestorm and storm currents and temperatures, and then revisiting the same area one and three days following passage. Given the broad spatial and temporal coverage of these data ($-R_{max}$ to $6R_{max}$), the Gilbert data will be used here to illustrate the salient features of the oceanic response. Flight-level meteorological observations and the near-surface wind stress distribution inferred from the Stepped Frequency Microwave Radiometer (SFMR),[23] were also obtained during Gilbert. The SFMR is a remote sensor in the microwave spectrum utilizing the extension of relationships between surface emissivity and near-surface wind stress. It has been found that surface emissivity is directly proportional to wind stress in tropical cyclones. The SFMR can be used jointly with a second remote sensor, C-band SCATterometer,[24] to acquire two almost independent near-surface wind speed measurements.

Typhoon 90 Experiment

As part of an ONR Accelerated Research Initiative (ARI) on *Tropical Cyclone Motion*, an international field experiment was conducted in the western Pacific Ocean between 1 August and 30 September 1990.[25] Four USSR research vessels acquired meteorological and oceanographic observations in a box pattern between 13 and 20°*N* and 126 to 141°*E*.[26] Of twelve typhoons that developed during the experiment, six passed through the experimental domain. Preliminary analyses of this extensive set of measurements indicate significant decreases in the surface salinity in regions directly influenced by precipitation. However, the advection of the salinity and temperature gradients could not be resolved because of the lack of concurrent velocity measurements at each hydrographic station. To assess these advective tendencies, future experiments should include either shipboard velocity profiling from an Acoustic Doppler Current Profiler[27] over the same space and time scales as the hydrographic observations, or the simultaneous deployments of AXCPs and Airborne eXpendable Conductivity and Temperature Depth profilers (AXCTD) to observe the 3-dimensional velocity, temperature and salinity structure over the scales of the atmospheric forcing.[28]

AIR-SEA PARAMETERS

One important parameter for estimating the expected baroclinic response depends on the internal Froude number (ration of the translation speed to the first baroclinic mode phase speed $\frac{U_h}{c_1}$). If the Froude number is less than unity, large mean currents are generated and upwelling is the dominant response because the tropical cyclone is stationary or moving slowly. For Froude numbers exceeding unity, the ocean response is *predominantly* baroclinic associated with upwelling and downwelling of the isotherms (isopycnals) and a spreading 3-dimensional wave wake. The non-dimensional storm scale ($\frac{2R_{max}}{\alpha_1}$), is a ratio of the scale of positive wind stress curl region to the baroclinic deformation radius (α_1) defined as the ratio of the first mode phase speed (c_1) and the Coriolis parameter (f). If this forcing scale is greater than this deformation radius, there will be a characteristic time scale for the phase and amplitude for each baroclinic mode to become out of phase with the other baroclinic modes in the mixed layer.[9] This time scale increases with mode number, which implies that the first mode will initially separate from the remaining baroclinic modes by virtue of the large phase speeds of about 3 m s^{-1}.

The key ratios for the barotropic response are the external Froude number (ratio of the translation speed to the barotropic mode phase speed $\frac{U_h}{c_0}$) and the nondimensional forcing scale. That is, the ratio of the atmospheric forcing scale to the deformation radius of the barotropic mode (α_0). The external Froude number is usually small, which has caused theoretical treatments to neglect the second-order terms (5-10%) involving the time evolution of the sea surface in the governing equations.[29] However, these second-order terms are necessary to balance the fields in the geostrophic adjustment process.[14] The nondimensional forcing scale depends on the scale of the applied wind field of the tropical cyclone, not just the $2R_{max}$ for the positive vorticity curl as in the baroclinic response. For broader and more complex structure storms, cyclonically-rotating winds of 10 m s^{-1} may extend to 8-10R_{max} as observed in Frederic,[30] thereby increasing the forcing scale from O(0.1) to O(0.5) or more. The implication here is that this barotropic response may be driven by the larger scale, near-surface wind field.

The observed air-sea parameters for tropical cyclones Frederic, Norbert, Josephine, Gloria and Gilbert are compared in Table 1. Frederic was a large storm with a well-defined eye, Norbert was a smaller, intense storm, while Josephine, Gloria and Gilbert were large storms with broad, flat wind profiles extending for several hundred kilometers. Using a wind-speed dependent formulation for the drag coefficient,[31] the maximum wind stress of Norbert was considerably larger than the other storms having a maximum value of 4.4 N m^{-2} at the primary radius of maximum winds. For Frederic and Gilbert, the maximum surface wind stress was about 3.3 N m^{-2} at 30 km and 45 km respectively, with a secondary wind stress maxima (2.4 N m^{-2}) at 90 *km* from the track of Gilbert. This double wind maxima may add difficulty in interpreting the oceanic response to a tropical cyclone within the context of a simple Rankine vortex.[19,32] Norbert and Josephine moved slowly whereas Frederic, Gloria and Gilbert moved faster as reflected in their Froude numbers exceeding two. In these cases, however, the oceanic response is expected to be *predominately* baroclinic associated with upweling and downwelling of the isopycnals in spreading 3-dimensional wakes. Since the observations were acquired at the mid-latitudes, the wavelength variations (Λ) are primarily due to the storm speeds.

THERMAL RESPONSE

Air-Sea Fluxes

Extensive hydrographic measurements of the oceanic response to tropical cyclones were acquired between 1975-1978 during the "Typhoon 75" (Tess), "Typhoon 78" (Virginia), and as part of the POLYMODE Experiment in the Atlantic in the wake of hurricane Ella.[6,33] The SSTs decreased by $5°C$ and large SST gradients were located at the edge of the wake region. In addition, the air temperatures decreased by 3-$5°C$ with a reduction in the wind speed, resulting in significant decreases in the sensible and latent heat fluxes.[35] These estimates indicate that the cool wake just behind the eye has a pronounced influence on the heat fluxes in the Atmospheric Planetary Boundary Layer necessary to sustain the tropical cyclone.[36]

One proven technique that resolves the air-sea fluxes in the APBL is the radome gust probe on research aircraft which is based upon the eddy correlation technique.[37] The mean and turbulent atmospheric variables are ensembled-averaged over a finite time interval to form the Reynolds stresses for heat, moisture and momentum as demonstrated in data acquired from the Genesis of Atmospheric Lows Experiment[38] and Frontal Air-Sea Interaction Experiment.[37,39] These Reynolds stresses are directly

TABLE 1

Air-Sea parameters and nondimensional scales in tropical cyclones Frederic,[79] Norbert,[84] Josephine,[84] Gloria[85] and Gilbert[88] that were measured extensively from current meter moorings and research aircraft.

Parameter	Frederic	Norbert	Josephine	Gloria	Gilbert
R_{max} (km)	30	34	52	46	45
τ_{max} (NAm^2)	3.3	4.4	3.4	3.4	3.3
U_h (m/s)	6.5	3.7	4.2	6.8	5.6
c_1 (m/s)	3.0	2.1	3.0	3.0	2.8
Fr	2.1	1.4	1.4	2.3	2
IP (d)	1.0	1.6	1.0	1.0	1.25
f ($s^{-1}x10^4$)	0.7	0.5	0.7	0.7	0.6
α_1 (km)	41	42	41	41	48
$(\frac{2R_{max}}{\alpha_1})$	1.5	1.1	2.5	2.2	1.9
Λ (km)	580	440	250	515	600
h (m)	30	40	55	45	35

proportional to the wind stress, sensible and latent heat fluxes, which are also important for the evaluation of the bulk transfer coefficients. Thus, the spatial variability of the air-sea fluxes in the APBL are critical to assess the ocean-atmosphere feedback mechanisms during these strong forcing events.

SSTs and Mixed Layer Depths

Numerical studies have shown that the maximum SST decreases and mixed layer depth increases occur on the right side of the storm track between 1-2R_{max}.[3,40] This bias is due to the nonlinear advection terms and to a lesser extent the asymmetries in the wind field.[40,41] Depending upon the intensity and speed of the tropical cyclone, there may also be a phase-locking between the rotating wind stress and the mixed layer velocity vector that induces a rightward bias.[3,42] Since this resonant excitation and wind field asymmetries are essentially linear effects, linear dynamics and observed wind fields are useful for investigating the impact of these mechanisms upon the oceanic response.

Observational studies of the thermal response became an integral part of the NOAA's Experimental Hurricane Research Program using AXBTs and the Airborne Infrared Radiometer Thermometer (AIRT).[2] An empirical model of upper ocean cooling was developed by stratifying 10 years of observations according to the U_h, τ_{max}, R_{max}, ocean temperature structure and mixed layer depths (MLD). Although in most of the documented cases, pre-storm conditions were not observed to quantify the total heat loss from the ocean to the atmosphere, the empirical model demonstrated a rightward bias in the SST decreases of 1-5°C depending on the storm speed (U_h) that followed a "crescent-shaped" pattern extending from the right-rear quadrant of the storm into the wake region. Recently, remotely-sensed SSTs from the NOAA Advanced Very High Resolution Radiometer (AVHRR) has also been found to be effective in monitoring the spatially-evolving SST patterns in the cold wakes of tropical cyclones.[43] The maximum SST response of 3.5°C was located about 70 km to the right of the track subsequent to the passage of hurricane Harvey (1981).

As shown in Figure 2, the objectively-analyzed SST fields derived from the hurricane Gilbert data illustrate the SST and MLD response to a tropical cyclone. Prestorm SSTs of about 29°C were nearly uniform over the domain without any evidence of a warm core eddy in the central Gulf of Mexico. During the storm, the SST cooling pattern began in advance of the eye with significant SST decreases beginning in back of the eye (Figure 2b). Farther back in Gilbert's wake, the observed SSTs acquired from the Wake I and Wake II experiments indicated a broad cooling pattern with a cool pool of water (25.5°C) located at about 1-2 R_{max}, which agrees with an empirical model.[2] The along-track displacement of this pool of cooler water may have been induced by the vertical motion associated with upwelling and downwelling cycles, which result from the divergence and convergence of the mixed layer currents. Thus, the net cooling induced by Gilbert was 3.5-4°C centered between 1-2 R_{max} on the right side of the track.

Based upon prestorm MLDs of 30-35 m, the cool wake was associated with deeper mixed layers of about 60 m induced by enhanced mixing processes due to a combination of surface wind stress and cooling,[4] and vertical current shear at the base of the mixed layer.[3] The passage of the hurricane core increased the MLD along the storm path. However, at about $\frac{\Lambda}{4}$ behind the eye, the mixed layer current vectors diverged from the track with relative maxima of about 1 m s^{-1}, which induced the upwelling of the cooler thermocline water underneath the storm track and decreased the MLD. Farther to the right (4-5 R_{max}), mixing effects were evidently enhanced by downwelling processes as the MLD increased to approximately 60 m. Although the presence of the eddy complicated the spatial variations of the MLDs, they were deeper (70 m) and farther to the right during the Wake I when the mixed layer currents diverged from the storm track as part of the next cycle of the near-inertial rotation, with relative maxima of about 1.4 m s^{-1}. During Wake II, the MLD pattern indicated a region of shallower water along the track of less than 30 m as a result of the convergence cycle of the mixed layer currents and the end of the upwelling or the beginning of the next downwelling cycle.

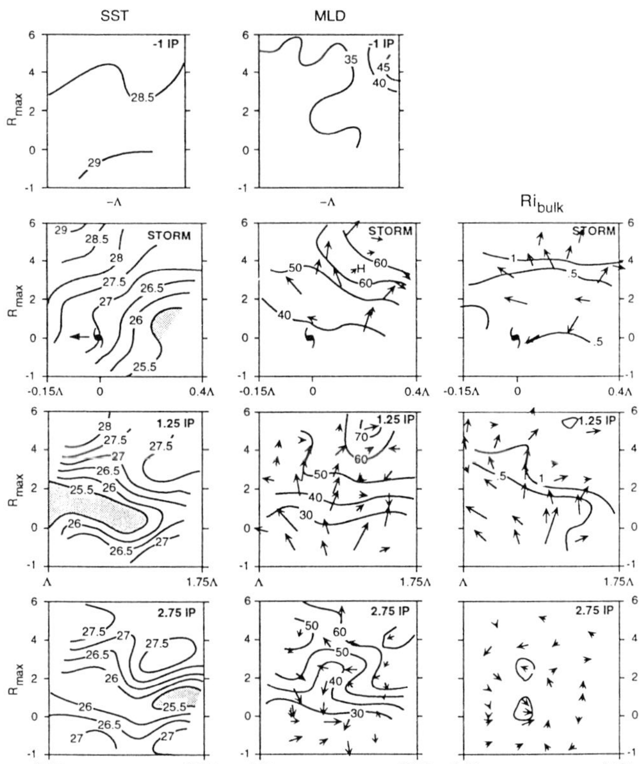

FIGURE 2. Objective analysis of SSTs (°C) (left panels), MLDs (m) (center panels), and the Ri_{bulk} numbers (right panels) derived from the a) Prestorm, b) Storm, c) Wake I and d) Wake II of hurricane Gilbert in the storm coordinate system of Λ (600 km) and R_{max} (45 km) for the along-track and cross-track distances. The location and direction of movement of Gilbert are depicted in the Storm frames with the hatched area in the Storm, Wake I and Wake II panels depicting regions of maximum cooling. The mixed layer currents are superposed on the MLDs (center panels) with maxima of 1.1, 1.4 and 1.0 m s^{-1} in Storm, Wake I and II, respectively. The differences between the mixed layer and thermocline currents are superposed on the Ri_{bulk} panels with maxima of 1, 1.5 and 0.6 m s^{-1} (from Shay et al., 1992).

Vertical Mixing Effects

To understand the observations acquired during hurricane Hilda,[17] mixing effects were included in a nonlinear model to account for the vertical transfer of momentum from the mixed layer into the thermocline.[44] The regions of enhanced vertical mixing were apparent over broad regions, whereas upwelling remained confined to the maximum wind regime in a stationary storm. Using a parameterization based upon the surface friction velocity ($u_\star = \sqrt{\tau_{max}\rho_o^{-1}}$), vertical advection associated with upwelling and downwelling effects were dominant underneath the storm whereas mixing processes were more pronounced beyond R_{max}.[4] Entrainment mixing has also been parameterized based upon vertical shear of the horizontal currents and a bulk Richardson number to simulate the mixed layer deepening induced by hurricane Eloise.[3] That is, when the bulk Richardson number drops below a critical value of 0.65 due to enhanced shear-instabilities, the upper ocean mixes causing the SST to decrease and the mixed layer to deepen. The model simulations based upon this parameterization agreed well with the observations at EB-10 acquired during the passage of Eloise in the Gulf of Mexico[3,42] and the ORTAH observations.[16]

During the passage of Gilbert, the estimated bulk Richardson numbers ranged from 0.2 to 1 between the storm track and the ocean eddy (0 - $4R_{max}$) as shown in Figure 2.[22] The vertical current shears remained large in the Wake I observations with pockets of low bulk Richardson numbers of 0.5-1.0 extending to 3 - $4R_{max}$. While it is difficult to separate the effects of shear versus stress-induced turbulent mixing during the Storm, the continued SST cooling and mixed layer deepening 1.2 IP following Gilbert were associated with a combination of the shear-induced mixing and vertical advection resulting from the upwelling and downwelling cycle. Maximum current differences between the mixed layer and thermocline (over vertical scales of 10-20 m) ranged from 0.6-1.0 m s^{-1} in regions of low Richardson numbers. However, in Wake II, the bulk Richardson numbers exceeded unity throughout the domain because of the deeper mixed layers. The vertical current shears would have to be substantially larger to induce further deepening by this mechanism.

Within the context of upwelling and downwelling of the isotherms (and isopycnals), there was considerable cross-track variability in the thermal structure between Wake I and II (Figure 3). Between -R_{max} and the storm track, the mixed layer decreased to about 20 m, whereas on the right side of the track (1 - $3R_{max}$), the divergence and convergence of the mixed layer currents induced an upwelling and downwelling of the isotherms, respectively. The observed mixed layer depths of 40-60 m in the Wake I experiment increased to 60-80 m during Wake II, which indicate higher temperatures at greater depths during Wake II due to the convergent mixed layer currents and the downwelling of the isotherms. These 20 m oscillations in the mixed layer depths were consistent with the expected amplitudes (9-10 m) of the isopycnal displacement based on the air-sea parameters ($\frac{\tau_{max}}{\rho_o f U h}$).[45] Based on the bulk Richardson number exceeding one in the warm core eddy ($5R_{max}$) with similar current shears observed in Wake I and II, variations in the temperature structure were due to the interactions of vertical mixing and the dynamically induced vertical and horizontal advection fields.

MOMENTUM RESPONSE

Given the fortuitous nature of encounters of moored arrays with tropical cyclones, many studies treated the ocean's momentum response using analytical expressions or numerical models. The ocean's momentum represents an evolving 3-dimensional response consisting of a broad spectrum of processes excited by strong winds such as surface waves, geostrophic currents and near-inertial motions. There may have also been instances when a pre-existing, geostrophically-balanced flow field of an oceanographic front or eddy has interacted with the hurricane-induced currents, thereby adding complexity to the momentum response. In this section, the focus will be on observations, and the relevance of analytical results and model simulations to the observed physical processes.

Surface Waves

The spectrum of surface wave motions forced by a tropical cyclone consists of a low-frequency swell peak and a higher-frequency wind sea propagating in different directions. At 240 km from the eye of hurricane Debby (1982), the 2 dimensional wave spectra from the Surface Contour Radar[46] indicated significant waves having spectral peaks at frequencies of 0.09 Hz (swell) and 0.12 (Hz) (wind sea). These low-frequency surface waves induce orbital velocities that decay exponentially with

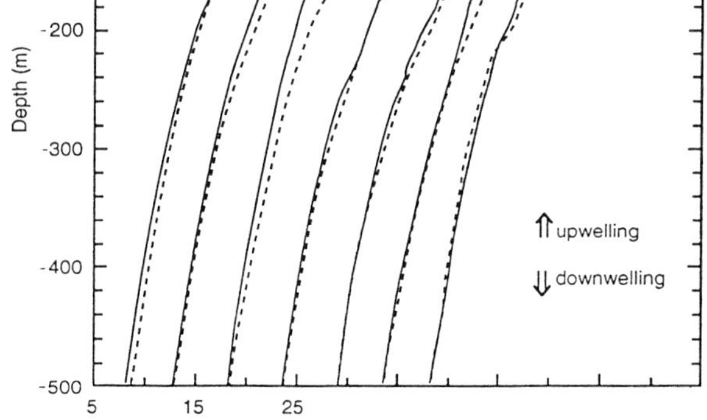

FIGURE 3. Cross-track temperature profiles (°C) from Wake I (solid) and Wake II (dashed) at 1.4 and 2.8 Λ. Each temperature profile is displaced by 4°C going from the -1 to about 5 R_{max} with a temperature scale in the lower left corner. The up and down arrows depict upwelling and downwelling of the isotherms in Wake I and II, respectively.

depth as shown in the observed velocity profiles acquired from Norbert, Josephine and Gloria. A three-layer, least-squares model was developed to isolate the effects of the surface wave-induced orbital velocities from the shear and steady current profiles in these storms.[15] These orbital velocity amplitudes ranged between 1-2 m s^{-1} due to low-frequency surface waves of 0.10 Hz, which corresponded to a 10 second wave period (wavelength of 156 m). In some cases, it has been found that the mean-wave induced currents were in the same direction as the steady mixed layer currents, which may modulate the vertical current shears in the layer, and hence vertical mixing processes.[20] To examine these scientific issues associated with the evolving wave spectra in a region of strong currents, a combined AXCP and SCR research flight was conducted in the Surface Wave Dynamics Experiment.[47] The addition of the spatially evolving 2-dimensional wave spectra provided key observations in resolving the orbital velocity amplitudes and directions of wave propagation detected by the AXCPs.

Velocity Structure

The orbital velocities isolated from the three-layer model[15] fits to the data were removed from the Gilbert current profiles as shown in Figure 4. At about $2R_{max}$, each successive velocity profile (in a storm coordinate system[16]), indicated an anti-cyclonic (clockwise) rotation of the current vector in the mixed layer starting at the storm center to 3 Λ due to the excitation of near-inertial motions. There was also a predominant clockwise rotation with depth as energy from the surface-intensified mixed layer propagated into the thermocline and the lower layers.[48] In the absence of strong baroclinic fronts and eddies, these forced oscillations have frequencies blue-shifted above f by 1-20%,[45] otherwise energy would be confined to the oceanic mixed layer when the wave frequency equals the inertial frequency. This frequency shift has been found to be proportional to the mixed layer Burger number,[45] which was about 1.03-1.04 for hurricane Gilbert based on a time series of current meter observations at 100 m.[49]

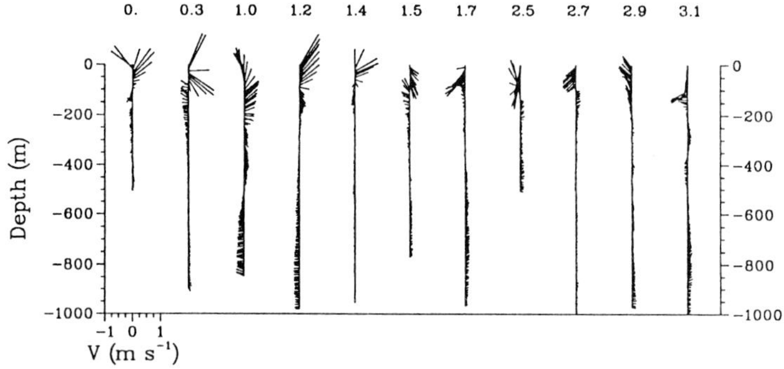

FIGURE 4. Observed along-section velocity structure (m s^{-1}) in the upper 1000 m excited by the passage of hurricane Gilbert at about $2R_{max}$ from the storm track. The times for each AXCP deployment are scaled by the inertial wavelength (600 km) given above each profile relative to the storm profile at $\Lambda = 0$. The velocity scale (m s^{-1}) is located in the lower left corner.

Vertical Structure

Using a two-layer ocean model based on linear dynamics,[5] a clear distinction was found between the baroclinic and barotropic response induced by the passage of a tropical cyclone. For the baroclinic response, the wind stress directly forces the mixed layer currents and the upwelling and downwelling of the isotherms to form a spreading 3-dimensional baroclinic ridge in the thermocline. The upwelling and downwelling cycle excites near-inertially rotating motions in the thermocline, which represents the propagation of energy from the surface intensified layer into the water column. By contrast, the barotropic response consists of a free-surface depression forced by the wind stress that is also part of the evolving 3-dimensional wave wake. The differential operators governing the baroclinic and barotropic modes are hyperbolic and elliptical depending on the ratio of the Froude number $\frac{U_h}{C_n}$, which allows the problems to be solved in terms of Bessel functions (J_o, K_o). Although the solution for the barotropic response applied to the steady-state solution, there may also be a small, but detectable time-dependent response driven by the gradients in the sea-surface depression.[13]

In a continuously-stratified fluid, all modes of oscillations are allowed as indicated by equation for the vertical structure of near inertial motions,

$$\frac{d^2\phi_n}{dz^2} + \frac{N^2(z)\phi_n}{c_n^2} = 0, \tag{1}$$

where ϕ_n is the nth eigenfunction of the vertical velocity, $x_n = (\sigma^2 - f^2)/K_n^2$ represents the phase speed of the nth mode ($K_n^2 = k^2 = l^2$ is the total horizontal wavenumber), σ is the wave frequency, and $N^2(z)$ is the buoyancy frequency derived from the vertical density gradients. The free surface boundary condition ($z = \eta$) requires pressure to be continuous across the air-sea interface,

$$\frac{d\phi_n}{dz} + \frac{gK_n^2}{\sigma^2 - f^2}\phi_n = 0, \tag{2}$$

whereas the bottom boundary condition is $\phi_n = 0$ at $z = -D$. An example of the vertical and horizontal velocity eigenfunctions ($u_n = d\phi_n/dz$) for the barotropic and first four baroclinic modes are shown in Figure 5 for an imposed free surface boundary condition. The kinematical boundary condition is $w = \frac{d\eta}{dt}$ that represents an oscillatory, non-zero vertical velocity at the sea surface associated with the barotropic mode. By contrast, the imposition of a rigid lid ($\phi_n = 0$) requires the barotropic mode of the vertical and horizontal velocities to vanish. The relative differences between the baroclinic modes for the vertical and horizontal velocity eigenfunctions estimated from (1) with and without a free surface boundary condition are 10^{-6} and 10^{-3}, respectively. Since these values are much less than unity, this implies that the baroclinic modes do not contribute to the free surface undulations. Notice that the first baroclinic mode amplitude in the thermocline (200 m) is directly out of phase wih the vertical velocity of the barotropic mode at the surface. That is, upwelling and downwelling zones in the thermocline associated with baroclinic processes will be in areas of downward and upward vertical velocities at the sea surface due to the barotropic response.[13]

Baroclinic Response

Several studies have simulated the ocean response to the passage of tropical cyclones using multi-layered[40,45,50,51] or multi-level models.[12,16,52] Besides the difference in a layer versus level treatment, the distinguishing characteristics between each model is the parameterization of vertical mixing. Using a multi-level primitive equation model with an embedded mixed layer based on turbulent kinetic energy closure[52,] and imposing a rigid lid, simulations of the baroclinic response to a model hurricane with the same characteristics of Frederic agreed well with the observed mixed layer currents.[56] However, the thermocline currents were significantly underestimated by 12-15 cm s^{-1} and decayed too quickly compared to the observations.[7] The simulated baroclinic response to hurricane Eloise using a layered model indicated that the amplitudes and phases of the simulated currents agreed well with the observed ocean currents at 53 m from EB-10.[45] These results demonstrated that the bulk Richardson number approach to parameterize vertical mixing provided realistic results.

To examine the baroclinic response using analytical expressions,[10] the wind stress terms are superposed onto the nth baroclinic modes of the horizontal velocity to form the forcing functions

$$X_n, Y_n = \frac{\tau^x, \tau^y}{\rho_o \int_{-D}^{0} \phi_{nz}^2 dz}, \tag{3}$$

where $\frac{d\phi_n}{dz} = \phi_{nz}$ and the mixed layer depth (MLD) is replaced by an equivalent forcing depth ($\int^0_{-D} \phi_{nz}^2 dz$).[54] Since the equivalent forcing depths increase with mode number, the projection of the wind stress onto the modes decreases, which is the reason why the low-order modes are excited more than the higher modes during the passage of atmospheric forcing events.[10]

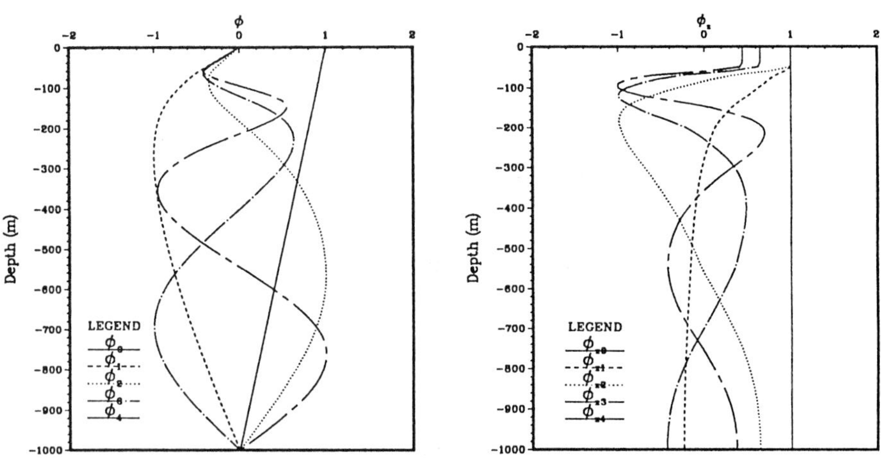

FIGURE 5. Normalized amplitudes of the vertical (left panel) and horizontal velocity eigenfunctions (right panel) for the barotropic mode (solid) and baroclinic modes 1 (dashed), 2 (dotted), 3 (chain-dotted) and 4 (chain-dashed) based on an $N^2(z)$ profile in the Gulf of Mexico.

As shown in Figure 6, the simulated mixed layer velocities based on a sum of the first ten forced baroclinic modes (see Figure 5) indicate the net divergence and convergence of the currents along the track with a maximum of 1.1 m s^{-1} at about R_{max}. This rightward displacement is due to the imposed wind field asymmetry using a Rankine vortex with similar parameters as Gilbert (see Table 1).[40] At the base of the mixed layer, the maximum vertical velocity is upward with an amplitude of 0.4 cm s^{-1} during the upwelling phase in the thermocline as the mixed layer currents diverge from the storm track. Over the next half cycle, vertical velocities are downward as the mixed layer currents converge towards the storm track. These alternating cells of upwelling and downwelling and the isopycnal displacements dynamically couple the mixed layer to the thermocline as part of the baroclinic near-inertial response excited by a translating tropical cyclone. Notice the qualitative agreement between the simulated and the observed mixed layer current vectors along $2R_{max}$ shown in Figure 4.

Barotropic Response

For the first time, moored subsurface current meter arrays have revealed that the hurricane-induced response may penetrate uniformly throughout the water column to the bottom as shown in Figure 1.[7,11] The near-bottom (457 m) current response induced by Frederic indicated near-inertial oscillations with amplitudes of about

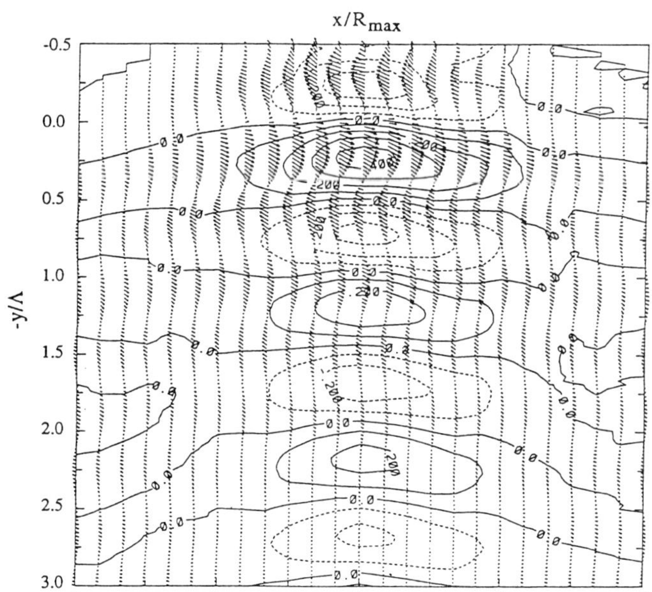

FIGURE 6. Simulated horizontal velocity structure (maximum of 1.1 m s^{-1}) superposed on the contours of the vertical velocities (maximum of 0.4 cm s^{-1} based upon first ten baroclinic modes. Along-track and cross-track distances are scaled in terms of Λ (600 km) and R_{max} (45 km) for Gilbert.

20 cm s^{-1}, whereas in hurricane Allen, near-inertial current amplitudes at 700 m were 15 cm s^{-1}. Even at the Ocean Thermal Energy Conversion mooring site at $5R_{max}$ from the track of Frederic, an impulse of about 5 cm s^{-1} was detected at 950 m. Barotropic oscillations have also been observed along the Australian margin during typhoons Ian and Victor from moored current meter arrays.[55] However, the prescribed bottom friction in their model prohibited the excitation of near-inertial motions because the e-folding time scale was 18 hours compared to the inertial period of 36 hours.

Based upon the discrepancy between the simulations and the observations in the thermocline and bottom layers during and subsequent to Frederic,[56] it was hypothesized that these differences were due to the neglect of the free surface and the barotropic response. A free-surface was incorporated into a multi-level primitive equation model capable of simulating both the baroclinic and barotropic ocean response to a hurricane.[12] In the free-surface model simulations, weak barotropic current oscillations of about 7 cm s^{-1} were found even after the removal of the hurricane forcing during the spin-down in a deep ocean.

The barotropic response to Frederic was examined using a combination of observations and simulations from an analytical and this multi-level primitive equation models with a free surface.[12,13] Both models indicated that the sea surface was depressed by about 20 cm from the undisturbed height with near-inertial oscillations of 3-5 cm in the numerical results (Figure 7). A depth-independent current oscillation of 0.11-0.13 m s^{-1} was found in the near-inertial wave band, which was consistent with the vertical velocity calculations at the sea surface and the observations. The spatial patterns of the depth-independent oscillations simulated by the numerical model followed the first baroclinic mode aplitudes with maximum values of about 0.45 ms^{-1} suggesting

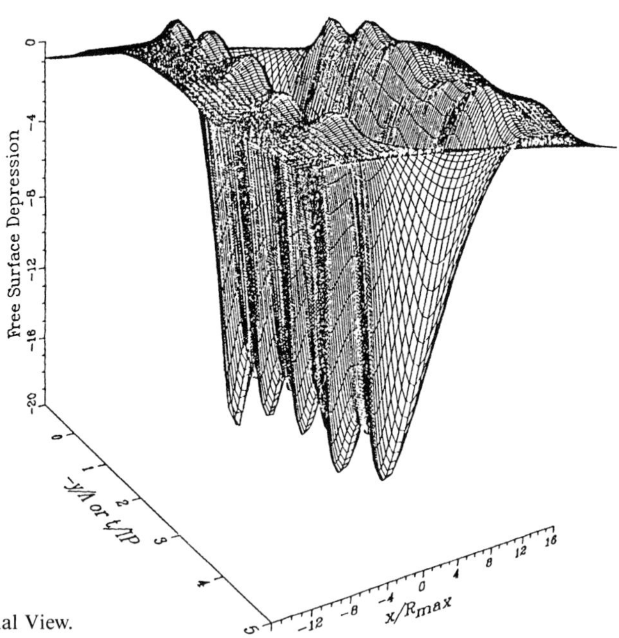

FIGURE 7.
Three Dimensional View.

interactions between the two modes. This depth-independent oscillation should not be confused with the barotropic geostrophic currents driven by the mean gradients in the sea surface depression that resembles an elongated racetrack like pattern.[13,29,55] The ratio of this barotropic amplitude (0.13 m s^{-1}) to the total mixed layer current (1.41 m s^{-1}) is about 0.09 compared to the value of the external Frounde number of 0.08. Thus, neglecting terms of O (0.1) in analytical models leads to the assumption of nondivergent flow ($\frac{d\eta}{dt}$ = 0), the elimination of the barotropic mode, and the conclusion that the sea surface oscillations are due to baroclinic, near-inertial motions.[29,57] This argument appears to violate the free surface boundary conditions given the value of unity for the barotropic vertical velocity compared to 10^{-6} for the baroclinic modal vertical velocities at the surface. The remaining science issues associated with the barotropic response are the temporal and spatial scales of the geostrophically-balanced depression, and the effect of abrupt bottom topography changes on the current oscillations.[51]

MESOSCALE INTERACTIONS

There have also been documented studies where tropical cyclones slowed down and became stationary for a period of time or interacted with pre-existing ocean features such as eddies and fronts. As a tropical cyclone slows down and becomes

FIGURE 7.
Plan View.

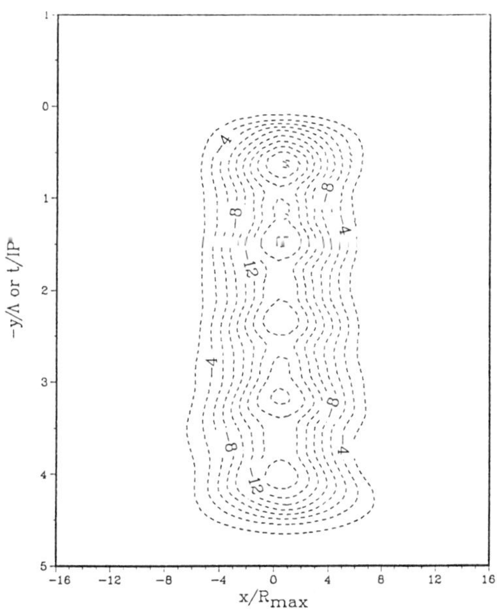

FIGURE 7. Three dimensional (left panel) and plan view (right panel) of the free surface height deviations (cm) relative to an undisturbed height simulated by a free surface model forced by a hurricane-like wind stress based on observed parameters from Frederic. A near inertial undulation of 3-5 cm was superposed on a broad sea surface depression with a maximum of 20 cm due to a barotropic current oscillation of about 10-11 cm s^{-1} in the wake of Frederic. The absissca and the ordinate are scaled by the inertial wavelength ($\Lambda = 580$ km) and the radius of maximum winds ($R_{max} = 30$ km) (from Shay et al., 1990).

stationary (Fr < 1) for more than an inertial period, strong geostrophically-balanced currents are excited due to the upwelling of cold water.[58] Typhoon Virginia was stationary for about 3 days, forcing a cyclonically-rotating cold-core eddy apparently with a ring structure and energy budget similar to those observed in the Gulf Stream.[34] Strong interactions between tropical cyclones Anita and Allen in the Gulf of Mexico with warm core eddies have also been observed.[2,11] The *RV Gyre* made extensive pre-storm XBT measurements in warm core eddy prior to the passage of hurricane Anita on 30 August 1977. Anita passed directly over this eddy at a speed of 2 m s^{-1} and induced extensive cooling that decreased the effect of the warm eddy. Hurricane Allen (1980) also passed over the edge of a warm core eddy in the western Gulf of Mexico;[11] however, the effect of the tropical cyclone on the eddy was not nearly as pronounced as in these other cases. Numerical simulations of the ocean response to Allen indicated a lack of correlation with the observations, which suggests that the background eddy field (via the ambient vorticity field) may have affected the hurricane-induced response.[51,59] Based on pre- and post-storm measurements, the Kuroshio current was weakened and displaced by the typhoon Mary (1968).[60] The oceanic response to typhoon Trix (1971) was investigated analytically by prescribing a pre-existing, geotrophically-balanced current field similar to the Kuroshio.[61] For translation speeds of less than 10 m s^{-1}, the interactions between the pre-existing and the typhoon-induced currents became increasingly more evident.

There was a warm core eddy located in the central Gulf of Mexico at about 5-6R_{max} to the right of the storm track during the passage of Gilbert.[49] In addition to the AVHRR images and the observed clockwise rotation of the current vectors from the AXCPs,[22] radar altimetry from the GEOSAT Exact Repeat Mission (not shown) detected this warm core eddy with an elevation of about 25 cm superposed on a depression skewed towards the right side of the storm track. This surface signature was consistent with observations from other field programs in the western Gulf of Mexico from previous years.[49] The physical mechanisms involved in the interactions between the tropical cyclone-induced response and pre-existing ocean features remains an important science issue requiring a combined observational and numerical approach.

SUMMARY

Recently acquired observations from research aircraft and vessels prior, during and subsequent to the passage of tropical cyclones are testament to the emerging technological capabilities that are now available. The acquisition of *in situ* observations are no longer fortuitous encounters with moored current meter moorings or drifting buoys. Rather, experiments can now be conducted from these platforms to address key scientific issues related to the oceanic response and coupled air-sea interactions in tropical cyclones. From a larger-scale perspective, remotely-sensed observations from the AVHRR and radar altimetry from the GEOSAT ERM (1986-1989) and the joint United States and France TOPEX/POSEIDON Mission offer effective methods to monitor SST changes and sea surface height anomalies in the directly-forced and evolving cold wakes of tropical cyclones on a Global basis.

These new types of observations provide the impetus to confirm or challenge previously held theories and notions about how the ocean responds to tropical cyclones. This blending of theories and models with the observations is critical to improve the understanding of these physical processes.

The salient features of the ocean's thermal and momentum response have been reviewed using a blend of observations, analytical expressions and numerical models in the directly forced and evolving wake regimes. The SST and MLD changes generally range from 1-5°C and 20-40 m depending upon the intensity and speed of the tropical cyclone and the underlying ocean stratification.[2] The empirical model of SST changes based on 10 years of observations agreed well with the observed patterns in Gilbert from the AVHRR and *in situ* observations.[22] The *rightward bias* of the SST and MLD response have been observed and modeled based on explanations of nonlinear interactions, wind field asymmetries,[40,41] and resonance excitation.[45] These effects have important consequences for feedback to the atmosphere by the air-sea fluxes, which may be important in the maintenance and intensification phases of tropical cyclones.

The upwelling and downwelling of the isopycnals in the thermocline are induced by the divergent and convergent near-inertial mixed layer currents. These cells tend to modulate vertical mixing processes associated with shear-instabilities and the lowering of the bulk Richardson number to less than unity. This does not rule out the importance of stress-induced turbulent mixing in the layer since there appears to be regimes within the oceanic response where each mechanism significantly contributes to the entrainment mixing. To investigate various types of vertical mixing parameterizations, numerical studies will now have to include the temperature and current profiles from AXCPs deployed in the ORTAH and Gilbert Experiments. That is, the numerical solution can no longer to tuned to agree with just a temperature profile, which places more rigorous constraints on the model physics to account for both the thermal and momentum balance. These profiles also have an important impact upon diagnostic studies to assess the relative roles of vertical mixing and advection on the thermal and momentum response over near-inertial time scales.

The strong atmospheric forcing and the horizontal pressure gradients induced by the isopycnal displacements excites near-inertially rotating currents in the mixed layer and thermocline. The divergent and convergent phases of the near-inertial currents occur over an along-track scale proportional to the product of the translation speed and the inertial period. The maximum near-inertial currents in the mixed layer have amplitudes of about 1-1.4 m s^{-1} with a large fraction (70%) of the observed near-inertial current response excited by tropical cyclones ascribed to linear dynamics and the first few baroclinic modes. While the low-order modes are more energetic by virtue of the superposition unto the wind stress and the replacement of the mixed layer depth by the equivalent forcing depth, the current shears at the base of the mixed layer are associated with higher-order baroclinic modes. The wind stress also forces a broad, undulating sea surface depression in the wake that is considerably larger than the wind stress curl region. This depression excites steady-state and near-inertially rotating currents, which are consistent with linear theories, numerical simulations, and *in situ* observations. While all modes are excited, it is indeed the gravest set of modes that significantly contributes to the ocean's momentum response.

Over the scales of the atmospheric forcing, the simultaneous deployment of aircraft expendables with measurements of the turbulent fluxes in the APBL[38,37] and remote sensing observations of ocean surface waves and winds by the NASA SCR,[46] C-SCAT[24] and SFMR[23] provide valuable measurements to understand the dynamical coupling between the two geophysical fluids when used in combination with coupled tropical cyclone - ocean models.[62] The satellite-based observations provide the larger scale context for these regional studies as well as monitoring the near- and far-fields on a global basis. This powerful combination of higher-resolution measurements with sophisticated numerical models will allow us to isolate the evolving 3-dimensional physical processes over regional to basin scales, which could only be inferred from single point measurements.

ACKNOWLEDGMENTS

The author gratefully acknowledges partial support from the National Science Foundation and the Office of Naval Research in the preparation of this manuscript.

REFERENCES

1. Palmen, E. 1948. On the formation and structure of tropical cyclones. *Geophysica.* 3:26-38.
2. Black, P.G. 1983. Ocean temperature changes induces by tropical cyclones. Ph.D. dissertation, The Pennsylvania State University, State College, PA, 278 pp.
3. Price, J.F. 1981. Upper ocean response to a hurricane. *J. Phys. Oceanogr.* 11:153-175.
4. Ellsberry, R.L., T. Fraim and R. Trapnell. 1976. A mixed layer model of the ocean thermal response to hurricanes. *J. Geophys. Res.* 81:1153-1162.
5. Geisler, J.E. 1970. Linear theory of the response of an ocean to a moving hurricane. *Geophys. Fluid Dyn.* 1:249-272.
6. Fedorov, K.N., A.A. Varfolomeyev, A.J. Ginsburg, A.G. Zatsepin, A. Yu. Krasnopevtsev, A.G. Ostrovskiy and V. Ye. Sklyarov. 1979. Thermal response of the ocean to the passage of hurricane Ella. *Oceanology.* 19:656-661.
7. Shay, L.K. and R.L. Elsberry. 1987. Near-inertial ocean current response to Hurricane Frederic. *J. Phys. Oceanogr.* 17:1249-1269.
8. Brink, K.H. 1989. Observations of the response of the thermocline currents to a hurricane. *J. Phys. Oceanogr.* 19:117-129.
9. Gill, A.E. 1984. On the behavior of internal waves in the wakes of storms. *J. Phys. Oceanogr.* 14:1129-1151.
10. Shay, L.K., R.L. Elsberry and P.G. Black. 1989. Vertical structure of the ocean current response to a hurricane. *J. Phys. Oceanogr.* 19:1249-1269.

11. Brooks, D. 1983. The wake of hurricane Allen in the western Gulf of Mexico. *J. Phys. Oceanogr.* 13:117-129.
12. Chang, S.W. 1985. Deep ocean response to hurricanes revealed by an ocean model with a free surface. *J. Phys. Oceanogr.* 15:1847-1858.
13. Shay, L.K., S.W. Chang and R.L. Elsberry. 1990. Free surface effects on the near-inertial current response to a hurricane. *J. Phys. Oceanogr.* 20:1405-1424.
14. Anthes, R.A. 1982. Tropical cyclones-their evolution, structure and effects. *Meteorlogical Monogram.* 19:American Meteorlogical Society. Boston, MA, 193 pp.
15. Sanford, T.B., P.G. Black, J. Haustein, J.W. Feeney, G.Z. Forristall and J.F. Price. 1987. Ocean response to hurricanes, Part I: Observations. *J. Phys. Oceanogr.* 17:2065-2083.
16. Price, J.F., T.B. Sanford and G.Z. Forristall. 1993. Forced stage response to a moving hurricane Part II. Dynamics and Modeling. *J. Phys. Oceanogr.* 24:233-260.
17. Leipper, D.F. 1967. Observed ocean conditions in hurricane Hilda. *J. Atmos. Sci.* 24:182-196.
18. Sanford, T.B., R.G. Drever, J.H. Dunlap and E.A. D'Asaro. 1982. Design, operation and performance of an expendable temperature and velocity profiler (XTVP). *Report APL-UW 8110, Applied Physics Lab.,* Univ. of Washington, Seattle, WA, 83 pp.
19. Black, P.G., R.L. Elsberry, L.K. Shay, R. Partridge and J.D. Hawkins. 1989. Hurricane Josephine surface winds and ocean response determined from air-deployed drifting buoys. *J. Atmos. Oceanic Tech.* 5:683-698.
20. Black, P.G., R.L. Elsberry and L.K. Shay. 1988. Airborne surveys of ocean current and temperature perturbations induced by hurricanes. *Adv. in Undersea Tech.* 16:51-60.
21. Voorhis, A.D. 1969. The horizontal structure of thermal fronts in the Sargasso Sea. *Deep-Sea Res.* 16:331-337.
22. Shay, L.K., P.G. Black, A.J. Mariano, J.D. Hawkins and R.L. Elsberry. 1992. Upper ocean response to hurricane Gilbert. *J. Geophys. Res.* 97:20,227-20,248.
23. Tanner, A., C.T. Swift and P.G. Black. 1987. Operational airborne remote sensing of wind speeds in hurricanes. *Proceedings of 17th AMS Conference on Hurricanes and Tropical Meteorology.* American Meteorological Society. Boston, Mass, 385-387.
24. McLaughlin, D.J., R.E. McIntosh, A. Pazmany, L. Hevizi and E. Boltniew. 1991. A C-Band scatterometer for remote sensing the air-sea interface. *IEEE Trans. Geosci. and Rem. Sensing.* 29:260-267.
25. Elsberry, R.L. 1989. ONR Tropical Cyclone Motion Research Initiative: Field Experiment Planning Workshop. NPS-63-89-002. Naval Postgraduate School, Monterey, CA, 93943, 79 pp.
26. Pudov, V., D., P.N. Svirkunov and V.P. Teslenko. 1991. *Proceedings 5th International Symposium on Tropical Meteorology.* USSR State Committee For Hydrometeorology. Institute of Experimental Meteorology, Obninsk, USSR, 23 pp.
27. Church, J.A., T.M. Joyce and J.F. Price. 1989. Current and density observations across the wake of typhoon Gay. *J. Phys. Oceanogr.* 19:259-265.

28. Hallock, Z.R. and W.J. Teague. 1990. SCTD Test: reliability and accuracy study (XTRAS). Naval Oceanographic and Atmospheric Research Laboratory Technical Note 69. Stennis Science Center, MS, 33 pp.
29. Ginis, I. and G. Sutyrin. 1993. Analytical and numerical studies of the hurricane-generated depth-averaged currents and sea surface effects. *Proceedings of the 20th Conference on Hurricanes and Tropical Meteorology.* American Meteorological Society. Boston, Mass, 574-578.
30. Powell, M.D. 1982. The transition of hurricane Frederic boundary layer wind field from the open Gulf of Mexico to landfall. *Mon. Wea. Rev.* 110:1912-1932.
31. Large, W.G. and S. Pond. 1981. Open ocean momentum flux measurements in moderate to strong winds. *J. Phys. Oceanogr.* 11:324-336.
32. Holland, G.J. 1980. An analytical model of the wind and pressure profiles in hurricanes. *Mon. Wea. Rev.* 108:1212-1218.
33. Pudov, V.D., A.A. Varfolomeyev and KN. Fedorov. 1978. Vertical structure of the wake of a typhoon in the upper ocean. *Oceanology.* 18:141-146.
34. Pudov, V.D. 1980. Mesostructure of the temperature and current velocity fields of a baroclinic ocean layer in the wake of typhoon Virginia. *Oceanology.* 20:8-13.
35. Ivanov, V.N. 1978. Certain results form the "Typhoon-75" expedition into the equatorial Pacific. *Soviet Meteorol. and Hydrol.* 1:8-16.
36. Frank, W.M. 1977. The structure and energetics of the tropical cyclone, II, dynamics and energetics. *Mon. Wea. Rev.* 105:1136-1150.
37. Friehe, C., W.J. Shaw, D.P. Rodgers, K.L. Davidson, W.G. Large, S.A. Stage, G.H. Crescenti, S.J.S. Khalsa, G.K. Greenhut and F. Li. 1991. Air-sea fluxes and surface layer turbulence around a sea surface temperature front. *J. Geophys. Res.* 95:8593-8610.
38. Bane, J.M. and K.E. Osgood. 1989. Gulf Stream air-sea interaction in the winter. *J. Geophys. Res.* 94:10,755-10,772.
39. Weller, R.A. 1991. Overview of the Frontal Air-Sea Interaction Experiment: a study of air-sea interaction in a region of strong oceanic gradients. *J. Geophys. Res.* 96:8501-8516.
50. Chang, S.W. and R.A. Anthes. 1978. Numerical simulations of the ocean's nonlinear, baroclinic response to translating hurricanes. *J. Phys. Oceanogr.* 8:468-480.
41. Greatbatch, R.J. 1983. On the response of the ocean to a moving storm: the nonlinear dynamics. *J. Phys. Oceanogr.* 13:357-367.
42. Martin, P.J. 1982. Mixed layer simulation of buoy observations taken during hurricane Eloise. em *J. Geophys. Res.* 87:409-427.
43. Stramma, L., P. Cornillon and J.F. Price. 1986. Satellite observations of sea surface cooling by hurricanes. *J. Geophys. Res.* 91:5031-5035.
44. O'Brien, J.J. 1967. The nonlinear response of a two-layer, baroclinic ocean to a stationary, axially symmetric hurricane: Part II. Upwelling and mixing induced by momentum transfer. *J. Atmos. Sci.* 24:208-215.
45. Price, J.F. 1983. Internal wave wake of a moving storm. Part I. Scales, energy budget and observations. *J. Phys. Oceanogr.* 13:949-965.

46. Walsh, E.J., D.W. Hancock III, D.E. Hines, R.N. Swift and J.F. Scott. 1985. Directional wave spectra measured by the surface contour radar. *J. Phys. Oceanogr.* 15:566-592.
47. Shay, L.K., E.J. Walsh and P.C. Zhang. 1994. Surface wave induced orbital velocities. *J. Atmos. and Oceanic Tech.* (in press).
48. Leaman, K.D. 1976. Observations of vertical polarization and energy flux of near-inertial waves. *J. Phys. Oceanogr.* 6:894-908.
49. Science Applications International Corporation. 1989. Gulf of Mexico physical oceanography program, Final Report: Year 5, Vol. II. *OCS Rep./MMS 89-0068.* Raleigh, NC, 325 pp.
50. Forristall, G.Z. 1980. A two-layer model for hurricane driven currents on an irregular grid. *J. Phys. Oceanogr.* 10:1417-1438.
51. Cooper, C. and J.D. Thompson. 1989. Hurricane generated currents on the outer continental shelf. Part 1: Model formulation and verification. *J. Geophys. Res.* 95:12,513-12,540.
52. Adamec, D., R.L. Elsberry, R.W. Garwood and R.L. Haney. 1981. An embedded mixed layer, ocean circulation model. *Dyn. Atmos. Oceans.* 6:69-96.
53. Garwood, R.W., Jr. 1977. An oceanic mixed layer model capable of simulating cyclic states. *J. Phys. Oceanogr.* 7:455-468.
54. Kundu, P. and R.E. Thomson. 1985. Inertial oscillations due to a moving front. *J. Phys. Oceanogr.* 15:1076-1084.
55. Hern, C.J. and P.E. Holloway. 1990. A three-dimensional barotropic model of the response of the Australian North West Shelf to tropical cyclones. *J. Phys. Oceanogr.* 20:60-80.
56. Hopkins, C. 1982. Ocean response to hurricane forcing. M.S. Thesis. Naval Postgraduate School. Monterey, CA 93943, 89 pp.
57. Cooper, C. and J.D. Thompson. 1989. Hurricane generated currents on the outer continental shelf. Part 1: Model sensitivity studies. *J. Geophys. Res.* 94:12,541-12,554.
58. Veronis, G. 1956. Parition of energy between geostrophic and nongeostrophic oceanic motions. *Deep Sea. Res.* 3:157-177.
59. Kunze, E. 1985. Near-inertial wave propagation in geostrophic shear. *J. Phys. Oceanogr.* 15:544-565.
60. Shuto, K. 1980. Influence of wind stress of a typhoon on the Kuroshio. *Kuroshio IV: Proceedings of the 4th Symposium for the Cooperative Study of Kuroshio and Adjacent Regions 14-17 Feb. 1979.* Japan Academy of Sciences, Tokoyo, Japan, 886-896.
61. Ichiye, T. 1977. Response of a two-layer ocean with a baroclinic current to a moving storm. Part I: Quasi-geostrophic Baroclinic Motion. *J. Oceanogr. Soc. Japan.* 33:151-160.
62. Chang, S.W. 1991. Development of a 3D coupled tropical cyclone-ocean model. *Proceedings of the 19th Conference on Hurricanes and Tropical Meteorology.* American Meteorological Society. Boston, Mass., 570-573.

The Oceans: Physical-Chemical Dynamics and Human Impact. Edited by S.K. Majumdar, E.W. Miller, G.S. Forbes, R.F. Schmalz and Assad A. Panah.

Chapter Nine

REMOTE SENSING OF INTERNAL WAVES IN THE OCEAN

JAMES H. CHURNSIDE and **ANDRE V. SMIRNOV**

National Oceanic and Atmospheric Administration
Environmental Research Laboratories
Boulder, CO 80303-3328

INTRODUCTION

Size is perhaps the most striking feature of the world's oceans. Any in situ measurements in this vast expanse are, quite literally, a drop in the ocean. Therefore, remote sensor measurements, especially those from satellites, are a very attractive alternative. Remote sensing of the oceans from the air, including photography, started in the 1950's.[1] Lidar, radar, radiometry, scatterometry, and synthetic aperture radar (SAR) were developed later. Most of these sensors have now been operated from satellites. These sensors can provide large quantities of information about conditions of the oceans and the atmosphere at the sea surface. Techniques for measuring sea surface height, sea surface temperature, ocean color, and ice cover are fairly well developed, and a number of fine books on these topics have been written.[2,3] Therefore, we do not include this information in the limited space available here.

Techniques using the sensors mentioned above are best for providing information about the surface layer of the ocean. Acoustic techniques are used for remote sensing of the interior of the ocean.[4] These depend on acoustic sources and receivers within the ocean, and perform poorly very near the surface because of surface reflections that make interpretation difficult. Thus, acoustic techniques tend to be complementary to the other techniques.

In this chapter, we discuss techniques for remote sensing of internal waves in the ocean. The emphasis is on attempts to observe physical processes in the interior of the ocean by measuring their effects on internal wave generation and propagation from above the surface. This is an exciting new field that promises to extend the usefulness of aircraft and satellite sensors.

IMPORTANCE OF INTERNAL WAVES

The most obviously important role of internal waves is in energy transfer and mixing in the ocean. The average energy density of internal waves is about 4000 J/m^2, which is about that of surface waves.[5] Thus, the role of internal waves in the dynamics of the ocean should be comparable to that of surface waves, although it is currently not as thoroughly understood.

Additionally, it should be possible to infer something of the profile of water density from a measurement of the dispersion of internal waves at several wavenumbers. As an example, consider an idealized two-layer model of the conditions found in a fjord during fresh-water runoff. We assume a thin layer of fresh water with thickness h and density ρ_{fresh} over a deep layer of salt water with density ρ_{salt}. The dispersion relationship for the lowest order mode is given by

$$\omega^2 = \frac{\Delta\rho}{\rho_{fresh}} ghk^2, \tag{1}$$

where ω is the frequency of the wave, $\Delta\rho$ is $\rho_{salt} - \rho_{fresh}$, g is the gravitational acceleration, and k is the magnitude of the horizontal wavenumber. This expression assumes long internal waves ($kh << 1$). Thus, the thickness of the layer can be estimated from measurement of the frequency and wavenumber of one internal wave. If the wave is being advected by a constant current, the current and the depth of the layer can be estimated from two different waves.

Under more realistic conditions, estimates of the vertical profiles of current and density can be made from measurements of the dispersion of several internal waves. In general, this is an ill-posed problem, and standard inversion techniques can be used. Also, one must make some assumptions about the relative influence of temperature and salinity on density in a particular situation to obtain either temperature or salinity profiles.

Another important role of internal waves is in sound propagation in the ocean. Surface waves, turbulence, tidal waves, inertial waves, and internal waves all cause fluctuations of propagating acoustic fields on different spatial and temporal scales. For sound that propagates along the sound channel without interacting with the surface or the bottom, internal waves can strongly affect the amplitude and phase.[6-8] This interaction is strong because the vertical displacement of the internal wave is greatest at the depth where the vertical gradient of density, and hence of sound speed, is the greatest. This produces the largest possible modulation of the sound speed. There has been some speculation that one consequence of this distortion of the ambient noise field in the ocean is that whales, which use the acoustic field for navigation, may become confused and end up beaching themselves. More research in this area is needed.

INTERNAL WAVE FIELDS

Internal waves are generated in the ocean by the vertical displacement of water within a density gradient. There are typically two regions of strong gradient that support such waves. The first is shallow, as in a seasonal thermocline from solar heating of the surface waters or a seasonal pycnocline from fresh-water runoff. In this region, internal waves dissipate relatively rapidly and the wave field is strongly dependent on local sources. Among the source mechanisms that have been extensively studied are motion of a ship through the water,[9] motion of water over an underwater obstruction,[10] nonlinear interactions with surface waves or current shears, and atmospheric disturbances at the surface.[11] Because they are near the surface, these waves generate relatively large modulations of surface roughness.

The other region is the permanent thermocline in the deep ocean. The distribution of internal waves associated with this gradient tends not to be dominated by local sources and sinks. In fact, measurements suggest a nearly universal spectrum of internal wave energy, with generally less energy at smaller scales. Garrett and Munk[5] proposed an energy spectrum of the form

$$E(k, l) = \frac{\frac{2}{\pi}\Lambda E N(z)\,\frac{1}{l_*}\,(\tau - 1)\left(1 + \frac{l}{l_*}\right)^{-\tau}}{N^2(z)\,k^2 + \Lambda^2 l^2}, \tag{2}$$

where k is the magnitude of the horizontal wavenumber, l is the vertical wavenumber, z is depth, Λ is the inertial frequency, $N(z)$ is the Brunt-Väisälä frequency, and $l_* = j_*\pi N(z)$. To fit available data, Garrett and Munk[5] proposed parameter values of $E = 6.3 \times 10^{-5}$, $\tau = 2.5$, and $j_* = 6$. The universality and shape of the spectrum suggest that it is caused by a myriad of weak interactions between current shears and existing internal waves such that an equilibrium is maintained.

The propagation of internal waves of any origin depends on the characteristics of the density gradient. In particular, propagation is described by the eigenvalue equation:[11]

$$\frac{d^2 W_n(z)}{dz^2} + \left[\frac{N^2(z)}{\omega^2} - 1\right] k^2 W_n(z) = 0, \tag{3}$$

where W_n is the n'th eigenmode of a disturbance in the vertical velocity that is propagating horizontally.

The Brunt-Väisälä frequency is given by

$$N(z) = \sqrt{-\frac{g}{\rho}\frac{\partial\rho}{\partial z}}, \tag{4}$$

where g is the gravitational acceleration and ρ is the density of the water, which depends on its temperature and salinity.

The appropriate boundary conditions are a hard surface ($W_n = 0$) at the bottom and a free surface at the sea surface. To first order, the second condition can be replaced with the condition of no displacement at the sea surface. The vertical displacement and horizontal velocity can be written, respectively, as

$$\zeta(\boldsymbol{r}, z, t) = \frac{i}{\omega} a_0 W_n(z) \exp[i(\boldsymbol{kr} - \omega t)] \tag{5}$$

and

$$u(\boldsymbol{r}, z, t) = \frac{i}{k} a_0 \frac{dW_n(z)}{dz} \exp[i(\boldsymbol{kr} - \omega t)], \tag{6}$$

where $\boldsymbol{r}$ is the 2-dimensional horizontal position, $\boldsymbol{k}$ is the horizontal wavenumber vector, and a_0 is the mode amplitude.

MANIFESTATIONS OF INTERNAL WAVES ON THE SEA SURFACE

Through nonlinear hydrodynamic interactions, internal waves affect surface waves of all sizes in the ocean, from capillary waves to large-scale swell. The vertical displacement of the surface induced by internal waves is negligible, and surface currents are induced by the vertical motion below. This creates regions of divergence a quarter wave in front of internal wave crests and convergence a quarter wave behind. It is this surface current that affects the surface waves, and several mechanisms have been proposed to explain this phenomenon.[12]

The first hypothesis was that surfactants were being swept toward the convergence zones. This would have a damping effect on small surface waves in those areas. However, the dominant effect is probably the direct modulation of surface waves by the internal-wave surface currents. This effect is easy to calculate. A small surface gravity wave with wavenumber K_0 will have a phase speed of $C = (g/K_0)^{1/2}$. If this wave then encounters a current of speed u in the same direction, the wavenumber will change to

$$K_1 = K_0 \left[\frac{1 + \left(1 + \frac{4u}{C}\right)^{1/2}}{2} \right]^{-1/2} . \tag{7}$$

At the same time, the amplitude will change to

$$A_1 = A_0 \left[\sqrt{\frac{K_0}{K_1}} \left(\sqrt{\frac{K_0}{K_1}} + 2\frac{u}{C} \right) \right]^{-1/2} , \tag{8}$$

where A_0 is the original amplitude. Thus, an opposing current (u negative) will decrease the wavelength and increase the amplitude of surface waves, and a following current will have the converse effects. The wavelength and amplitude effects act together to either increase or decrease the rms wave slope, or surface roughness.

For surface waves propagating in the same direction as an internal wave, the result will be an increase in surface roughness above the crest of the internal wave and a decrease in roughness above the trough. If the propagation is in the opposite direction, that relationship will be reversed. If there is some other angle between the internal and surface waves, only that component of the surface wave in the axis of the propagation of the internal wave is affected. As an example, we note that the current at any point on the surface, induced by an internal wave in the two-layer model discussed previously, is

$$u = -\zeta\left(\frac{g}{h}\frac{\Delta\rho}{\rho_{fresh}}\right)^{1/2}, \tag{9}$$

where ζ is the displacement of the pycnocline directly below the point of interest. Thus, the amplitude of internal waves can be inferred from measurements of the surface wave roughness.

There are other, more complex interactions that also probably have an influence. In the regions where the induced surface currents oppose the surface wave propagation, surface wave steepness is increased, which can lead to a higher density of breaking waves. This will increase the surface roughness there. It will also increase the levels of turbulence near the surface in those regions, which increases the attenuation of short surface waves and tends to decrease surface roughness there.

Another interaction is a resonant interaction in which some coherent combination of surface waves propagates with the same velocity as an internal wave, and energy can be transferred from the internal wave over long distances. The simplest case is a three-wave interaction between two surface waves, *S*1 and *S*2, and an internal wave. The resonance condition is

$$\begin{aligned} \boldsymbol{K}_{S1} - \boldsymbol{K}_{S2} &= \boldsymbol{k} \\ \omega(\boldsymbol{K}_{S1}) - \omega(\boldsymbol{K}_{S2}) &= \omega(\boldsymbol{k}) \end{aligned} . \tag{10}$$

These conditions can be satisfied for decimeter surface waves, which have group velocities comparable to the phase velocities of internal waves. They cannot generally be satisfied for the longer surface waves that contain most of the wind wave energy.

Four-wave resonance involving three surface waves and a single internal wave is not possible. There is no combination of the four wavenumbers and frequencies that satisfy conservation equations similar to Eq. (10). For longer waves, a five-wave resonance between four surface waves and an internal wave can be significant.[13]

Much more work is necessary to understand all of the interactions between surface waves and internal waves. One major obstacle is that there is only one good set of field data[14] with which to compare theoretical models.

REMOTE SENSING TECHNIQUES

Perhaps the most direct technique for remote measurement of internal waves is one in which the vertical motion of the water is tracked. Figure 1 presents a typical example from an acoustic sounder mounted on a ship traveling up Knight Inlet in British Columbia.[15] Modulation of the depth of scattering layers in the ocean by internal waves is evident in the figure. Similar scattering layers have been observed using airborne lidar.[16] To date, no clear internal waves have been reported in lidar data, but they should be observable by lidar.

Another acoustic technique relies on variations in sound speed created by propagation of internal waves. When acoustic energy propagates through a random field of internal waves, the amplitude and phase of the acoustic wave are distorted by the sound speed perturbations.[7,8] Measurements of the statistics of a received acoustic signal can be used to obtain information about the internal wave field between the source and the receiver.[17,18] In this approach, it is generally the spectrum of the internal waves that is of interest. Measured statistics of the acoustic waves include phase and amplitude variance and covariance. Analysis of spatiotemporal covariance can also be used to measure the current across the path.[17]

Perhaps the most common remote sensing of internal waves is by direct visual observation or photography. Figure 2 shows photographs of slicks, or areas with no high-frequency surface waves, taken from the air over Knight Inlet in British Columbia. These can generally be observed in moderately calm conditions (wind speeds of less than 7-8 m/s but not zero). When the winds are much stronger, surface conditions are dominated by the wind waves, and internal-wave modulations are swamped. If there is no wind to ripple the surface, internal waves are also not observable. Internal-wave-induced slicks can be discriminated from other types by their phase speed.

Each point on the surface reflects light to the observer; the amount depends on the position of the observer, the angle of the surface, and the illumination incident from the specular direction. Thus, an increase in surface roughness will increase or decrease the average brightness of an area, depending on these factors. Most commonly for observers near the surface, an increase in roughness produces an increase in brightness because light reaches the observer from nearer the solar direction in the sky. The surface roughness is commonly described in terms of the mean-square slope, which can be expressed[14] by

$$\delta^2 = \delta_0^2\left(1 + 2\frac{\alpha u}{c}\right), \tag{11}$$

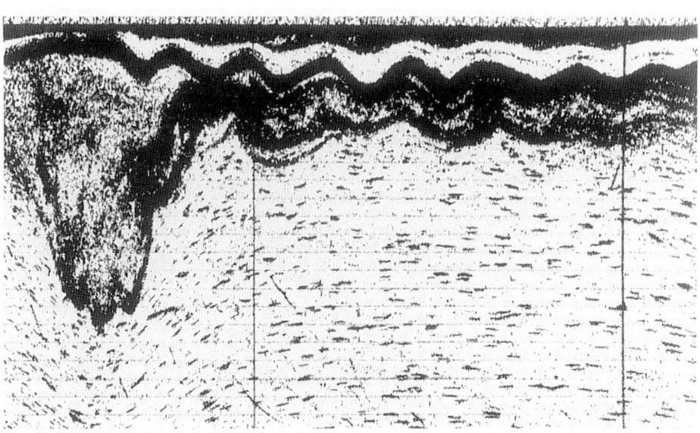

FIGURE 1. Acoustic echosounder record of internal waves in Knight Inlet, British Columbia.

where δ_0^2 is the mean-square slope when $u = 0$, c is the phase speed of the internal wave, and α is a dimensionless coefficient that depends on the wind velocity and c. The wavenumber and frequency of internal waves are easily obtained from optical images, especially if they contain reference objects such as ships or coastlines. Amplitude measurements are more difficult. Directional wave spectra for surface waves have been inferred from optical images with some success;[19,20] but these techniques have yet to be applied to internal wave studies.

Radiometric techniques, in both the infrared and microwave portions of the spectrum, can also be used for remote sensing of internal waves.[21] The relationships between internal waves and thermal emission from the surface are very complex. However, the dominant mechanism is through surface roughness.[22] One effect is that the emitting surface area within the field of view of a radiometer increases as the surface roughness increases. Also, the heat flux and temperature gradient near the surface are strongly affected by surface roughness. Since the observed thermal emission originates from a surface layer with a thickness of about a 0.1 to 1 mm in the microwave and about 10 μm in the infrared, the emitting temperature is

FIGURE 2. Optical images of internal waves in Knight Inlet, British Columbia. The top image was taken from a near-grazing view angle, and the bottom was taken looking straight down. The top image in Figure 2 was reproduced from Farmer, D.M. and J.D. Smith, 1978. Nonlinear Internal Waves in a Fjord, pp. 465-493. In: J. Nihoul (Ed.) *Hydrodynamics of Estuaries and Fjords.* Elsevier Scientific Publishing, Amsterdam, The Netherlands.

strongly affected by surface temperature gradients. Another effect is that the emissivity of the surface will be affected by the presence of surfactants, which are concentrated at internal wave convergence zones.

To date, the most useful remote sensing technique for the study of internal waves has been microwave radars that view the surface well away from the vertical. In this geometry, specular reflections from the surface can generally be neglected, and the primary interaction is Bragg scattering. This is a coherent addition of contributions from many waves when

$$K = \kappa(\sin\theta + \sin\phi), \tag{12}$$

where K is the surface wave wavenumber, κ is the microwave wavenumber, and θ and Φ are the angles of the incident and reflected beams to the normal of the surface without the wave. For the backscatter configuration typical of radar, θ and Φ are the same.

Generally, the intensity of the radar return depends on the amplitude of the Bragg-matched surface wave and not on overall surface roughness. However, breaking surface waves can generate facets that are perpendicular to the radar create brief specular reflections. In that case, internal waves can be observed through their interaction with breaking waves.

Several radar configurations have been used. The most straightforward is a real-aperture imaging radar. Figure 3 is a typical image from an airborne system; it was taken in the New York Bight with a near-grazing view angle. The bright spikes in the image are thought to be breaking waves. In order to increase image resolution, especially from space, SARs have also been used. Figure 4 is a typical satellite SAR image, also of the New York Bight. The wind speeds at the surface were 7-8 m/s when these two radar images were obtained.

A third type is Δk-radar,[23] which transmits two microwave frequencies such that the resulting microwave interference pattern on the surface matches the surface

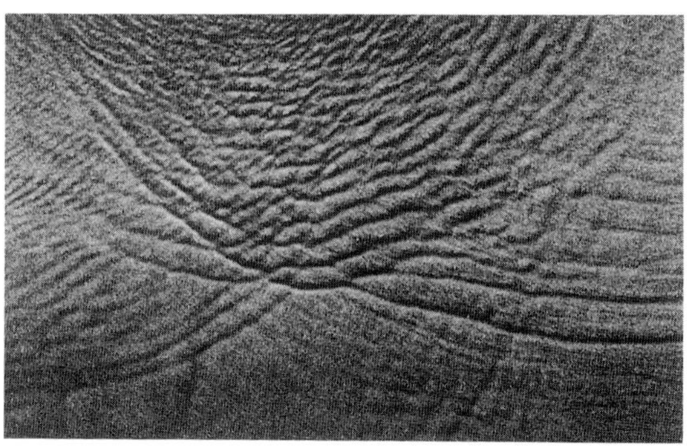

FIGURE 3. Real aperture radar image of internal waves in the New York Bight. The airborne radar was horizontally polarized with a wavelength of 2.25 cm and an incidence angle of 70°. The scale of the image is 1 cm = 2.7 km.

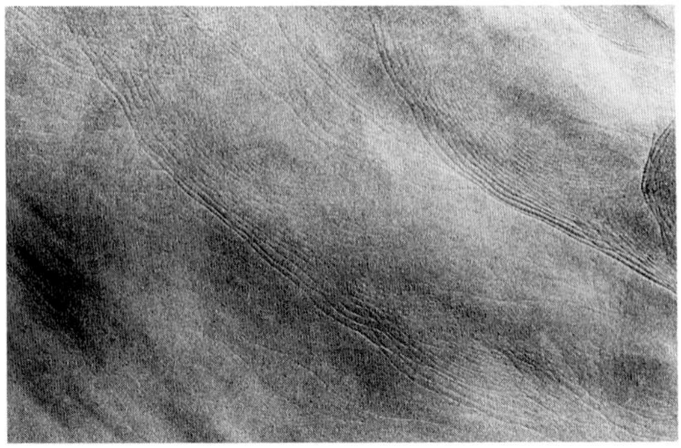

FIGURE 4. SAR image of internal waves in the New York Bight. The radar, operating from the ERS-1 satellite, was horizontally polarized with a wavelength of 5.7 cm and an incidence angle of 23°. The scale of the image is 1 cm = 13.3 km.

pattern of a particular internal wave. A large signal is generated when that particular internal wave is present, and no signal is obtained when it is not.

A theory describing the perturbation of the surface height spectrum by internal wave surface currents has been used to calculate the expected backscatter modulations associated with an internal wave. Gasparovic et al.[24] gave the equation for the wave-induced current $u(x,z)$ in the form

$$u(x,z) = -\zeta c\, sech^2[k(x-ct)]\frac{\partial W_1}{\partial z}. \tag{13}$$

The surface strain rate is obtained by computing a horizontal derivative of the velocity at the surface, with the result

$$\frac{\partial u}{\partial x}\Big|_{z=0} = 2\zeta kc\frac{\partial W_1}{\partial z}\Big|_{z=0} sech^2[k(x-ct)]\tanh[k(x-ct)], \tag{14}$$

where x is the surface position in the scattering plane.

The radar cross section is predicted by the Bragg model to be proportional to the value of the spectral density function of sea surface waves at the Bragg wavenumber. Under spatially homogenous and stationary conditions, the spectral density is the equilibrium density produced by the local winds, since short surface waves are strongly attenuated. Caponi, et al.[25] extended this model to allow for spatial modulation of the backscatter cross section by spatial variations in surface current. To first order, the spatial and temporal modulation of the radar backscatter cross section is given by

$$\frac{\Delta\sigma(\kappa_B, \mathbf{r}, t)}{\sigma_0(\kappa_B)} = \frac{\Delta F(\kappa_B, \mathbf{r}, t)}{F_0(\kappa_B)}, \tag{15}$$

where $\Delta\sigma$ is the change in the radar cross section from its unmodulated value of σ_0, ΔF is the change in the surface-wave spectral density from its unmodulated value of F_0, κ_B is the Bragg wavenumber, **r** is position on the surface, and t is time.

DISCUSSION

A variety of techniques exist for remote sensing of internal waves in the ocean. The most useful of these depend on the surface manifestations of internal waves. Just as a Wilson cloud chamber discloses the paths of energic particles, the sea surface discloses the paths of internal waves to observers above the surface. In this analogy, the sea surface ripple plays the role of the supersaturated atmosphere in the cloud chamber. In the Wilson cloud chamber, particle tracks are measured to provide information about the nuclear processes that created the particles. In a similar manner, we feel that it should be possible to use internal wave measurements to provide information about the dynamics of the interior of the ocean. Most of efforts in remote sensing of internal waves to date have concentrated on internal wave detection. Efforts to use these techniques to obtain information about other processes in the ocean have just begun.

ACKNOWLEDGEMENTS

We thank Dr. David Farmer and Ms. Grace Kamitakahara-King of the Canadian Institute of Ocean Sciences for the images of Knight Inlet. The images of the New York Bight were taken as part of a 1992 experiment sponsored by the U.S. Department of Defense.

REFERENCES

1. Ewing, G. 1950. Slicks, Surface Films and Internal Waves. *J. Marine Res.* 9:161-187.
2. Robinson, I.S. 1985. *Satellite Oceanography.* Ellis Horwood, Ltd., Chichester, pp. 455.
3. Gower, J.F.R. 1981. *Oceanography from Space.* Plenum Press, New York, pp. 978.
4. Munk, W.H. and C. Wunsch. 1979. Ocean Acoustic Tomography: A Scheme for Large Scale Monitoring. *Deep-Sea Res.* 26:123-161.
5. Garrett, C. and W. Munk. 1975. Space-Time Scales of Internal Waves: A Progress Report. *J. Geophys. Res.* 80:291-297.
6. Mooers, C.N.K. 1975. Sound Velocity Perturbations Due to Low Frequency Motions in the Ocean. *J. Acoust. Soc. Amer.* 57:1067-1075.
7. Desaubies, Y.J.F. 1978. On the Scattering of Sound by Internal Waves in the Ocean. *J. Acoust. Soc. Am.* 64:1460-1469.
8. Flatte, S.M. and R.B. Stoughton. 1986. Predictions of Internal-Wave Effects on Ocean Acoustic Coherence, Travel-Time Variance, and Intensity Moments for Very Long-Range Propagation. *J. Acoust. Soc. Am.* 84:1414-1424.
9. Watson, G., R.D. Chapman and J.R. Apel. 1992. Measurements of the Internal Wave Wake of a Ship in a Highly Stratified Sea Loch. *J. Geophys. Res.* 97:9689-9703.

10. Farmer, D.M. and J.D. Smith. 1980. Tidal Interaction of Stratified Flow with a Sill in Knight Inlet. *Deep-Sea Res.* 27A:239-254.
11. Phillips, O.M. 1977. *The Dynamics of the Upper Ocean, 2nd ed.* Cambridge University Press, New York, pp. 336.
12. Gargett, A.E. and B.A. Hughes. 1972. On the Interaction of Surface and Internal Waves, *J. Fluid Mech.* 52:179-191.
13. Mityagina, M.I., V.G. Pungin, A.V. Smirnov and V.S. Etkin. 1991. Changes in the Energy-Bearing Region of the Sea Surface Wave Spectrum in an Internal Wave Field Based on Remote Observation Data. *Izvestiya, Atmospheric and Oceanic Physics* 27:925-929.
14. Hughes, B.A. and H.L. Grant. 1978. The Effect of Internal Waves on Surface Wind Waves. 1. Experimental Measurements. *J. Geophys. Res.* 83,C1:443-454.
15. Hughes, B.A. and T.W. Dawson. 1988. Joint Canadian-U.S. Ocean Wave Investigation Project: An Overview of the Georgia Strait Experiment. *J. Geophys. Res.* 93:12219-12234.
16. Hoge, F.E., C.W. Wright, W.G. Krabill, R.R. Buntzen, G.D. Gilbert, R.N. Swift, J.K. Yungel and R.E. Berry. 1988. Airborne Lidar Detection of Subsurface Oceanic Scattering Layers. *Appl. Opt.* 27:3969-3977.
17. Farmer, D.M. and S.F. Clifford. 1986. Space-Time Acoustic Scintillation Analysis: A New Technique for Probing Ocean Flows. *IEEE J. Ocean. Eng.* OE-11:42-50.
18. Uscinski, B.J. 1986. Acoustic Scattering by Ocean Irregularities: Aspects of the Inverse Problem. *J. Acoust. Soc. Am.* 79:347-355.
19. Stillwell, D. Jr. and R.O. Pilon. 1974. Directional Spectra of Surface Waves from Photographs. *J. Geophys. Res.* 79:1277-1284.
20. Lubard, S.C., J.E. Krimmel, L.R. Thebaud, D.D. Evans and O.H. Shemdin. 1980. Optical Image and Laser Slope Meter Intercomparisons of High-Frequency Waves. *J. Geophys. Res.* 85:4996-5002.
21. Volkov, Yu. A. and Yu. M. Kuftarkov. 1980. Thermal Effect of Internal Gravity Waves on the Free Surface of the Ocean. *Izves. Atmos. Ocean. Phys.* 16:795-798.
22. Wu, S.T. and A.K. Fung. 1972. A Nonlinear Model for Microwave Emissions and Backscattering from the Sea Surface. *J. Geophys. Res.* 71:5917-5929.
23. Palmer, A.J. 1991. Surface Current Mapping Performance of Bistatic and Monostatic Δk Radars. *IEEE Trans. Geosci. Remote Sens.* 29:1014-1016.
24. Gasparovic, R.F., J.R. Apel and E.S. Kasischke. 1988. An Overview of the SAR Internal Wave Signature Experiment. *J. Geophys. Res.* 93:12304-12316.
25. Caponi, E.A., D.R. Crawford, H.C. Yuen and P.G. Saffman. 1988. Modulation of Radar Backscatter from the Ocean by a Variable Surface Current. *J. Geophys. Res.* 93:12249-12263.

The Oceans: Physical-Chemical Dynamics and Human Impact. Edited by S.K. Majumdar, E.W. Miller, G.S. Forbes, R.F. Schmalz and Assad A. Panah.

Chapter Ten

GLOBAL METEOROLOGICAL TELECONNECTIONS ASSOCIATED WITH EL NIÑO/SOUTHERN OSCILLATION EVENTS

DANIEL J. LEATHERS
Center for Climatic Research
Department of Geography
University of Delaware
Newark, DE 19716

INTRODUCTION

El Niño/Southern Oscillation (ENSO) events are complex phenomena that play an important role in the interannual variability of the global climate system. El Niño events are the oceanic component of the ENSO. A canonical El Niño event generally begins with the development of positive sea surface temperature (SST) anomalies along the coasts of Ecuador and Peru in December, with the anomalous warmth in this area reaching peak intensity in the spring.[1] Often, the warming of the oceanic temperatures along the coast of South America is associated with a large-scale warming of the tropical Pacific, with the area of anomalous warmth emcompassing the region from South America to the international dateline.[2] Temporally, the area of positive temperature anomalies migrates westward from the South American coast from the spring through the summer, reaching its peak areal extent during the subsequent winter season.[1]

The Southern Oscillation (SO), the atmospheric component of ENSO, involves a redistribution of atmospheric mass in the tropics and subtropics of the Pacific basin. This mass redistribution induces changes in surface air pressure, wind and temperature fields, cloudiness and precipitation within the tropics. In addition, this large-scale adjustment of the tropical atmosphere has been linked to changes in other meteorological variables across the planet. General characteristics of the SO have been known since the pioneering work of several authors early in this century.[3,4,5,6]

It was not until much later that Bjerknes[7] recognized the coupled nature of the atmosphere and ocean in the tropical Pacific and was able to hypothesize the physical linkages between the El Niño and the Southern Oscillation phenomena. At present, the interaction between the two is considered to be so synergistic in nature that they are referred to together as El Niño/Southern Oscillation events.

Teleconnections are statistical associations among meteorological variables that are separated by large distances.[8,9] In the case of ENSO events, a large number of studies have shown that ocean/atmosphere changes occurring in the tropical Pacific basin have ramifications that extend across the globe. Although there is still much discussion on the manner in which the changes in the tropical Pacific are "connected" to other areas, most studies suggest that large-scale variations in mid-tropospheric geopotential height patterns accompany ENSO events in the middle latitudes. Specifically, a mid-tropospheric thermal anomaly results from the release of latent heat in areas of anomalous convection that form in the environs of the central equatorial Pacific during most ENSO episodes. This localized heating anomaly is believed to lead to the propagation of Rossby waves into the mid-latitudes of the Northern and Southern Hemispheres perturbing the hemispheric geopotential height field.[10,11] Alternatively, in the tropics, this same anomalous heating source leads to a massive readjustment in the east-west Walker circulation,[12] the specifics of which will be discussed below.

It is the purpose of this chapter to review the nature of meteorological teleconnections which are associated with ENSO events. ENSO events are arguably the most intensely researched topic of the last two decades within the atmospheric sciences. Literally hundreds of papers detail all aspects of the phenomena. Therefore, it is impossible to review all the work that has been completed to date. Thus, papers cited within this chapter represent a small sampling of the material that has been published on this topic. The chapter will be partitioned geographically. First, tropical teleconnections will be discussed, with an emphasis placed on precipitation anomalies occurring throughout the region. Mid-latitude teleconnections will be addressed next, with a discussion of meteorological changes that occur in the Northern and Southern Hemispheres during ENSO events. Finally, a brief discussion of ENSO forecasting efforts will be presented.

TROPICAL TELECONNECTIONS

The Walker Circulation

Bjerknes[7] was the first to use the term Walker circulation in describing the large-scale east - west atmospheric circulation that spans the equator. Within the Pacific basin, the average atmospheric circulation along the equator includes an intense area of convection centered in the area of Indonesia. This center of large-scale upward motion in the atmosphere, and its associated center of mid-tropospheric latent heat release, is a major source of energy within the global climate system. This region of upward motion is complemented by a region of strong subsidence in the eastern portions of the Pacific basin, to the west of South America. This vertical motion field induces relative low surface pressures in the western Pacific and relative high pressure in the eastern portion of the basin. Between these two centers, easterly winds dominate at the surface with a westerly return flow at upper tropospheric levels.[13,14] The area of large-scale ascent in the troposphere is associated with very high SSTs at the surface in the Indonesian region. In fact, this portion of the Pacific Ocean has the highest SSTs found on the planet. The area of subsidence is located over the relatively cool waters of the eastern equatorial Pacific. Bjerknes[7] believed that the Walker circulation was a result of this SST gradient. However, more recent work suggests a synergistic relationship between the surface wind field in the tropics and the SST distribution.[2]

During ENSO events, the region of highest SSTs are located near the international dateline in concert with the area of strongest convection. This results in a decrease in convection in the western Pacific and a decrease in precipitation. The areas near the international dateline, and farther to the east, receive greater precipitation as a result of enhanced convective activity. In addition, the surface and upper tropospheric wind fields over the Pacific basin are modified. Westerly wind anomalies are prevalent at the surface (resulting in a decrease in the magnitude of the normal low-level easterly flow), while the upper-level west to east circulation is also weakened. This realignment of the Walker circulation in the Pacific basin results in fluctuations of the tropical circulation across the globe. Moreover, the movement of the area of strongest convection to the central Pacific results in a mid-tropospheric heating anomaly in that region that initiates changes in extratropical large-scale geopotential height fields, especially in the winter hemisphere.[15]

Tropical Precipitation Anomalies During ENSO

Many studies have established that important changes in tropical rainfall patterns occur during ENSO events.[16,17] Following the convention of Rasmusson and Carpenter,[1] in the discussion to follow, year (-1) will refer to the period prior to the initial December/January warming along the South American coast, year (0) will refer to the twelve month period in which SST anomalies peak along the South American coast and begin their migration across the equatorial Pacific and year (+ 1) will refer to the period following the peak SST warming in the central Pacific basin. Much of the information to be presented in this chapter is derived from the work of Ropelewski and Halpert[16] who investigated precipitation anomalies associated with ENSO events using monthly data from across the globe. Figure 1 is a composite map indicating areas across the planet that evidence diverse changes in meteorological variables associated with ENSO events, while Table 1 indicates the nature of those changes. This map is an expansion of Figure 21 from Ropelewski and Halpert.[16]

Within the Pacific basin, the perturbation of the Walker circulation by ENSO results in extremely dry conditions for the area of the Maritime continent including Indonesia, New Guinea and the surrounding islands (see area A, Figure 1). Specifically, the time period from the summer of year (0) to the spring of year (+ 1) is much dryer than normal as a result of the movement of the most intense convection away from the area.[16,18] This has often led to extreme drought conditions in these areas producing far-reaching impacts on the societies involved.[19] The same perturbation in the Walker circulation brings positive precipitation anomalies to the islands of the central Pacific; specifically from the summer of year (0) to the summer of year (+ 1; see area B, Figure 1).

The northeastern portion of South America is prone to alternate periods of flood and drought (see area C, Figure 1). Caviedes[20] noted that droughts in this region

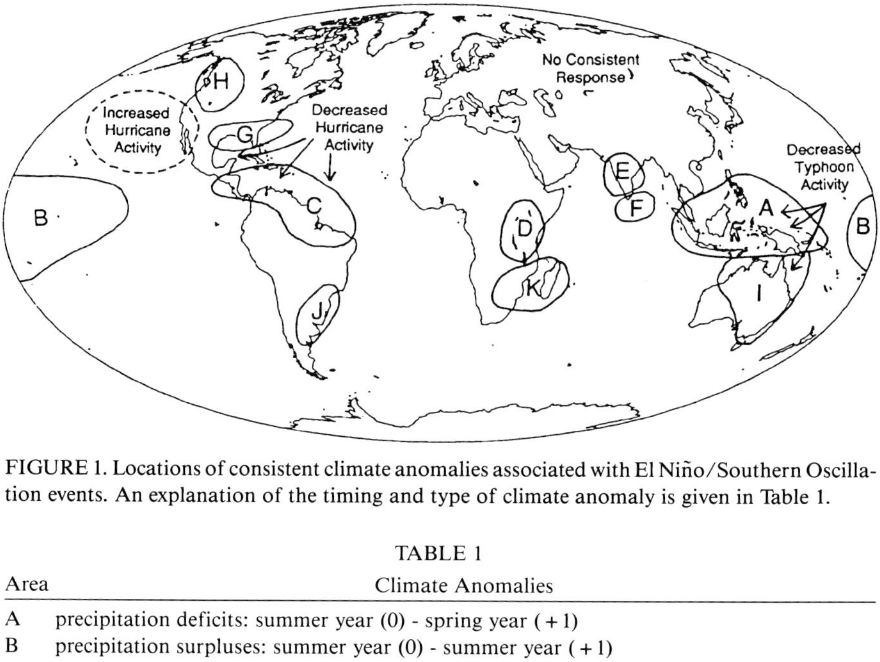

FIGURE 1. Locations of consistent climate anomalies associated with El Niño/Southern Oscillation events. An explanation of the timing and type of climate anomaly is given in Table 1.

TABLE 1

Area	Climate Anomalies
A	precipitation deficits: summer year (0) - spring year (+ 1)
B	precipitation surpluses: summer year (0) - summer year (+ 1)
C	precipitation deficits: summer year (0) - spring year (+ 1)
D	precipitation surpluses: autumn year (0) - spring year (+ 1)
E	precipitation deficits: June year (0) - September year (0) {deficits occur within the summer monsoon season}
F	precipitation surpluses: autumn year (0) - January year (+ 1)
G	precipitation surpluses: autumn year (0) - spring year (+ 1) temperatures below normal: autumn year (0) - spring year (+ 1)
H	temperatures above normal: autumn year (0) - spring year (+ 1)
I	precipitation deficits: summer year (0) - spring year (+ 1)
J	precipitation surpluses: autumn year (0) - spring year (+ 1)
K	precipitation deficits: November year (0) - spring year (+ 1)

often occur during times of positive SST anomalies along the coast of Ecuador and Peru, while several other investigators have studied the relationship between rainfall in this region and canonical ENSO events.[21,22,23] In general, a rainfall deficiency is noted in this area from the summer of year (0) to the late spring of year (+ 1).[16] The relationship between ENSO and drought is very robust and is likely to be a result of the large-scale perturbation of the Walker circulation.[24]

Just to the north in the subtropics, an area encompassing portions of Central America and the southern Caribbean evidences a consistent precipitation anomaly with ENSO events (see area C, Figure 1). Rogers[25] and Ropelewski and Halpert[16] have studied the precipitation/ENSO relationship in this region. Both investigations suggest that summer precipitation during year (0) of ENSO is below normal across this area, similar to conditions at the time in northeast South America. The physical mechanisms underlying this relationship are not well understood but are probably tied in some way to the mechanisms that are responsible for precipitation anomalies that occur in northeast South America at the same time.

Along the eastern margin of the Atlantic basin, eastern equatorial Africa also evidences a strong association between precipitation and ENSO (see area D, Figure 1). In this case, the relationship is one of above normal precipitation between the autumn of year (0) and the late spring of year (+ 1). The relationship is quite robust, and is linked to an equatorward movement of a branch of the Intertropical Convergence Zone (ITCZ) in response to modifications of the Walker circulation.[16]

Although the Indian subcontinent is not wholly contained within the tropics, relationships between summer monsoonal precipitation and ENSO will be addressed in this section because fluctuations in the global equatorial Walker circulation are a likely mechanism linking the two phenomena (see area E, Figure 1). The summer southwest monsoon is the main precipitation producer for most of the Indian subcontinent. The rainy season typically lasts from June to September, during which 80% of the annual precipitation falls over a large area of the country. Thus it is clear that a failure of the monsoon, or a change in its timing, can have dramatic impacts on the population.[26]

Walker and Bliss[5] were the first to formally link precipitation anomalies in India to large-scale atmospheric changes in the Pacific basin. Since that time a number of authors have investigated the relationship using diverse methodologies and data sources.[27,28,29,16,17] In general, the evidence indicates a weakening of the summer monsoon circulation during ENSO years which leads to negative precipitation anomalies. All studies point to a tendency for monsoonal rains over the subcontinent to be below normal during the summer monsoon season of year (0). In fact, the four driest monsoon seasons on record correspond to strong ENSO events as identified by Quinn et al.[30] Ropelewski and Halpert[16] found a general decrease in monsoonal rains over the subcontinent from June to September of year (0), with twenty-one of 26 ENSO events responding in this general manner. It is important to note that several other large-scale forcing mechanisms may be related to the strength and timing of the summer monsoon. These additional factors include the extent and depth of Eurasian snow cover,[31] 40 to 50 day atmospheric oscillations in the tropics[32,33] and regional circulation anomalies during the period just before the onset of the monsoon.[34] In extreme southern India and Sri Lanka, a different relationship is found (see area F, Figure 1). In this region, the winter monsoon rainfall during year (0) of an ENSO event is characterized by above normal precipitation in 21 of 26 ENSO episodes.[16]

Impacts on Tropical Cyclones

Tropical cyclone activity is also affected by ENSO events (see Figure 1).[35,36,37,38,39] In the Atlantic basin, Gray[35] has shown that strong westerly winds at upper tropospheric levels act to inhibit the formation of tropical storms and hurricanes through the enhancement of vertical wind shear. These adverse upper tropospheric conditions are prevalent during ENSO years. Observational evidence suggests that the number of hurricanes in the Atlantic basin is greatly reduced during year (0) of ENSO.[40] In the northwest Pacific/Australian basin Dong[41] has shown that cyclone activity is significantly reduced during ENSO events while Hastings[39] has shown that the primary area of cyclone development moves from the northeast coast of Australia toward the international dateline. Interestingly, tropical cyclone activity in the northeast Pacific basin (coastal Mexico) seems to intensify during ENSO years and tropical cyclones are likely to be more numerous in this region.[40] In addition, cyclones that form in the northeast Pacific basin during ENSO years are likely to move farther west than normal, in some cases as far as the Hawaiian Islands.[40]

NORTHERN HEMISPHERE TELECONNECTIONS

Mid-Tropospheric Height Patterns

Perturbations of extratropical mid-tropospheric height fields are a well known consequence of ENSO events. This fact has been verified by theoretical and modeling works[11,15,42,43,44,45] and by observational studies.[7,46,47,48,49] Most investigations have shown that perturbations in the geopotential height fields are a result of a mid-tropospheric heating anomaly that develops in association with anomalous convection in the portion of the equatorial Pacific basin near the international dateline.[50] The heating anomaly is associated with large releases of latent heat in the area of anomalous convection. Rossby waves are able to disperse away from the mid-tropospheric equatorial heating anomaly, into mid- and high-latitude areas, in regions of mean upper-level westerly winds which are generally found in the winter hemisphere.[11]

Arguably the most persistent pattern of mid-tropospheric height anomalies associated with ENSO events is the Pacific/North American (PNA) teleconnection pattern[8,9] (Figure 2). This pattern of mid-tropospheric height anomalies is characterized by an anomalous trough in the Aleutian region, a strong ridge over the Rocky mountains and a deeper than normal trough over eastern North America.[9]

This pattern of variation has been recognized as the strongest mode of interannual variability during the Northern Hemisphere winter[51] and it occurs on time scales ranging from days to seasons. Yarnal and Diaz[52] have shown that the PNA pattern occurs 54% of the time during the winter months that span year (0) and year (+ 1) of an ENSO event.

Leathers et al.[9] have constructed a climatology of the correlation between an index that represents the strength and phase of this teleconnection pattern and surface temperature and precipitation across the United States. They found strong correlations with both variables during the winter season. Strong PNA patterns result in lower than normal temperatures across the southeast United States and positive temperature anomalies across the Pacific northwest. Precipitation correlations are not strong, however, there is a tendency for reduced precipitation amounts in both the intermountain west and the Ohio river valley during the winter season.

While the positive phase of the PNA pattern is quite common during ENSO events, it is important to note that the extratropical response to ENSO events across North America varies greatly from event to event. Thus, a response to ENSO in the form of a positive PNA pattern is only one of a suite of geopotential height perturbations that lead to changes in meteorological parameters across the North American continent. This can be seen by inspection of Figure 3, which shows 500 mb height departures associated with January and February months during year (+ 1) of ENSO events. An indication of a PNA-like pattern is clear, however, the departure pattern indicates a mixing of diverse circulation modes. Thus, the mid-latitude response to an ENSO episode is dependent upon the specific characteristics of a given event.

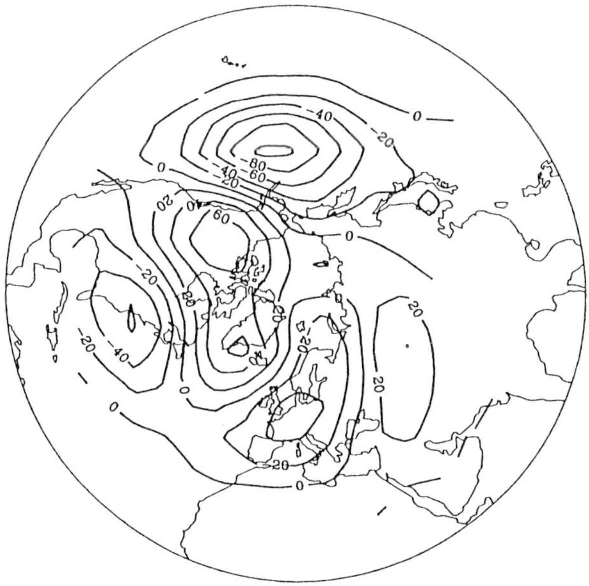

FIGURE 2. January/February 500 mb height anomalies (gpm) associated with values of the Pacific/North American Teleconnection Index greater than 1.0.

The ENSO Signal in North America

Across the North American continent several areas experience both temperature and precipitation anomalies during ENSO events. Ropelewski and Halpert[53] used monthly temperature and precipitation data to investigate the association between ENSO and the surface climate. For temperature, they found a pattern of anomalies consistent with the positive phase of the PNA teleconnection pattern during the time period from December of year (0) to March of year (+ 1) (see areas G and H, Figure 1). This pattern is represented by lower than normal temperatures across the southeast United States and higher than normal temperatures over the northwest United States and southwest Canada.

Although the precipitation response was not as strong, consistent ENSO related anomalies were found over the Gulf Coast region of the United States and the Great Basin.[53,54] Positive precipitation anomalies were found in stations near the Gulf Coast from October of year (0) to March of year (+ 1). This precipitation increase is likely to be the result of an active sub-tropical jet and increased baroclinicity in the region, which may be associated with the positive phase of the PNA pattern. The increased baroclinicity may lead to the development of more storms in the gulf with a stronger than normal subtropical jet enhancing the moisture advection into the area. In the Great Basin, positive precipitation anomalies are found from April of year (0) to October of year (0). This relationship is difficult to explain; however, it is robust with 9 of 11 ENSO events associated with above normal precipitation in the area.[16]

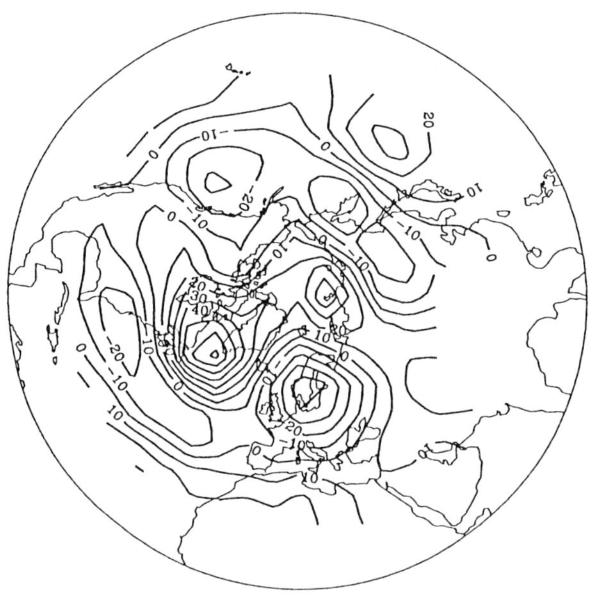

FIGURE 3. January/February 500 mb height anomalies (gpm) associated with year (+ 1) of ENSO events during the period 1947 through 1989.

Schonher and Nicholson[55] have shown that a relationship exists between ENSO events and precipitation in California. This association varies according to the strength and position of the SST anomalies that are associated with a given ENSO episode. In comparing California rainfall to the ENSO classification scheme of Fu et al.,[56] the authors found that ENSO events characterized by large positive temperature anomalies east of the international dateline correspond with abnormally high rainfall amounts in California, especially in the southern portions of the state. During events in which SSTs are of a moderately anomalous character, the response in California is one of normal to slightly dry conditions. Finally, when the ENSO event is characterized by weak SST anomalies that are mainly confined to the eastern Pacific, near the South American coast, the response in California is one of widespread drought. Moreover, the results indicated that the specific ENSO response depends upon the exact location within the state, with the strength of all relationships decreasing to the north.

SOUTHERN HEMISPHERE TELECONNECTIONS

Ropelewski and Halpert[16] have identified three regions within the southern Hemisphere that evidence consistent ENSO related precipitation departures. In general, the entire continent of Australia experiences a decrease in precipitation during ENSO events (see area I, Figure 1). Specifically, northern and eastern sections of the continent experience below normal precipitation from the autumn of year (0) to the spring of year (+1). In northern Australia, the unusually dry conditions are related to a disruption of the typically wet winter monsoon season. However, this explanation does not explain unusually dry conditions along the entire latitudinal extent of eastern Australia. The central and southern portions of the continent (including Tasmania) do not show as strong or consistent a pattern of deficient rainfall as the north and east. However, precipitation is usually below normal in the central desert regions from the spring of year (0) through the spring of year (+1), while the extreme southeast and Tasmania evidence precipitation deficits mainly during the summer of year (0).

On the eastern side of the Pacific basin, southern Brazil, Uruguay and northern Argentina experience wetter than normal conditions from the autumn of year (0) through the early spring of year (+1; see area J, Figure 1).[16] Although the exact cause of the precipitation enhancement is not known with assuridy, it is believed to be related to a modification of the upper-level westerly wind field in the area.[16]

In southeastern Africa a decrease in rainfall from November of year (0) to the late spring of year (+1) is noted (see area K, Figure 1).[16] This area of southern Africa experiences dry conditions, during ENSO events, at the same time that eastern Equatorial Africa experiences above normal precipitation. This pattern of precipitation anomalies is related to an equatorward shift of the ITCZ that brings enhanced wetness to the equatorial regions while regions to the south are left abnormally dry.

THE USE OF GEOPHYSICAL AND STATISTICAL MODELS IN ENSO FORECASTING

The utility of an accurate forecast of an ENSO event is clear given the ubiquitous nature of the resulting teleconnections. There have been several efforts to model the ENSO phenomena using standard geophysical modeling techniques.[57,58,59,60,61,62] These efforts have been conducted in order to diagnose the relevant physics involved in the ENSO phenomena, and for use in the potential projection of ENSO events. All of these models have met with varying success in the simulation of the most important characteristics of ENSO events; primarily SST and wind field anomalies. In addition, several authors have used statistical modeling techniques in order to study the underlying mechanisms of ENSO and its possible predictability.[63,64]

At present, the Climate Analysis Center (CAC) of the National Oceanic and Atmospheric Administration (NOAA) publishes monthly projections from both a statistically based model[64] and a geophysical model developed by Cane et al.[58] and Barnett et al.[62] The statistical model of Barnston and Ropelewski[64] uses canonical correlation analysis (CCA) in order to forecast ENSO fluctuations. This model is used operationally to forecast the 3-month mean SST across diverse areas of the Pacific basin. The model can be used to project SST conditions for a number of different periods into the future. The CCA uses sea-level pressure data from a large portion of the globe and SST data from the tropical Pacific as predictor variables. The method has shown considerable skill at predicting the SST condition in the eastern portion of the Pacific basin. In fact, it accurately predicted the most recent ENSO event that began during the winter of 1991-92.[64]

The geophysical model of Cane et al.[58] and Barnett et al.[62] employs a coupled atmosphere-ocean model to simulate the ENSO phenomena. Specifically, in a predictive mode, the model will project SST conditions in the eastern portion of the Pacific basin encompassing the area from 90°W to 180°W longitude. These projections are made for lead times of 0 to 15 months. This model has also shown skill, above persistence, in forecasting SST conditions several seasons in advance.

DISCUSSION

Much of the interannual variability of the global climate system is associated with ENSO events. The material presented above indicates that meteorological teleconnections associated with these episodes are scattered across the globe, and include precipitation and temperature anomalies and changes in the patterns of extratropical and tropical cyclogenesis, all of which are a result of global-scale perturbations of the atmospheric circulation.

There is significant evidence to suggest that ENSO events have been an important component of global climatic variability during the last several hundred years.[30,65] However, additional evidence indicates that the relationship between ENSO events and some meteorological teleconnections are not stable over time.[66,67] This fact would suggest that ancillary changes within the climate system modulate the relationship between ENSO events and the meteorological response in certain areas of the planet over long time periods. Thus, the relationship between ENSO events and regional climates may be very different in a world in which the climatological mean conditions are dissimilar from today (i.e. a doubled CO_2 environment). Another important point to note is that the magnitude and spatial distribution of the SST anomalies during a specific event play an important role in the exact specification of regional climate perturbations.[55,56] Even with these potential problems in mind, in the future an increased accuracy in the prediction of ENSO events may be used, along with the relationships discussed above, to infer, at least in a gross manner, the timing and magnitude of important regional climate anomalies. Such "predictions" may help to ameliorate the societal disruption that can result from large climatic fluctuations.

REFERENCES

1. Rasmusson, E.M. and T.H. Carpenter. 1982. Variations in Tropical Sea Surface Temperature and Surface Wind Fields Associated with the Southern Oscillation/El Niño. *Mon. Wea. Rev.* 110:354-384.
2. Trenberth, K.E. 1991. General Characteristics of El Niño - Southern Oscillation, pp. 13-42. In: M. Glantz, R.W. Katz and N. Nicolls (Ed.) *Teleconnections Linking Worldwide Climate Anomalies.* Cambridge University Press, New York, NY, pp. 535.
3. Walker, G.T. 1923. Correlation in Seasonal Variations of Weather, IX: A Further Study of World Weather. *Mem. Indian. Meteor. Dept.* 24:275-332.
4. Walker, G.T. and E.W. Bliss. 1930. World Weather IV. *Mem. Roy. Meteor. Soc.* 3:81-95.
5. Walker, G.T. and E.W. Bliss. 1932. World Weather V. *Mem. Roy. Meteor. Soc.* 4:53-84.
6. Walker, G.T. and E.W. Bliss. 1937. World Weather VI. *Mem. Roy. Meteor. Soc.* 4:119-139.
7. Bjerknes, J. 1969. Atmospheric Teleconnections from the Tropical Pacific. *Mon. Wea. Rev.* 97:163-172.
8. Wallace, J.M. and D.S. Gutzler. 1981. Teleconnections in the Geopotential Height Field during the Northern Hemisphere Winter. *Mon. Wea. Rev.* 109:784-812.
9. Leathers, D.J., B. Yarnal and M.A. Palecki. 1991. The Pacific/North American Teleconnection Pattern and United States Climate. Part I: Regional Temperature and Precipitation Associations. *J. Climate* 4:517-528.
10. Horel, J.D. and J.M. Wallace. 1981. Planetary Scale Atmospheric Phenomena Associated with the Southern Oscillation. *Mon. Wea. Rev.* 109:813-829.
11. Webster, P.J. 1981. Mechanisms Determining the Atmospheric Response to Sea Surface Temperature Anomalies. *J. Atmos. Sci.* 38:554-571.
12. Lim, H. and C.-P. Chang. 1983. Dynamics of Teleconnections and Walker Circulations Forced by Equatorial Heating. *J. Atmos. Sci.* 40:1897-1915.
13. Krishnamurti, T.N. 1971. Tropical East-West Circulations During the Northern Summer. *J. Atmos. Sci.* 28:1342-1347.

14. Krishnamurti, T.N., N. Kanamitsu, W.J. Koss and J.D. Lee. 1973. Tropical East-West Circulations During the Northern Winter. *J. Atmos. Sci.* 30:780-787.
15. Webster, P.J. 1982. Seasonality in the Local and Remote Response to Sea Surface Temperature Anomalies. *J. Atmos. Sci.* 39:41-52.
16. Ropelewski, C.F. and M.S. Halpert. 1987: Global and Regional Scale Precipitation Patterns Associated with the El Niño/Southern Oscillation. *Mon. Wea.* Rev. 115:1606-1626.
17. Kaladis, G.N. and H.F. Diaz. 1989. Global Climatic Anomalies Associated with Extremes in the Southern Oscillation. *J. Climate* 2:1069-1090.
18. Berlage, H.P. 1957. Fluctuations in the General Atmospheric Circulation of more than One Year, their Nature and Prognostic Value. *Mededlingen en Verhandelingen No. 69.* Koninklijk Nederlands Meterologische Institut, Staatsdrukkerijs-Gravenhage, Nederlands.
19. Malingreau, J.-P. 1987. The 1982-83 Drought in Indonesia: Assessment and Monitoring, pp. 11-18. In M. Glantz, R. Katz and M. Krenz (Ed.) *The Societal Impacts Associated With The 1982-83 Worldwide Climate Anomalies.* Environmental and Societal Impacts Group, National Center for Atmospheric Research, Boulder, CO, pp. 105.
20. Caviedes, C.N. 1973. Secas and El Niño: Two Simultaneous Climatic Hazards in South America. *Proceedings of American Geography* 5:44-49.
21. Markham, C.G. and D.R. McLain. 1977. Sea Surface Temperature Related to Rain in Ceara, Brazil. *Nature* 265:320-323.
22. Hastenrath, S. and L. Heller. 1977. Dynamics of Climatic Hazards in Northeast Brazil. *Quart. J. Roy. Meteor. Soc.* 103:77-92.
23. Hastenrath, S., M.-C. Wu and P.-S. Chu. 1984. Toward the Monitoring and Prediction of Northeast Brazil Droughts. *Quart. J. Roy. Meteor. Soc.* 110:411-425.
24. Kousky, V.E., M.T. Kagano and I.F.A. Cavalanti. 1984. A Review of the Southern Oscillation: Oceanic-Atmospheric Circulation Changes and Related Rainfall Anomalies. *Tellus* 36A:490-502.
25. Rogers, J.C. 1988. Precipitation Variability Over the Caribbean and the Tropical Americas Associated with the Southern Oscillation. *J. Climate* 1:172-182.
26. Kaladis, G.N. and S.K. Sinha. 1991:ENSO, Monsoon and Drought in India, pp. 431-458. In: M. Glantz, R.W. Katz and N. Nicholls (Ed.) *Teleconnections Linking Worldwide Climate Anomalies.* Cambridge University Press, New York, NY, pp. 535.
27. Angell, J.K. 1981. Comparison of the Variations in Atmospheric Quantities with Sea Surface Temperature Variations in the Equatorial Eastern Pacific. *Mon. Wea. Rev.* 109:230-243.
28. Shukla, J. and D. Paolino. 1983. The Southern Oscillation and Long-Range Forecasting of Summer Monsoon Rainfall over India. *Mon. Wea. Rev.* 111:1830-1837.
29. Mooley, D.A. and B. Parthasarathy. 1983. Indian summer Monsoon and El Niño. Pageoph 121:339-352.
30. Quinn, W.H., D.O.L. Zoph, K.S. Short and R.T.W. Kuo Yang. 1978. Historical Trends and Statistics of the Southern Oscillation, El Nino and Indonesian Droughts. *Fishery Bulletin* 76:663-678.

31. Hahn, D. and J. Shukla. 1976. An Apparent Relationship Between Eurasian Snow Cover and Indian Monsoon Rainfall. *J. Atmos. Sci.* 33:2461-2463.
32. Lau, K.-M. and P.H. Chan. 1985. Aspects of the 40-50 Day Oscillation During the Northern Winter as Inferred from Outgoing Longwave Radiation. *Mon. Wea. Rev.* 113:1889-1909.
33. Webster, P.J. 1987. The Variable and Interactive Monsoon pp. 269-30. In: J.S. Fein and P.L. Stephens (Ed.) *Monsoons.* John Wiley and Sons, New York, NY.
34. Mooley, D.A., B. Partyhasarathy and G.B. Pant. 1986. Relationships Between All-India Summer Monsoon Rainfall and Location of Ridge at 500 mb level Along 75°E. *J. Clim. Appl. Meteor.* 25:633-640.
35. Gray, W.M. 1984a. Atlantic Seasonal Hurricane Frequency, Part I: El Niño and 30 mb Quasi-biennial Oscillation Influences. *Mon. Wea. Rev.* 112:1649-1668.
36. Gray, W.M. 1984b. Atlantic Seasonal Hurricane Frequency, Part II: Forecasting its Variability. *Mon. Wea. Rev.* 112:1669-1683.
37. Nicholls, N. 1984. The Southern Oscillation, Sea Surface Temperature, and Interannual Fluctuations in Australian Tropical Cyclone Activity. *J. Climatology* 4:661-670.
38. Nicholls, N. 1985. Predictability of Interannual Variations of Australian Seasonal Tropical Cyclone Activity. *Mon. Wea. Rev.* 113:1143-1149.
39. Hastings, P.A. 1990. Southern Oscillation Influence on Tropical Cyclone Activity in the Australian/South-West Pacific Region. *Int. J. Climatology* 10:291-298.
40. Gray, W.M. and J.D. Sheaffer. 1991. El Niño and QBO Influences on Tropical Cyclone Activity pp. 257-284. In: M. Glantz, R.W. Katz and N. Nicholls (Ed.) *Teleconnections Linking Worldwide Climate Anomalies.* Cambridge University Press, New York, NY pp. 535.
41. Dong, K. 1988. El Nino and Tropical Cyclone Frequency in the Australian Region and the Northwest Pacific. *Australian Meteorological Magazine* 36:219-226.
42. Hoskins, B.J. and D.J. Karoly. 1981. The Steady Linear Response of a Spherical Atmosphere to Thermal and Orographic Forcing. *J. Atmos. Sci.* 38:1179-1196.
43. Simmons, A.J. 1982. The Forcing of Stationary Wave Motion by Tropical Diabatic Heating. *Quart. J. Roy. Meteor. Soc.* 108:503-534.
44. Simmons, A.J., J.M. Wallace and G.W. Branstator. 1983. Barotropic Wave Propagation and Instability, and Atmospheric Teleconnection Patterns. *J. Atmos. Sci.* 1363-1392.
45. Held, I. and I.-S Kang. 1987. Barotrophic Models and the Extratropical Response to El Nino. *J. Atmos. Sci.* 44:3576-3586.
46. Namias, J. 1969. Seasonal Interactions Between the North Pacific Ocean and the Atmosphere During the 1960s. *Mon. Wea. Rev.* 97:173-192.
47. Namias, J. 1976: Some Statistical and Synoptic Characteristics Associated with El Niño *J. Phys. Oceanogr.* 6:130-138.
48. Lau, K.-M. and J.S. Boyle. 1987. Tropical and Extratropical Forcing of the Large-Scale Circulation: A Diagnostic Study. *Mon. Wea. Rev.* 115:400-428.

49. Livezey, R.E. and K.C. Mo. 1987. Tropical-Extratropical Teleconnections During the Northern Hemisphere Winter. Part II: Relationships Between Monthly Mean Northern Hemisphere Circulation and Proxies for Tropical Convection. *Mon. Wea. Rev.* 115:3115-3132.
50. Rasmusson, E.M. 1991. Observational Aspects of ENSO Cycle Teleconnections pp. 309-343. In: M. Glantz, R.W. Katz and N. Nicholls (Ed.) *Teleconnections Linking Worldwide Climate Anomalies.* Cambridge University Press, New York, NY, pp. 535.
51. Barnston, A.G. and R.E. Livezey. 1987. Classification, Seasonality and Persistence of Low-Frequency Atmospheric Circulation Patterns. *Mon. Wea. Rev.* 115:1083-1126.
52. Yarnal, B.M. and H.F. Diaz. 1986. Relationships Between Extremes of the Southern Oscillation and the Winter Climate of the Anglo-American Pacific Coast. *J. Climatology* 6:197-219.
53. Ropelewski, C.F. and M.S. Halpert. 1986. North American Precipitation and Temperature Patterns Associated with the El Niño Southern Oscillation *(ENSO) Mon. Wea. Rev.* 114:2352-2362.
54. Douglas, A.V. and P.J. Englehart. 1981. On a Statistical Relationship Between Autumn Rainfall in the Central Equatorial Pacific and Subsequent Winter Precipitation in Florida. *Mon. Wea. Rev.* 109:2377-2382.
55. Schonher, T. and S.E. Nicholson. 1989. The Relationship Between California Rainfall and ENSO Events. *J. Climate* 2:1258-1269.
56. Fu, C., H.F. Diaz and J.O. Fletcher. 1986. Characteristics of the Response of SST in the Central Pacific Associated with Warm Episodes of the Southern Oscillation. *Mon. Wea. Rev.* 114:1716-1738.
57. Cane, M.A. and S.E. Zebiak. 1985. A theory for El Niño and the Southern Oscillation. *Science* 228:1085-1087.
58. Cane, M.A., S.E. Zebiak and S.C. Dolan. 1986. Experimental Forecasts of El Niño. *Nature* 321:827-832.
59. Zebiak, S.E. and M.A. Cane. 1987. A model El Niño-Southern Oscillation. *Mon. Wea. Rev.* 115:2262-2278.
60. Battisti, D.S. 1988. The Dynamics and Thermodynamics of a Warming Event in a Coupled Tropical Atmosphere-Ocean Model. *J. Atmos. Sci.* 45:2889-2919.
61. Schopf, P.S. and M.J. Suarez. 1988. Vacillation in a Coupled Ocean-Atmosphere Model. *J. Atmos. Sci.* 45:549-566.
62. Barnett, T., N. Graham, M.A. Cane, S.E. Zebiak, S. Dolan, J.J. O'Brien and D. Legler. 1988. On the Prediction of the El Nino of 1986-1987. Science 241:192-196.
63. Barnett, T.P. 1981. Statistical Prediction of North American Air Temperature from Pracific Predictors. *Mon. Wea. Rev.* 109:1021-1041.
64. Barnston, A.G. and C.F. Ropelewski. 1992. Prediction of ENSO Episodes Using Canonical Correlation Analysis. *J. Climate* 5:1316-1345.
65. Hamilton, K. and R.R. Garcia. 1986. El Niño/Southern Oscillation Events and their Associated Midlatitude Teleconnections 1531-1841. *Bull. Amer. Meteor. Soc.* 67:1354-1361.
66. Ramage, C.S. 1983. Teleconnections and the Siege of Time. *J. Climatology* 3:223-231.
67. Elliot, W.P. and J.K. Angell. 1988. Evidence for Changes in Southern Oscillation Relationships During the Last 100 Years. *J. Climate* 1:729-737.

The Oceans: Physical-Chemical Dynamics and Human Impact. Edited by S.K. Majumdar, E.W. Miller, G.S. Forbes, R.F. Schmalz and Assad A. Panah. © 1994, The Pennsylvania Academy of Science.

Chapter Eleven

TIDAL ENERGY DISSIPATION: Observations from Astronomy, Geodesy, and Oceanography

RICHARD D. RAY
Hughes STX
Goddard Space Flight Center, Code 926
Greenbelt, MD 20771

INTRODUCTION

Tidal friction is a classic subject in geophysics, and most people are aware of the basic astronomical implications: that the friction removes energy from the earth's rotation, lengthening the day by some 2.3 ms each century; that the lost angular momentum passes to the moon, which slowly recedes from the earth at about 3.8 cm y^{-1}.

Since at least 1920, when the late Sir Harold Jeffreys calculated the amount of tidal energy being dissipated, geophysical folklore has held that friction in the shallow seas is the principal physical mechanism responsible for the tidal accelerations. Folklore may well be true, but modern evidence as yet provides no strong quantitative support. In fact, as we shall see, recent dissipation estimates from many suggested mechanisms vary widely; the nature and location of tidal energy dissipation remains one of the outstanding problems in the scientific study of the tides.

In contrast, estimates over the past decade of the total amount of this dissipation, or corresponding estimates of the secular acceleration of the moon, all derived from a variety of sources, have converged in a most satisfactory way. Modern methods of space geodesy, especially laser ranging to the moon and to artificial satellites, have considerably clarified older, sometimes discordant, estimates based on historical astronomical observations. These estimates confirm that tidal friction, whatever its physical mechanism, is a controlling factor in the evolution of the moon's orbit and the earth's rotation. Scenarios for the history of this evolution have also been clarified,[1,2] but because of the nature (or uncertainty) of the dissipation mechanism, the history of the lunar orbit cannot be accurately parameterized in time. Tidal studies have therefore failed to provide strong constraints on theories of the lunar origin[3]—always since G.H. Darwin a long-sought goal. It is certainly clear, however, that the present lunar acceleration is abnormally large, because when integrated backwards, it implies an earth-moon close approach in much too recent time. Dissipation theories must account for this.[4]

This chapter reviews recent progress on these questions. Emphasis is given to summarizing the estimates of the secular rate of tidal energy dissipation and the concomitant secular accelerations of the lunar mean motion and earth rotation. The last part of this chapter considers the variety of suggested mechanisms by which this tidal energy is dissipated; it thereby provides the primary link to the rest of this book.

More detailed introductions to this subject may be found in the famous monographs of Munk and MacDonald[5] and Lambeck.[6]

TIDAL MECHANICS

The standard diagram depicting the geometry of tidal friction is shown in Figure 1. The familiar two-sided bulge of the principal semidiurnal tide is shown delayed relative to the lunar transit, the result of a dissipative response to the tidal potential. The gravitational forces on the offset bulges produce a torque acting to slow the earth's rotation, while the bulges' counter-torque increases the moon's angular momentum, forcing it to recede from the planet with reduced angular velocity. In this form, the consequences of tidal friction are a gradual lengthening of both the day and the month.

Figure 1 is a fair representation of the solid-earth body tide, although with much exaggerated lag angle, but it is a considerable distortion of the ocean tide whose complex dynamics—amphidromes, local resonances, etc.—precludes simple description, particularly by an equilibrium tidal potential. As stressed by Lambeck,[6] however, if one inserts a spherical harmonic expansion of the ocean tide into a

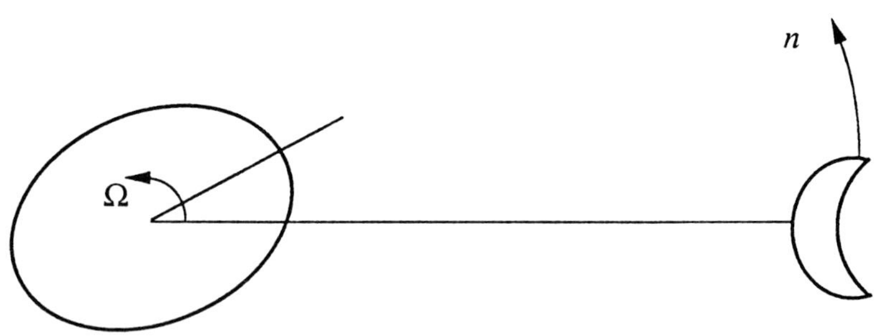

FIGURE 1. Tidal friction delays the earth's tidal bulge relative to the lunar transit, while $\Omega > n$ sweeps the bulge ahead of the moon.

global integral describing the tidal torque on the earth, one finds that only terms of the same degree and order as the tidal potential contribute to the integral. Thus in discussions of tidal energetics and associated effects on the earth's spin and lunar orbit, the simple picture of a lagged ellipsoid as in Figure 1 remains a valid one.

Before examining observational evidence, it is helpful to establish some relationships between variables of interest. The lunar mean motion n and lunar semimajor axis a are related through Kepler's law: n^2a^3 = constant, or (overdot denotes time derivative)

$$\dot{n} = -\frac{3}{2}\left(\frac{n}{a}\right)\dot{a}. \tag{1}$$

The relationship to the earth's rotation rate Ω is established through conservation of angular momentum, which in the direction of the earth's spin, can be written

$$C\Omega + \mu a^2(1 - e^2)\, n \cos I = \text{constant}$$

or, with (1),

$$C\dot{\Omega}_m = \mu a^2\,[\tfrac{1}{3}\dot{n}\cos I + ne\dot{e}\cos I + n\dot{I}\sin I], \tag{2}$$

where C is the earth's polar moment of inertia, μ is the earth/moon reduced mass, and e and I are the eccentricity and mean inclination of the lunar orbit. The subscript on $\dot{\Omega}$ denotes the change caused by lunar tides only. We have neglected a term arising from the slow lunar spin.

The secular change in E, the earth's rotational energy $\frac{1}{2}C\Omega^2$ plus the earth/moon orbital kinetic and potential energy $-\frac{1}{2}\mu n^2a^2$ is

$$\dot{E} = C\Omega\dot{\Omega} - \mu n a^2\dot{n}/3, \tag{3}$$

where $\dot{\Omega}$ now denotes total change caused by all tides. The second term is roughly 30 times smaller than the first, and their difference—between the rotational energy lost by the earth and the small energy gained by the moon—must be dissipated into heat. The rotational acceleration $\dot{\Omega}$ implies a rate of change in the length of day of $-\dot{\Omega}/\Omega$.

The astronomical consequences of these equations, simplified by $e = I = 0$ but including lunar spin, have been elegantly summarized by Counselman,[7] who gives diagrams showing that any planet-satellite system evolves toward one of three states: the satellite orbit decays inward until the satellite is destroyed; the orbit decays outward toward escape; or the satellite's orbit and planet's spin approach stable synchronism. Which of these outcomes is attained depends on the satellite: planet mass ratio and on the initial spins and orbital angular velocity. The earth-moon system is in the third category: the moon will continue to recede from a more slowly spinning earth until synchronism is reached, at which time the length of day and lunar month will both be approximately 50 present days. Solar tides, however, will continue to slow the earth's rotation, and will subsequently cause the lunar mean motion to exceed the rotation rate Ω. This reverses Figure 1, so that the tidal bulge lags (rather than leads) the moon; at that point, the moon will begin a slow spiral back toward the earth. Such, in fact, is the present state of Phobos, which orbits Mars 3⅓ times each Martian day, with positive secular acceleration[8] (and with expected catastrophic demise within 40 million years).

OBSERVATIONAL EVIDENCE

In addition to its secular deceleration from tidal friction, the earth's rotation displays complex variations across nearly all time scales. Shorter period variations—interannual to sub-daily—have recently received considerable attention,[9] an outgrowth especially of new campaigns of space-geodetic observations. Longer-term variations, by their nature, require older data. In a comprehensive discussion of astronomical determinations of $\dot{\Omega}$, Stephenson and Morrison[10] show that "decadal" variations extend to at least century scales, and that historical eclipses imply a mean rate of change in the length of day of 1.4 ms cy^{-1} during the past thousand years, but 2.4 ms cy^{-1} during the preceding thousand.

Thus it is clear that mechanisms other than tidal friction affect secular trends in Ω. Core-mantle coupling is usually invoked for decadal or longer changes, while a reduction in the earth's polar moment of inertia, probably from high-latitude rebound under melted Pleistocene ice sheets, has been observed[11] and must also affect Ω. Studies of tidal friction must therefore rely primarily on observations of $\dot{n}$ and $\dot{E}$, not $\dot{\Omega}$. A possible exception draws on discoveries in paleontology.

Paleontological

The fossil remains of corals, bivalve molluscs, and stromatolites may perhaps provide the most direct evidence for large, long-term changes in the length of day. Both living animals and their fossilized remains display growth-related banding on an annual cycle (at least in higher latitude climates), with a finer, usually daily, growth pattern superimposed. Counts of the fine-scale markings within the annual bands give the number of days in a year, after averaging to allow for year-to-year environmental variations. In an influential 1963 paper, Wells[12] reported that fossilized corals showed roughly 400 days per year during the Devonian, about what might be extrapolated from current estimates of $\dot{\Omega}$.

Many of the subsequent studies in this field have generated considerable controversy. Lambeck[6] reviews the status as of 1980, and while attempting some quantitative acceleration estimates, he emphasizes the many complicating factors involved. Richardson's recent summary[13] of experiments on living bivalves is even more pessimistic. He reports that bivalves once thought to have strictly daily patterns can sometimes have tidal patterns (twice per day, or less, depending on tidal ranges) or patterns of no apparent rhythm, and he concludes that it is premature to use bivalve fossils to infer the number of paleodays per year.

The study of rhythmically bedded fine-grain sediments, so-called "rhythmites," are prone to some of the same difficulties, although Williams[14] has obtained reasonable estimates from some Precambrian formations. We will refrain from quantitative interpretations of these data and instead turn to $\dot{n}$.

Astronomical

The first person to suggest that the moon is subject to an anomalous acceleration was Edmond Halley in 1695. His suggestion, in one of the early *Philosophical Transactions,* appears as a rather cryptic note near the end of a paper primarily devoted to historical and philological matters; it was apparently based on discrepancies he had noticed between the observed and computed times of four historical eclipses, recorded circa A.D. 900 by al-Battānī. Within half a century, further study of ancient and medieval eclipses by Dunthorne, Lalande, Tobias Mayer, and others had established the anomalous (apparent) acceleration as some $+20'' \text{ cy}^{-2}$ (arcsec per century²; this older literature customarily quotes $\frac{1}{2}\dot{n}$). The subsequent history, involving Laplace, Adams, Delaunay, and others, is well known[15]; improvements to the lunar theory reduced the acceleration to about $10'' \text{ cy}^{-2}$, which Ferrel and Delaunay attributed to the effects of tidal friction.

Ancient and medieval eclipses were studied in the first quarter of this century by Fotheringham, in the 1970s by the late R. Newton, and, more recently, by Stephenson and colleagues.[10,16] Although it is possible to use eclipse data to deduce both $\dot{n}$ and $\dot{\Omega}$, the two variables are highly correlated in the analysis. This, and unreliable or misinterpreted historical reports, likely accounts for some of the spurious values quoted over the years by different investigators—see the tabulation by Lambeck.[17] Stephenson and Morrison argue that it is preferable to adopt $\dot{n}$ from other measurements (see below) and to use historical eclipse records to study $\dot{\Omega}$. A reliable uncoupling of $\dot{n}$ and $\dot{\Omega}$ is more likely if solar, lunar, and planetary observations are subjected to a combined analysis, and this generally requires telescope-based observations (i.e., post-1600).

Any astronomical observation (e.g., of the occultation of a star by the moon) is recorded in Universal Time UT, which is regulated by the earth's rotation. A computed prediction of such an event follows from gravitational theory, which is parameterized by the more uniform Terrestrial Dynamical Time TDT (formerly Ephemeris Time). Many observations gathered over several centuries will generate a time series of observed timing discrepancies $\Delta T = \text{TDT} - \text{UT}$. Building on the work of Newcomb and others, Spencer Jones[18] convincingly showed that plots of ΔT (or equivalently, longitude discrepancies divided by mean motion) for the sun, Mercury, and Venus all lie approximately on the same curve (see also Figure 5.2 of Lambeck[6]). The lunar ΔT, however, shows an additional offset quadratic in time. These facts are consistent with ΔT being caused by variations in earth rotation, with the moon displaying an additional quadratic through its secular acceleration, which is determined via

$$\dot{n} = n \frac{d^2}{dt^2} [\Delta T_{Moon} - \Delta T_{Sun}].$$

This procedure was followed by Morrison and Ward[20] when they analyzed observations spanning three centuries of (a) transits of Mercury across the solar disk, thus defining ΔT_{Sun}, and (b) lunar occultations of stars, thus defining ΔT_{Moon}. They simultaneously solved for several corrections to the orbital elements of Mercury and the sun, including the relativistic correction to Mercury's perihelion. For the lunar acceleration, they found

$$\dot{n} = (-26 \pm 2)'' \text{ cy}^{-2}.$$

Lunar occultations observed since 1955 can be analyzed directly in terms of atomic time, without the confusing influences of earth rotation. Early attempts to do this were unreliable, and some rather wild values were reported. They were hindered by a number of systematic errors in the lunar ephemeris and the star catalogs and by the shortness of the time series. After twenty years, for example, n displaces the moon by only about 0.5″, which was about the typical error in any one occultation observation. Using data through 1980, Van Flandern[19] found $\dot{n} = (-21.4 \pm 2.6)''$ cy^{-2}. It would seem that a reanalysis of these data would be timely, since the intervening years should considerably help to determine the quadratic offset in lunar position.

Lunar Laser Ranging

Laser ranging to retroreflectors on the lunar surface date from only August 1969, but this short time series is more than compensated for by the extremely high precision of the measurements, which is now below 5 cm. This sensitivity has stimulated improved models of earth orientation, the lunar orbit (including relativistic effects), and the lunar physical librations. It is often thought that the ranging precision establishes the limit in the direct measurement of $\dot{a}$, but in fact the measurements are much more sensitive to quadratic changes in mean anomaly, or in the time of perigee passage, induced by $\dot{n}$. In 1988, Newhall et al.[21] reported $\dot{n} = (-24.9 \pm 1.0)''\ cy^{-2}$, but a more recent analysis, based on 23 years of range data, gives[22] $\dot{n} = (-25.9 \pm 0.5)''\ cy^{-2}$, in remarkable agreement with the Morrison-Ward value. Kepler's law then implies $\dot{a} = 3.8 \pm 0.1$ cm y^{-1}. Because of the quadratic time dependence in the secular changes of the orbit, the standard error in $\dot{n}$ will continue to fall rapidly (another decade should give errors $< \pm 0.2''\ cy^{-2}$).

Because $\dot{n}$ is induced by tidal constituents dependent on the lunar orbit's inclination to the earth's equator, a complete 18.6-y cycle of the lunar node around the ecliptic creates modulations which allow[23] separation of the diurnal and semidiurnal tidal contributions to $\dot{n}$. Williams et al.[22] now report a (lumped) diurnal contribution of $(-3.9 \pm 0.4)''\ cy^{-2}$ and a semidiurnal contribution of $(-22.4 \pm 0.6)''\ cy^{-2}$, with an additional $+0.4''\ cy^{-2}$ attributed to (model dependent[24]) energy dissipation in the moon.

Owing to equation (3), $\dot{n}$ alone does not establish the total energy dissipation $-\dot{E}$ in the tides nor the corresponding rotational acceleration $\dot{\Omega}$. For these, and for the contributions of the solar tides, we must turn to direct estimations of the tides.

Tides from Satellite Tracking

The solid-earth and ocean tides affect the earth's gravitational potential and thereby induce perturbations in the orbits of artificial satellites. As an example, the dominant periodic tidal signal in the orbit of the *Starlette* satellite is a 91-day 1100-millarcsec oscillation in the longitude of its node, caused by the K_1 body tide. Precise tracking—and especially laser ranging to the passive geodetic satellites *Starlette, Lageos,* and *Ajisai*—allows these perturbations to be directly analyzed for amplitudes and phases of the inducing tides.[17] Some modern analyses[25] combine the tidal determination with a large-scale adjustment of spherical harmonic coefficients of the geopotential, with certain harmonics expressed as static gravitational terms plus oscillations at known tidal frequencies. The satellite perturbations with periods longer than 1 day are generated only by prograde waves with spherical harmonics of order m equal to the tidal species,[17] and primarily of degree n even, although odd degree terms perturb the eccentricity. And since perturbations from high degree harmonics decay rapidly with altitude, only low degree terms are effective. Satellite tracking studies therefore determine the amplitudes D^{ν}_{nm} and phase lags ψ^{ν}_{nm} of constituent waves of form

$$h^{\nu}_{nm}(\varphi, \lambda, t) = D^{\nu}_{nm} P^{m}_{n}(\sin\varphi) \cos(A_{\sigma} + \sigma t + m\lambda - \tfrac{\pi}{2}\delta_{m,1} - \psi^{\nu}_{nm})$$

with m = tidal species, $n = 2, 3, \ldots, 6$ or so, φ is latitude, λ longitude, σ the tidal speed of constituent ν, and A_{σ} the equilibrium argument at the origin of time t. Without additional information, one cannot separate the degree-2 ocean-tide terms from those of the body tide, so the usual procedure is to adopt an earth model with nominal Love number k_2 that is strictly real (i.e., no phase lag, implying no body dissipation). Residual perturbations not accounted for by the model are then attributed to the ocean tide and its loading, with the latter removed by adopting a set of real loading numbers k'_n. Strictly real Love numbers are a good first approximation for the solid tide (see below).

Implications from the tidal solutions of two recent analyses of satellite tracking data are covered in the following section. One by Lerch et al.,[25] denoted GEM-T3S, is a combined gravity/tide solution using optical, Doppler, and laser tracking data to 31 different satellites. The other, by Cheng et al.,[26] is an analysis of laser tracking to *Starlette,* using several very long arcs of duration one or more years.

Hydrodynamic and Altimetric Ocean-Tide Models

Since the early computer experiments of Hansen and Pekeris, several generations of numerical ocean-tide models have appeared, each incorporating successive refinements in physical modeling (e.g., realistic tidal loading) and improved deep-sea observations (e.g., satellite altimetry). In this section, we examine the implications for tidal dissipation of two global models, one hydrodynamic and one altimetric; these two models are generally considered the most comprehensive and accurate.[27] Within the next few years, however, many new models will likely appear as a result of the Topex/Poseidon satellite altimeter mission.

The well-known model of Schwiderski[28] consists of the 8 major short-period, as well as 3 long-period, constituents on a global 1° × 1° grid. The model was derived by an early form of data assimilation, which incorporated the known tidal constants at over 2000 coastal and island sites, plus a few bottom-pressure sites, into a numerical solution of the Laplace tidal equations. The model of Cartwright and Ray[29] was derived from two years of *Geosat* altimeter data by a strictly empirical response-type analysis, which defined the tidal admittances across the diurnal and semidiurnal tidal bands. The model is near-global (no Arctic) on $1° \times 1.475°$ grid, the odd longitude spacing being the distance between *Geosat* groundtracks. A crucial step in the analysis of *Geosat* data is the correction of errors in the radial position of the satellite, for which see the original references. Since the altimeter is sensitive only to the geocentric tide, the ocean tide (relative to the seabed) can be extracted only by assuming an earth model with Love number h_2 and loading coefficients h'_n. Analogous to the satellite tracking procedures, these coefficients were assumed strictly real, i.e., no solid-earth dissipation. Comparisons of the Schwiderski and Cartwright-Ray global tide charts show[27] many areas of general agreement, but also some large open-ocean regions with height differences greater than 10 cm, and much larger differences on the continental shelves.

Direct numerical quadrature of global or near-global models such as these can be used to generate complete sets of spherical harmonic coefficients D_{nm}, ψ_{nm} for the tidal elevations, n and m being limited only by the grid resolutions. The tidal effect on the lunar orbit can then be found be writing down the gravitational potential of the tide in terms of the D_{nm}, ψ_{nm}, converting coordinates to Keplerian elements, and inserting the transformed potential into the Lagrange planetary equations. This has been worked out by Lambeck.[17] As in the case of torque and dissipation, only the degree $n = 2$ terms, with order m equal to tidal species, are effective. The tide-induced secular change in the lunar semimajor axis a is found to be

$$\dot{a} = \sum_{\nu} A_{\nu} D^{\nu}_{nm} \sin (\psi^{\nu}_{nm} + \tfrac{\pi}{2} \delta_{m,1}), \tag{4}$$

summing over all lunar constituents ν of species $m = 0, 1, 2$. Similar expression hold for the secular rates in lunar mean motion $\dot{n}$, eccentricity $\dot{e}$, and mean inclination $\dot{I}$, where coefficients such as A_{ν} can be found in tabulations by several authors but here are evaluated* directly from Eq. (27) of Lambeck.[17] In Table 1, these secular rates are separated into contributions from each of the major constituents, for the Cartwright-Ray ocean tide model. The K_1 and K_2 tides, which have no secular effect on the lunar eccentricity or mean motion, are split into their lunar and solar components. Table 1 neglects a few minor constituents that affect the lunar acceleration—primarily the fortnightly Mf and the nodal line in O_1; these two lines partly cancel in their effect on $\dot{n}$, and all neglected terms should affect the total $\dot{n}$ by less than 1%.

The acceleration of the lunar mean motion is dominated by the principal semidiurnal tide M_2, the contribution of the principal diurnal tide O_1 being nearly an order of magnitude smaller. The increase in the lunar eccentricity is primarily due to N_2. Since N_2 originates from a modulation of the principal tide caused by the varying earth-moon distance, e and N_2 each act to reinforce the other. The same is true of the smaller Q_1.

*As secular rates from various investigations converge, the importance of consistent constants for A_{ν}, etc. grows more acute. Here, Lambeck's R is taken as the earth's equatorial radius, rather than the mean radius, in agreement with the Cartwright-Tayler tidal potential and with orbit analyses by the Goddard group.[25] The loading number $k'_2 = k_2 - h_2$ is taken from Wahr's frequency-dependent theory, and the mean density of the ocean is taken as 1035 kg m.$^{-3}$

Several methods for computing the rate of tidal energy dissipation are considered in the next section, but the most reliable is to ignore dissipation mechanisms and to compute the rate of working of tidal forces on the complete earth-ocean-atmosphere system, as considered by Platzman.[30] In keeping with the assumption of real Love numbers and neglecting temporarily the atmosphere, dissipation occurs only in the ocean and must equal the rate of working of tidal forces on the ocean minus the flux of energy into the solid earth. (The latter is disposed of elastically by the earth working against the tidal forces.[31]) Analogous to (4), the dissipation is found to take the form

$$-\dot{E} = \sum_{\nu} \dot{\epsilon}_{\nu} D^{\nu}_{nm} \sin(\psi^{\nu}_{nm} + \tfrac{\pi}{2}\delta_{m,1}),$$

with coefficients ϵ_{ν} tabulated by Cartwright and Ray[29] (except their coefficient for N_2 should be 0.185, not 0.199). The major constituent terms in $\dot{E}$ are listed in Table 1 for the Cartwright-Ray model, with M_2 obviously dominant.

The rotational accelerations for the lunar tides in Table 1 follow from Eq. (2). Generally, $\dot{\Omega}$ is dominated by the $\dot{n}$ term in (2), but the $\dot{e}$ is important for N_2 while $\dot{I}$ is significant for M_2 and O_1 and is the only contributor for K_1^L and K_2^L. For the solar tides, $\dot{\Omega}$ has been deduced from $\dot{E}$ via the first term in (3); the secular changes to the earth's orbit are insignificant and only the earth's rotational energy $\frac{1}{2}C\Omega^2$ diminishes.

The totals from Table 1 are reproduced in Table 2 along with similar totals from the other tide models considered, plus the several independent determinations of $\dot{n}$. The $\dot{\Omega}$ and $\dot{E}$ totals include a correction for the S_2 atmospheric tide, which acts to accelerate the earth's rotation:[29] $\dot{\Omega} = +55''\ \mathrm{cy}^{-2}$, $\dot{E} = +0.16$TW. Driven by insolational heating, S_2 is the only atmospheric tide of consequence in the energy budget; the gravitational air tides account for no more than about 10GW.[32]

The tabulations of $\dot{n}$ show a very satisfying convergence of estimates, excepting the older Van Flandern and Schwiderski estimates. Agreement in the satellite-based

TABLE 1

Secular Rates Implied by the Cartwright-Ray Model

Tide	$\dot{n}$ $''\ \mathrm{cy}^{-2}$	$\dot{e}$ $10^{-19}\ \mathrm{s}^{-1}$	$\dot{I}$ $10^{-19}\ \mathrm{s}^{-1}$	$\dot{\Omega}$ $''\ \mathrm{cy}^{-2}$	$\dot{E}$ TW
M_2	−20.56	−0.34	−2.60	−916	−2.54
N_2	−1.30	4.76	−0.11	−39	−0.10
O_1	−2.84	−0.05	1.82	−63	−0.17
Q_1	−0.18	0.67	0.08	−3	−0.01
K_1^L	—	—	−2.27	−66	−0.19
K_1^S	—	—	—	−31	−0.09
K_2^L	—	—	−0.27	−8	−0.02
K_2^S	—	—	—	−4	−0.01
S_2	—	—	—	−155	−0.44
P_1	—	—	—	−10	−0.03
Total	−24.88	5.04	−3.35	−1294	−3.61

$1''\ \mathrm{cy}^{-2} = 4.868 \times 10^{-25}\ \mathrm{s}^{-2}$; 1TW $= 10^{12}$W

$\dot{\Omega}$ estimates is somewhat misleading, since there are solar and lunar differences that have partly cancelled; the $\dot{n}$ comparisons are more representative. The latest lunar-ranging value is slightly higher than previous reports, but quotes a lower uncertainty ($\pm 0.5''$ cy^{-2}), comparable to quoted uncertainties in the satellite-tracking studies. The altimeter model is probably less accurate since it is sensitive to subtle tidal-height cancellations in the numerical quadrature of the D_{22} harmonic. The more recent tabulated estimates must be considered in general agreement, and it would be premature to attribute differences to physical mechanisms such as a varying gravitational constant.

An $\dot{\Omega}$ acceleration of $-1240''$cy^{-2} corresponds to a rate of change in the length of day of 2.26 ms cy^{-1}. As noted previously, there are also long-term nontidal changes to Ω. The one usually attributed to glacial rebound and manifested as a changing second harmonic in the geopotential (nominally[11] $\dot{J}_2 = -3 \times 10^{-11}$ y^{-1}), accelerates the earth by about 300″ cy^{-2}, giving $\dot{\Omega}_{total} = -940''$ cy^{-2}. Now when the lunar acceleration is measured in UT, i.e., taken relative to the earth's rotation, we find a long-term apparent acceleration of

$$\dot{n} - \frac{n}{\Omega}\dot{\Omega}_{total} = -25 + 35 = +10''\text{cy}^{-2}.$$

It is rather startling to observe that this was the number commonly accepted in the mid-nineteenth century, based on the discrepancy between Dunthorne's early eclipse analysis and the lunar theory as rectified by John Couch Adams.

DISSIPATION MECHANISMS

The argument that most of the 3.5TW of tidal power is dissipated in the shallow seas is, to some extent, one of elimination, and we begin with the solid earth.

The specific dissipation function Q^{-1} of the solid earth is reasonably well established for seismic and free-oscillation frequencies. Obtaining reliable estimates for solid-earth tidal dissipation rests on extrapolating Q to diurnal and semidiurnal frequencies. Zschau[33] has attempted this by combining seismic-based Earth models

TABLE 2

Estimates of Secular Acceleration and Energy Dissipation

Year	Author	Method	$-\dot{n}$ ″cy^{-2}	$-\dot{\Omega}$ ″cy^{-2}	$-\dot{E}$ TW
1975	Morrison-Ward[20]	Post-1670 occultations	26		
1981	Van Flandern[19]	Post-1955 occultations	21.4		
1981	Schwiderski[28]	Hydrodynamic	19.44	964	2.68
1988	Newhall et al.[21]	Lunar laser ranging	24.9		
1990	Cheng et al.[26]	Satellite tracking	25.12	1240	3.50
1991	Cartwright-Ray[29]	Satellite altimetry	24.88	1240	3.45
1992	Lerch et al.[25]	Satellite tracking	25.33	1245	3.46
1992	Williams et al.[22]	Lunar laser ranging	25.9		

with observational constraints from the period of the Chandler wobble, which at approximately 435 d is some 8 d longer than a purely elastic wobble. Under the assumption that Q is a smooth function of frequency, Zschau shows that the observed Chandler period implies an M_2 solid-earth dissipation rate between 0.07 and 0.14TW, with "most probable" being 0.12TW. This corresponds to a solid-earth phase lag of about 0.2°. Similar estimates can be deduced from the complex Love numbers of Wahr and Bergen,[34] which were also constrained by the Chandler wobble. If the pole tide (the ocean's response to the Chandler wobble) is nonequilibrium, which has occasionally been reported, the earth-tide dissipation rate will necessarily be less than 0.12TW.

Far greater phase lags, and hence greater dissipation, have been inferred from some past gravity-tide measurements, but the recent careful study by Baker et al.[35] of gravity measurements at five European sites contradicts this. It now seems likely that many of the earlier reports of larger phase lags were corrupted by instrument miscalibrations and poor ocean-loading corrections.

Although Zschau's rate of 0.12TW for M_2 is four times greater than an earlier estimate he had made (which neglected mantle dispersion and frequency dependence in Q), the latest estimate still confirms that the solid earth is a minor contributor to the total tidal dissipation. (It seems large enough, however, that satellite tracking and altimeter studies should move toward a more rigorous separation of solid and fluid tides.)

Turning to the oceans, one may evaluate the dissipation resulting from frictional stress on the ocean bottom and show by simple order-of-magnitude estimates that the deep-ocean regions provide an inadequate energy sink. Frictional stress is usually taken as a quadratic in velocity: $.002\rho v^2$. Typical deep-ocean tidal currents rarely exceed 3 cm s^{-1}, so dissipation per area must be less than $.002\rho v^3$ = .06mW m^{-2}, for a global total ~ 10GW. In contrast, currents in the shallow seas often exceed 30 cm s^{-1}; with a total area ~3 × 10^{13} m^2, this gives a dissipation of ~ 2TW, certainly the right order of magnitude. Jeffreys and Heiskanen attempted more detailed estimates from the shallow seas, obtaining 1.1 and 1.9TW respectively, but this method's dependence on the cube of poorly known velocities limits its usefulness.

More reliable estimates of shelf dissipation rest on calculating the energy flux out of the deep oceans. This is dependent only on the first power of the current and its phase relative to tidal height, and is independent of the actual stress parameterization or even the dissipation mechanism; it would include dissipation, for example, through interactions with ordinary breaking waves, a mechanism once suggested by Jeffreys. Miller[36] used this flux method to estimate the M_2 dissipation in shelf and shallow seas as 1.4-1.7TW (±50%). He attributed about 15% to the Bering Sea alone, and this is now considered too large. But Miller also neglected* the seas round Antarctica, and dissipation in the ice-covered Weddell and Ross Seas may be significant. In fact, Doake[37] has suggested that flexing of Antarctic ice shelves by the underlying ocean tide may alone dissipate 2TW (±50%) of tidal power. There is hardly any observational evidence for this, although a few spot

*I thank C. Le Provost for noticing this.

values of large tidal "ages" and abnormal nonlinear tides[38,39] do suggest large dissipation in the area. A reliable estimate of the energy flux across the 60°S parallel would be of some value, and ongoing satellite altimeter missions may help to define this.

Conversion of tidal energy into baroclinic modes, either at the continental slopes or across smaller-scale abyssal features, represents another possible energy sink. Mechanisms for internal-tide dissipation are themselves far from clear, but the important point is that energy, once converted to baroclinic motion, is permanently lost to the barotropic tide. There is general agreement[40] that generation of internal tides at the continental slopes is an insignificant sink; typical scattering fluxes are 100W m^{-1}, for a global total of ~ 15GW. In a more detailed analysis of all significant shelf regions, Baines[41] found 14.5GW for M_2, 2.7GW for S_2. Scattering by deep-ocean bottom topography, however, appears more important. Munk[42] and Bell[43] developed theoretical models for internal tide generation and got energy fluxes implying global dissipations of 0.5 and 0.3TW, respectively. Sjöberg and Stigebrand,[44] extending a model they previously used to study mixing in fjords, recently made a more detailed global calculation by considering the generation of progressive waves (no reflection) across a fairly realistic bathymetric model, with specified climatological density stratification and Schwiderski's barotropic tidal heights. For M_2, they get 1.3TW—more than half the required rate according to Table 1. Important dissipation areas (see their Figure 6) are the mid-ocean ridges in the North Atlantic and western Indian Oceans and the vast region of rugged topography across the western Pacific. Interestingly, Bell had constrained his model with measured bathymetric spectra from the eastern North Pacific, an area that Sjöberg and Stigebrand figure to be relatively less important for dissipation.

When Bell published his work, the required astronomical rate appeared to be 5-6TW, so his 0.3TW looked insignificant. Now that modern space geodesy has established 2.5TW for M_2, conversion to internal tides appears more important. The Sjöberg-Stigebrand estimate of 1.3TW, if it holds up, is more than adequate for closing the energy gap, and, in addition, suggests that dissipation in the deep ocean may be important in numerical models of the barotropic tides.

SUMMARY

Tidal energy dissipation, primarily in the oceans, acts to slow the earth's rotation, lengthening the day by some 2.3ms each century. The lost angular momentum is transferred to the moon, whose mean motion is decelerated by 26″ cy^{-2}. Table 2 summarizes the recent satisfactory concordance in these estimates. Miller's figure of 1.4-1.7TW for energy dissipation in the shallow seas accounts for more than half the required M_2 amount, and, based on suggestions of large dissipation around Antarctica, it may be an underestimate. In addition, if the recent Sjöberg-Stigebrand open-ocean model is realistic, then conversion to baroclinic modes more than makes up for the remaining tidal dissipation.

ACKNOWLEDGEMENT

It is a pleasure to thank D.E. Cartwright and G.W. Platzman for many extended discussions on tides and tidal energetics.

REFERENCES

1. Goldreich, P. 1966. History of the lunar orbit. *Rev. Geophys.,* 4:411-439.
2. Mignard, F. 1982. Long time integration of the moon's orbit, p. 67-91. In: P. Brosche and J. Sündermann (ed.) *Tidal Friction and the Earth's Rotation.* Springer-Verlag, Berlin.
3. Boss, A.P. and S.J. Peale. 1986. Dynamical constraints on the origin of the moon, pp. 59-101. In: W.K. Hartmann (ed.) *The Origin of the Moon.* Lunar & Planetary Institute, Houston.
4. Hansen, K.S. 1982. Secular effects of oceanic tidal dissipation on the moon's orbit and the earth's rotation. *Rev. Geophys. Space Phys.,* 20:457-480.
5. Munk, W.H. and G.J.F. MacDonald. 1960. *The Rotation of the Earth.* Cambridge U. Press.
6. Lambeck, K. 1980. *The Earth's Variable Rotation.* Cambridge U. Press.
7. Counselman, C.C. 1973. Outcomes of tidal evolution. *Astrophys. J.,* 180:307-314.
8. Jones, D.H.P., A.T. Sinclair and I.P. Williams. 1989. Secular acceleration of Phobos confirmed from positions obtained on La Palma. *Mon. Not. R. Astr. Soc.,* 237:15-19.
9. Hide, R. and J.O. Dickey. 1991. Earth's variable rotation. *Science,* 253:629-637.
10. Stephenson, F.R. and L.V. Morrison. 1984. Long-term changes in the rotation of the earth: 700 B.C. to A.D. 1980. *Phil. Trans. R. Soc. Lond., A313:47-70.*
11. Yoder, C.F. et al. 1983. Secular variation of earth's gravitational harmonic J_2 from *Lageos* and nontidal acceleration of earth's rotation. *Nature,* 303:757-762.
12. Wells, J.W. 1963. Coral growth and geochronometry. *Nature,* 197:948-950.
13. Richardson, C. 1990. Tidal rhythms in the shell secretion of living bivalves, p. 215-26. In: P. Brosche & J. Sündermann (ed.) *Earth's Rotation from Eons to Days.* Springer-Verlag, Berlin.
14. Williams, G.E. 1989. Late Precambrian tidal rhythmites in South Australia and the history of the Earth's rotation. *J. Geol. Soc. Lond.,* 146:97-111.
15. Kushner, D. 1989. The controversy surrounding the secular acceleration of the moon's mean motion. *Arch. Hist. Exact Sci.,* 39:291-316.
16. Stephenson, F.R. and S.S. Said. 1989. Non-tidal changes in the earth's rate of rotation as deduced from medieval eclipse observations. *Astron. Astrophys.,* 215:181-189.
17. Lambeck, K. 1977. Tidal dissipation in the oceans: astronomical, geophysical, and oceanographic consequences. *Phil. Trans. R. Soc. Lond,* A287:545-594.
18. Spencer Jones, H. 1939. The rotation of the earth, and the secular accelerations of the sun, moon, and planets. *Mon. Not. R. Astr. Soc.,* 99:541-558.
19. Van Flandern, T.C. 1981. Is the gravitational constant changing? *Astrophys. J.* 248:813-816.

20. Morrison, L.V. and C.G. Ward. 1975. An analysis of the transits of Mercury; 1677-1973. *Mon. Not. R. Astr. Soc.,* 173:183-206.
21. Newhall, XX, J.G. Williams and J.O. Dickey. 1988. Earth rotation from lunar laser ranging, pp. 159-164. In: A. Babcock and G. Wilkins (Ed.) *The Earth's Rotation and Reference Frames for Geodesy and Geodynamics.* Kluwer, Dordrecht.
22. Williams, J.G., XX Newhall and J.O. Dickey. 1992. Diurnal and semidiurnal tidal contributions to lunar secular acceleration (abstract). *Eos,* Fall Suppl. 43:126.
23. Williams, J.G., W.S. Sinclair and C.F. Yoder. 1978. Tidal acceleration of the moon. *Geophys. Res. Lett.,* 5:943-946.
24. Yoder, C.F. 1981. The free librations of a dissipative Moon. *Phil. Trans. R. Soc. Lond.,* A303:327-338.
25. Lerch, F.J. et al. 1992. Geopotential models of the earth from satellite tracking, altimeter, and surface gravity observations: GEM-T3 and GEM-T3S, *NASA Tech. Memo. 104555,* Goddard Space Flight Center, 118 pp.
26. Cheng, M.K., C.K. Shum, R.J. Eanes and B.E. Schutz. 1990. Long-period perturbations in Starlette orbit and tide solution. *J. Geophys. Res.* 95:8723-8736.
27. Ray, R.D. 1993. Global ocean tide models on the eve of Topex/Poseidon. *IEEE Trans. Geosci. Remote Sens.,* 31:355-364.
28. Schwiderski, E.W. 1981. Global ocean tides: Atlas of ocean tidal charts and maps: Parts I to IX, NSWC Tech. Repts., Naval Surface Weapons Center, Dahlgren, Va.
29. Cartwright, D.E. and R.D. Ray. 1991. Energetics of global ocean tides from Geosat altimetry. *J. Geophys. Res.,* 96:16897-16912.
30. Platzman, G.W. 1984. Planetary energy balance for tidal dissipation. *Rev. Geophys. Space Phys.,* 22:73-84.
31. Platzman, G.W. 1985. The role of earth tides in the balance of tidal energy. *J. Geophys. Res.,* 90:1789-1793.
32. Platzman, G.W. 1991. An observational study of energy balance in the atmospheric lunar tide. *Pure Appl. Geophys.,* 137:1-33.
33. Zschau, J. 1986. Tidal friction in the solid earth: constraints from the Chandler wobble period, p. 315-344. In: A. Anderson & A. Cazenave (Ed.) *Space Geodesy and Geodynamics.* Academic.
34. Wahr, J.M. and Z. Bergen. 1986. The effects of mantle anelasticity on nutations, earth tides, and tidal variations in rotation rate. *Geophys., J.R. Astr. Soc.,* 87:633-668.
35. Baker, T.F., R.J. Edge and G. Jeffries. 1991. Tidal gravity and ocean loading in Europe. *Geophys. J. Int.,* 107:1-11.
36. Miller, G.R. 1966. The flux of tidal energy out of the deep ocean. *J. Geophys. Res.,* 71:2485-2489.
37. Doake, C.S.M. 1978. Dissipation of tidal energy by Antarctic ice shelves. *Nature,* 275:304-5.
38. Cartwright, D.E. 1979. Analyses of British Antarctic Survey tidal records. *Br. Antarc. Sur. Bull.,* 49:167-179.
39. Pedley, M., J.G. Paren, J.R. Potter. 1986. The tidal spectrum underneath Antarctic ice shelves. *J. Geophys. Res.,* 91:13001-13009.

40. Huthnance, J.M. 1989. Internal tides and waves near the continental shelf edge. *Geophys. Astrophys. Fluid Dyn.,* 48:81-106.
41. Baines, P.G. 1982. On internal tide generation models. *Deep-Sea Res.,* 29:307-338.
42. Munk, W.H. 1966. Abyssal recipes. *Deep-Sea Res.,* 13:707-730.
43. Bell, T.H. 1975. Topographically generated internal waves in the open ocean. *J. Geophys. Res.,* 80:320-327.
44. Sjöberg, B. and A. Stigebrand, 1992. Computations of the geographical distribution of the energy flux to mixing processes via internal tides. *Deep-Sea Res.,* 39:269-291.

The Oceans: Physical-Chemical Dynamics and Human Impact. Edited by S.K. Majumdar, E.W. Miller, G.S. Forbes, R.F. Schmalz and Assad A. Panah.

Chapter Twelve

OCEAN SURFACE TEMPERATURE VARIATION AND CHANGE

MICHAEL A. PALECKI
Meteorology/Climatology Program
Department of Geography
University of Nebraska
Lincoln, Nebraska

INTRODUCTION

The oceans of Earth are of singular importance in governing climate and making this world hospitable to life. The interactions that take place at the atmosphere/sea surface boundary transfer momentum, heat and mass; these exchanges are crucial to the Earth's energy balance, momentum budget and hydrologic cycle. The state of the ocean surface is a product of these complex interactions with the atmosphere, simultaneously impacting the atmospheric state and responding to atmospheric forcing. One of the most important variables describing the state of the ocean surface is temperature. Sea surface temperatures, or SSTs, are consequential from the standpoint of energy transfers to the atmosphere, which are a function of the marine air/SST difference. Equally important, however, is the recognition that the time change of SST patterns reflects the dynamic processes of climate variation and change which impact human society. Understanding the nature of ocean surface temperature variability and change is the goal of this chapter.

While sailors and naturalists have been observing the oceans for thousands of years, the formality of data collection by most ocean-going vessels was not realized until the 1850s. Matthew Maury, the superintendent of the US Hydrologic Office, organized an international conference to codify standard observational practices, including the collection of SST measurements.[1] Since this time, observations have been collected by ships of opportunity, as well as by oceanographic research vessels. Many of these early records have survived to the present; of course, the number of ship observations available for analysis and their spatial coverage has increased substantially over time. The difficulties with creating a uniform and comparable set of SST observations for the period since the 1950s provides a daunting challenge. For example, the Comprehensive Ocean-Atmosphere Data Set (COADS) contains over 80 million observations.[2,3] Simple changes in ship type (small wooden to large metallic), ship velocity (slow to fast), and SST measurement system (wooden buckets to engine water intake tubes) create biases in the data that change over time. Correction procedures are complex, using geographically varying estimates of ship types and measurement types.[4] These have been discussed in excellent reviews by Folland et al.[5] and Maul.[6] The SST data are usually presented to the scientific community in gridded or binned formats, with all necessary corrections already completed. The data used in deriving the maps in this chapter are from the Global Ocean Surface Temperature Atlas (GOSTA) project described by Bottomley et al.[7] These SST data are given as monthly mean anomalies on a 5° latitude by 5° longitude grid covering the ocean areas of the world.

The mean patterns of seasonal SSTs are shown in Figure 1. The analysis domain in all figures is limited to regions north of 45°S, due to the lack of data in the southern oceans. All seasons show a general meridional temperature gradient in both hemispheres; however, a closer inspection indicates circulation dynamics at work. During winter and spring (Figures 1A, 1B), the tropical warm pool is at its smallest extent. Distinctly cooler oceanic regions extend equatorward on the western margins of South America and Africa in all months, as ocean currents bring cold water into tropical latitudes. Upwelling occurs where ocean currents turn away from these coastlines, further cooling the surface water. This is also apparent west of the United States, especially during summer and fall (Figures 2C, 2D). The eastern continental margins, on the other hand, are flanked predominantly by warm currents, such as the Gulf Stream in the Atlantic. Where these warm waters meet cooler waters, very strong temperature gradients tend to form; these regions have important affects on the overriding atmospheric circulation.[8,9] It should be noted that the regions of larger temperature gradient in the mid-latitudes of both hemispheres are also very sensitive to atmospheric forcing.

In the next section, the focus will be placed on ocean surface temperature variability. The foremost mode of variation in the global oceans occurs over three to five year cycles in the equatorial East Pacific, with its extreme warm mode termed El Niño and its extreme cold mode termed La Niña. Other important modes of variation occur in the tropical Atlantic and in the extratropics; these will be discussed briefly. Following this section, the topic of long-term ocean surface temperature change will be explored. While the oceans have been warming during the Twentieth Century, some regions have changed more than others. Finally, the issue of potential future climate change will be discussed in light of the findings presented.

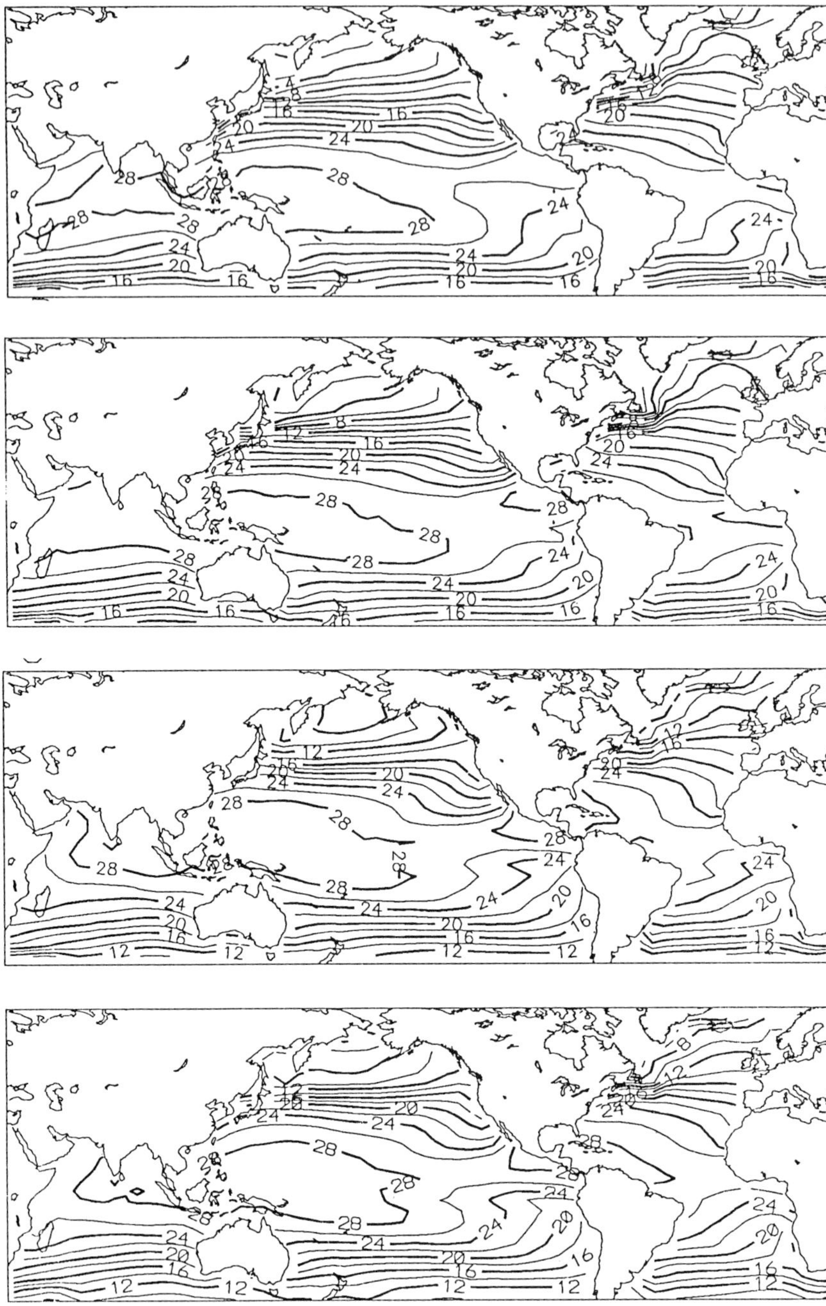

FIGURE 1. Seasonal mean SST patterns, 1951-1980. Values calculated by the GOSTA project[7]: (A) winter, (B) spring, (C) summer, (D) fall.

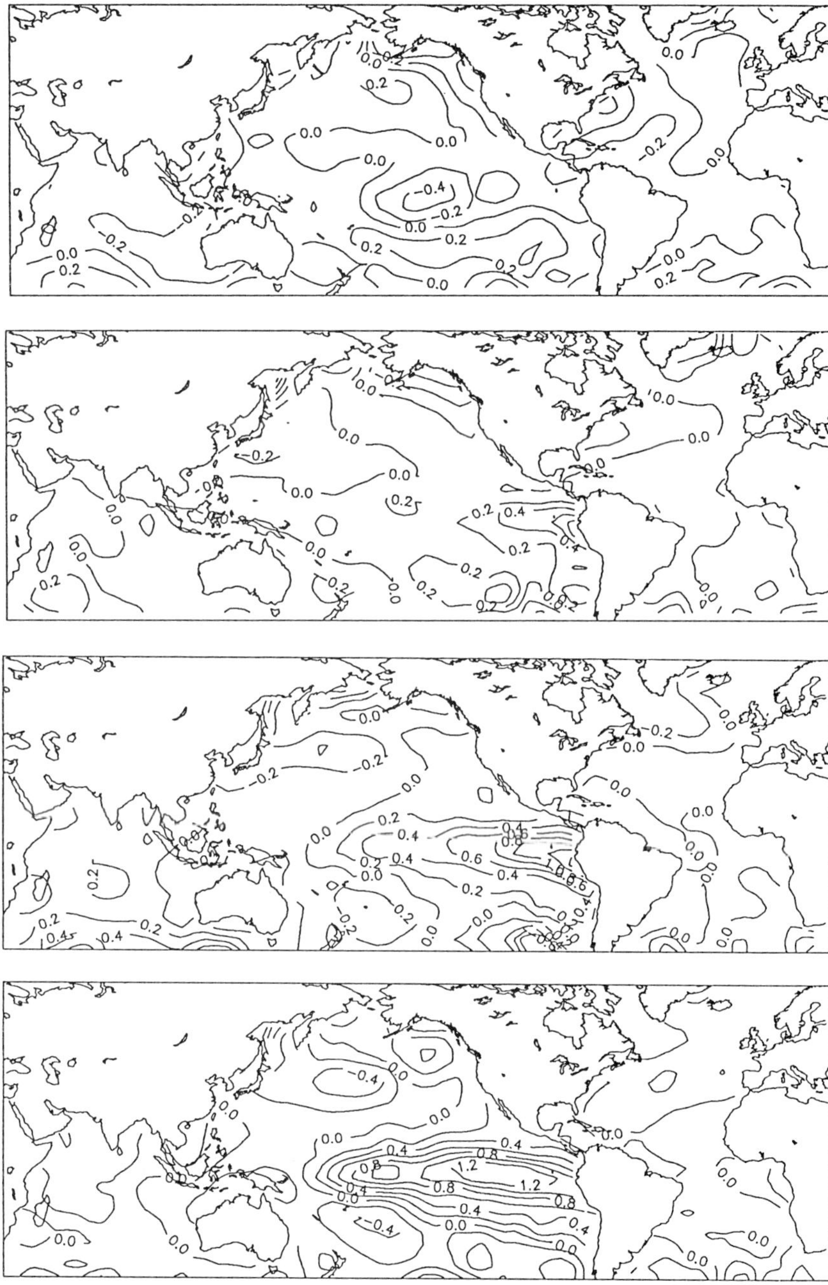

A

B

C

D

FIGURE 2. Composite El Niño event, created from all strong or medium events since World War II (see text for details): (A) winter Year 0, (B) spring Year 0, (C) summer Year 0, (D) fall Year 0, (E) winter Year +1, (F) spring Year +1, (G) summer Year +1, (H) fall Year +1.

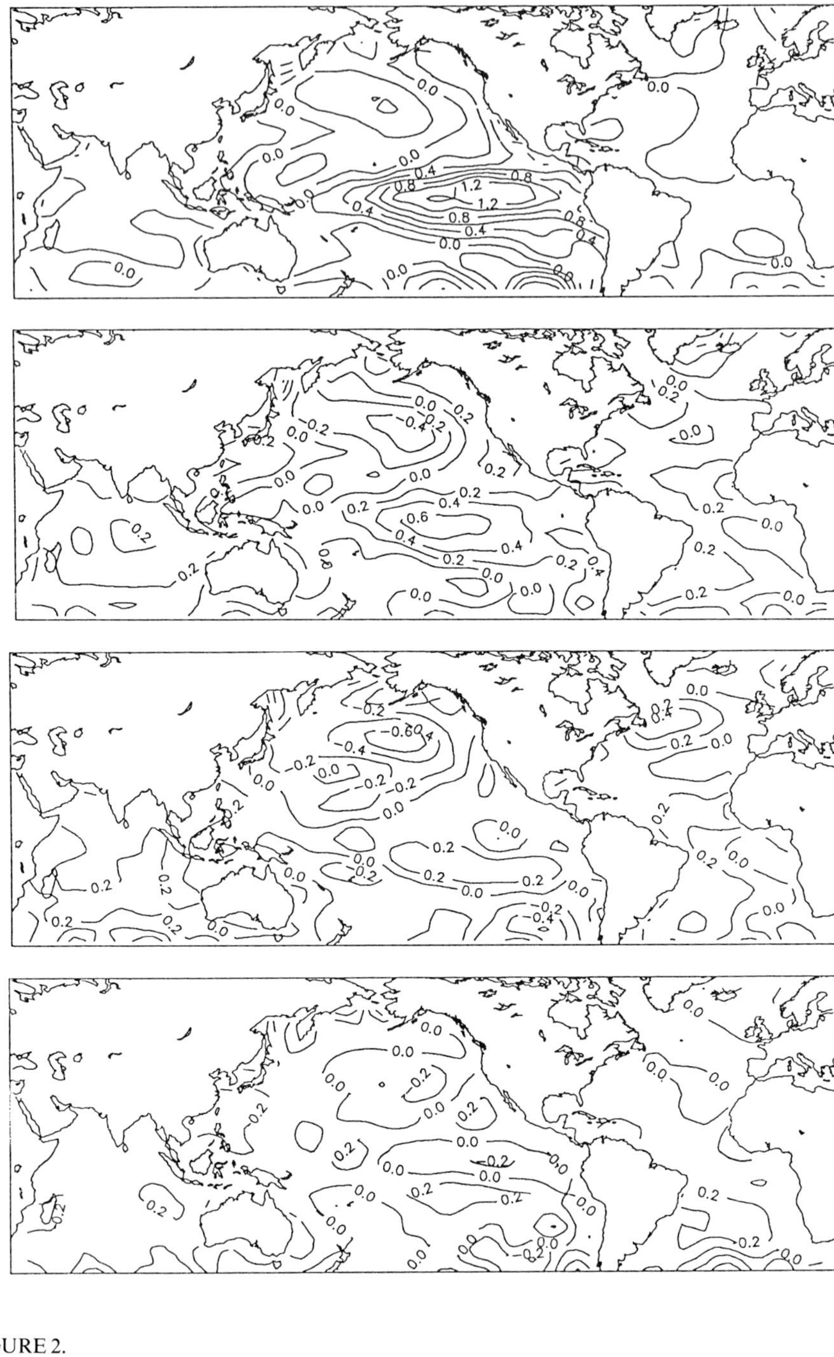

E

F

G

H

FIGURE 2.

OCEAN TEMPERATURE VARIABILITY

El Niño/La Niña

The earliest documentation of the phenomenon of El Niño in the coastal waters of Peru is recorded in the logs of a Captain Colnet in 1795.[10] This warming of the coastal waters of Peru has probably been known about through the entire period of human habitation, due to its effect on local environmental conditions. These years were referred to as "*anos de abundancia*", or years of abundance, by the late Nineteenth Century Peruvians, as plentiful rains occurred on the desert coasts of southern Peru.[11] However, we now know that these ocean warmings are not merely the coastal events observed near Peru, but instead are part of a warming anomaly stretching across the entire eastern equatorial Pacific Ocean when at its maximum extent. This was not recognized until an El Niño occurred in 1957-1958, during the intensive global observations of the International Geophysical Year.[11,12,13] The El Niño is readily apparent in SST records for the region from 150° to 90°W and 5ö°S[14]; Figure 3 displays the monthly average SST for this region during the Twentieth Century. The time series of the record is punctuated by the occurrences of approximately 21 separate El Niños of varying intensities,[15] including the most recent one in 1992. In examining this time series, it is readily apparent that large negative temperature anomalies also develop in this region; these are the La Niña events. While originally thought to be a "normal" state between El Niños, the La Niña is now recognized as a climate anomaly that may be as severe as the El Niño event, but opposite in its physical attributes.[16]

The El Niño and La Niña events are the oceanic manifestations of the most important coupled ocean-atmosphere mode in the tropics. The atmospheric component of this system is known as the Southern Oscillation[17]; it is most commonly represented as a sea level pressure difference between the western and eastern Pacific equatorial zone. It is this east/west pressure gradient across the Pacific that helps to establish typically strong trade winds across the tropical Pacific. In fact, a time series of the Southern Oscillation (not shown) would correlate very highly with the east equatorial Pacific SST record shown in Figure 3. The changes in the Southern

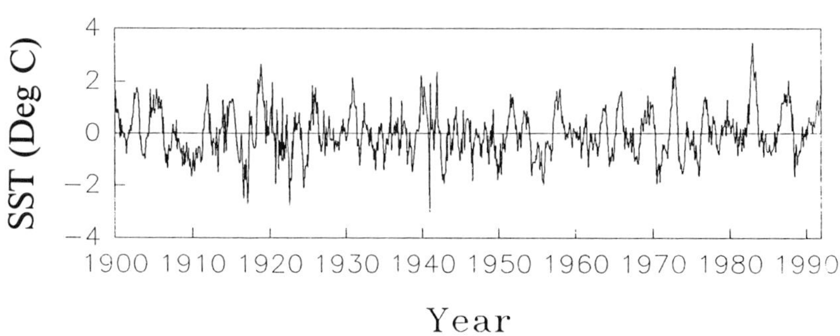

FIGURE 3. Time series of east equatorial Pacific SSTs, 1900-1991.

Oscillation represent variations in the structure of zonally-oriented tropical circulation. Normally, the strong trade winds flow into a region of lower pressure in the western Pacific; rising air at this location then reaches upper levels and spreads in the equatorial plane. Some of the air spreads north/south, joining the Hadley circulation, while some continues to flow in the equatorial zone. This air descends where surface conditions near the equator are cooler and upward motions are suppressed. The eastern Pacific, with its cold surface waters, is a preferred place for air to descend. The subsiding air can then join the trade winds to complete the circulation. This atmospheric phenomenon is known as the Walker Circulation.[13] The changes in strength of the Walker circulation are represented by the Southern Oscillation. Leathers[18] describes the atmospheric manifestations of the Southern Oscillation elsewhere in this book. Here, the discussion will be limited to the changes recognized in the oceanic surface environment.

Many aspects of El Niño and La Niña behavior have been isolated through the use of compositing techniques.[19,20] In this approach, individual events are identified in the historical record and averaged together to reveal the central tendencies of an anomaly type. The composite El Niño event presented in Figure 2 was derived by averaging the seasonal departures of SST for every moderate or strong event since World War II.[21] This period was chosen because the SST data have undergone virtually no corrections, and are more densely distributed than earlier time periods.[5] A total of nine substantial El Niños have been composited, occurring in the years 1951, 1953, 1957, 1965, 1969, 1972, 1976, 1982, and 1986. The eight composite panels represent the seasonal mean SST departures for the year of peak warming and the following year, conventionally referred to as Year 0 and Year +1.[19] An El Niño event usually lasts from 1 to 2 years from beginning to end. Despite its irregular periodicity, the El Niño composites display very strong patterns, due mainly to the tendency of the period of maximum warming to phase lock with the seasonal cycle.[15] However, it should be noted that each individual event has its own characteristics of areal coverage and timing, so the composite anomalies are generally not as strong as those of individual events.

The winter of year 0 (Figure 2A) indicates temperatures that are below normal in the central equatorial Pacific. The subtropics to the south, however, display above normal temperatures that are associated with enhanced convection along the South Pacific Convergence Zone (SPCZ). This is thought to be a precursor to the El Niño event.[16,20,22] In many El Niño years, temperatures along the coast of Peru begin to warm in late December (thus the name El Niño, or the Christ Child); however, this warming is usually not clear in the composite until spring (Figure 2B). Warming near the coast of Peru reaches levels greater than 1°C very quickly, and starts to spread westward toward the dateline. There are no other distinct anomalies elsewhere in the global oceans, except for the SPCZ warm pool, which is shrinking in size. During this time period, trade winds are greatly reduced, upwelling of cold water along the coast has ceased, and local warming of the surface water is taking place.

By summer of Year 0, the El Niño is intensifying rapidly (Figure 2C), and the trade winds have collapsed over a wide area. A return flow of warm water from the west equatorial Pacific takes place, as the dynamic pressure of the trade winds is no longer holding sea levels higher in the west than the east. The combination of local warming and warm water advection leads to the eastward extension of the climatological warm pool (Figure 1C), which effectively shifts the preferred region of convection to the central and eastern equatorial Pacific location of the warm SST anomaly. Important changes are also occurring in the mid-latitudes of both the northern and southern Pacific, where substantial cooling is taking place. This mid-latitude cooling reaches its nadir at the time of maximum El Niño strength during the fall of Year 0 (Figure 2D). The dynamic link between the two regions is related to selective intensification of the Hadley circulation at the longitude of anomalous convection; this reorganization causes the subtropical jet stream maximum to extend further east in the North Pacific, altering the basic state of the extratropical circulation. Associated changes in surface air circulations then lead to the formation of the cool pool in the North Pacific.

By the winter of Year +1 (Figure 2E), the El Niño has reached its mature stage. While a considerable warm pool still exists east of 180°W, the waters off the coast of Peru are beginning to cool. The North Pacific remains cool throughout Year +1, with most negative anomalies occurring in the summer of Year +1 (Figure 2G). By this time, though, the El Niño warming in the central equatorial Pacific has reached the end of its life cycle. Interestingly, some fairly substantial SST departures can be seen in the Atlantic Ocean basin during spring and summer of Year +1 (Figure 2F, 2G), with warming from the equator southward. Finally, by the winter of Year +1 (Figure 2H), cooler than normal ocean surfaces are seen in the east equatorial Pacific; in some cases, an El Niño is followed immediately by a La Niña.

The composite La Niña event shown in Figure 4 is derived from seven events,[15] centered in 1949, 1954, 1964, 1970, 1973, 1975, and 1988. The La Niña SST patterns are generally similar spatially to the El Niño patterns but are of opposite sign. The winter pattern of Year 0 (Figure 4A) illustrates the high likelihood for a La Niña event to follow immediately after an El Niño event. By spring, however, cooling below normal has commenced on the Peruvian coastline (Figure 4B). During this period, SSTs are warmer than normal in the west Pacific, leading to an enhancement of normal convection and a lowering of sea level pressure. The stronger pressure gradient results in increased trade wind velocity, further enhancing upwelling in the east Pacific and advecting this cold water further west than normal by summer (Figure 4C). The maximum cooling occurs in the central equatorial Pacific during the fall (Figure 4D), as the eastern edge of the climatological west Pacific warm pool (Figure 1D) is displaced westward.

In the winter of Year +1 (Figure 4F), substantial warming is taking place in the mid-latitudes of both hemispheres in the Pacific. The warming in the region of the SPCZ indicates the end of the La Niña event, as it often portends the return to El Niño conditions. Still, the cool pool in the central equatorial Pacific remains large through the spring (Figure 4F). In the Atlantic tropics, cool conditions develop as the Pacific La Niña ends; this is analogous to the warming that occurs in the tropical Atlantic at the end of an El Niño event. By summer and fall of Year +1 (Figures 4G, 4H), the La Niña event has dissipated.

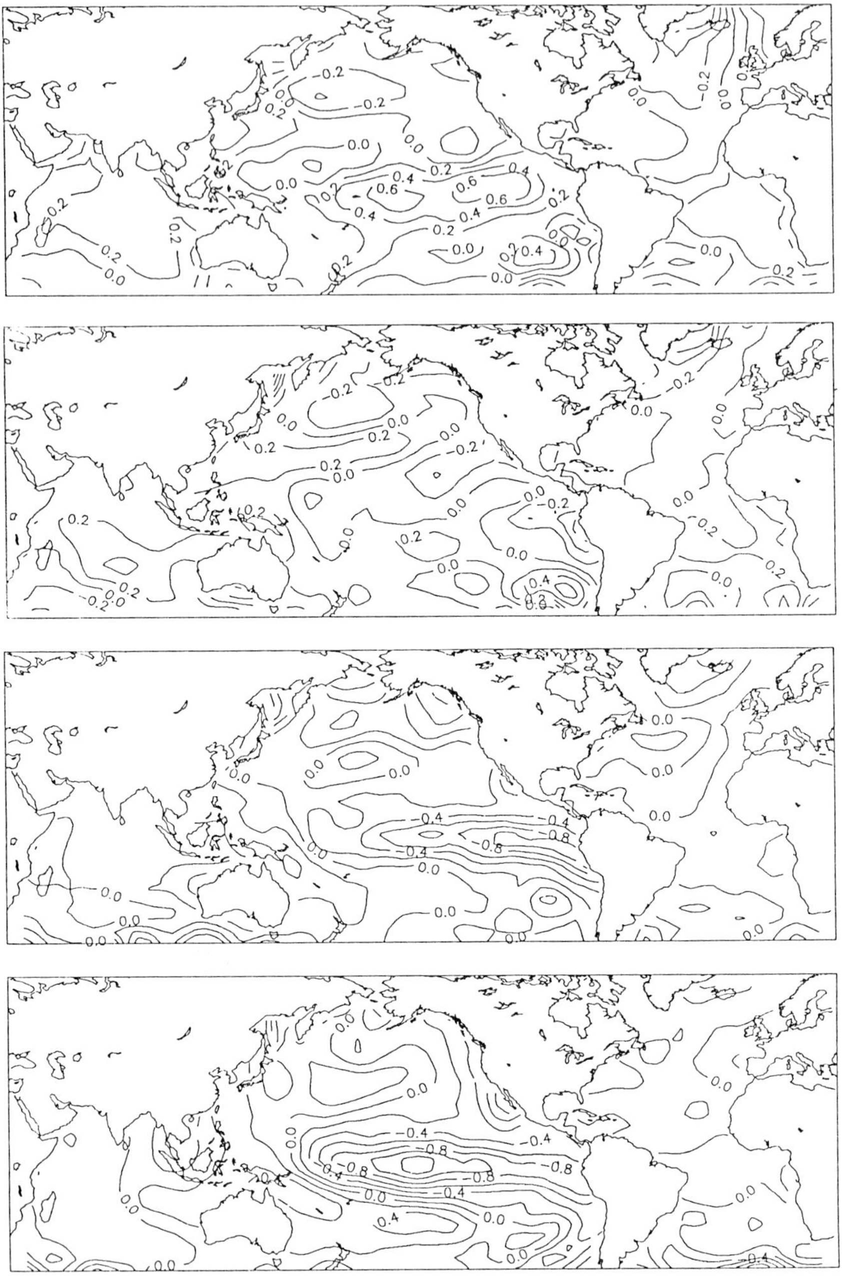

FIGURE 4. Composite La Niña event, created from all strong or medium events since World War II (see text for details): (A) winter Year 0, (B) spring Year 0, (C) summer Year 0, (D) fall Year 0, (E) winter Year +1, (F) spring Year +1, (G) summer Year +1, (H) fall Year +1.

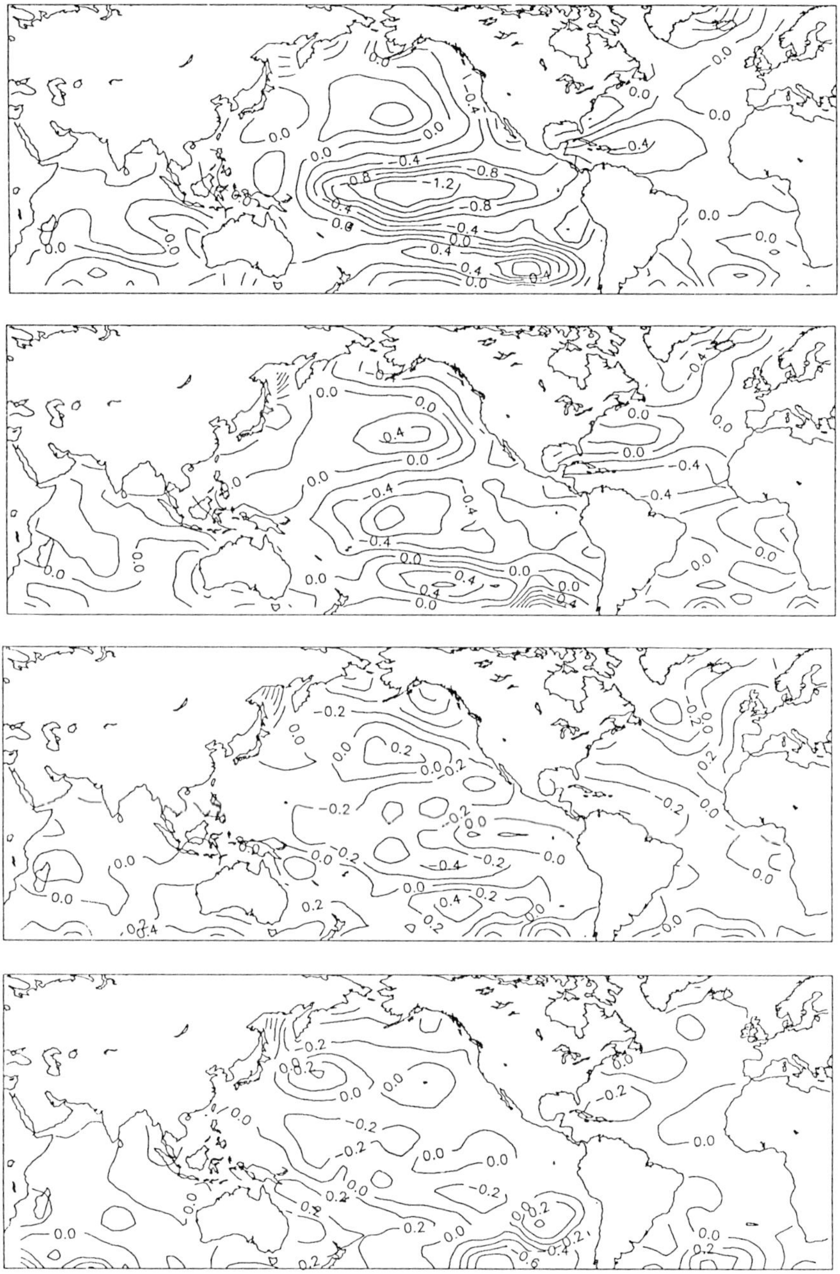

E F G H

FIGURE 4.

El Niño and La Niña events are two phases of the most important single mode of global SST variation.[23] These events also have influence on oceans elsewhere, especially in the mid-latitudes of the Pacific and the tropical Atlantic. However, the linkage of the tropical Pacific climate state to SST variability elsewhere is not always straightforward, as other sources of climate variation can also be visited upon these regions. The SST variability in the Atlantic and the Northern Hemisphere extratropics will be discussed below.

The Tropical Atlantic

The variability of the tropical Atlantic region is very important for the determination hydrologic conditions in the Sahel[24] and northeastern Brazil.[25] Any perceived linkage of Atlantic SSTs to the occurrence of El Niño events seems to be unsteady through time;[11] although the composite La Nina demonstrates a distinct cooling in the north tropical Atlantic during Year +1 (Figure 4E, 4F). This region north of the equator corresponds to the second mode of tropical Atlantic SST variance as determined by rotated principal component analysis.[26] The first mode of SST variance is located in the Atlantic south of west Africa. These two patterns (not shown) represent more than 60 percent of the variance of SST in the tropical Atlantic region. The presumed existence of a dipolar pattern of simultaneous variance straddling the latitude of the Intertropical Convergence Zone (ITCZ) has been shown to be an artifact of the unrotated principal component method.[26]

As in the case of El Niño/La Niña, periodic warmings and coolings in the Atlantic appear to be related to changes in the local Walker circulation.[27,28] However, unlike in the Pacific basin, the Walker circulation cell in the tropical Atlantic is very compact from west to east; in anomalous situations, convection centers tend to be displaced meridonally rather than zonally. The overall correlation of the SSTs in the tropical Atlantic to an index of the Southern Oscillation reveals statistically significant simultaneous correlations only to the north of the equator, where regressions do not explain much more than 20 percent of the variance in most places.[29] It is possible that lag relationships between eastern equatorial Pacific SSTs and Atlantic SSTs may reveal stronger correlations, as tropical Atlantic warming (cooling) often follows peak El Niño warming (La Niña cooling) by a season or two in the composites shown in Figure 2 and 4.

The Northern Hemisphere Extratropics

The principal nonseasonal modes of global SSTs have been identified by Hsuing and Newell.[23] As is expected, the first mode is related to El Nino/La Nina events, with a pattern of highly positive values in the east equatorial Pacific, and highly negative values in the mid-latitudes of the north central Pacific and southwest Pacific. A center with highly positive values is also found in the extreme eastern Pacific near the west coast of the United States. This is a pattern seen very clearly in the composites, such as Figures 2D or 4E. The second eigenvector pattern (not shown) consists of positive values in both the north Pacific and northeast Atlantic, and may be related to a general downward temperature tendency since the 1950s.

The third mode, while dominated by a general positive pattern in the Southern Hemisphere, also demonstrates positive/negative dipoles oriented west to east in both the northern Atlantic and northern Pacific oceans.

The first and third modes have been identified in other studies focusing on the north Atlantic or north Pacific basins. Namias and colleagues have found that the north Pacific modes are highly related to atmospheric circulation anomalies, both overhead and downwind.[30,31,32,33,34] The first north Pacific basin mode is highly persistent, and is coincident with the Pacific/North American Pattern of mid-tropospheric heights during winter;[34,35,36] this atmospheric pattern is associated with the amplitude of the normal trough-ridge-trough pattern in this region. In the north Atlantic, a north/south dipole forms the first mode of SST variation and is oriented from the northwest Atlantic to the central subtropical Atlantic. This is substantially related to the atmospheric circulation mode known as the North Atlantic Oscillation,[35] representing a dipole pattern of inverse air pressure relationships between the subpolar and subtropical areas of the north Atlantic.

While the ocean system is fully coupled with the atmosphere in the middle and higher latitudes, atmospheric circulation anomalies often lead extratropical SST anomalies. Wallace and Jiang[37] examined the most substantial modes of north Pacific SST and sea level pressure variability and found that changes in the atmospheric state preceded changes in the oceanic state. However, once the SST anomaly fields are established, persistence of the atmospheric pattern supporting the SST pattern is large. The propensity for SST anomaly patterns to change in the extratropics is actually more closely related to activity in the east coast baroclinic zones of Asia and North America.[36] Major SST tendency patterns are associated with the West Pacific Oscillation and West Atlantic Pattern in the mid-troposphere, which are found straddling the jet stream maxima, extracting energy from the baroclinic waves passing through the jet. When these patterns are strong, the atmospheric circulation at the surface tends to cool SSTs to the north and south of the southern node of the atmospheric pattern, where the westerlies to the north and trade winds to the south are enhanced. Warming may occur in the subtropics at this time. The opposite result is found when the atmospheric circulation patterns are weaker than normal. The actual locations of cooling or warming SSTs may be shifted off the center of the atmospheric anomaly; the flux of sensible heat from the ocean tends to be larger to the east and west of the atmospheric circulation center, as these are the regions of preferred warm or cold air advection.[38]

In summary, the picture of global ocean temperature variability is very complex. The dominant El Niño/La Niña mode has been subject to a great deal of scrutiny, and the state of knowledge regarding these events has improved greatly over the last decade. Efforts to predict the occurrences of these events have been refined substantially, but there is still much that is not known. The state of knowledge of extratropical SST variations has also improved, especially with the advent of satellite technology capable of measuring SSTs in remote areas. Short term climate predications based on real-time SST data are now operational at the United States Climate Analysis Center.[39] Much of what has been learned about ocean surface temperature variations can also be applied to the question of long-term climate change, which is addressed in the next section.

OCEAN TEMPERATURE CHANGE

The interpretation of long-term historical records of ocean temperature is fraught with difficulties. Changes in spatial coverage, ship type, and measurement techniques have required the application of empirical corrections in order to compare data over time. The general correspondence between the results of several research groups indicates that a fairly robust signal for global oceanic warming exists after 1900. There is considerably more divergence in the findings derived from the Nineteenth Century data, so the analyses shown here encompass the post-1900 period.

Figures 5A and 5B display the Northern Hemisphere and Southern Hemisphere ocean surface temperature change from 1900-1989, as derived by the United Kingdom Meteorological Office.[5] Both hemispheres have undergone substantial warming during the Twentieth Century. However, the Southern Hemisphere

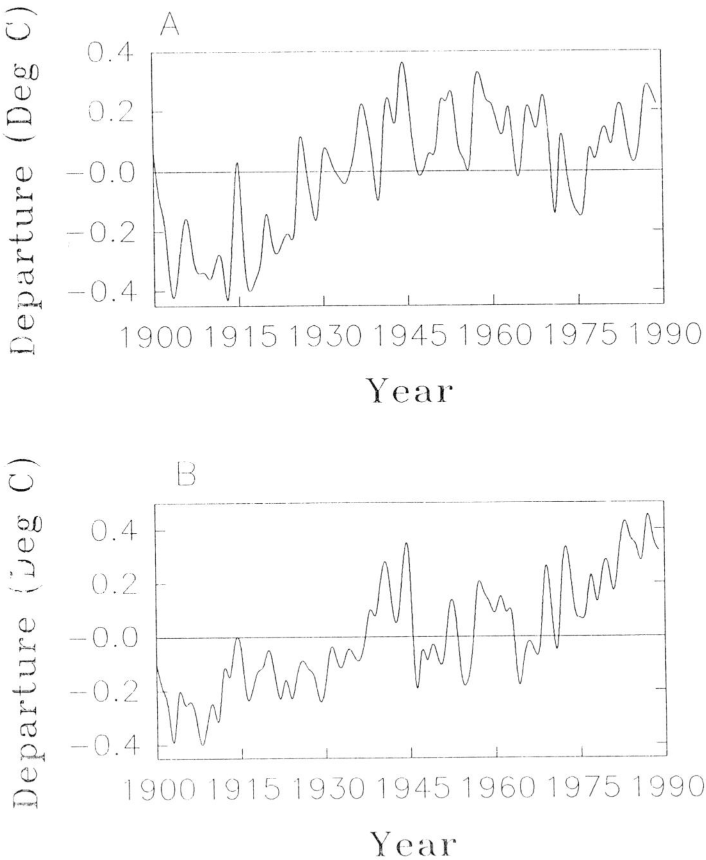

FIGURE 5. Hemispheric SST departures, 1900-1989. Values from Folland et al.[5]: (A) Northern Hemisphere, (B) Southern Hemisphere.

warming has been larger and has trended more steadily upward than the Northern Hemisphere record. In the north, temperatures rose rapidly until the 1940s, and have since been fairly level, except for a markedly cooler period centered in the early 1970s. The oceans in each hemisphere are behaving in a manner similar to land surface air temperatures (not shown), but are lagging behind by 5 to 10 years due to the thermal properties of the ocean.

The patterns of seasonal SST trends during the Twentieth Century are shown in Figure 6. These maps display the slope of the least squares regression between each SST time series and year, multiplied by 100 in order to present the trends as °C per century. Grid points that were missing more than two-thirds of the data between 1900 to 1925 were not included in the regression analysis. The resulting patterns indicate a highly differentiated response to climate change has been taking place during the last 92 years. During all seasons of the year, an extraordinary cooling has been taking place in the North Atlantic, south of Greenland. Yet the remainder of the Atlantic has been undergoing a substantial warming, especially at the high latitudes of the Southern Hemisphere (Figure 6A-6D). Meanwhile, in the north Pacific, a very strong warming is taking place during the winter, spring, and fall, with a slightly reduced warming during the summer. A bit further south, temperature change is very small in the central north Pacific in all seasons except for the fall, when the entire north Pacific has undergone a sizable warming (Figure 6D). In the tropical Pacific, SST trends generally are weakly positive, but diminish to near zero in the Southern Hemisphere subtropics. Interestingly, warming is strongest in the tropics during winter in the equatorial east Pacific (Figure 6A), and may indicate a change in the frequency or intensity of El Niño events. The far southern Pacific is warming consideraly during all seasons, with the possible exception of spring. Finally, the Indian ocean displays the same distinctive temperature change pattern in each season: warming near southern Asia, very little if any warming in the central Indian Ocean, and strong warming at more poleward latitudes.

The causes of the differences in trend between the Atlantic and Pacific oceans are not clear. Changes in the frequency and magnitude of El Niño/La Niña events may swamp any underlying trends in the central Pacific, while secular changes in the North Atlantic Oscillation may be leading to cooler conditions in the north Atlantic, south of Greenland. Note that the furthest northern reaches of the Atlantic show a warming much like that in the northernmost Pacific; at the same time, the southern poleward latitudes are undergoing substantial warming. These sub-polar zones may have enhanced warming due to ice-albedo-temperature feedbacks that are expected to be important near the seasonal sea ice margin. Finally, the entire south Atlantic is warming rapidly without regard for the processes holding subtropical south Pacific warming to very small trends.

CONCLUSIONS

The temperature patterns of the ocean surfaces are produced through complex interactions between the atmosphere and ocean. These arrangements impact human society on all time scales, from our day-to-day weather to the long-term trends that

FIGURE 6. Global SST trends, 1900-1991. The values displayed are the slope of the linear regression between SSTs and years, multiplied by 100 to provide the results in terms of °C per century: (A) winter, (B) spring, (C) summer, (D) fall.

may result from an enhanced greenhouse effect. The major mode of oceanic surface temperature variability is typified at its extremes by El Niño/La Niña events. As shown in the chapter by Leathers,[18] extraordinary powerful atmospheric climate anomalies associated with the El Niño/La Niña cycle may affect lands from the tropics to the high latitudes. In addition, the ordinary coupled behaviors of the oceans and atmospheres in the tropical Atlantic and northern extratropics also produce land surface climate variability.

By studying these modes of ocean surface temperature variability, more can be learned about processes that lead to long-term changes in global climatic conditions. Foremost among these is the possibility that secular changes in the magnitude and frequency of certain *atmospheric* circulation modes may be reflected in trends of surface temperature.[40] Unfortunately, simple hemispheric temperature averages tend to become an overriding concern to policy makers, even though they mask the striking patterns of spatially-distributed SST trends. If the spatial patterns of SST trends become the foci of investigation, then one can ask fundamental questions regarding the origins of the observed global-scale warming. For instance, do they represent an enhanced greenhouse effect climate change? While the observed high latitude SST warming may meet some of the expectations of models predicting potential climate changes, it is hard to establish proof with one type of climate variable and location. The cooling trend in the North Atlantic deep water formation zone is a powerful reminder that ocean temperatures may also be subject to processes involving the deep ocean circulation. In addition, the south Atlantic has not been recognized previously as a region in which to expect preferential warming under an enhanced greenhouse effect. At present, the patterns shown in Figure 6 provide more questions than answers regarding the future path of the climate state of the oceans.

REFERENCES

1. Perry, A.H. and J.M. Walker. 1977. *The Ocean-Atmosphere System.* Longman, New York, NY, pp. 160.
2. Slutz, R.J., S.J. Lubker, J.D. Hiscox, S.D. Woodruff, R.L. Jenne, D.H. Joseph, P.M. Steurer and J.D. Elms. 1985. *COADS Comprehensive Ocean-Atmosphere Data Set - Release 1.* Climate Research Program, ERL, Boulder, CO. pp. 39 + appendices.
3. Woodruff, S.D., R.J. Slutz, R.J. Jenne and P.M. Steurer. 1987. A Comprehensive Ocean-Atmosphere Data Set. *Bull. Am. Met. Soc.* 68:1239-1250.
4. Folland, C.K. and D.E. Parker. 1989. Observed variations of sea surface temperature, pp. 31-52. In: *NATO Advanced Research Workshop on Climate-Ocean Interaction, Oxford, UK, 26-30 September 1988.* Kluwar Academic Press, New York.
5. Folland, C.K., T.R. Karl and K. Ya. Vinnikov. 1990. Observed climate variations and change, pp. 194-238. In: J.T. Hougton, G.J. Jenkins and J.J. Ephraums (Ed.) *Climate Change: The IPCC Scientific Assessment.* Cambridge University Press, Cambridge, UK. pp. 365.

6. Maul, G.A. 1992. Temperature and sea level change, pp. 78-112. In: B.G. Levi, D. Hafemeister and R. Scribner (Ed.) *Global Warming: Physics and Facts. AIP Conference Proceedings 247.* American Institute of Physics, New York. pp. 310.
7. Bottomley, M., C.K. Folland, J. Hsiung, R.E. Newell and D.E. Parker. 1990. *Global Ocean Surface Temperature Atlas (GOSTA).* HMSO, London. pp. 20 + 313 plates.
8. Harnack, R.P. and A.J. Broccoli. 1979. Associations between sea surface temperature gradient and overlying mid-tropospheric circulation in the north Pacific region. *J. Phys. Ocean.* 9:1232-1242.
9. Lazante, J.R. 1983. A further assessment of the association between sea surface temperature gradient and the overlying mid-tropospheric circulation. *J. Phys. Ocean.* 13:1971-1974.
10. Zuta, S., D. Enfield, J. Valdivia, P. Lagos and C. Blandin. 1980. Physical aspects of the 1972-73 El Niño phenomenon, pp. 11-62. In: *Proceedings of the Workshop on the Phenomenon known as 'El Niño', Guayaquil, Ecuador, 4-12 December 1974.* UNESCO, Belgium. pp. 284.
11. Philander, S.G. 1990. *El Niño, La Niña, and the Southern Oscillation.* Academic Press, San Diego, CA. pp. 293.
12. Bjerknes, J. 1966. A possible response of the atmospheric Hadley circulation to equatorial anomalies of ocean temperature. *Tellus* 18:820-829.
13. Bjerknes, J. 1969. Atmospheric teleconnections from the equatorial Pacific. *Mon. Wea. Rev.* 97:163-172.
14. Reynolds, R.W. 1988. A real-time global sea surface temperature analysis. *J. Clim.* 1:75-86.
15. Diaz, H.F. and G.N. Kiladis. 1992. Atmospheric teleconnections associated with the extreme phases of the Southern Oscillation, pp. 7-28. In: H.F. Diaz and V. Markgraf (Ed.) *El Niño: Historical and Paleoclimatic Aspects of the Southern Oscillation.* Cambridge University Press, Cambridge, UK. pp. 476.
16. van Loon, H. and D.J. Shea. 1985. The Southern Oscillation. Part IV: The development of warm and cold events. *Mon. Wea. Rev.* 113:2063-2074.
17. Walker, G.T. 1923. Correlation in seasonal variations in weather. VIII. A preliminary study of world weather. *Mem. Indian Meteorol. Dep.* 24:75-131.
18. Leathers, D.J. 1994. Global meteorological teleconnections associated with El Niño/Southern Oscillation events. In: S.K. Majumdar, E.W. Miller, G.S. Forbes, R.F. Schmalz, and A.A. Panah (Ed.) *The Oceans: Physical-Chemical Dynamics and Human Impact.* The Pennsylvania Academy of Science Publication, Easton, PA.
19. Rasmusson, E.M. and T.H. Carpenter. 1982. Variations in tropical sea surface temperature and surface wind fields associated with the Southern Oscillation/ El Niño. *Mon. Wea. Rev.* 110:354-384.
20. Meehl, G.A. 1987. The annual cycle and interannual variability in the tropical Pacific and Indian Ocean regions. *Mon. Wea. Rev.* 115:27-50.
21. Quinn, W.H. 1992. A study of Southern Oscillation-related climatic activity for A.D. 622-1900 incorporating Nile River flood data, pp. 119-149. In: H.F. Diaz and V. Markgraf (Ed.) *El Niño: Historical and Paleoclimatic Aspects of the Southern Oscillation.* Cambridge University Press, Cambridge, UK. pp. 476.

22. van Loon, H. and D.J. Shea. 1987. The Southern Oscillation. Part VI: Anomalies of sea level pressure on the Southern Hemisphere and of Pacific sea surface temperature during the development of a warm event. *Mon. Wea. Rev.* 115:370-379.
23. Hsuing, J. and R.E. Newell. 1983. The principal nonseasonal modes of variation of global sea surface temperature. *J. Phys. Ocean.* 13:1957-1967.
24. Lamb, P.J. and R.A. Peppler. 1992. Further case studies of tropical Atlantic surface atmospheric and oceanic patterns associated with Sub-Saharan drought. *J. Clim.* 5:476-488.
25. Hastenrath, S. and L. Greischar. 1993. Circulation mechanisms related to northeast Brazil rainfall anomalies. *J. Geophys. Res.* 98:5093-5102.
26. Houghton, R.W. and Y.M. Tourre. 1992. Characteristics of low-frequency sea surface temperature fluctuations in the tropical Atlantic. *J. Clim.* 5:765-771.
27. Philander, S.G.H. 1986. Unusual conditions in the tropical Atlantic ocean in 1984. *Nature* 322:326-238.
28. Mechoso, C.R. and S.W. Lyons. 1988. On the atmospheric response to SST anomalies associated with the Atlantic warm event during 1984. *J. Clim.* 1:422-428.
29. Hameed, S., K.R. Sperber and A. Meinster. 1993. Teleconnections of the Southern Oscillation in the tropical Atlantic sector in the OSU coupled upper ocean-atmosphere GCM. *J. Clim.* 6:487-498.
30. Namias, J. 1969. Seasonal interactions between the north Pacific Ocean and the atmosphere during the 1960s. *Mon. Wea. Rev.* 97:173-192.
31. Namias, J. and R.M. Born. 1970. Temporal coherence in North Pacific sea-surface temperature patterns. *J. Geophys. Res.* 75:5952-5955.
32. Namias, J. 1978. Multiple causes of the North American abnormal winter 1976-77. *Mon. Wea. Rev.* 106:279-295.
33. Namias, J. 1986. Persistence of flow patterns over North America and adjacent ocean section. *Mon. Wea. Rev.* 114:1368-1383.
34. Namias, J., X. Yuan and D.R. Cayan. 1988. Persistence of north Pacific sea surface temperature and atmospheric flow patterns. *J. Clim.* 1:682-703.
35. Wallace, J.M. and D.S. Gutzler. 1981. Teleconnections in the geopotential height field during the Northern Hemisphere winter. *Mon. Wea. Rev.* 109:784-812.
36. Wallace, J.M., C. Smith and Q. Jiang. 1990. Spatial patterns of atmosphere-ocean interaction in the northern winter. *J. Clim.* 3:990-998.
37. Wallace, J.M. and Q. Jiang. 1986. On the observed structure of the interannual variability of the atmosphere/ocean climate system, pp. 17-43. In: H. Cattle (Ed.) *Atmospheric and Oceanic Variability.* Royal Meteorological Society, Bracknell, Berkshire, UK. pp. 182.
38. Cayan, D.R. 1992. Latent and sensible heat flux anomalies over the northern oceans: Driving the sea surface temperature. *J. Phys. Ocean.* 22:859-881.
39. Barnston, A.G. and R.E. Livezey. 1989. An operational multifield analog/anti-analog prediction system for United States seasonal temperatures. Part II: Spring, summer, fall and intermediate 3-month period experiments. *J. Clim.* 2:513-541.
40. Palecki, M.A. and D.J. Leathers. 1993. Northern Hemisphere extratropical circulation anomalies and recent January land surface temperature trends. *Geophy. Res. Let.* 20:819-822.

The Oceans: Physical-Chemical Dynamics and Human Impact. Edited by S.K. Majumdar, E.W. Miller, G.S. Forbes, R.F. Schmalz and Assad A. Panah. © 1994, The Pennsylvania Academy of Science.

Chapter Thirteen

GULF STREAM THERMAL ANOMALIES: Local Dynamics and Implications for the Gyre Circulation

LEWIS M. ROTHSTEIN and **CRAIG GILMAN**

Graduate School of Oceanography
Narragansett Bay Campus
University of Rhode Island
Naragansett, RI 02882-1197
and
Center for Earth and Ocean Research
University of Victoria
Victoria British Columbia

INTRODUCTION

Classical linear dynamical views of oceanic western boundary currents in general, and the Gulf Stream in particular, teach us that these currents exist because of basin-scale wind stress curl patterns and an ambient planetary vorticity gradient field, with some form of frictional decay invoked (historically along the western boundary or the bottom) to balance the input of energy from the wind (e.g. Stommel, 1948). This view dynamically involves the entire subtropical gyre and assigns the narrow western boundary currents the role of returning poleward the much broader interior equatorward mass transport. Although compelling in its simplicity, this inherently rather passive character of the western boundary current competes with ideas that assign a much more active role to this boundary layer; one in which the western boundary current actively influences the interior circulation and hence itself. Even though this non-linear feedback is undoubtedly more realistic, our modern theories of the large scale circulation (Luyten, et al., 1983; Rhines and Young, 1982) still implicitly assign a passive role to the western boundary current. We here present some rather compelling observations and numerical model results that may help us to better understand the Gulf Stream's role in determining its own fate. Our approach here is with a process-oriented numerical modelling study, firmly rooted in the set of observations that initially motivated the study. We attempt to represent the observations with a set of "reduced physics" equations that, we hope, include the fundamental physics of the process we observe. This enables isolation of the essential dynamics in ways which will help us to better understand the mechanisms involved.

The story is organized as follows: We first briefly describe a set of observations that suggest ways in which the Gulf Stream communicates with its surroundings. The visual signatures of this interaction are satellite observations of synoptic scale thermal anomalies which propagate along the Stream's major axis. It is useful to translate these surface thermodynamic signatures into potential vorticity anomalies, and we do so with the aid of additional observations. The relevant local dynamics are observationally described and then investigated beyond that by studying a numerical simulation of the observations. The model helps fill in the picture, allowing for a more comprehensive dynamical discussion of the local dynamics. We conclude with a rather speculative discussion of the possible importance that these synoptic scale interactions may have on the gyre scale circulation.

OBSERVATIONS: SYNOPTIC SCALE THERMAL AND POTENTIAL VORTICITY ANOMALIES

Since Gulf Stream eddies transport large quantities of momentum, potential vorticity and heat, much study has been directed towards the dynamics of these features at different stages of their life span. These include the conditions for eddy formation (Robinson et al., 1988; Pratt and Stern, 1986), the self-propagation of isolated vortices (Nof, 1981, Smith and O'Brien, 1983), and the interaction of an eddy with a jet, including the eventual 'capture' of an eddy (Smith and Davis, 1989; Stern and Flierl, 1987). Our goal here is to supplement these studies with a recent set of observations (Gilman and Rothstein, 1994; hereafter GR) and with a numerical simulation of a mean flow — mesoscale eddy interaction. Of specific interest is the impact that a Gulf Stream/eddy interaction has on fluid parcel pathways in the Gulf Stream, the eddy, and neighboring waters.

In a study using five years of thermal infrared (IR) satellite images, a sequence of mesoscale processes was repeatedly observed that depicted the evolution of a feature that we have labeled a Gulf Stream 'warm surge' (GR). The sequence begins with an interaction between a cold core eddy and the Gulf Stream between Florida and Cape Hatteras. This interaction results in the Gulf Stream surface layers being detrained out of the Stream and advected cyclonically around the eddy. The sea surface temperature (SST) of the Gulf Stream downstream of the interaction decreases 1-3°C (over a few days) due to this sudden loss of the warm surface layer upstream. The Stream/eddy detrainment process eventually weakens, and the warm surface water flowing through the Florida Straits resumes its normal flow past Cape Hatteras. This return of warm water (the 'warm surge') is easily visible on the IR images as a 'jet' with a sharp leading edge that is often confined to the cyclonic (Slope Water) side of the Stream (Figure 1). The phase of upstream detrainment is labeled as the 'cold phase' of the event (since downstream SST decreases), and likewise the return of the warm surface layer is referred to as the 'warm phase'. It should be kept firmly in mind that we were able to identify 9-11 of these warm surge events in *each* year of the 5 year study. The ubiquity of these events will play a role in our speculations about their possible implications for the gyre circulation.

Vertical profiles of the Gulf Stream during warm surge events indicate substantial vertical stretching and shrinking of the Gulf Stream's surface layers down to the top of the main thermocline (GR). Richardson (1979) observed similar stretching in the surface layers of a cold core eddy when Gulf Stream water was detrained into the eddy. The upper water column of the Gulf Stream is likewise affected when the surface layers are first detrained and subsequently return several days later. In addition, independent velocity sections across the Stream during the warm surge

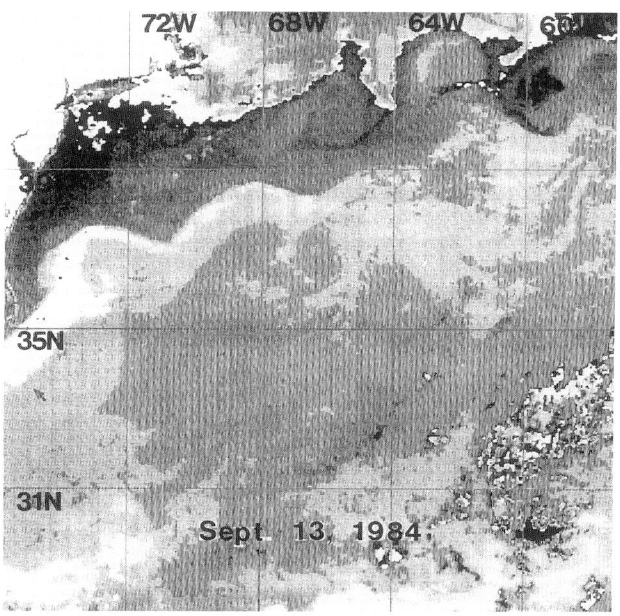

FIGURE 1. Thermal IR image from September 13, 1984. The coldest water is shown by the darkest regions with increasing brightness representing increasing temperature. A 4°x4° grid is shown on the image. A cold eddy had previously coalesced with the Gulf Stream east of Cape Hatteras (lower left hand corner), evident by the detrainment of warm Gulf Stream water around the eddy. The coalescence is here weakening, with a warm surge extending to 71°W visible on the cyclonic side of the Gulf Stream as a narrow band of warm water.

event indicate a substantial changes in the relative vorticity of the Stream's upper water column. The alteration in both the layer thickness and the relative vorticity indicate a rapid change in the potential vorticity of the Gulf Stream. Thus, the warm surge event represents not only a thermal anomaly pattern (rapid cooling followed by rapid warming of the Stream's surface), but also a potential vorticity anomaly pattern. GR reported that in the velocity core of the Stream the layer potential vorticity decreased by more than 50% during the cold phase of the surge event.

When the surface layers are detrained out of the Gulf Stream, conservation of mass implies that these layers must be replenished. The obvious candidates for replenishment include the eddy, the Gulf Stream's subsurface layer (Subtropical Underwater (SUW)), the Sargasso or Slope Water surface waters. Unfortunately, it is not possible to hydrographically distinguish the source because: 1) the SUW and the surface Sargasso waters near the Gulf Stream have the same T-S properties (T = 23-20°C, S = 36.5-36.7°/$_{oo}$) (Watts, 1983; Worthington, 1976; Fuglister, 1960); and 2) the T-S properties of each specific eddy are unknown. (The eddy originally consists of Gulf Stream water, however it is likely to undergo substantial mixing during it's propagation through the Sargasso (Richardson, 1983; Ring Group, 1981) making it difficult to identify this as the source. Studies have shown that cold core eddies do significantly change their surface temperature and salinity structure over time (Ring Group, 1981)). However, satellite images and cross-stream velocity sections presented in GR do indicate an enhanced entrainment of water from the Sargasso Sea thus replacing the Stream's detrained surface layer and suggesting that this is the replenishment source. Figure 2, from May 11, 1983, shows the leading edge of a 25°C warm surge 'jet' (extending just downstream of 73°W), with 23°C water downstream of the jet. The northern edge of the Gulf Stream is a continuous front, however the southern edge is discontinuous implying an exchange of surface water between the Stream and the Sargasso (also evident by the filamented structure in the region). A cross-stream velocity section obtained on May 10 at the PEGASUS transect line shown on Figure 2 more directly indicates entrainment of water from the Sargasso into the Stream throughout the upper several hundred meters of the water column (see GR for details). The cross-stream velocities on May 10 increase four-fold (from the Sargasso into the Stream) over the mean values obtained by averaging sixteen PEGASUS transects. Unfortunately, no other velocity sections were available during other warm surge events, thus no comprehensive conclusions could be made concerning the horizontal exchange between the Gulf Stream and neighboring waters, although the present example is compelling.

Several other questions were raised by the observational study of GR in which it became evident that a numerical simulation would greatly aide the investigation. Because the satellite observations occur only once a day at best it is not possible to continously observe the local dynamics of the Stream/eddy interaction that lead to the formation of the warm surge. Also, because there were no direct velocity measurements in the interaction area, or in the Sargasso and Slope water further downstream, it was not possible to continuously monitor downstream detrainment or entrainment of waters below the surface during a warm surge event. However, these issues can be further investigated by numerical analyses which give a continuous depiction of the Stream/eddy interaction and its consequences on the circulation of neighboring waters.

For the present set of observations, a simple yet accurate model of a Gulf Stream — cold core eddy interaction was needed to specifically investigate our main hypothesis: that upstream detrainment of warm Gulf Stream surface waters by Stream/eddy interactions is responsible for the observed downstream cold/warm thermal anomaly sequence seen in the IR images. Specifically, the model had to be designed to address some fundamental questions put forth from the observations. These questions include: a) Upon detrainment, where do the warm surface waters go? Do these warm waters merely circulate round the eddy or will they permanently be ejected into the Sargasso Sea?; b) What is the source of the cooler water which replenishes the detrained warm surface waters? Is there only horizontal advection from the cold eddy itself or is advection from the Sargasso Sea and Slope Water important?; c) Is the observed modification of the Gulf Stream's available potential energy (APE) strong enough to further feed back on the sequence of events leading to eddy detrainment of Stream warm waters, i.e., can the alterations to the APE by the local dynamics be significant enough to cause the Stream to become unstable downstream of the original interaction region?; and finally d) What are the Lagrangian pathways of various marked fluid particles before, during and after a detrainment event?

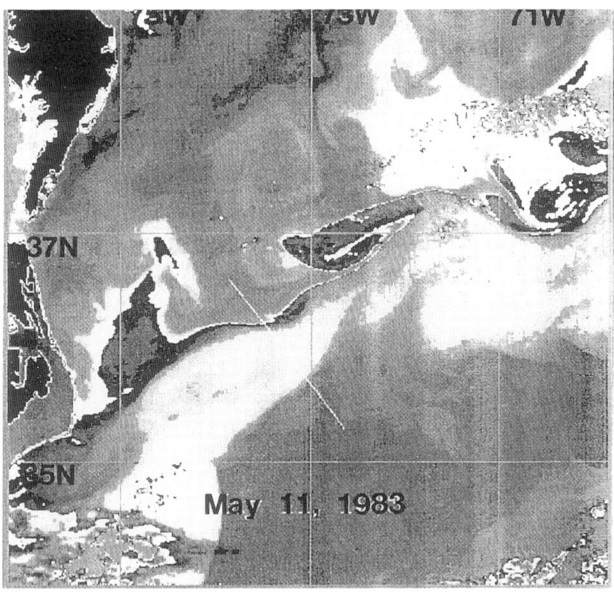

FIGURE 2. IR image from the Cape Hatteras region on May 11, 1983. Temperature representation is the same as Figure 1. The leading edge of a warm surge is seen just upstream of a discontinuity in the southern wall of the Gulf Stream. The PEGASUS transect line is indicated.

We next present a brief description of the model that we have chosen to highlight these issues and present the results of a local dynamics experiment.

NUMERICAL EXPERIMENTS

Model Physics

A nonlinear 'stacked' shallow water model was employed for this study consisting of four dynamically active layers each of constant density above a deep quiescent lowest layer (4½ layers), with parameters set to accurately model the Gulf Stream. (The model was originally developed by Soiland (1991) to simulate the Lagrangian motion of particles in an idealized (Gulf Stream) jet and to study the dynamical adjustments these particles experiences as they travel though meander crests and troughs.) The fluid is presumed hydrostatic and the Boussinesq approximation is presumed valid. The Gulf Stream is *specified* as a strong meandering jet with propagating and growing finite amplitude meanders. A Gaussian cyclonic eddy is introduced south of the Stream and self-propelled toward the jet on a β-plane to initiate the eddy/Stream interaction. Lagrangian particles (or floats) are marked and their pathways are followed before, during and after the Stream/eddy interaction. The model consists of thermodynamically inactive fluid, thereby relinquishing any hope of simulating the SST evolution of the cold/warm surge. However, this is the simplest first approximation to dynamically simulate the rather complex mechanical interaction processes and to isolate horizontal advective processes without the added complications of diapycnal mixing across the active layers and the role of mixed layer turbulence in determining SST (processes that no doubt are important in the real system). Nevertheless, these are important omissions and suggest some obvious extensions to the present effort.

The model domain is rectangular and was specified, for a local dynamics experiment, as approximately 450km longitude x 1000km latitude. In all of the experiments, the model is configured as a 'channel' with periodic east-west boundary conditions (the last and first points in the x-direction are considered as neighboring points) and open north-south boundaries. The layer and density structure of the model was initialized to represent an indealized Gulf Stream and a cyclonic eddy. The Gulf Stream is modelled as a meandering 'thin jet', specified by a Stream axis and cross-stream structure. The Stream axis was defined as a sinusoid with the cross-stream and density structure based on Gulf Stream observations (Soiland, 1991). The eddy is specified as a cyclonic Gaussian eddy. The ratio of the eddy size to the first internal Rossby radius (R_d) is fixed at unity. Adjustable variables include the meander wavelength and amplitude, the eddy pressure distribution (i.e. velocity), and the initial Stream/eddy separation distance.

The depth of each layer is initially specified with constant values far to the north and south of the jet and the eddy. The cross-stream thicknesses resemble a hyperbolic tangent, obtained as a fit of the mean PEGASUS temperature section across the Gulf Stream (see Figure 3 from Halkin and Rossby, 1985). The model height field (and the actual Gulf Stream isopycnals) slope downward by 600-700m from the northern side of the Stream to the Sargasso side. Since the warm surge phenomena appeared in the observations to significantly affect only the upper 50-200m, the upper layer of the model was initialized to represent this layer. The second layer was initialized as a thick layer (600m), so that the initial height of the eddy (i.e. vertical displacement of the layer) could be adjusted. From the mean PEGASUS temperature profile, the interfaces between numerical layers best represent the 20, 14, 10 and 7°C isotherms. The initial density of each layer in the model was chosen to represent the observed average Gulf Stream values with an upper layer density initialized at σ_t = 26.0. The intermediate layers are initialized to give reasonable density jumps between layers, with the 20-14°C layer represented as σ_t = 26.5, the 14-10°C layer as σ_t = 27.0, and the 10-7°C as σ_t = 27.5. The deep inactive layer is represented as σ_t = 28.0, slightly higher than actual Gulf Stream values, but this was needed in order to increase the velocities in the lowest active layer to reasonable values (Soiland, 1991). A more detailed description of the model (including important information about the numerical issues encountered and the techniques that were used) and a complete derivation of the governing equations can be found in Soiland, 1991.

The potential vorticity (q), q_y and APE fields were monitored during all the

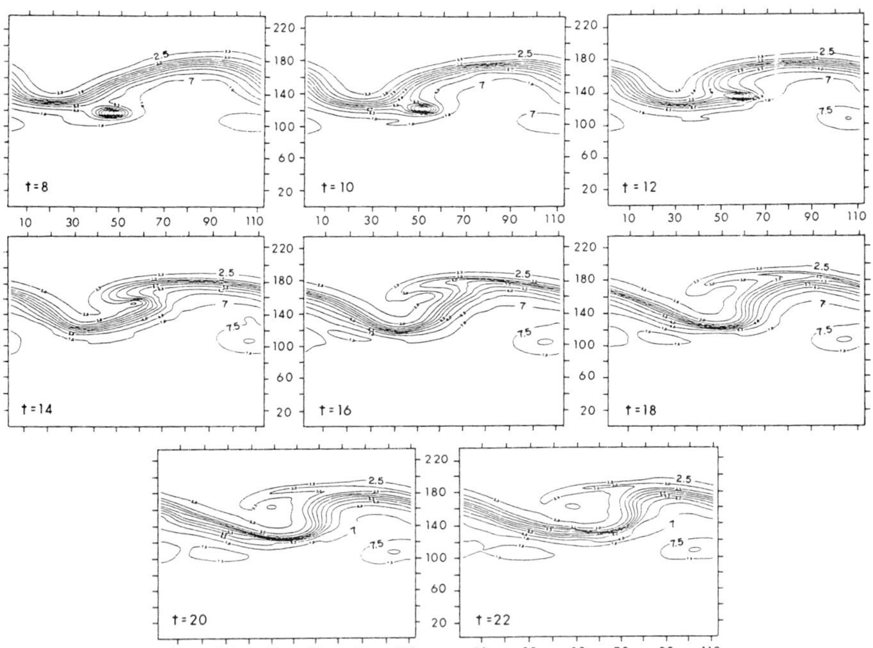

FIGURE 3. The evolution of the (layer 2) height field for Experiment 2 beginning on Day 8 and finishing on Day 22. The eddy begins to interact with the Stream on day 8, is rapidly advected downstream, and has been completely absorbed by day 16.

experiments. Both Eulerian and Lagrangian mean circulations proceed along constant q lines in a system with no forcing or dissipation (McDowell, et al., 1982) and f/h estimates in the open ocean are accurate depictions of the geostrophic contours which then define the flow field. In the Gulf Stream one must also account for the relative vorticity which is not negligible. Throughout the experiment the spatial structure and time evolution of the q field will be monitored to obtain information on the pathway and the stability of the flow. Baroclinic, barotropic or mixed instabilities are assumed to be responsible for Gulf Stream meander growth and eddy formation, with baroclinic instability thought to be the most important mechanism (Dewar and Bane; 1989; Johns, 1988; Ikeda and Apel, 1981). The mean APE is the energy source for conversion to eddy kinetic energy in a baroclinic instability, thus the end result of the instability process is a decrease in total APE and an increase in total kinetic energy. A necessary condition for baroclinic instability is that the potential vorticity gradient of the basic current must vanish on a line in the meriodional plane (Charney and Stern, 1962). Therefore both energy fields and q_y will be calculated throughout the experiments.

The model has been purposefully designed to follow Lagrangian trajectories of marked water parcels (hereafter 'floats'). The model isn formulated in an Eulerian framework, but Lagrangian information may be extracted from the Eulerian field by simply keeping track of a limited number of marked fluid parcels (Soiland, 1991). Also, the dynamic properties of these marked water parcels can be followed with time. The floats may move vertically within each layer, and since the fluid moves as a Taylor column within each layer, the floats can be thought of as tracking fluid columns. Since there is no assumed communication vertically between the layers in this study the floats cannot be exchanged between layers.

RESULTS

Since we are applying the numerical experiments to Gulf Stream studies, the text often refers to the model jet as the 'Gulf Stream', the region north of the jet as the 'Slope Water', and the region south of the jet as the 'Sargasso Sea', even though the experiments are a simplification of this system. The experiments emphasize local dynamics of the jet/eddy interaction by placing the majority of the floats in the jet and only a few floats outside the jet. By monitoring the Stream floats, we hope to study the Gulf Stream flow pattern in the cold and warm phases of the surge event as observed in the satellite imagery.

There are a number of previous studies that provided insight into the design of our experiment. Using a 2-layer primitive equation model to simulate an eddy interaction with an ocean jet, Smith and Davis (1989) found no tendency for eddies to coalesce with a non-meandering jet or a meander of opposite sign vorticity. Coalescence occurred only in experiments where a collision occurred between an eddy and a like sign meander. Thus in our experiment the cyclonic eddy is propelled towards a cyclonic meander because, while complete coalescence was not necessary for the formation of a warm surge event, it was observed in the majority of events.

Furthermore, Stern and Flierl (1987) indicated eddy capture occurs when the eddy-jet separation distance is less than a non-dimensional length scale l, defined as the square root of the vortex circulation divided by the potential vorticity of the shear flow. Smith and Davis (1989) obtained a quantitative comparison of the non-dimensional parameters of Stern and Flierl (1987) and reported l to be approximately $70km$ with a jet and eddy e-folding width equal to a first internal Rossby radius R_d of $40km$. Our experiment design incorporated the same parameters (jet width, eddy radius, potential vorticity magnitude in the jet) as in Smith and Davis (1989).

Since a comparison of the Lagrangian pathways of parcels in experiments with and without a Stream/eddy interaction was warranted, we ran experiments with initial jet/eddy separation distances of $l = R_d$ and $l = 2R_d$. In experiments with $l = 2R_d$ no interaction occurred, and in all experiments with $l = R_d$ there was an interaction, in qualitative agreement with the studies discussed above.

The initial jet structure is designed to resemble that observed in the North Atlantic, with a Stream maximum at the north wall of the Stream, decreasing q southward across the Stream, and generally constant q values found to the north and south of the jet (McDowell, et al., 1982). There is a q maximum at the center of the eddy due to the thinness of the layer there.

Our first experiment was a control experiment in which no Stream/eddy interaction occurred and we simply summarize the salient points here without detail. During this 24 day experiment, the eddy underwent β-induced westward propagation at a rate of $2.5km/day$. This speed is consistent with observed estimates of both the flow velocities of the recirculation on both sides of the Gulf Stream (up to $4km/day$) and the maximum westward phase speed for the first mode baroclinic Rossby waves of $4km/day$ (Worthington, 1976; McWilliams and Flierl, 1979). The meander amplitude increased slightly at a rate of $1km/day$ during the model run.

All the floats initially placed in the jet remained in the jet for the duration of this control experiment; likewise with the floats placed in the eddy. A simple kinematic mechanism for ejecting floats out of the meandering jet was investigated by Bower (1991) where it was found that floats with particle speeds lower then the meander propagation speed tended to be ejected. Since our objective is to study entrainment/detrainment during eddy interactions, floats in this experiments were placed towards the center of the jet (explaining why all the floats remained in the Stream) and this strategy will be used in all subsequent experiments, so that the majority of floats will not be expelled from the jet *before* an interaction begins.

The q field underwent no substantial alteration in structure or magnitude during this control experiment, and q_y did not indicate the potential for baroclinic instability in any of the layers, as the gradient remained positive at all locations within the jet. The APE of the system remained constant during the entire experiment. To summarize, during this control experiment with no Stream/eddy interaction the Stream floats remained in the Stream, three floats from the surrounding waters were entrained into the Stream (following the observational pattern reported in Bower and Rossby, 1989), and the dynamical fields were not altered.

The initial conditions for the main interaction experiment were identical to those in the control except now the eddy is initially placed closer to the jet (within one

Rossby radius). The evolution of the layer 2 height field beginning on day 8 and extending to the end of the experiment is presented in Figure 3. (In all figures, the height field is contoured in hundreds of meters, with both axes divided into model grid points, each point representing $4km$ (also true for all remaining figures). All q values have been non-dimensionalized (using the local value of the Coriolis parameter).) On day 8 the eddy begins to be absorbed into the jet, evident by the height contours (which here approximate streamlines) encircling the eddy. For the next several days, as the eddy attaches to the jet, it is rapidly (10-12km/day) advected downstream. Between days 12-14, the eddy is rapidly absorbed into the jet, followed by a breaking of the meander crest just downstream of the interaction. By day 16, the eddy has been completely absorbed and a large 'shingle' (Chew, 1981) has appeared on the meander crest which is continuously stretched longitudinally as the meander propagates downstream.

The interaction in this experiment quite accurately depicts a mode of interaction described observationally by Richardson (1983). Figure 4 shows the coalescence of eddy Bob with the Gulf Stream (Richardson, 1980). The temporal and spatial

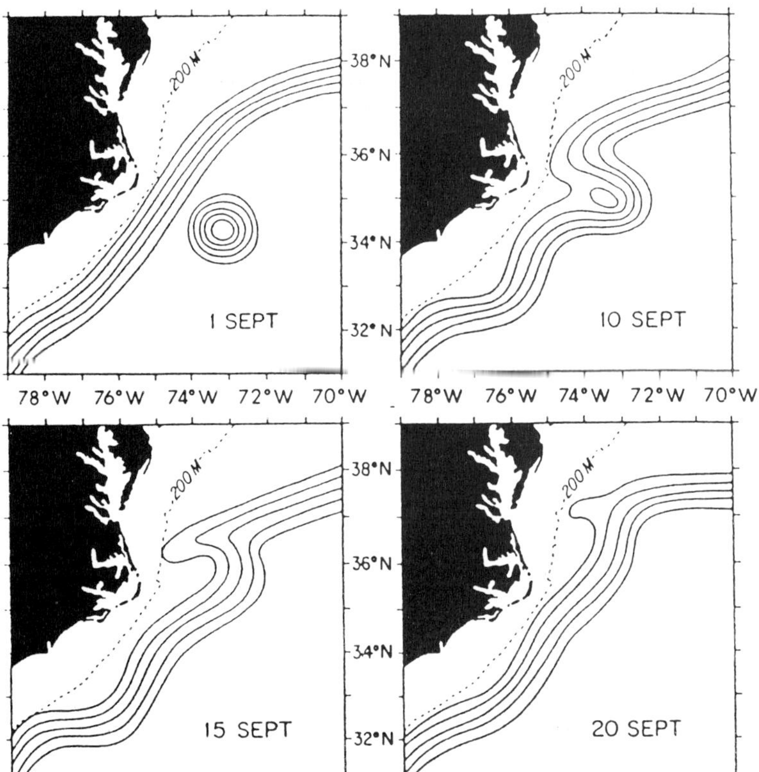

FIGURE 4. Schematic diagram showing the coalescence of ring Bob with the Gulf Stream in September 1977. Contours indicate depth inferred from XBT surveys, buoy trajectories and infrared images (from Richardson, 1980).

evolution of the height contours closely resembles the pattern observed in the numerical experiment as the eddy merges with the current and then the meander breaks back. The shingle is not observed to be continuously stretched out with time as in our experiment, but this could be due to the impinging continental shelf which appears to restrict the growth of this feature.

To depict the evolution of the Gulf Stream flow during the eddy interaction, a series of float lines were placed completely across the Stream (in the north-south direction) at increasing upstream distances from the interaction site. Thus, each line of floats will enter the interaction site as the interaction evolves and an evolution of the Lagrangian pathways of these Stream parcels can be followed. Each plot in Figure 5 shows the trajectories for a line of floats that entered the interaction site on a) day 8, b) day 10, c) day 11, d) day 13, and e) day 15. Before the interaction began all Stream float trajectories resembled the control experiment with the floats remaining evenly distributed across the Stream as they propagate downstream. As the interaction began, floats on the cyclonic side of the Stream were detrained into the eddy and several were subsequently ejected into the Sargasso Sea (Figure 5a). One of the detrained floats is seen to re-enter the Stream and resume its propagation down the jet. The floats that were not detrained remained evenly distributed across the Stream as they propagate downstream. As the interaction proceeds, parcels on the anti-cyclonic side of the Stream continue to be detrained into the eddy and ejected into the Sargasso. Interestingly, the non-detrained parcels become increasingly confined to the cyclonic side of the Stream (Figure 5b and 5c). A similar flow pattern was repeatedly observed in the Gulf Stream satellite imagery during the cold phase of the surge event (Figure 1) as downstream of the interaction the warm Gulf Stream surface water is visible as a narrow band 'squeezed' to the cyclonic side of the Stream. Eventually, as the eddy is advected downstream and absorbed, the detrainment of parcels diminishes and more of the detrained parcels re-enter the Stream rather than being ejected into the Sargasso. Also, the confinement of Stream parcels to the anticyclonic side is diminished (Figure 5d and 5e) as the floats are more evenly distributed across the Stream.

As in the control experiment, floats were also initially placed in the Slope Water and the Sargasso Sea to investigate any entrainment of neighboring waters to replenish the Stream's detrained waters. Similar to the control, the Slope Water floats were entrained into a meander only when a meander crest propagated towards (and caught up to) the float. The Sargasso floats, however, were greatly influenced by the Stream/eddy interaction. Five of the seven Sargasso floats were entrained into the Stream, beginning two days after the eddy interaction began (Figure 6). The two floats placed initially upstream of the interaction zone were advected south of the eddy, and towards the jet, but were not entrained. The earlier a float was entrained into the Stream (all were entrained between Day 10-13) the closer to the north wall of the Stream it eventually reached, indicating a greater cross-stream velocity as it entered the Stream. The dashed lines on Figure 6 indicate the Stream position on Day 8, however the meander is growing northward with time, thus the trajectories may not penetrate as far to the cyclonic side of the Stream as shown on the figure. However the tendency to move across the Stream is clear. Thus, several days after the interaction begins, the Stream (downstream of the interaction) is comprised of Sargasso Sea water on it's anticyclonic side (Figure 6) and Gulf Stream water that propagated past the interaction on it's cyclonic side (Figure 5c). Again, this pattern corresponds to the observational data (satellite and cross-stream velocity) presented in GR that inferred an enhanced entrainment of Sargasso Sea water during the cold phase of the surge event.

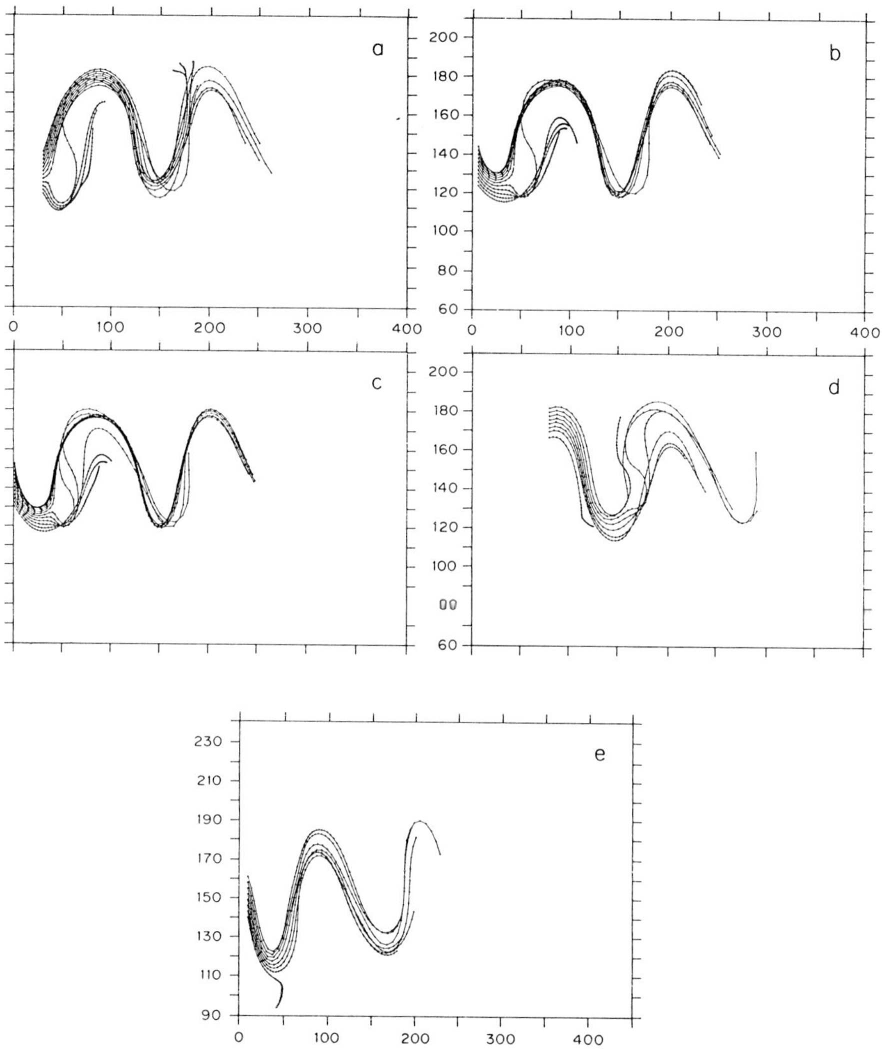

FIGURE 5. Trajectories for separate lines of floats initially placed across the Stream. The floats entered the Stream/eddy interaction zone on a) day 8, b) day 10, c) day 11, d) day 13, and e) day 15.

The alteration in the structure of the meander is also clearly evident in the evolution of the q and ∂_q/∂_y fields. Before the interaction the q isopleths are evenly distributed across the Stream (as in the control experiment), however as the interaction proceeds the q isopleths become closely packed on the cyclonic side of the jet (Figure 7a). Again, this is evidence of entrainment of Sargasso waters into the Stream. Since the fluid parcels conserve potential vorticity in the model, parcels initially in the Sargasso will retain their initial q value of (non-dimensional) 1. By day 16, these parcels incorporate the entire anticyclonic side of the Stream downstream of the interaction, and have 'squeezed' the Gulf Stream parcels ($q > 1$) into a narrow band on the cyclonic side. The q gradient is substantially altered (Figure 7b) with several reversals in sign within the jet indicating a potential for baroclinic instability. Also, the APE of the system did decrease by 4.1% during the 24-day experiment (not shown).

In summary, during this interaction experiment the Stream's flow is divided near the interaction site into parcels that are 1) detrained around the eddy and then often ejected into the Sargasso or 2) constricted into a narrow band on the cyclonic side of the jet. Following detrainment out of the jet, entrainment into the jet from the Sargasso occurs downstream of the interaction, all consistent with the satellite observation.

SUMMARY AND DISCUSSION OF RESULTS

It has been known for many years that cyclonic eddies interact and coalesce with the Gulf Stream off the southeastern coast of the United States. The impact that these interactions have on an eddy's structure has been observationally well

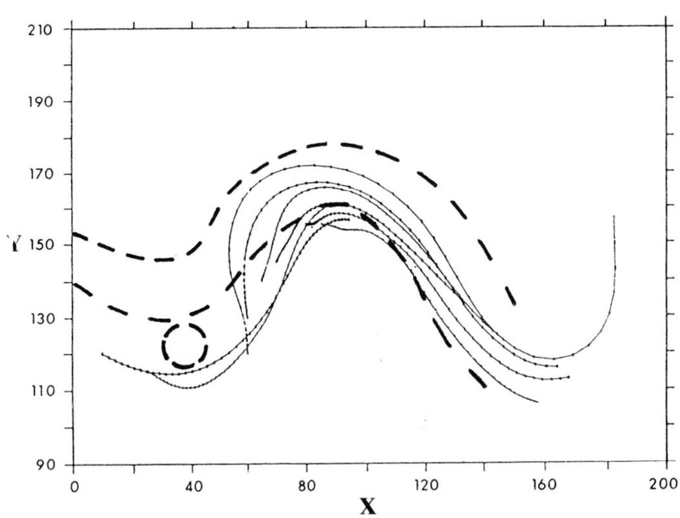

FIGURE 6. Float trajectories (tic marks every six hours) for the eight floats initially placed to the south of the jet on Day 8 for Experiment 2. The dashed lines indicate the approximate position of the Stream and eddy when the floats were released.

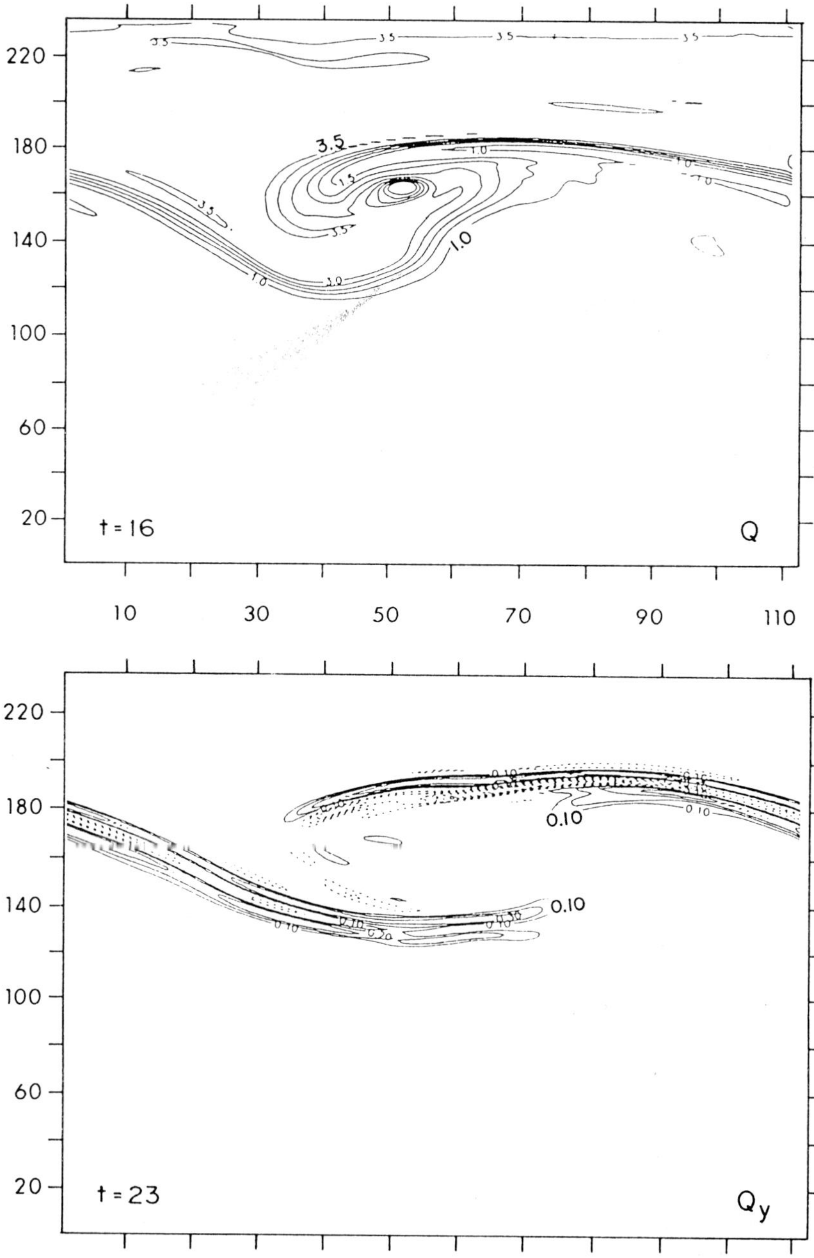

FIGURE 7. The a) Day 16 potential vorticity field for layer 2, Experiment 2. Downstream of the interaction the q isopleths become tightly confined to the cyclonic side of the Stream. b) Day 23 meridional gradient of potential vorticity field for layer 2 at the end of Experiment 2.

documented (Richardson, 1983; Richardson et al., 1979). It has also been known that these interactions must have a significant impact on the Gulf Stream's structure and dynamical properties. Richardson et al., 1979 (in a study of cyclonic eddy Allen) reported, 'these (cyclonic) ring interactions could be important in the redistribution of water, heat, and energy within the Gulf Stream system.' Observational analyses (GR) and numerical experiments presented here indicate that a Stream/cyclonic eddy interaction not only alters the eddy's and Stream's physical and dynamic structure, but also has significant impact on the neighboring oceanic water masses.

The numerical experiments have given insight into several questions raised by the observational analyses of a warm surge (GR). First, during the model detrainment of Gulf Stream waters by the cyclonic eddy, a significant amount of water is ejected into the Sargasso Sea. This process, observed in satellite studies over several years, causes warm surface water to be mixed into the Sargasso or exchanged eddy to eddy. The dynamics of ejection of the detrained Gulf Stream water out of the eddy could be analogous to the ejection of fluid parcels out of Gulf Stream meanders. Bower and Rossby (1989) showed that cross-stream displacement of RAFOS floats (and thus expulsion of the floats) is correlated to trajectory curvature of Gulf Stream meanders. As parcels approach cyclonic bends in the Stream they downwell and move offshore along sloping density surfaces. Very sharp changes in path curvature along parcel trajectories induce larger vertical and lateral motions which, in turn, leads to the expulsion of the parcel (Bower and Rossby, 1989). In an eddy, the extremely sharp changes in the path curvature could easily lead to the ejection of parcels under the same mechanism.

The ejection of Stream water into the Sargasso has important consequences for the warm surge event. If the detrained water does not merely circulate round the eddy and re-enter the Stream, then another source must replenish the surface layer of the Gulf Stream. Both the numerical experiments and the observations data imply that a major replenishing source is horizontal advection from the Sargasso Sea. One of the main differences between numerical experiments in which a Stream/eddy interaction did or did not occur was the enhanced entrainment of Sargasso water, beginning approximately two days after the onset of the upstream detrainment of Gulf Stream water.

The analyses indicate that Gulf Stream parcels in the center of the Stream and on the anticyclonic side are detrained into the eddy, while parcles on the cyclonic side become confined into a narrow band near the north wall of the Stream. This pattern is seen observationally in Figure 1, as upstream of the interaction site the Gulf Stream surface waters of 25°C or greater (represented as white) are approximately 80*km* wide, while downstream of the interaction these waters comprise only a narrow 20*km* band. Thus, during the period when the Gulf Stream's surface layers are being ejected into the Sargasso (the cold phase), the Stream's surface layer (downstream of the interaction) is comprised of Sargasso water (on its anticyclonic side) and Gulf Stream water that propagated past the interaction (on its cyclonic side). As the interaction weakens and the eddy is absorbed, more parcels that were detrained into the eddy re-enter the Stream (Figure 5), until eventually the Stream resumes its uninterrupted flow.

The numerical and observational results suggest that the interaction does not enhance entrainment of Slope Water into the Stream. Slope Water entrainment occurred in all experiments only when a forward propagating meander crest caught up to a float in the Slope Water. (A process also observed in the Gulf Stream with RAFOS floats (Bower and Rossby, 1989)). Also, the satellite images show no alteration of the north wall of the Stream (i.e. it remains a continuous front) throughout the warm surge event, while the southern edge does not remain continuous. This is evident in Figure 2, where it appears that a 'window' has opened on the southern edge of the Stream indicative of Stream/Sargasso exchange. Thus, all indications are that the Sargasso, not the Slope water, is the replenishing source of the Gulf Stream's surface layer. Finally, implicating the Sargasso as the major replenishing source for the detrained Gulf Stream water is physically realisitic since it is the closest water mass near the Stream parcels that are detrained (i.e. the Stream's anticyclonic side). This result may have implications on the gyre circulation, to be discussed in the next section.

Parcels initially placed in the eddy (but outside the core) entered the Stream and propagated downstream. However, parcels placed in the core (within 20*km* of the center) of the eddy were often ejected north of the Stream. This occurred as the eddy was being absorbed and an 'open meander' briefly formed around the eddy and parcels in the core could be freely exchanged with the Slope Water. Thus, these parcels completed a circuit round the Gulf Stream system, as originally they were north of the Stream, subsequently trapped in the cold core of the eddy, and eventually deposited back north of the Stream during the interaction.

The model was designed to isolate horizontal advection processes, however vertical advection (i.e. strong horizontal divergence) of Gulf Stream water is arguably important. The next step in numerically simulating this complex mesoscale interaction process might be to include diapycnal mixing across the active layers. By including a thermodynamically active fluid new hypotheses on the role of vertical advection could be addressed. For example, one possibility is that the cross-stream divergence in the Stream's surface layers is balanced by vertical advection of deeper Gulf Stream water. The loss of geostrophic momentum, caused by the interaction, would result in a local decrease of the sea surface height in the interaction zone. This in turn would cause a geostrophic surface flow from the Stream into the eddy (upstream of the interaction), and a flow from the eddy into the Stream (downstream of the interaction). Thus, the Stream's surface water in the cold phase could plausibly be comprised of: first, both eddy water and upwelled Gulf Stream water; next, mostly upwelled Gulf Stream water (since there is not an infinite supply of eddy water); and finally, upwelled water and surface Sargasso water as horizontal advection becomes important. With the inclusion of active thermodynamics and diapycnal mixing processes in the numerical analyses this hypothesis, and others, can be further investigated. Also, the warm surge event, which involves rapid cooling and warming of the Stream's surface layers, could significantly affect the mixed layer and heat exchange with the atmosphere and warrants further investigation with a detailed regional mixed layer model.

While the focus of this study was on Lagrangian pathways of fluid parcels, dynamical Eulerian fields were also monitored. The Eulerian fields, consistent with the Lagrangian pathways, showed a similar alteration in the Gulf Stream's structure after the eddy interaction. The q field (a tracer of the flow field) indicated that the Stream's flow, downstream of the interaction, is confined to a narrow band on the cyclonic side of the Stream (Figure 7a). In a separate experiment initialized with a multiple meander of the Gulf Steram (not presented here), meanders downstream of the interaction indicate meander 'breaking', while the upstream meanders did not. The Stream also evolved towards conditions necessary for baroclinic instability. For all 24 day model runs without a Stream/eddy interaction the q field, it's meridional gradient, and the energy balance remained unchanged.

In summary, we have isolated and identified plausible Lagrangian pathways of parcels in the Gulf Stream system due to a Stream/eddy interaction. Gulf Stream parcels entering the interaction site may 1) propagate past the interaction site, however they become confined into a narrow band near the north wall, or 2) become detrained into the eddy and subsequently be ejected into the Sargasso or re-enter the Stream. Downstream of the interaction there is enhanced entrainment of parcels from the Sargasso into the Gulf Stream, but there is no enhancement of parcels entering the Stream from the north. The system is incomplete without explicit accounting of vertical advection and mixed layer physics, however by isolating horizontal advection the numerical analysis has supported many of the tendencies seen in the observational data.

EFFECT ON THE GYRE SCALE CIRCULATION: CONJECTURE

We here offer a number of ideas, based upon published dynamical hypotheses, of the possible implications of the warm surge events for the gyre scale circulation. Keep in mind the ubiquity of the phenemenon we report (i.e. 9-11 warm surge events per study year).

The warm surge events and their impact on the Gulf Stream/subtropical Gyre system can be viewed in several dynamically complimentary ways. They may be thought of as (1) q anomalies in the Gulf Stream, (2) paired cold/warm anomalies, (3) indicators of rapid changes in the energy balance of the Stream, and/or (4) a method of injecting warm water into the Sargasso Sea. Each viewpoint has different dynamical implications for the Gulf Stream System and will be discussed in turn.

Significant layer variations of the Gulf Stream's upper waters are observed during a warm surge event, thus the events can be thought as of q anomalies in the Gulf Stream. The inclusion of PEGASUS velocity data emphasizes that the Gulf Stream's absolute vorticity structure is also substantially altered during an event. Quasi-geostrophic numerical models have shown that in a double gyre system (designed to represent the subtropical/subpolar gyres) the cross-stream dissipation of vorticity is an important sink for anomalous potential vorticity. Thus parcels of anomalous q within the western boundary current can lose (negative) vorticity by dissipative meandering of the jet between the gyres (Lozier and Riser, 1989). These model cross-gyre exchanges are often triggered by vorticity anomalies being swept up through the model western boundary current, with these anomalies controlling both model meander growth, overshooting of the model current and hence the dissipative meadering mechanism. Cessi et al. (1987) numerically indicated that the anomalous values of low q carried northward by the western boundary current may also help drive the recirculation gyre south of the Gulf Stream.

Recent studies indicate that cooling (whether along-stream or cross-stream) is a possible mechanism for controlling the pathway of the Stream (Huang, (1990); Nof, (1983, 1986); Adamec and Elsberry, (1985)). These studies reported on *seasonal* shifts of the Stream under the mechanism of seasonal *atmospheric* cooling. For example, as a current enters a region of atmospheric cooling its density changes and new pressure gradients, which were not present upstream, are generated. These pressure gradients drive a flow in a direction perpendicular to the current upstream, and as a result of this flow the interface steepens and the current is displaced (Nof, 1983). Our observations indicate intense cooling of the Stream *due to ocean advective processes* on the order of *days*. Such transient events, which significantly alter the dynamic and thermodynamic structure of the Gulf Stream, may plausibly play a significant role in defining the Stream's pathway, stability and heat budget.

Another possibility for setting Stream pathways is determined by the build up and release of APE, i.e. by changing the steepness of density surfaces and it's effect on the potential vorticity balance. By stretching water columns the APE field is altered thus creating a possible energy source for baroclinic instability that could result in enhanced meandering or eddy formation. The data indicate that the upper layers of the Gulf Stream undergo vertical stretching and shrinking during the various phases of the event which alter the Stream's APE (GR). One possibility, that again needs to be investigated with numerical simulations, is that during the cold phase the near surface isotherms are stretched to replenish the detrained surface layer thus causing an increase in the Stream's APE. This may possibly serve as an energy reservoir for baroclinic instability manifesting itself as enhanced Gulf Stream meandering.

Finally, the observation of warm surge events may play a key role in a recent theory of the mid-latitude circulation. Cushman-Roisin (1987) assigns control of the entire Subtropical Gyre to dynamical (wind) action and thermodynamical (heat-flux) reaction. In that quasi-geostrophic model, the potential vorticity input from the wind stress curl is balanced by atmospheric cooling at the western boundary which resets the potential vorticity of the gyre and thus closes the circulation. A key feature of that theory (for which he did not have supporting observations) is that the atmospheric cooling is triggered by *an injection of anomalous amounts of warm water into the Sargasso Sea* (Cushman-Roisin, 1987), thus suggesting an important atmosphere/ocean feedback. A warm surge event could be a mechanism for rapidly injecting these anomalous warm waters into the Sargasso Sea (via eddy detrainment of Gulf Stream water), and because of the frequency of this process (9-11 events per year) could play a role in maintaining the potential vorticity balance of the entire Subtropical Gyre.

Again, many of these questions can be investigated with the aid of numerical models. Future research should implement numerical experiments to investigate the subsequent downstream perturbations induced by the interaction. We hope to address the question of whether the surges do affect the large scale circulation, and if so, toward an understanding of the dynamics involved.

ACKNOWLEDGEMENTS

The Office of Naval Research funded these efforts under contract NA00014-90-J-1874 and we are most grateful for their support.

REFERENCES

Adamec, D. and R.L. Elsberry. 1985. Response of an intense oceanic current system to cross-stream cooling events, *J. Phys. Ocean.,* **15**, 273-287.

Bower, A.S. 1991. A simple kinematic mechanism for mixing fluid parcels across a meandering jet. *J. Phys. Oceanogr.,* **21**, 173-180.

Bower, A.S., and H.T. Rossby. 1989. Evidence of cross-frontal exchange processes in the Gulf Stream based on isopycnal RAFOS float data, *J. Phys. Ocean.,* **19**, 1177-1190.

Cessi, P., G. Ierley, and W. Young. 1987. A model of the inertial recirculation driven by potential vorticity anomalies, *J. Phys. Ocean.,* **17**, 1640-1652.

Charney, J.G. and M. Stern. 1962. On the stability of internal baroclinic jets in a rotating atmosphere, *J. Atmos. Sci.,* **19**, 159-172.

Chew, F. 1981. Shingles, spin-off edches and an hypothesis. *Deep Sea Research,* **28**, 379-391.

Cushman-Roisin, B. 1987. On the role of heat flux in the Gulf Stream-Sargasso Sea Subtropical Gyre system, *J. Phys. Ocean.,* **17**, 2189-2202.

Dewar, W.K., and J.M. Bane. 1989. Gulf Stream Dynamics. Part II: Eddy Energetics at 73 W, *Journal of Physical Oceanography,* **19**, 1574-1587.

Fuglister, F. 1960. North Atlantic Atlas, W.H.O.I. Atlas Series, Vol. 1, 1960.

Gilman, C. and L. Rothstein. 1994. Gulf Stream warm surges: Thermal anomalies resulting from cold core ring interactions. *J. Geophys. Res.,* in press.

Halkin, D. and T. Rossby. 1985. The structure and transport of the Gulf Stream at 73°W, *J. Phys. Ocean.,* 1439-1452.

Huang, R.X. 1990. Does atmospheric cooling drive the Gulf Stream recirculation, *J. Phys. Ocean.,* **20**, 750-757, 1990.

Ikeda, M., and J.R. Apel. 1981. Mesoscale eddies detached from spatially growing meaders in an eastward flowing jet using a two-layer quasi-geostrophic model, *J. Phys. Ocean,* **11**, 1638-1661, 1981.

Johns, W.E. 1988. One-dimensional baroclinically unstable waves on the Gulf Stream potential vorticity gradient near Cape Hatteras, *Dynamics of Oceans and Atmospheres,* **11**, 323-350.

Lozier, M. Susan, and S.C. Riser. 1989. Potential vorticity dynamics of boundary currents in a quasi-geostrophic ocean, *J. Phys. Ocean.,* **19**, 1373-1396.

Luyten, J.R., J. Pedlosky and H. Stommel. 1983. The ventilated thermocline, *J. Phys. Ocean,* **13,** 292-309.

McDowell, S., Rhines, P., and T. Keefer. 1982. North Atlantic potential vorticity and its relation to the general circulation, *J. Phys Ocean.,* **12,** 1417-1436.

McWilliams, J.C., and G.R. Flierl. 1979. On the evolution of isolated non-linear vortices, **9,** 1155-1182.

Nof, D. 1986. Comments on "The response of intense ocean current systems entering regions of strong cooling" *J. Phys. Ocean.,* **16,** 996-999.

Nof, D. 1983. On the resopnse of ocean currents to atmospheric cooling, *Tellus,* **35a,** 60-72.

Nof, D. 1981. On the beta-induced movements of isolated baroclinic eddies, *J. Phys Ocean.,* **11,** 1662-1672.

Pedlosky, J. 1979. *Geophysical Fluid Dynamics,* Springer-Verlag, 710 pp.

Pratt, L.J. and M. Stern. 1986. Dynamics of potential vorticity fronts and eddy detachment, *J. Phys. Ocean,* **16,** 1101-112004.

Reid, R.O., Elliot, B.A., and D.B. Olson. 1981. Available Potential Energy: a Clarification, *J. Phys. Ocean,* **11,** 15-29.

Rhines, P.B. and W.R. Young. 1982. A theory of the wind-driven circulation. I. Mid-ocean gyres, *J. Mar. Res.,* **40,** (supplement), 559-596.

Richardson, P. 1983. "Gulf Stream Rings" in *Eddies in Marine Science* (A.R. Robinson, ed.) 565 pp.

Richardson, P. 1980. Gulf Stream Ring Trajectories, *J. Phys. Ocean,* **10,** 90-104.

Richardson, P., C. Maillard, and T.B. Sanford. 1979. The physical structure and life history of cyclonic Gulf Stream ring Allen, **84,** 7727-7741.

Ring Group. 1981. Gulf Stream Cold-Core rings: Their physics, chemistry and Biology, *Science,* **212,** 1091-1100.

Robinson, A.R., M.A. Spall, and N. Pinardi. 1988. Gulf Stream simulations and the dynamics of ring and meander processes, *J. Phys. Ocean,* **18,** 1811-1853.

Smith, D.C. and Davis, G.P. Jr. 1989. A numerical study of eddy interaction with an ocean jet, *J. Phys. Ocean,* **19,** 975-986.

Smith, D.C. and J.J. O'Brien. 1983. The interaction of a two-layer isolated mesoscale eddy with bottom topography, *J. Phys. Ocean,* **13,** 1681-1697.

Soiland, H. 1991. Numerical studies of Lagrangian motion in a meandering jet, Univ. of Rhode Island Grad. School of Oceanography Ph.D. thesis, 184 pp.

Stern, M.E., and Flierl, G.R. 1987. On the interaction of a vortex with a shear flow, *J. Geophys. Res.* **92,** 10733-10744.

Stommel, H. 1948. The westward intensification of wind-driven ocean currents, *Trans. Amer. Geophysical Union,* **29,** 202-206.

Watts, D.R. 1983. "Gulf Stream variability" in *Eddies in Marine Science* (A.R. Robinson, ed.) 565 pp.

Worthington, L.V. 1976. On the North Atlantic circulation, *Oceanogr. Stud.,* **6,** 1-110.

The Oceans: Physical-Chemical Dynamics and Human Impact. Edited by S.K. Majumdar, E.W. Miller, G.S. Forbes, R.F. Schmalz and Assad A. Panah. © 1994, The Pennsylvania Academy of Science.

Chapter Fourteen

EL NIÑO SOUTHERN OSCILLATION: Causes and Effects

E. WILLARD MILLER

Professor of Geography and Associate Dean for Resident Instruction (Emeritus)
College of Earth and Mineral Sciences
The Pennsylvania State University
University Park, PA 16802

INTRODUCTION

It has long been known that the coastal areas of Peru and Ecuador are subject to violent changes in weather conditions. This phenomenon was first recorded by the Spanish conquistador, Francisco Pizarro in 1525. Because the cold current is replaced by a warm current at times during the Christmas season, the anomalous weather behavior came to be known as the El Niño, the Christ Child.

The scientific study of the El Niño began in the late nineteenth century when Hildebrandsson (1897) discovered evidence of the El Niño/Southern Oscillation in the form of an out-of-phase relationship between the surface pressure anomalies at Sydney, Australia and Buenos Aires, Argentina. Five years later Lockyer and Lockyer confirmed these findings and indicated the El Niño period to be about 3.8 years. By 1902 these authors indicated that the El Niño phenomenon influenced weather worldwide.

In the 1920s and 1930s, Sir Gilbert Walker prepared a series of articles that related the Southern Oscillation to El Niño (ENSO). In 1932, Walker and Bliss wrote:

> When pressure is high in the Pacific Ocean it tends to be low in the Indian Ocean from Africa to Australia; these conditions are associated with low temperatures in both areas, and rainfall varies in the opposite direction to pressure. Conditions are related differently in winter and summer, and it is therefore necessary to examine separately the seasons December to February and June to August.

Little attention was paid to the El Niño effect in the world between the 1930s and 1960s. In 1966, Berlage provided evidence of the strong relationship between the interannual surface pressure variations at Djakarta, Indonesia and the Sea Surface Temperature (SST) at Puerto Chicama on the Peruvian coast. In the same year, Bjerknes postulated that the weakening of the equatorial easterlies across the eastern half of the Pacific that occurs in association with negative swings of the Southern Oscillation occurs at a time when the pressure rises in Australia and Indonesia and falls over the tropical southeast Pacific. This, in turn, reduces equatorward upwelling in the eastern Pacific and therefore leads to an increase of SST along the Peru and Ecuador coasts. This concept is now accepted as fundamental to the ENSO. Since then, studies by Kidson (1975), Trenberth (1976), Barnett (1988) and many others have confirmed the relationship conceived by earlier scientists.

DATA COLLECTION

It has long been recognized that the ocean possesses a large and active reservoir of heat that determines many of the climatic patterns on the earth. Nevertheless, this has been a paucity of oceanographic weather data. With the growing interest in the El Niño/Southern Oscillation phenomenon and the recognition that the weather patterns associated with these oscillations are global in character. Studies now show that the physical mechanisms underlying the statistical correlation demonstrates that the combined El Niño/Southern Oscillation (ENSO) phenomenon is a major factor in the global interannual climatic variability. There is now recognition that an understanding of the conditions in the oceans and atmospheric boundary layers of the tropical oceans is of the utmost importance to predicting weather on the land.

In 1979, the National Oceanographic and Atmospheric Administration (NOAA) initiated a research program identified as the Equatorial Pacific Ocean Climate Studies (EPOCS) to investigate the oceanic and atmospheric interaction associated with the ENSO phenomenon. The goal of the EPOCS program is to provide an improved understanding of ENSO in order to develop the capability to simulate tropical Pacific Ocean and atmospheric conditions and to predict various aspects of the evolution of these conditions. Because of the interaction between ocean-atmosphere systems in ENSO development, it is necessary to understand the relationship between the coupled system. To understand these interrelationships, data must be available. Initial progress was secured by separately considering the effects of SST anomalies on the atmosphere and the effects of wind perturbations on the ocean. Since its inception, scientific guidance and planning for EPOCS have been provided jointly by atmospheric and oceanographic researchers. This close cooperation assures a coordinated approach to the coupled ocean-atmosphere problems.

The major ENSO event of 1982-1983 created a worldwide interest in collecting data follwed by an analysis of the climatic events. EPOCS played a major role in documenting these events and critically comparing observations with theoretical simulations.

To further the scientific research, the international Tropical Ocean Global Atmosphere (TOGA) program was established in 1984 and the EPOCS program is now coordinated within the broader TOGA framework. Thus, it is part of the World Climate Reseach Program. To implement the program, a network of instruments provide data to the Global Telecommunications System including expendable bathytheromographs released by voluntary observation ships, tide gauges on numerous islands, and in the central Pacific, where there are neither islands nor commercial ships, an array of moorings with thermistor instruments, and in a few places current meters.

The TOGA program has concentrated its efforts in the Southern Oscillation of the Pacific. As a response to the new and additional data a numerical model of the tropical Pacific, capable of realistic simulation of oceanic conditions, has been completed. Each month TOGA produces maps on surface and subsurface temperatures and wind conditions. These monthly oceanic maps and the counterpart of daily weather maps reveal the evolution, month by month, of the evolution of El Niño conditions. This system is now being extended to the global oceans.

CHARACTERISTICS OF A COMPOSITE OF THE EL NIÑO/ SOUTHERN OSCILLATION

The El Niño/Southern Oscillation phenomenon occurs on an irregular basis and may last for one to three years. From 1925 to 1992 there have been at least 14 ENSOs (1925-26, 1929-31, 1940-41, 1943, 1948, 1951, 1953, 1957-58, 1965, 1969, 1972-73, 1974, 1982-83, 1991-92). Each El Niño is distinct and has varying influences on the weather of the world.

In order to understand the El Niño phenomenon, Rasmusson and Carpenter (1982) have developed a composite El Niño within the time span 1950-73. This composite representation averaged the strongest six El Niño events that occurred during that period. This composite reveals the strong relationship in the pattern of Sea Surface Temperatures (SST) and the wind anomaly for each event. Their study presented evidence of (1) strong positive SST anomalies extending from the Galapagos Islands eastward to the South American coast during January-May and weaker but spatially coherent warming of the SST extending across the entire equatorial Pacific eastward of 170°E peaking around October, (2) heavy rainfall along the Peru-Ecuador coast and extending over the central equatorial Pacific, and (3) a pronounced negative swing of the SO.

The composite El Niño indicates a sequence of events that characterize the general phenomenon (Wooster and Fluharty, 1985). It is revealed that the El Niño event is preceded by a rather long period of stronger than average easterlies in the western Pacific which depresses the thermocline and increases its level in the western Pacific relative to the eastern Pacific. The Sea Level Temperature (SST) is slightly warmer than normal in the western Pacific and colder than normal in the eastern Pacific. The Sea Level Temperature (SST) is slightly warmer than normal in the western Pacific and colder than normal in the eastern Pacific. In September of the year preceding the warming, the easterlies decrease west of the dateline. Warming of the surface layer off South America begins in December and increases to a maximum value in June or July. The warming is accompanied by an increase in sea level along the west coast of South America. Westerly wind anomalies appear in the central Pacific in the northern summer of the preceding year. The positive SST anomalies propagate westward along the equator from the eastern boundary so that by the end of the first year the maximum SST anomaly shifts to the central Pacific. The secondary peak in SST off South America follows a sharp decrease in the northern spring of the second year and a return to below normal SST. Warm SST persists in the central Pacific until the middle of the second year. By then the winds return to strong easterlies and the event has run its course.

While similarities of the El Niño/Southern Oscillation are evident for most ENSOs, there are also some contrasts. The 1957-58 El Niño differed from the others in that there were positive anomalies preceding the event and the SSTs did not return to normal level during the first year. It has been noted by Rasmusson and Carpenter that the components Niño SST signal off South America can be considered as an amplification of the annual signal because of the sharp rise at the beginning of the onset year coincides with the second heating cycle.

THE EL NIÑO/SOUTHERN OSCILLATION OF 1982-83

The El Niño of 1982-83 was notable for its influence of weather worldwide. The first evidence of the 1982-83 ENSO appeared in the late northern spring of 1982 in the western Pacific when westerly wind replaced the easterly trade winds and positive Sea Surface Temperatures (SST) became evident (Wooster and Fluharty, 1985). Because of the length and strong worldwide effects of this El Niño, it is used as a case study.

Temperatures

The Sea Surface Temperature (SST) in the eastern Pacific started to increase in the northern summer of 1982 and by December 1982 anomalies of more than 3.5°C were observed at 120°W. The maximum SST anomaly of more than 4.0°C in the Peruvian coast was recorded in June 1983. This was an extremely large anomaly. As a result, a vast tongue of warm water stretched for 8,000 miles across the Pacific Ocean along the equator. This rise in temperature preceded the annual heating system by five or six months.

Changes in the upper ocean thermal and current structure on the equator was measured by moors at 95° and 110°W (Halperin et al. 1983). At 95°W the thermocline depth described by the 20°C isotherm began decreasing in mid-August 1981 and deepened until mid-May 1982 when it was about 50-60 meters deeper than in August 1981. After a second rise, the thermocline in August 1982 again deepened rapidly to a depth of 150 meters at 109°30′W and 125 meters at 95°W by mid-December (Halperin et al. 1983). The warming occurred first at 109°W and moved eastward. The thermocline deepening was arrested in January 1983 and a rapid rise in depth in the 20°C isotherm occurred. Upper ocean temperature remained warm and the surface mixed layer was anomalously thick although the deeper thermocline isotherm (particularly at 110°W) was near the pre-event level.

Rainfall

During the El Niño of 1982-83, heavy and repetitive convectional rainstorms devastated coastal Ecuador and Peru. Normally the easterlies of the region produce a desert environment. With the disappearance of the easterlies and the development of the westerlies, the moisture laden winds brought rainfall to the coastal mountains.

Goldberg, Tisnado and Scofield (1987) have analyzed the precipitation regime for 66 stations within the Chira-Piura area of northwestern Peru from January 22 to April 30, 1983. During this period 72 rainfall producing convection systems were analyzed. Fifty percent of these systems lasted in excess of 12 hours. Most of the storms were of large meteorological size and penetrated to high altitudes. Most of the storms reached maximum intensity during the evening and nighttime hours. This suggests convective outflow boundaries, low-level winds, radiative cooling of cloud types, and orographic effects from the Andes.

In a comparison of an El Niño versus non El Niño years, the difference in the amount of precipitation is striking. In a non El Niño year, rainfall totals about 250 mm. In an El Niño year, the rainfall totals about 4,000 mm with rainfall exceeding 100 mm on many days.

Sea Level

Associated with other physical change were the large rises in sea level. For example, the sea level on the Galapagos Islands began to increase in August 1982 (Hayes, 1982). It continues to rise and by January 1983 the increase was nearly 0.5 meters. During this period the Sea Surface Temperatures (SST) increased by about 5°C. The central Pacific experienced a similar sea level rise (Lukas et al. 1984). This type of correlation provides evidence of the importance of equatorial changes in the initial oceanic response to western and central Pacific wind changes. Numerical model simulations indicate that the total response is more complicated than simply a summation of these waves (Cane, 1984).

In January 1983, sea level fell rapidly on the Galapagos Islands. This has been interpreted as an upwelling Kelvin wave in response to the return of strong southeast trade winds in the western Pacific in December 1982 (Philander, 1989). A subsequent sea level rise in March 1983 was in response to the local westerly wind anomaly which developed in the eastern Pacific in early 1983. A second sea level peak was not observed in the central Pacific.

Currents

During the 1982-83 El Niño/Southern Oscillation there was a dramatic change in the surface and subsurface equatorial currents of the Pacific Ocean. From July until November 1982 the subsurface current was similar to its normal structure. Beginning in mid-December 1982 the subsurface current began to change rapidly. Its speed decreased and reversed direction changing from an eastward moving to a westward subsurface current. By August 1983, the current returned to its normal position.

The movement of the equatorial surface current also experienced changes during the various stages of the 1982-83 El Niño. From January until July 1982, the current movements were essentially normal. By September the flow had weakened and was less zonal. By December the surface current had become weakly eastward over a broad expanse of the eastern Pacific. From January to March 1983, an intense westward flow occurred near the equator. East of 130°W the flow reversed in April-May 1983 and an eastward equatorial current prevailed. The conclusion of the El Niño in July-August 1983 was accompanied by a return to a normal westward flow. An intense Equatorial Front north of the equator and associated meridional shear led to the most pronounced mesoscale eddy activity ever reported. These changes in motion appear to be examples of this current's instability. During El Niño warming such motions appeared to be suppressed, presumably because of a weakening of the meridional gradient.

Summation

The uniqueness of the 1982-1983 appears to be: (1) the event began in the late northern spring and not the late northern fall of the typical ENSO, (2) the SST variation was at least twice as large as normal variations, (3) the SST anomaly moved from west to east instead of east to west, and (4) the westerly wind anomalies penetrated much further east than in the typical ENSO. This penetration was so extreme that there were westerly winds at the equator in the eastern Pacific.

Economic Livelihood Effects of the El Niño

The altered weather patterns of 1982-93 in the equatorial Pacific had a major effect on the economy of the region. As typical examples, the fishing industry of Peru and Ecuador and the guano fertilizer industry of the Galapagos Islands are used as case studies.

Peruvian and Ecuadorian Fishing Industry.

In normal years, fishing is a major industry off the coast of Ecuador and Peru. The Peruvian anchovy and sardine industry is of the greatest importance to the economy of the Peruvians. Under normal climatic conditions, the coasts are bathed by a cold current from the south that is laden with phytoplankton (diatoms) and zooplankton (copepods). These are the right food sources for these fish.

Alamo and Bauchon (1987) analyze the effect on sardine populations when a change in the currents during the El Niño occurs. Under normal conditions the north-flowing current and the earth's rotation push the warm surface waters westward and the colder, deeper waters high in nutrients flow along the coasts providing food for the fish. When the El Nino appears the warm water moves southward and the cold current with its nutrients disappears. The plankton decrease and the fish disappear. The quantitative analysis of the stomach contents of the Peruvian sardine and other fish indicate that the fish displayed a clear tendency to change their diet as a consequence of the El Niño event in respect to normal years. The plankton biomass decrease caused a lack of food for the fish and resulted in their near disappearance.

As a consequence, the Peruvian and Ecuadorian fishing industry was devastated during the 1982-83 El Niño. At Chimbote, the fishing center for the anchovy industry, the fishing fleet was idle. The fish-meal plant was idle. A small production of sardines remained, but only sufficient to feed the local population. In contrast to the decrease in sardines and anchovies, the warm current was favorable to shrimp. The production of shrimp increased from 5 tons per week to over 75 tons.

Sea Bird Devastation. The eastern equatorial Pacific coast and islands abound with unique seabirds - the guanoz, cormorant, Peruvian bobby and pelican. These birds are economically important because they produce an organic fertilizer, known as guano, which has been used in agriculture since the time of the Incas (Tovar, Guillen and Cabrera, 1987).

In order to determine the effects of the El Niño, Tovar, Guillen and Cabrera measured the bird population before and after the 1982-83 El Niño. In 1981-82, the population size prior to the El Niño during the reproductive cycle was 5.75 million birds (47% booby, 46% guanoz, 7% pelican) but during the El Niño 1982-83 up to March 1984, the population declined to 2.74 million (55% guanoz, 38% booby, 7% pelican). The data indicated that during July 1983 guanoy birds deserted their islands and coastal locations, leaving only 0.12 million birds. The exodus was caused basically by a lack of food. In addition, the El Niño event of 1982-83 caused the death of at least 2.8 million chicks. By 1985-86 cycle the bird population had increased to 3.16 million, primarily due to the reappearance of anchovies.

The devastation of the marine bird population occurred across the Pacific Ocean. For example, it is estimated that 17 million birds were forced from Christmas Island.

Worldwide Weather Influence

In many places throughout the world weather anomalies were experienced during 1982-83. Diverse trends were experienced with heavy precipitation in some areas causing floods and related damage. In other areas drought conditions were equally devastating to the economy of the region.

Worldwide Drought. A large number of areas experienced unprecedented drought conditions. In the western Pacific areas of Central America and Mexico dry conditions reduced crop yields and desert-like grazing lands severely reduced the animal population. It was estimated the drought during 1982-83 reduced the agricultural income by over $600 million in Mexico and Central America.

In the eastern Pacific and Indian Oceans severe drought occurred in Indonesia, the Philippines, India, Sri Lanka, Southern Africa and was most devastating in Australia. Of a total agricultural damage of $4,400 million, $1,000 million occurred in Southern Africa, $750 million in the Philippines and Indonesia and $2,500 million in Australia. In Indonesia at least 340 deaths were attributed to starvation. In India water shortages raised fears of jaundice outbreaks and other epidemics and in Southern Africa drought reduced crop production by 40 to 70 percent, spreading disease and malnutrition.

In Australia, the driest of the continents, a high pressure ridge normally reduces precipitation, but occasionally it breaks and rain comes in from the Indian Ocean. In 1982-83 this did not occur. As a consequence, eastern Australia became a complete desert. The sheep industry was devastated as well as the wheat yields. Many sheep flocks were reduced by 75 percent. Vast fires swept through the dry semidesert and desert regions.

Worldwide Floods. In contrast to the drought-stricken areas, there were a number of regions where vast devastation occurred due to flooding. On the coastal areas of Ecuador and Peru, the rains devastated the leading banana areas and the coastal rice regions were flooded. Thousands of houses were badly damaged or destroyed. Diseases destroyed the cattle and chickens of the area. Thousands of people from the rural areas migrated to the cities. In Guayaquil a great slum, known as Guasmo developed. Many of the shanties rose above the flood waters. Sanitation was abysmal with flooded toilet pits spreading disease. Mosquitos and other insects bred on the stagnant pools causing outbreaks of typhus, salmonella infections and typhoid fever reached epidemic proportions. The flooding in coastal Equador and northern Peru is estimated to have caused $1,000 million in damages and at least 600 deaths.

Rains of flooding proportion occurred throughout the equatorial Pacific. A rare hurricane slammed into Hawaii. There were six tropical storms that struck the French Polynesian islands. On Tahiti a hurricane had not been experienced in 75 years, but the storm of 1982-83 destroyed 1,500 homes and roofs were torn off an additional 6,000. Many persons from the outer French islands sought refuge in Tahiti despite discouragement of the government. The estimated damage to the French Polynesian islands was set at $280 million.

In the United States flooding occurred in the Gulf and West Coasts. In March 1983, a Pacific storm generated a tornado in the Watts area of Los Angeles. Storms loosened a mountainside that slid into a valley creating a lake and destroying Thistle, Utah. The exceptional storm caused $1,100 million damage to agriculture, roads and buildings in western United States and $1,270 million additional damage to the Gulf Coast. At least 65 persons died in the United States from flooding.

CONCLUSION

The non periodic El Niño/Southern Oscillation illustrates that the world's weather is tied into a single system. Vast anomalies are created throughout the world when these phenomena occur. Although the El Niño has long been known to develop equatorial waters of the Pacific Ocean off the coasts of Peru and Ecuador, the understanding of its effect on worldwide climate is still at a development stage. While increased data collection has begun, it will be decades before the mechanism that triggers the El Niño and its accompanying Southern Oscillation will be completely understood. And even more time will be needed to understand how these phenomena affect world weather systems.

SELECTED REFERENCES

Aceituno, P. 1992. El Niño, the Southern Oscillation, and ENSO: Confusing Names for a Complex Ocean-Atmosphere Interaction. *Bulletin of the American Meteorological Society.* 73(April):483-485.

Alamo V. Alejandro and Marilu Bauchon. 1987. Changes in the Food and Feeding of the Sardine (Sardinops sagax sagax) During the Years 1980-1984 off the Peruvian Coast. *Journal of Geophysical Research.* 87(C13):14,411-14,416.

Alarcon, Pedro Antonio de. 1903. *El Niño de la Bola.* New York: American Book Company. 278 p.

Arkin, P.A., J.D. Kipman and R.W. Reynolds. 1983. *The 1982-83 El Niño/Southern Oscillation Event Quick Look Atlas.* Washington, DC: Climate Analysis Center.

Arkin, P.A. and J.E. Janowiak. 1987. The Global Climate for December 1985 - February 1986: Conflicting ENSO Signals Observed in the Equatorial Pacific. *Monthly Weather Review.* 115:297-316.

Avaria, Sergio and Pablo Munoz. 1987. Effects of the 1982-83 El Niño on the Marine Phytoplankton off Northern Chile. *Journal of Geophysical Research.* 87(C13):14,369-14,382.

Barnett, T.P., N. Graham, M. Cane, S. Zebiak, S. Dolan, J. O'Brien and D. Leger. 1988. Prediction of the El Niño of 1986-87. *Science.* 241:192-196.

Barnston, Anthony, G. 1987. The Global Climate for March-May 1986: Continued Uncertainty About the Possible Onset of an ENSO Episode. *Monthly Weather Review.* 115:317-339.

Battisti, D.S. 1988. Dynamics and Thermodynamics of a Warming Event in a Coupled Tropical Atmosphere-Ocean Model. *Journal of the Atmospheric Sciences.* 45(October 15) 2889-2919.

Berlage, H.P. 1966. The Southern Oscillation and World Weather. *Koninklijk Nederlands Meteorologisch Instituut Mededelinger en Verhandelingen.* 88:152 pp.

Bjerknes, J.A. 1966. A Possible Response of the Atmosphere Hadley Circulation to Equatorial Anomalies of Ocean Temperature. *Tellus.* 18:820-829.

__________. 1961. "El Niño" Study Based on Analysis of Ocean Temperatures 1935-57. *International American Tropical Tuna Commission Bulletin.* 5(3): 219-271.

Boer, G.J. 1989. Concerning the Response of the Atmosphere to a Tropical Sea Surface Temperature Anomaly. *Journal of the Atmospheric Sciences.* 46(July1):1898-1921.

Brooks, C.E.P. and H.W. Braby. 1921. The Clash of the Trades in the Pacific. *Quarterly Journal of the Royal Meteorological Society.* 47:1-13.

Canby, Thomas. 1984. El Niño's III Wind. *National Geographic.* 165:144-184.

Cane, M.A. 1984. Modeling Sea Level During El Niño. *Journal of Physical* Oceanography. 14(12):1864-1874.

__________. 1983. Oceanographic Events During El Niño. *Science.* 222:1189-1195.

__________ and S.E. Zebiak. 1985. A Theory for El Niño and the Southern Oscillation. *Science.* 228:1085-1087.

Carrasco, S. and H. Santander. 1987. The El Niño Event and Its Influence on the Zeroplankton of Peru. *Journal of Geographic Research.* 87(C13):14,405-14,410.

Chao, B.F. 1989. Length of Day Variations Caused by El Niño-Southern Oscillation and Quasi-Biennial Oscillation. *Science.* 243(February 17):923-925.

Chen, W.Y. 1983. The Climate of Spring 1983 - A Season with Persistent Global Anomalies Associated with El Niño. *Monthly Weather Review.* 111:2371-2384.

Cucalon, E. 1987. Oceanographic Variability off Ecuador Associated with an El Niño Event in 1982-1983. *Journal of Geophysical Research.* 92(C13):14,309-14,322.

Deser, Clara and John M. Wallace. 1987. El Niño Events and Their Relation to the Southern Oscillation: 1925-1986. *Journal of Geophysical Research.* 92(C13): 14,187-14,196.

El Niño: An American Geophysical Union (AGU) Chapman Conference. *Journal* of Geophysical Research. Section C, Oceans. 92:14,187,14,471. (29 chapters).

Enfield, D.B., R.L. Smith and S.P. Hayes. 1983. Sea Level Variability at the Eastern Equatorial Pacific Boundary in 1982. *Tropical Ocean Atmosphere Newsletter.* No. 21.

Equatorial Pacific Ocean Climate Studies (EPOCS), Progress and Plans. 1985. Rockville, MD: U.S. Department of Commerce, National Oceanic and Atmospheric Administration, Environmental Research Laboratories. 90 pp.

Firing, E., R. Lucas, J. Sadler and K. Wyrtki. 1983. Equatorial Undercurrent Disappears During the 1982-83 El Niño. *Science.* 227:1121-1123.

Fraedrich, K. 1988. El Niño/Southern Oscillation Predictability. *Monthly Weather Review.* 116(May):1001-1012.

Glantz, Michael H., Richard W. Katz and Neville Nicholls, eds. 1991. *Teleconnections Linking Worldwide Climate Anomalies, Scientific Basis and Societal Impact.* New York: Cambridge University Press. 535 pp.

Glynn, P.W., ed. 1990. *Global Ecological Consequences of the 1982-83 El Niño-Southern Oscillation.* Amsterdam, NY. Elsevier. 563 pp.

Goldberg, R.A., G. Tisnado and R.A. Scofield. 1987. Characteristics of Extreme Rainfall Events in Northwestern Peru During the 1982-83 El Niño. *Journal of Geophysical Research.* 92(C13):14,225-14,241.

Graham, N.E. and W.B. White. 1988. The El Niño Cycle: A Natural Oscillation of the Pacific Ocean-Atmosphere System. *Science.* 240(June3):1293-1302.

Gunther, E.R. 1936. A Report on Oceanographical Investigations in the Peru Coastal Current. *Discovery Report.* 13:107-276.

Halperin, D., S.P. Hayes, A. Leetmaa, D.V. Hansen and S.G.H. Philadner. 1983. Oceanographic Observations of the 1982 Warming of the Tropical Eastern Pacific. *Science.* 221:1173-1174.

Hamilton, K. and R.R. Garcia. 1986. El Niño/Southern Oscillation Events and Their Associated Midlatitude Teleconnections. *Bulletin of the American Meteorological Society.* 67:1354-1361.

Harrison, D.E. 1987. Monthly Mean Island Surface Winds in the Central Tropical and El Niño Events. *Monthly Weather Review.* 115(December):3133-3145.

Hayes, S.P. 1982. A Comparison of Geostrophic and Measured Velocities in the Equatorial Undercurrent. *Journal of Marine Research.* 40(Suppl):219-229.

Held, I.M. et al. 1989. Transients and the Extratropical Response to El Niño. *Journal of the Atmospheric Sciences.* 46(January 1):163-174.

__________. and I.S. Kang. 1987. Barotropic Models of the Extratropical Response to El Niño. *Journal of the Atmospheric Sciences.* 44(December 1):3576-3586.

Hildebrandsson, H.H. 1897. *Quelques Recherches sur les Entres D'Action de l' Atmosphere. K. Svenska Vetens-Akad. Handl.* 29:33 pp.

Hileman, B. 1993. Global Warming Trend Interrupted in 1992. *Chemical & Engineering News.* 71(January 25):7.

Huyer, Adriana, Robert L. Smith and Theresa Paluszklewicz. 1987. Coastal Upwelling off Peru During Normal and El Niño Times, 1981-1984. *Journal of Geophysical Research.* 92(C13):14,297-14,308.

Kerr, R.A. 1992. Unmasking a Shifty Climate System. *Science.* 255(March 20): 508-1510.

__________. 1990. The Climate System as a Ticking Clock. *Science.* 249(September 14):1246-1247.

Kidson, J.W. 1975. Tropical Eigenvector Analysis and the Southern Oscillation. *Monthly Weather Review.* 103:187-196.

Kleeman, R. 1991. A Simple Model of the Atmospheric Response to ENSO Sea Surface Temperature Anomalies. *Journal of the Atmospheric Sciences.* 48(January 1):3-18.

Kousky, V.E. 1987. The Global Climate for December 1986 - February 1987: El Niño Returns to the Tropical Pacific. *Monthly Weather Review.* 115(November): 2822-2838.

Leetmaa, A. 1989. The Interplay of El Niño and La Niña. *Oceanus.* 32(Summer): 30-34.

__________ and J. Witte. 1985. *El Niño Atlas 1982-83.* Nova University Press.

Lockyer, N. and W.J.S. Lockyer. 1902. On Some Phenomena Which Suggest a Short Period of Solar and Meteorological Changes. *Proceedings of the Royal Society of London.* 70:500.

__________ and __________. 1902b. On the Similarity of the Short-Period Pressure Variation Over Large Areas. *Proceedings of the Royal Society of London.* 71:134-135.

__________ and __________. 1904c. The Behavior of the Short-Period Atmospheric Pressure Variation Over the Earth's Surface. *Proceedings of the Royal Society of London.* 73:457-470.

Lukas, R., S.P. Hayes and K. Wyrtki. 1984. Equatorial Sea Level Response During the 1982-83 El Niño. *Journal of Geophysical Research.* 89:10,425-10,430.

Martinez Mediero, Manuel. 1991. *Las Largas Vacaciones de Oliveira Salazar y El Niño de Belen.* Madrid: El Publico, Centro de Documentacion Teatral, 189 pp. [in Spanish]

McPhaden, M.J. and J. Picaut. 1990. El Niño-Southern Oscillation Displacements of the Western Equatorial Pacific Warm Pool. *Science.* 250(December 7): 1385-1388.

Mears, E.G. 1943. The Ocean Current "The Child". *Smithsonian Institute Annual*

Report. 245-251.

Murphy, R.D. 1926. Oceanic and Climatic Phenomena Along the West Coast of South America During 1925. *Geographical Review.* 16:26-54.

Neelin, J.D. 1990. A Hybrid Coupled General Circulation Model for El Niño Studies. *Journal of the Atmospheric Sciences.* 47(March 1):674-693.

1986. *El Niño Phenomenon and Fluctuations of Climate:* Lectures Presented at the Thirty-Sixth Session of the WMO Executive Council. Geneva, Switzerland: Secretariat of the World Meteorological Organization. 46 pp.

1992. El Niño Still on the Prowl. *Science.* 256(April 7):312-313.

1980. *Proceedings of the Workshop on the Phenomenon Known as "El Niño", Guayaquel, Ecuador, 4-12 December 1974.* Organized within the International Decade of Ocean Exploration (IDOE) by the Intergovernmental Oceanographic Commission (IOC) . . . [et al.] Paris: Unesco, 284 pp.

Ochoa, N. and O. Gomez. 1987. Dino flagellates as Indicators of Water Masses During El Niño, 1982-1983. *Journal of Geophysical Research.* 87(C13):14,355-14,368.

Owen, J.A. and T.N. Palmer. 1987. The Impact of El Niño on an Ensemble of Extended-Range Forecasts. *Monthly Weather Review.* 115(September):2103-2117.

Parker, B.B. 1991/1992. Sea Level as an Indicator of Climate and Global Change. *Marine Technology Society Journal.* 25(Winter):13-24.

Philander, S.G. 1992. El Niño. *Oceanus.* 35(Summer):56-61.

__________. 1990. *El Niño, La Niña and the Southern Oscillation.* San Diego: Academic Press, 289 pp.

__________. 1989. El Niño and La Niña. *American Scientist.* 77(September-October):451-459.

Quinn, William H., Victor T. Neal and Santiago E. Anunez de Mayolo. 1987. El Niño Occurrences Over the Past Four and a Half Centuries. *Journal of Geophysical Research. Section C. Ocean.* 92:14,449-14,463.

Quiroz, R.S. 1983. The Climate of the "El Niño" Winter of 1982-83 - Season of Extraordinary Climatic Anomalies. *Monthly Weather Review.* 111:1685-1706.

Rasmusson, E.M. and T.H. Carpenter. 1982. Variations in Tropical Sea Surface Temperatures and Surface Wind Field Associated with the Southern Oscillation/El Niño. *Monthly Weather Review.* 110:354-384.

__________ and J.M. Wallace. 1983. Meteorological Aspects of the El Niño/Southern Oscillation. *Science.* 222:1195-1202.

Ribera Chevremont, Evaristo. 1950. El Niño de Arcilla. Novela. San Juan, PR: Biblioteca de Autores Puertorriquenos, 145 pp.

Richey, J.E., et al. 1989. Amazon River Discharge and Climate Variability: 1903-1985. *Science.* 246(October 6):101-103.

Schott, G. 1931. Der Peru-Strom und Seine Nordlichen Nachbargebiete in Normaler und Anormaler Ausbildung. *Ann. Hydrogr. Mar. Meterol.* 59:161-252.

Shang, E.C. and Y.Y. Wang. 1992. On the Possibility of Monitoring El Niño by Using Modal Ocean Acoustic Tomography. *Journal of the Acoustical Society of America.* 91(January):136-140.

Tovar, H., V. Guillen and D. Cabrera. 1987. Reproduction and Population Levels of Peruvian Guano Birds, 1980-1986. *Journal of Geophysical Research. Section C.*

Ocean. 92:14,455-14,445-446.

Trenberth, K.E. 1976. Spatial and Temporal Variations of the Southern Oscillation. *Quarterly Journal of the Royal Meteorological Society.* 102:639-653.

__________, G.W. Branstater and P.A. Arkin. 1988. Origins of the 1988 North American Drought. *Science.* 242:1640-1645.

Trillmich, Fritz and Kathryn A. Ono, eds. 1991. *Pinnipeds and El Niño.* New York: Springer-Verlag, 293 pp.

Troup, A.J. 1965. The Southern Oscillation. *Quarterly Journal of the Royal Meteorological Society.* 102:639-653.

Wagner, H.J. 1985. The Global Climate for March-May 1985: Strong Phase Reversed in May Ocean from Two Years Ago. *Monthly Weather Review.* 114:500-523.

Wakata, Y. and E.S. Sarachik. 1991. Unstable Coupled Atmospher-Ocean Basin Modes in the Presence of a Spatially Varying Basic State. *Journal of the Atmospheric Sciences.* 48(September 15):2060-2077.

Walker, G.T. 1923. Correlation in Seasonal Variation of Weather, VIII: A Preliminary Study of World Weather. *Mem. Indian Meteorological Department.* 24:75-131.

__________. 1924. Correlation in Seasonal Variations of Weather, IX: A Further Study of World Weather. *Mem. Indian Meteorological Department.* 24:275-332.

__________. 1928. World Weather II. *Mem. Royal Meteorological Society.* 2:97-106.

__________ and E.W. Bliss. 1930. World Weather IV. *Mem. Royal Meterological Society.* 3:81-95.

__________ and __________. 1932. World Weather V. *Mem. Royal Meteorological Society.* 4:53-84.

__________ and __________. 1937. World Weather Vi. Mem. Royal Meteorological *Society.* 4:119-139.

Wang, B. 1992. The Vertical Structure and Development of the ENSO Anomaly Mode During 1979-1989. *Journal of the Atmospheric Sciences.* 49(April 15):698-712.

Wang, X.L. and T. Murakami. 1988. Intraseasonal Disturbance Activity Before, During and After the 1982-83 ENSO. *Journal of the Atmospheric Sciences.* 45(December 15):3754-3770.

Wells, L.E. 1990. Holocene History of the El Niño Phenomenon as Recorded in Flood Sediments of Northern Coastal Peru. *Geology.* 18(November):1134-1137.

Wilkerson, F.P., R.D. Dugdale and R.T. Barber. 1987. Effects of El Niño on New, Regenerated and Total Production in Eastern Boundary Upwelling Systems. *Journal of Geophysical Research.* 87(C13):14,347-14,354.

Wolf, W.L. and R.B. Smith. 1987. Lenght-of-Day Changes and Mountain Torque During El Niño. *Journal of the Atmospheric Sciences.* 44(December):3656-3660.

Wooster, Warren S. and David L. Fluharty, eds. 1985. *El Niño North: Niño Effects in the Eastern Subarctic Pacific Ocean.* Seattle: Washington Sea Grant Program, University of Washington, 312 pp.

Wyrtki, K. 1984. The Slope of the Sea Level Along the Equator During the 1982/1983 El Niño. *Journal of Geophysical Research.* 89:10,419-10424.

Zebiak, Stephen E. and Mark A. Cane. 1987. A Model El Nino-Southern Oscillation. *Monthly Weather Review.* 115:2262-27788.

The Oceans: Physical-Chemical Dynamics and Human Impact. Edited by S.K. Majumdar, E.W. Miller, G.S. Forbes, R.F. Schmalz and Assad A. Panah. © 1994, The Pennsylvania Academy of Science.

Chapter Fifteen

HEAT AND MOISTURE TRANSFER

JOSEPH CHI
University of the District of Columbia
Washington, DC 20008

INTRODUCTION

Most of the radiant energy absorbed by the sea and its atmosphere is transformed into internal energy and stored in the sea water. As a consequence, the immediate source of the energy which drives the marine atmosphere may be identified with the air boundary layer over the sea surface and specifically with its internal energy distribution. Transport of heat, moisture and momentum may bring about subsequent energy transformations in the marine atmosphere. Much of our understanding of processes in the marine planetary boundary layer (MPBL) has come from careful observation and sound theoretical modeling. Recent advances in turbulence modeling and computational techniques have made it possible to simulate simultaneous effects of numerous parameters on dynamics of the MPBL. An objective of this article is to present a turbulent theory and numerical procedure for predicting effects of heat, moisture and momentum transfer on dynamics of marine atmospheres.

For this objective, a turbulence model and governing equations for heat, moisture and momentum transfer in the MPBL will be formulated. A flexible finite-element numerical procedure and computer algorithm will be described. In conclusion, results of predictions will be presented and discussed. As far as possible, confidence of the theory will be established by comparison of predictions with published data from observations.

TURBULENCE MODEL AND GOVERNING EQUATIONS

Conservation Equations for Mean Flows

In order to simulate dynamics of atmospheric air in the MPBL, conservation equations must first be formulated. The coordinates are chosen so that x- and y-axes are directed towards east and north, respectively, and z-axis is vertically upward. When Boussinesq approximation is used, the conservation equations for momentum, enthalpy and moisture of atmospheric air in northern hemisphere become:

$$\frac{\partial U}{\partial x}+\frac{\partial V}{\partial y}+\frac{\partial W}{\partial z}=0 \tag{1}$$

$$\frac{\partial U}{\partial t}+U\frac{\partial U}{\partial x}+V\frac{\partial U}{\partial y}+W\frac{\partial U}{\partial z}-fV=-\frac{\partial P}{\rho\partial x}-\frac{\partial\overline{u^2}}{\partial x}-\frac{\partial\overline{uv}}{\partial y}-\frac{\partial\overline{uw}}{\partial z} \tag{2}$$

$$\frac{\partial V}{\partial t}+U\frac{\partial V}{\partial x}+V\frac{\partial V}{\partial y}+W\frac{\partial V}{\partial z}+fU=-\frac{\partial P}{\rho\partial y}-\frac{\partial\overline{uv}}{\partial x}-\frac{\partial\overline{v^2}}{\partial y}-\frac{\partial\overline{vw}}{\partial z} \tag{3}$$

$$\frac{\partial W}{\partial t}+U\frac{\partial W}{\partial x}+V\frac{\partial W}{\partial y}+W\frac{\partial W}{\partial z}=-\frac{\partial P}{\rho\partial y}-\frac{\partial\overline{uw}}{\partial x}-\frac{\partial\overline{vw}}{\partial y}-\frac{\partial\overline{w^2}}{\partial z}+\beta gS_v \tag{4}$$

$$\frac{\partial H}{\partial t}+U\frac{\partial H}{\partial x}+V\frac{\partial H}{\partial y}+W\frac{\partial H}{\partial z}=-\frac{\partial\overline{uh}}{\partial x}-\frac{\partial\overline{vh}}{\partial y}-\frac{\partial\overline{wh}}{\partial z} \tag{5}$$

$$\frac{\partial R}{\partial t}+U\frac{\partial R}{\partial x}+V\frac{\partial R}{\partial y}+W\frac{\partial R}{\partial z}=-\frac{\partial\overline{ur}}{\partial x}-\frac{\partial\overline{vr}}{\partial y}-\frac{\partial\overline{wr}}{\partial z} \tag{6}$$

where the upper-case dependent variables represent the mean fields and the corresponding lower-case variables represent turbulent fluctuations of the same variables. Thus, U, V and W are mean velocity components in the x, y and z directions and at the time t. P is the mean pressure, H the mean moist specific enthalpy which is defined as $(S + LR_v)$, and R the mean total moisture mixing ratio which is defined as $(R_v + R_l)$. S is the dry specific enthalpy which is defined as $C_p[T - (T_o - gz/C_p)]$, L the water latent heat of vaporization, C_p the constant pressure specific heat, T the temperature, T_o the standard temperature, R_v the water vapor mixing ratio, R_l the liquid water mixing ratio, g the gravitational acceleration, f the Coriolis parameter, S_v the virtual dry specific enthalpy which is defined as $S(1 + 1.609R_v - R)$, ϱ the standard density, ν the kinematic viscosity, and β the buoyancy coefficient which is defined as $[1/(C_p T_o)]$. The second-order correlations in these equations represent the mean turbulent fluxes of momentum, enthalpy and moisture fluxes. Transport equations for turbulence are discussed in the following section.

Turbulence Transport Equations

The turbulence transport theory has been investigated by several authors, for example, see Harlow and Hirt (1969), Lewellen and Teske (1973), Mellor and Yamada (1974) and Chi (1977). Within the marine planetary boundary layers, variations of the second order correlations are dominant in the vertical direction. Transport equations for the second-order correlations may be written as follows:

$$\frac{\partial\overline{u^2}}{\partial t}=\frac{\partial}{\partial z}(Aq\lambda\frac{\partial\overline{u^2}}{\partial z})-2\overline{uw}\frac{\partial U}{\partial z}-B\frac{q}{\lambda}(\overline{u^2}-\frac{q^2}{3})-D\frac{q^3}{3\lambda} \tag{7}$$

$$\frac{\partial\overline{v^2}}{\partial t}=\frac{\partial}{\partial z}(Aq\lambda\frac{\partial\overline{v^2}}{\partial z})-2\overline{vw}\frac{\partial V}{\partial z}-B\frac{q}{\lambda}(\overline{v^2}-\frac{q^2}{3})-D\frac{q^3}{3\lambda} \tag{8}$$

$$\frac{\partial\overline{w^2}}{\partial t}=\frac{\partial}{\partial z}(3Aq\lambda\frac{\partial\overline{w^2}}{\partial z})+2\beta\overline{ws_v}-B\frac{q}{\lambda}(\overline{w^2}-\frac{q^2}{3})-D\frac{q^3}{3\lambda} \tag{9}$$

$$\frac{\partial\overline{uv}}{\partial t}=\frac{\partial}{\partial z}(Aq\lambda\frac{\partial\overline{uv}}{\partial z})-\overline{uw}\frac{\partial V}{\partial z}-\overline{vw}\frac{\partial U}{\partial z}-B\frac{q}{\lambda}\overline{uv} \tag{10}$$

$$\frac{\partial\overline{uw}}{\partial t}=\frac{\partial}{\partial z}(2Aq\lambda\frac{\partial\overline{uw}}{\partial z})+\beta g\overline{us_v}-(\overline{w^2}-C\overline{q^2})\frac{\partial U}{\partial z}-B\frac{p}{\lambda}\overline{uw} \tag{11}$$

$$\frac{\partial\overline{vw}}{\partial t}=\frac{\partial}{\partial z}(2Aq\lambda\frac{\partial\overline{vw}}{\partial z})+\beta g\overline{vs_v}-(\overline{w^2}-C\overline{q^2})\frac{\partial V}{\partial z}-B\frac{p}{\lambda}\overline{vw} \tag{12}$$

$$\frac{\partial\overline{uh}}{\partial t}=\frac{\partial}{\partial z}(Aq\lambda\frac{\partial\overline{uh}}{\partial z})-\overline{uw}\frac{\partial H}{\partial z}-\overline{wh}\frac{\partial U}{\partial z}-E\frac{q}{\lambda}\overline{uh} \tag{13}$$

$$\frac{\partial \overline{vh}}{\partial t} = \frac{\partial}{\partial z}(Aq\lambda \frac{\partial \overline{vh}}{\partial z}) - \overline{vw}\frac{\partial H}{\partial z} - \overline{wh}\frac{\partial V}{\partial z} - E\frac{q}{\lambda}\overline{vh} \tag{14}$$

$$\frac{\partial \overline{wh}}{\partial t} = \frac{\partial}{\partial z}(2Aq\lambda \frac{\partial \overline{wh}}{\partial z}) - \overline{w^2}\frac{\partial H}{\partial z} + \beta g \overline{hs_v} - E\frac{q}{\lambda}\overline{wh} \tag{15}$$

$$\frac{\partial \overline{h^2}}{\partial t} = \frac{\partial}{\partial z}(Aq\lambda \frac{\partial \overline{h^2}}{\partial z}) - 2\overline{wh}\frac{\partial H}{\partial z} - F\frac{q}{\lambda}\overline{h^2} \tag{16}$$

$$\frac{\partial \overline{ur}}{\partial t} = \frac{\partial}{\partial z}(Aq\lambda \frac{\partial \overline{ur}}{\partial z}) - \overline{uw}\frac{\partial R}{\partial z} - \overline{wr}\frac{\partial U}{\partial z} - E\frac{q}{\lambda}\overline{ur} \tag{17}$$

$$\frac{\partial \overline{vr}}{\partial t} = \frac{\partial}{\partial z}(Aq\lambda \frac{\partial \overline{vr}}{\partial z}) - \overline{vw}\frac{\partial R}{\partial z} - \overline{wr}\frac{\partial V}{\partial z} - E\frac{q}{\lambda}\overline{vr} \tag{18}$$

$$\frac{\partial \overline{wr}}{\partial t} = \frac{\partial}{\partial z}(2Aq\lambda \frac{\partial \overline{wr}}{\partial z}) - \overline{w^2}\frac{\partial R}{\partial z} + \beta g \overline{rs_v} - E\frac{q}{\lambda}\overline{wr} \tag{19}$$

$$\frac{\partial \overline{r^2}}{\partial t} = \frac{\partial}{\partial z}(Aq\lambda \frac{\partial \overline{r^2}}{\partial z}) - 2\overline{wr}\frac{\partial R}{\partial z} - F\frac{q}{\lambda}\overline{r^2} \tag{20}$$

$$\frac{\partial \overline{hr}}{\partial t} = \frac{\partial}{\partial z}(Aq\lambda \frac{\partial \overline{hr}}{\partial z}) - \overline{wr}\frac{\partial H}{\partial z} - \overline{wh}\frac{\partial R}{\partial z} - F\frac{q}{\lambda}\overline{hr} \tag{21}$$

where q is the turbulent fluctuations velocity defined as:

$$q = \sqrt{\overline{u^2} + \overline{v^2} + \overline{w^2}} \tag{22}$$

λ is a characteristic length which may be calculated by the equation:

$$\lambda = \frac{\kappa z \delta}{\delta + 10\kappa z} \tag{23}$$

and the length scale δ is defined as follows:

$$\delta = \frac{\int_0^\infty qzdt}{\int_0^\infty qdz} \tag{24}$$

In equations 7-21, the time-derivative terms on the left-hand side model the transient variation of the turbulent flux. The second-order derivative terms on the right-hand side model the turbulent diffusion. While the production of turbulence due to buoyancy is modeled by terms with buoyant coefficient β, the production of turbulence due to friction is modeled by the products of second-order correlations and gradient of mean variables. Terms with coefficients B, C, E and F represent the turbulent redistribution. The turbulent dissipation is modeled by terms with coefficient D.

Evaluations of the coefficients $\varkappa$, A, B, C, D, E and F have been made by Lewellen and Teske (1973), Mellor and Yamada (1974) and Chi (1991). Table 1 shows their respective results:

TABLE 1

Model Coefficients for Turbulence

Authors	$\varkappa$	A	B	C	D	E	F
Lewellen & Teske (1973)	0.36	0.25	0.60	0.000	0.146	0.42	0.27
Mellor & Yamada (1974)	0.35	0.23	0.43	0.056	0.133	0.43	0.25
Chi (1991)	0.35	0.21	0.46	0.053	0.132	0.44	0.23

It can be seen from this table that the three sets of values spread at about the order of accuracies as the empirical data from which these coefficients were derived.

In conclusion, we have formulated governing equations 1-6 for conservation of mass, momentum, enthalpy and moisture of atmospheric air in the MPML and established correlation equations 7-21 for turbulent transportation of mass, momentum, enthalpy and moisture fluxes. The coefficients $\varkappa$, A, B, C, D, E and F for correlation equations have been reviewed, and their values equal to 0.35, 0.21, 0.46, 0.053, 0.132, 0.44 and 0.23 will be used in this article.

FINITE-ELEMENT EQUATIONS AND PROGRAM ALGORITHMS

In the previous section, partial differential equations 1-21 governing conservations of mass, momentum, enthalpy and moisture and transportations of turbulent stresses and fluxes have been derived. In this section of the paper, a finite-element procedure firstly will be used to derive from equation 1-21 a set of ordinary differential equations which may be used to simulate dynamics of atmospheres in the MPBL. Program algorithms then will be developed for coding a graphics-oriented computer program for simulating dynamics of atmospheres in the MPBL.

Finite-Element Equations

The conservation and turbulent transport equations 1-21 derived in section 2 belong to a class of equations in mathematical physics for dealing with initial value convection/diffusion problems. They can be written in the following general form:

$$\frac{\partial \alpha}{\partial t}+U_j\frac{\partial \alpha}{\partial x_j}-\frac{\partial}{\partial x_j}(\gamma\frac{\partial \alpha}{\partial x_j})+\phi=0, \quad \textit{Within domain } \Omega \tag{25}$$

and their boundary and initial conditions may be written as:

$$\gamma\frac{\partial \alpha}{\partial n}+a\alpha+b=0, \quad \textit{On boundary } \Gamma \tag{26}$$

$$\alpha(x,y,z,0)=\alpha_0(x,y,z), \quad \textit{At initial time zero} \tag{27}$$

where α (x,y,z,t) are mean values of velocity components, enthalpy, moisture and turbulent correlations; ϕ (x,y,z,t) values are sources/sinks for α; the Four items in equation 25 describe the system's inertia, convection, diffusion and source/sink strength, respectively; and 'a' and 'b' values are specified to enforce the appropriate boundary conditions in equation 26.

The standard Galerkin finite element method is employed for the discretization of spatial variables. The flow domain to be simulated is divided into small sub-domains called finite elements. Multiplying both sides of equation set 25 by a weighing function ω and integrating over the finite element Ω_e, the weighted residual equation set can be derived:

$$\int_{\Omega_e}\omega[\frac{\partial \alpha}{\partial t}+U_j\frac{\partial \alpha}{\partial x_j}-\frac{\partial}{\partial x_j}(\gamma\frac{\partial \alpha}{\partial x_j})+\phi]d\tau=0 \tag{28}$$

Integrating the second-order term in equation 28 by parts and substituting equation 26 for normal gradients at the element's boundary result in the following equation:

$$\int_{\Omega_e}[\omega\frac{d\alpha}{dt}+\gamma\frac{\partial \omega}{\partial x_j}\frac{\partial \alpha}{\partial x_j}+U_j\omega\frac{\partial \alpha}{\partial x_j}+\phi]d\tau+\oint_{\Gamma_e}(a\omega\alpha+b\omega)d\sigma=0 \tag{29}$$

The weighing functions ω and dependent variables α (x,y,z,t) are interpolated in each element Ω_e as follows:

$$\omega_e(x,y,z)=\psi_n(x,y,z)\Omega_n \tag{30}$$

$$\alpha_e(x,y,z,t)=\psi_n(x,y,z)\mathrm{A}_n(t) \tag{31}$$

where ψ_n denotes the interpolation functions at the nth node of each finite element, and A_n and Ω_n are the element nodal values of the variable α and the weighing function ω, respectively.

The Galerkin procedure of substituting equations 30 and 31 into equation 29 leads to the following finite-element equation:

$$C_{mn}\frac{dA_n}{dt}+(K_{mn}+U_{mn}+H_{mn})A_n+L_n=0 \tag{32}$$

where C_{mn}, K_{mn}, U_{mn} and H_{mn} are the mth-row and nth-column elements of square matrices which describe distributions of the finite element's capacitance, diffusivity, convectiveness and surface flux, respectively, and L_n is the nth element of a load vector for the finite-element source/sink terms; they are defined individually as follows:

$$C_{mn}=\int_{\Omega_e}\psi_m\psi_n d\tau \tag{33}$$

$$K_{mn}=\int_{\Omega_e}\gamma\frac{\partial\psi_m}{\partial x_j}\frac{\partial\psi_n}{\partial x_j}d\tau \tag{34}$$

$$U_{mn}=\int_{\Omega_e}U_j\psi_m\frac{\partial\psi_n}{\partial x_j}d\tau \tag{35}$$

$$H_{mn}=\int_{\Gamma_e}a\psi_m\frac{\partial\psi_n}{\partial n}d\sigma \tag{36}$$

$$L_n=\int_{\Omega_e}\phi\psi_n d\tau+\int_{\Gamma_e}b\psi_n d\sigma \tag{37}$$

Assemblying the finite-element equation 32 over the whole flow field yields a set of ordinary differential equations:

$$[C]\frac{d\{A\}}{dt}+[K+U+H]\{A\}+[L]=0 \tag{38}$$

where the vector {A} contains all node point variables lying within the flow domain Ω; [L] is the global load vector; and the matrices [C], [K], [U] and [H] represent global distributions of capacitance, diffusiveness, convection and surface conductance, respectively.

Solution Procedure and Program Algorithms

The finite-element procedure described in section *Finite-Element Equations* has transformed the initial-elliptical boundary value partial differential equations 1-21 into a large-order system of the ordinary differential equation set 38. Solution of (38) may start from a Taylor series expansion of $\{A\}$ about time t_n:

$$\{A\}_{n+1} = \{A_n\} + \Delta t[\theta \frac{\partial \{A\}_{n+1}}{\partial t} + (1-\theta)\frac{\partial \{A\}_n}{\partial t}] \tag{39}$$

where θ is an implicitive coefficient. Multiplying both sides of equation 39 by capacitance matrix [C], equation 39 may be re-written as:

$$\{F(A_{n+1})\} = [C][\{A\}_{n+1} - \{A_n\} - (\Delta T)\theta \frac{\partial \{A\}_{n+1}}{\partial t} - (\Delta t)(1-\theta)\frac{\partial \{A\}_n}{\partial t}] \tag{40}$$

With expressions for time derivatives provided by equation 38, equation 40 becomes:

$$\begin{aligned} \{F\} &= [C]_{n+1}(\{A\}_{n+1} - \{A\}_n) \\ &+ \Delta t\theta([K+U+H]_{n+1}\{A\}_{n+1} + \{L\}_{n+1}) \\ &+ \Delta t(1-\theta)([K+U+H]_n\{A\}_n + \{L\}_n) \end{aligned} \tag{41}$$

The equation set 41 may be solved for $\{A\}_{n+1}$ values at time t_{n+1} from the known $\{A\}_n$ values at time t_n by a Newton's iterative process, and a computer program has been written to facilitate the process. Listed below is a step-by-step algorithm which has been used in coding the program:

Step 1. Set $\{A\}_{n+1}$ equal to $\{A\}_n$.

Step 2. Evaluate element matrices and load vectors by equations 33-37; and assemble them to form the global matrices and load vector $[C]_{n+1}$, $[K]_{n+1}$, $[U]_{n+1}$, $[H]_{n+1}$, $\{L\}_{n+1}$, respectively.

Step 3. Evaluate error vector $\{F\}_{n+1}$ from equation (41).

Step 4. Evaluate the Jacobian matrix [J(F)] by the equation:

$$[J(F)]_{n+1} = [C]_{n+1} + \Delta t\theta[K+U+H]_{n+1} \tag{42}$$

Step 5. Evaluate $\{\delta A\}_{n+1}$values from the equation:

$$[J(F)]_{n+1}\{\delta A\}_{n+1} = -\{F\}_{n+1} \tag{43}$$

Step 6. Update the $\{A\}_{n+1}$ values using the equation:

$$\{A\}_{n+1,new} = \{A\}_{n+1,old} + \{\delta A\}_{n+1} \tag{44}$$

Step 7. Calculate the root-mean-square discrepancy μ by the equation:

$$\mu = \sqrt{\frac{\Sigma(\delta A)^2}{N}} \tag{45}$$

Step 8. If μ is less than or equal to the preset convergent criterion ϵ, go to step 10 else go to step 9.

Step 9. Let $\{A\}_{n+1}$ equal to $\{A\}_{n+1,new}$ and repeat steps 1-8.

Step 10. Output $\{A\}_{n+1,new}$ values, and let $\{A\}_n$, $[K+U+H]_n$ and $\{L\}_n$ equal to $\{A\}_{n+1,new}$, $[K+U+H]_{n+1,new}$ and $\{L\}_{n+1,new}$.

Step 11. If end of the simulation period has not been reached, increase the time by Δt and go to step 1. Else stop the program.

In conclusion, a finite-element computer program has been written to calculate heat, moisture and momentum transfer in the MPBL. In addition, the computer program has been integrated with a graphics-oriented software PATRAN to faciliate users' interactive generation of models for simulating dynamics of atmospheres in the MPBL. Several examples of the simulation results are presented and discussed in the following section.

RESULTS AND CONCLUSION

The graphics-oriented finite-element computer program described in *Finite-Element Equations and Program Algorithms* has been tested to confirm its accuracy and to simulate dynamics of atmospheres in the MPBL. Several examples of the simulation are discussed below.

Simulation Procedure and Accuracy

To illustrate the simulation procedure and its accuracy, we in this section will start from a simple 1-D transient heat conduction problem with known boundary temperatures and followed by a 2-D rotating-flow problem. Examples for the more involved problems of the MPBL will be presented in *Stratified Atmospheric Layers* and *Heat, Moisture and Momentum Transfer in Vortex Flows.*

Sucec (1985) considered an infinite slab of thickness 0.3 meter. Initially because of an electrical current, the slab is in a steady state with a temperature distribution in Degrees C given by the equation:

$$T = 100 + 400\sin(\frac{\pi x}{0.3}) \tag{46}$$

However, at time t = 0, the electric current is shut off, thus causing a trasient in the temperature, while the slab faces continue to be held at 100°C by a coolant. The thermal diffusivity of the material is constant at $2.84 x 10^5 m^2/s$. We would like to find the temperature within the slab as a function of position and time by using the present software.

A model for this problem has been created. Figure 1 shows comparisons of predicted temperature distributions with exact solutions for the slab at time equal to 0, 150, 300, 450, and 750 seconds, respectively.

To illustrate flow studies using the present software, a very popular two-dimensional driven cavity problem (e.g., see Bozeman and Dalton, 1973; Kawahara and Okamoto 1976; and Chi, 1990) will be used as an example. Figure 2-A shows a sketch of the driven cavity with the nondimensional length by height being one by one. The cavity fluid is initially at rest, and the top lid of the cavity is started impulsively with a velocity which is used as the reference velocity. Reynolds number for the cavity flow is assumed to be 100. At the test run, the flow domain is divided

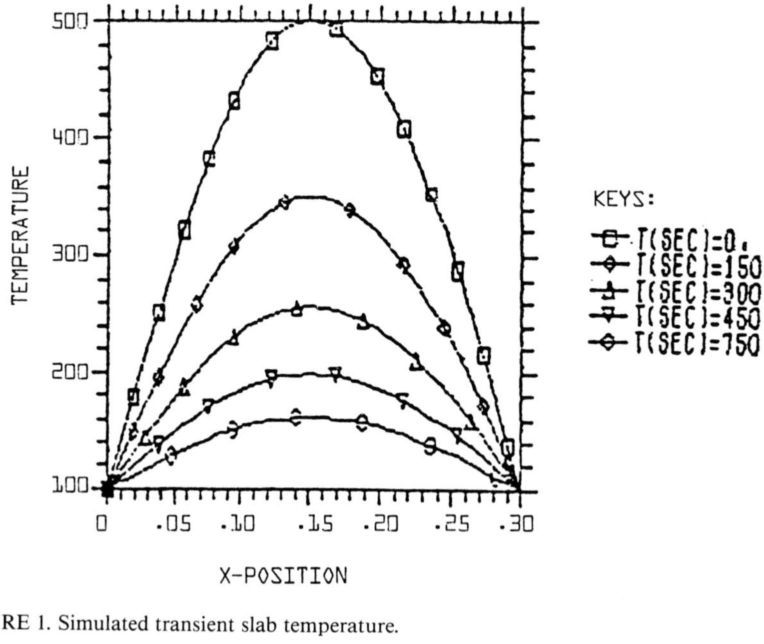

FIGURE 1. Simulated transient slab temperature.

into a mesh of 20 by 20. Figures 2A to 2E shows snapshots of the flow field at the nondimensional time τ equal to 0.0, 0.4, 1.0, 3.0 and 5.0, respectively. It can be seen in these snapshots that, as time advances from 0 to 3, penetration of the flow depth increases. As τ advances from 3.0 onwards, there are little further changes. That is, steady state of the flow has reached at τ equal 3.

Stratified Atmospheric Layers

The atmospheric boundary layer in the region outside of the surface roughness but still close to the earth/sea surface is the most extensively studied example of a turbulent shear layer incorporating the influence of stratification on the dynamics of turbulence. Following Monin -Obukhov (1953), some success has been achieved by Businger, Wyngaard, Izumi and Bradley (1971) and Wyngaard et al. in experimentally obtaining universal functions that describe the dependence of the mean turbulence characteristics of height, shear stress and heat flux. Thus, this layer provides a good test for the present turbulent model.

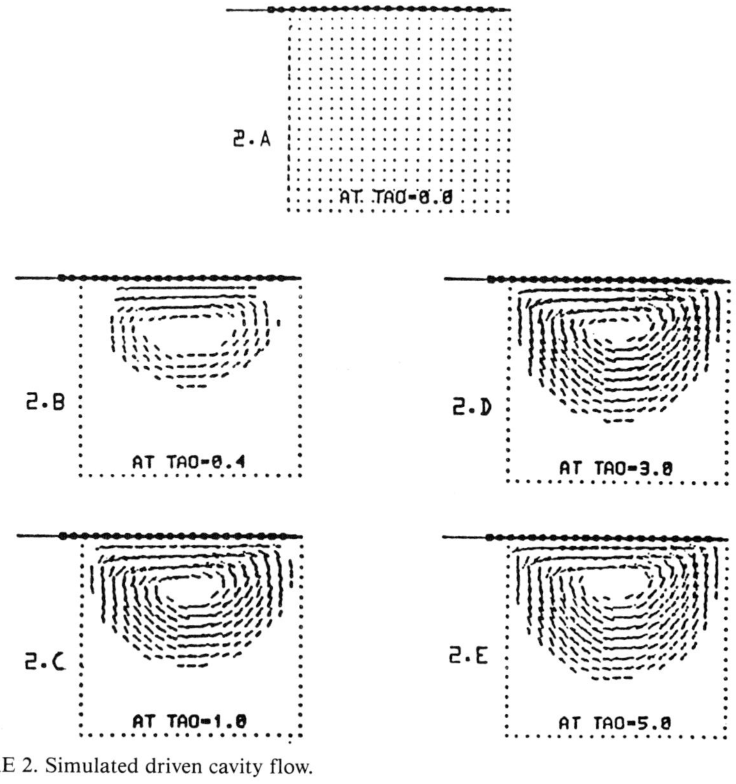

FIGURE 2. Simulated driven cavity flow.

Without loss of generality, we can orient our coordinate system so that $\overline{uw}$ is equal to zero. From a simple scale analysis it may be determined that the Coriolis terms cannot be important in this thin layer. In addition, we will introduce the potential temperature Θ which is equal to $[T - T_o + (gz/C_p)]$, redefine the coefficient β equal to $1/T_o$ instead of $1/(C_pT_o)$ and assume the absence of liquid water. The resulting set of simplified turbulent transport equations can now be written as:

$$\frac{\partial \overline{u^2}}{\partial t} = \frac{\partial}{\partial z}\left(Aq\lambda \frac{\partial \overline{u^2}}{\partial z}\right) - 2\overline{uw}\frac{\partial U}{\partial z} - B\frac{q}{\lambda}\left(\overline{u^2} - \frac{q^2}{2}\right) - D\frac{q^3}{3\lambda} \tag{47}$$

$$\frac{\partial \overline{v^2}}{\partial t} = \frac{\partial}{\partial z}\left(Aq\lambda \frac{\partial \overline{v^2}}{\partial z}\right) - B\frac{q}{\lambda}\left(\overline{v^2} - \frac{q^2}{2}\right) - D\frac{q^3}{3\lambda} \tag{48}$$

$$\frac{\partial \overline{w^2}}{\partial t} = \frac{\partial}{\partial z}\left(3Aq\lambda \frac{\partial \overline{w^2}}{\partial z}\right) + 2\beta g\overline{w\theta} - B\frac{q}{\lambda}\left(\overline{w^2} - \frac{q^2}{2}\right) - D\frac{q^3}{3\lambda} \tag{49}$$

$$\frac{\partial \overline{uw}}{\partial t} = \frac{\partial}{\partial z}\left(2Aq\lambda \frac{\partial \overline{uw}}{\partial z}\right) + \beta g\overline{u\theta} - \left(\overline{w^2} - C\overline{q^2}\right)\frac{\partial U}{\partial z} - B\frac{q}{\lambda}\overline{uw} \tag{50}$$

$$\frac{\partial \overline{u\theta}}{\partial t} = \frac{\partial}{\partial z}\left(Aq\lambda \frac{\partial \overline{u\theta}}{\partial z}\right) - \overline{uw}\frac{\partial \Theta}{\partial z} - \overline{w\theta}\frac{\partial U}{\partial z} - E\frac{q}{\lambda}\overline{u\theta} \tag{51}$$

$$\frac{\partial \overline{w\theta}}{\partial t} = \frac{\partial}{\partial z}\left(2Aq\lambda \frac{\partial \overline{w\theta}}{\partial z}\right) - \overline{w^2}\frac{\partial \Theta}{\partial z} + \beta g\overline{\theta^2} - E\frac{q}{\lambda}\overline{w\theta} \tag{52}$$

$$\frac{\partial \overline{\theta^2}}{\partial t} = \frac{\partial}{\partial z}\left(Aq\lambda \frac{\partial \overline{\theta^2}}{\partial z}\right) - 2\overline{w\theta}\frac{\partial \Theta}{\partial z} - F\frac{q}{\lambda}\overline{\theta^2} \tag{53}$$

Equations 47-53 may be integrated to large t to yield the following steady-state normalized functions for shear stress ϕ_M, heat flux ϕ_H, shear production of shear stress $\phi_{S,m}$, shear production of heat flux $\phi_{S,H}$, buoyant production of shear stress $\phi_{B,}$, and buoyant production of heat flux $\phi_{B,H}$:

$$\phi_M = \frac{\kappa z}{(-\overline{uw})^{1/2}} \frac{\partial U}{\partial z} \tag{54}$$

$$\phi_H = \frac{\kappa z(-\overline{uw})^{1/2}}{-\overline{w\theta}} \frac{\partial \Theta}{\partial z} \tag{55}$$

$$\phi_{S,M} = \frac{\kappa z\overline{w^2}}{(-\overline{uw})^{3/2}} \frac{\partial U}{\partial z} \tag{56}$$

$$\phi_{S,H} = \frac{\kappa z\overline{w^2}}{(-\overline{w\theta})(-\overline{uw})^{1/2}} \frac{\partial \Theta}{\partial z} \tag{57}$$

$$\phi_{B,M} = \frac{\kappa z\beta g(-\overline{u\theta})}{(-\overline{uw})^{3/2}} \tag{58}$$

$$\phi_{B,H} = \frac{\kappa z\beta g(-\overline{\theta^2})}{(-\overline{uw})^{1/2}(-\overline{w\theta})} \tag{59}$$

In Figures 3-8, predicted values of these normalized functions are plotted versus the dimensionless stability parametric value of (z/L) which is defined by the expression:

$$(z/L) = \frac{\kappa z\beta g(-\overline{w\theta})}{(-\overline{uw})^{3/2}} \tag{60}$$

For comparison, shown in the same figures are experimental data of Businger, Wyngaard, Izum and Bradley (1971), Wyngaard and Cote (1971) and Wyngaard, Cote and Izum (1971). Excellent agreement between simulation and experiments can be seen in these figures.

Heat, Moisture and Momentum Transfer in Vortex Flows

The vortex-flow situation as shown in Figure 9 occurs frequently in marine atmospheres. It represents many challenging problems which are of interests to researchers in areas of heat, moisture and momentum transfer; turbulence modeling; strong influences of the gravitational, Coriolis and centrifugal forces on the

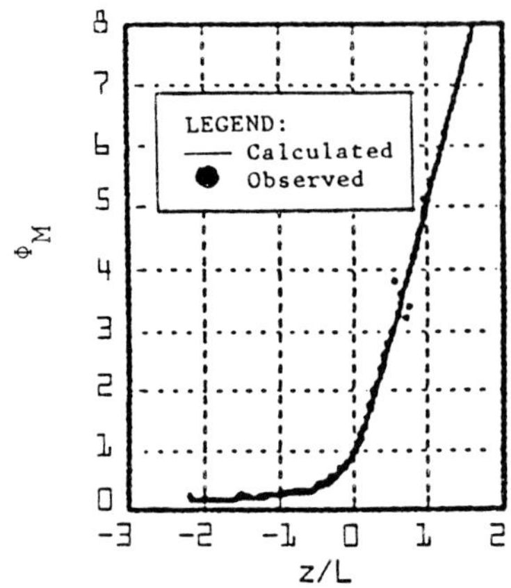

FIGURE 3. Turbulent shear stress.

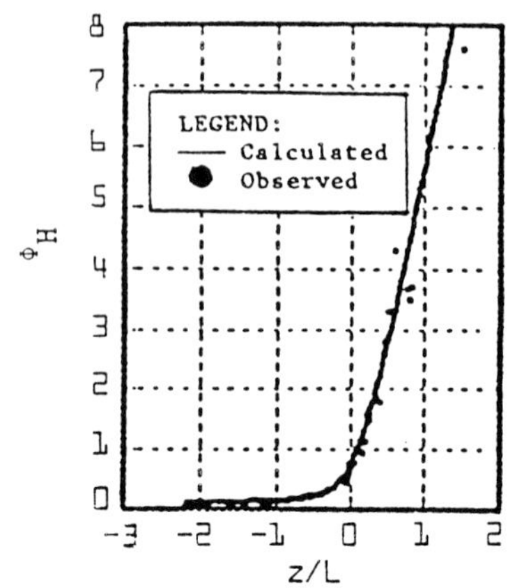

FIGURE 4. Turbulent heat flux.

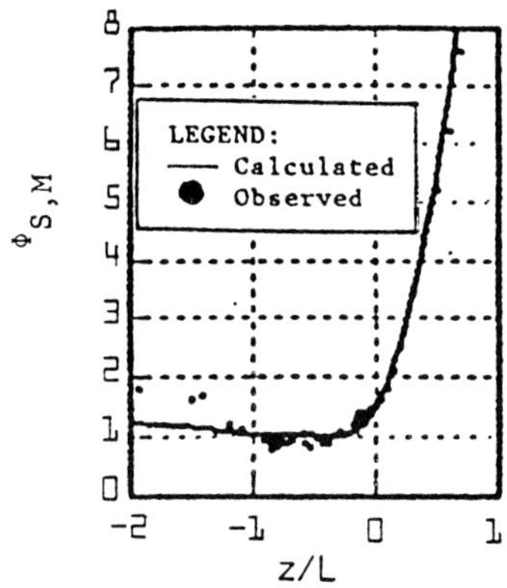

FIGURE 5. Shear production of turbulent shear stress.

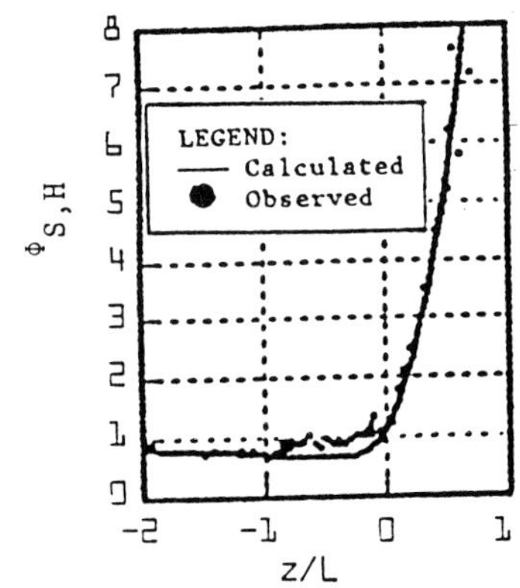

FIGURE 6. Shear production of turbulent heat flux.

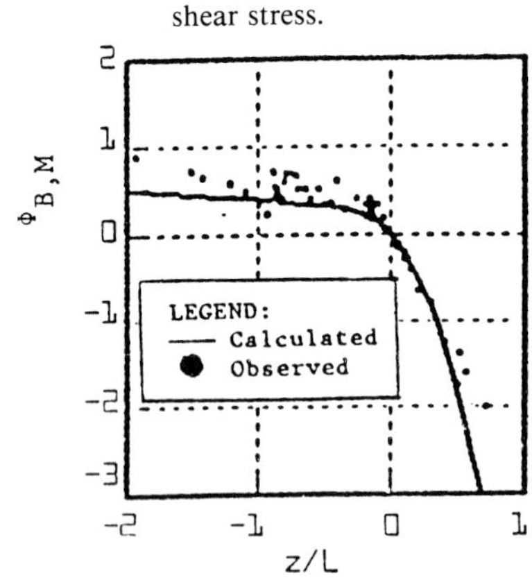

FIGURE 7. Buoyant production of turbulent shear stress.

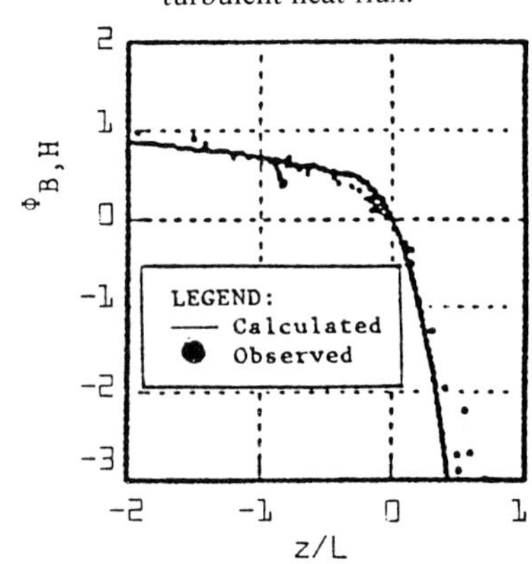

FIGURE 8. Buoyant production of turbulent heat flux.

MPBL; and effects of air-water interface roughness and entrainment of liquid droplets in marine atmospheres.

Several solutions of vortex flow have been obtained by Chi et al. (1969, 1977 and 1987). For illustration, we will discuss in this section a set of solutions for the MPBL flow under a main vortex as specified in the following table:

TABLE 2
Main Vortex at an Altitude of One Meter

U and V distribution at z = 1 km	Figure 10
Maximum tangential velocity V_{max} at z =.1 km	52 m/s
Radius at which V_{max} occurs	40 km
Air temperature at z = 1 km	273°C
Coriolis force coefficient	$1x10^{-5}$ s^{-1}
Saturated air temperature at z = 0	288°C
Gradients of U, V, H, R, q^2 and $\epsilon$with respect to z at z = 1 km	0.00

The simulation results for this sample vortex MPBL are plotted in Figures 11-15. The MPBL tangential velocity V and radial velocity U at several radial positions are plotted in Figure 11 and 12, respectively. The MPBL specific enthalpy $(H - H_\infty)$ and moisture ratio $(R - R_\infty)$ values are plotted versus height in Figures 13-14, respectively. Distribution of the turbulence velocity square q^2 in the MPBL are plotted in Figure 15.

Several interesting marine vortex boundary layer characteristics can be observed from these figures. Figure 11 shows that at the region far above the sea surface tangential velocity dominates. The balance force for this region is, therefore, a balance of the radial pressure gradient and combination of the centrifugal and Coriolis forces. Near the sea surface, retardation of the gangential velocity is seen in Figure 11. This retardation is accompanied by a reduction in centrifugal and Coriolis forces. The balance of pressure, centrifugal and Coriolis forces is thereby destroyed. The flow in this region is then characterized by the entrainment of flow

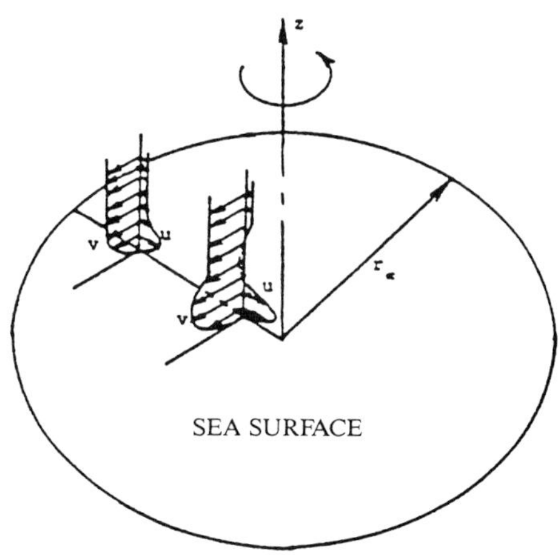

FIGURE 9. Schematic diagram of a votex flow over the sea surface.

to the boundary layers as indicated by the large induced radial velocity shown in Figure 12. Figures 13 and 14 show large enthalpy and moisture values at vicinity of the sea surface and their decreased values at high altitude. Advection and diffusion of the turbulent energy can be observed from distributions of q^2-values shown in Figure 15.

CONCLUSION

In conclusion, a turbulence model and governing equations for conservation of heat, moisture and momentum in the marine planetary boundary layer have been derived. A finite-element procedure has been developed to transform the governing partial differential equations for turbulence, and heat, moisture and momentum

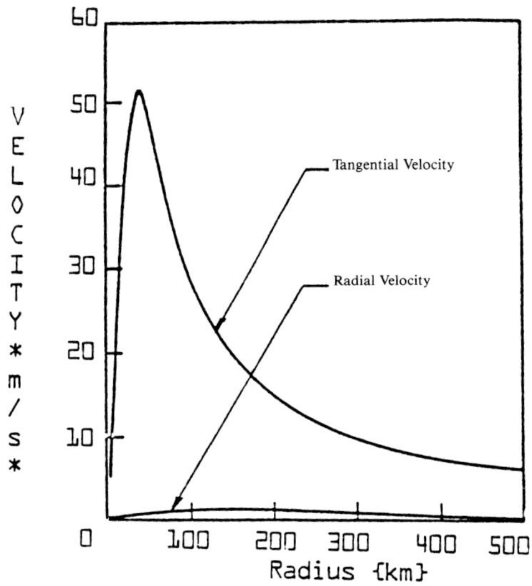

FIGURE 10. A votex tangential and radial velocities V and -U

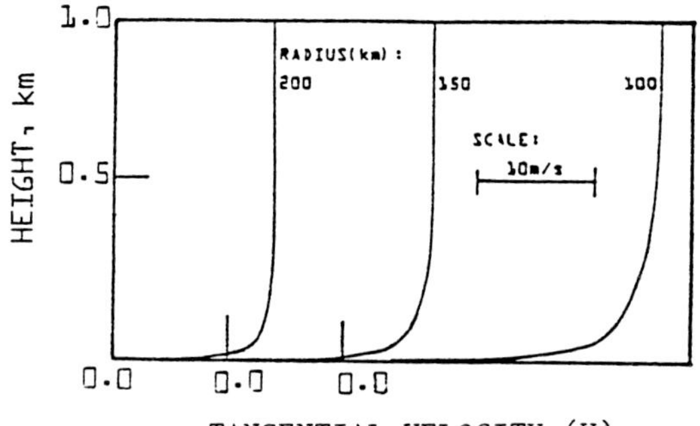

FIGURE 11. Tangential velocity versus height at different radii.

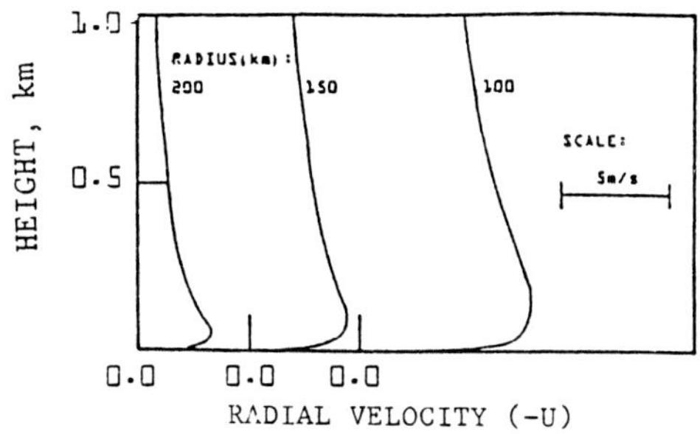

FIGURE 12. Radial velosity height.

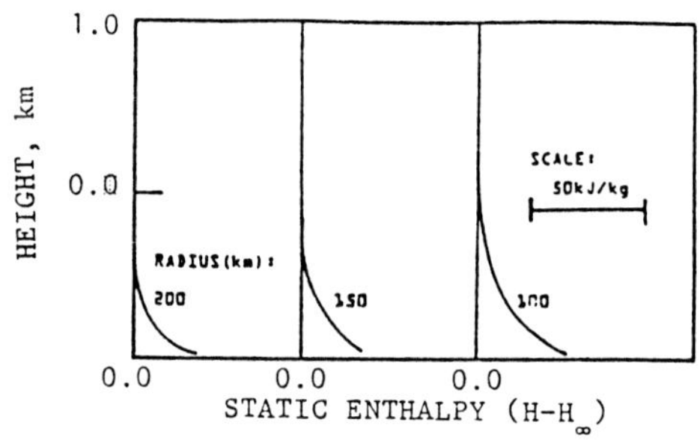

FIGURE 13. Static enthalpy vs height.

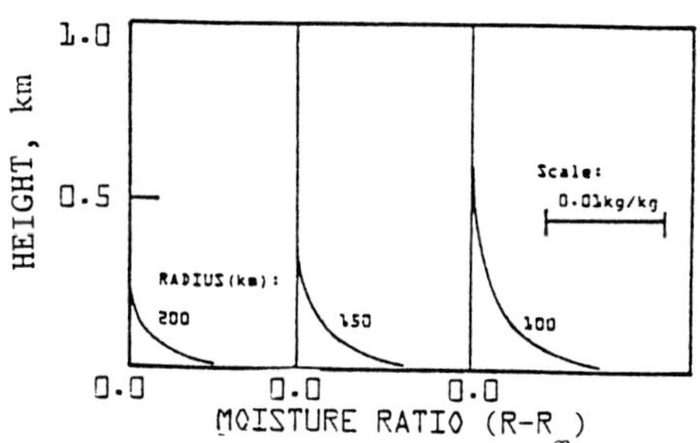

FIGURE 14. Moisture ratio vs height.

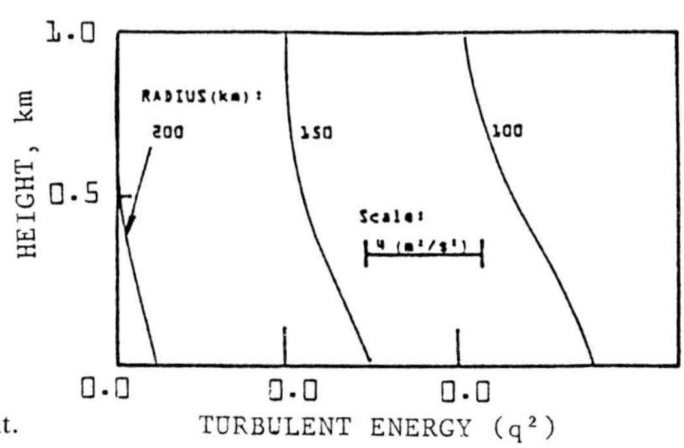

FIGURE 15. Turbulent energy vs height.

transfer into a large set of ordinary differential equations. Computer algorithms have been developed to solve the resulting set of ordinary differential equations by a Newton's iterative method. Several simulations have been made to solve specific problems of heat transfer, fluid flows and dynamics of marine atmospheres. Confidence in the theory has been established by comparing simulation results with exact solutions for heat transfer and published experimental data for stratified atmospheric layers.

ACKNOWLEDGEMENT

This work has been partially supported by the Office of Naval Research; the author wishes to express his appreciation for its support.

REFERENCES

1. Bozeman, J.D. and C. Dalton. 1973. Numerical study of viscosity in a cavity. J. Comp. Physics, 12, 348-363.
2. Businger, J.A., J.C. Wyngaard, Y. Izumi and E.F. Bradley. 1971. Flux profile relationships in the atmospheric surface layer. J. Atmos. Sic., 28, 181-189.
3. Chi, J. 1977. Numerical analysis of turbulent end-wall boundaries of intense vortices. J. Fluid Mech., 82, 209-222.
4. Chi, J. 1987. Heat, moisture and momentum transfer in turbulent vortex flows over the water surface. Proc. ASME/JSME Thermal Engineering Joint Conference, 2, 627-633.
5. Chi, J. 1989. A finite element computer program for dynamic simulation of thermal and fluid systems. Proc. ASEE Annual Conference, 89, 788-790.
6. Chi, J. 1990. A graphics-oriented finite-element computer program for simulation studies of rotating flows. Proc. ASEE Annual Conference, 90, 1436-1439.
7. Harlow, F.H. and C.W. Hirt. 1969. Generalized transport equations of anisotropic turbulence. Los Alamos Sci. Lab., Rept. No. LA-4086.
8. Kawahara, M. and T. Okamoto. 1976. Finite element analysis of steady flow of viscous fluid using stream functions. Proc. ASCE, 47, 123-135.
9. Lewellen, W.S. and M. Teske. 1973. Prediction of the Monin-Obukhov similarity functions from an invariant model of turbulence. J. Atmos. Sci., 30, 1340-1345.
10. Mellor, L.F. and T. Yamada. 1974. A hierarchy of turbulence closure models for planetary boundary layers. J. Atmos. Sci., 31, 1791-1806.
11. Monin, A.S. and A.M. Obukhov. 1953. Dimensionless characteristics of turbulence in the atmospheric surface layer. Dokl. Akad. Nauk SSSR, 93, 223-226.
12. Suces, J. 1985. HEAT TRANSFER, Wm. C. Brown Publishers, Dubuque, Iowa. Wyngaard, J.C. and O.R. Cote. 1971. The budgets of turbulent kinetic energy and temperature variance in the atmospheric surface layer. J. Atmos. Sci., 28, 190-121.
13. Wyngaard, J.C., O.R. Cote and Y. Izumi. 1971. Local free convection, similarity, and the budgets of shear stress and heat transfer. J. Atmos. Sci., 28, 1171-1182.

The Oceans: Physical-Chemical Dynamics and Human Impact. Edited by S.K. Majumdar, E.W. Miller, G.S. Forbes, R.F. Schmalz and Assad A. Panah. © 1994, The Pennsylvania Academy of Science.

Chapter Sixteen

TROPICAL CYCLONE MOTION
How Can We Predict the Potentially Disastrous Storm Track?

RUSSELL L. ELSBERRY

Department of Meteorology
Naval Postgraduate School
Monterey, CA 93943-5000

BACKGROUND

Although most people think of tropical ocean regions in terms of warm and pleasant weather, the tropical cyclone that forms over these oceans is one of the most destructive weather phenomena in the world. During May 1991, more than 135,000 people were killed when a tropical cyclone struck Bangladesh. Disastrous events such as this have led to the island native saying — "Typhoon come, all die." During August 1992, over $20 billion damage was sustained in southern Florida as Hurricane Andrew came ashore. Although the Bangladesh storm and Andrew had approximately the same intensity, the difference in loss of lives is dramatic as only about 20 people were killed as a direct result of Andrew. In high technology nations, the property damage may be high but the loss of lives has recently been kept small. In the developing nations of the tropics, the equivalent dollar amount of the damage may be smaller, but the impact on the fragile economy can be disastrous. Inadequate construction practices or materials may contribute to the total destruction of homes and business structures. Lack of flood control measures can add to the damage because of the heavy precipitation from the tropical cyclone. Entire coastal villages are frequently wiped out as the rise of the ocean surface in shallow coastal areas (called the storm surge) brings ashore a layer of water that may be 10 m high.

Tropical cyclone is the general name for a counterclockwise (in Northern Hemisphere) rotating storm in the tropics. Regional names such as typhoon in the western North Pacific and hurricane in the Atlantic and eastern North Pacific refer to the most intense stage in which the maximum wind speed must exceed 75 mph (32 m s^{-1}) and may be up to 200 mph (85 m s^{-1}). For convenience in public warnings, a name (selected from an alphabetized list alternating between male and female) is given the storm when the maximum winds near the center exceed gale force (41 mph or 17 m s^{-1}). Less intense stages are called tropical depressions or waves in the tropical easterlies.

A typical life cycle of an Atlantic hurricane is summarized in Figure 1. As the tropical storm moves into the region of highest sea-surface temperatures in the western oceans, the maximum winds increase and the highest winds move closer to the center of the low pressure. The size of the storm, which is defined here as the radius from the center at which 15-17 m s^{-1} winds are found, typically increases during this stage. Gale force winds in large typhoons may extend 700 miles from the center, but only extend 70 miles in a midget typhoon. On the average, the tropical cyclone reaches maximum intensity at the time the storm motion changes from westward to eastward (called recurvature). The tropical cyclone then is moving into an environment that is less favorable as the increasing westerly winds aloft tend to shear off the top of the storm and the sea-surface temperatures decrease. Although the central intensity normally decreases, the size of the storm increases as the gale-force winds expand.

The tropical cyclone track prediction problem is to forecast the movement of the central region of high winds, heavy precipitation, and large ocean surface waves. Although the tropical cyclone center position is the focus of the warning messages, the danger area depends as well on the storm structure, especially the size. Errors in the track prediction may result in some areas not having adequate time to evacuate or prepare, or may cause unncessary preparations in other areas that will not experience damage. Because these track errors grow linearly with length of forecast interval (e.g., 100 miles at 24 h and 300 miles at 72 h), the tropical cyclone forecaster must first set a larger "watch area" than is likely to be affected. The actual "warning area" will be posted over a smaller area as the storm approaches and the likely path becomes better known.

Great benefits are to be gained by decreasing tropical cyclone track forecast errors. First, unnecessary preparations and evacuations (e.g., from oil rigs offshore or from low-lying coastal regions) can be avoided. Second, those areas that are going to be struck are given longer warning times in which to prepare. Third, public confidence in the warning system will be increased by fewer incidents of "crying wolf" unnecessarily, and thus the public will be more motivated to respond to the warnings and make the proper decision for the actual threat.

The focus of this chapter is an assessment of what is being done to improve tropical cyclone prediction. The first section will describe recent advances in understanding tropical cyclone motion.[1] The second section will describe the observing system and some new technology to be applied to the tropical cyclone track prediction problem. Finally, the role of the meteorological forecast in the overall warning system will be addressed to emphasize that the track forecast is just one aspect of disaster preparedness. Education, planning and advance preparations are equally important components in avoiding loss of life and minimizing damage from tropical cyclones.

ADVANCES IN UNDERSTANDING TROPICAL CYCLONE MOTION

As indicated above, tropical cyclones usually form and move westward in the trade winds between the broad subtropical (around 30° lat.) high pressures and the lower pressure region near the equator (Figure 2). In the western North Pacific where three-eighths of all the world's named tropical cyclones form, this low pressure trough is deeper, so that the trade winds to the north are stronger. Furthermore, the deep trough pulls air from the Southern Hemisphere into a band of strong westward winds on the equatorward side. It is within this favorable environment of westward winds on the poleward side and eastward winds on the equatorward side that contributes to the counterclockwise (in the Northern Hemisphere) rotation of tropical cyclones. The energy source for the development is provided by water vapor previously evaporated from the warm tropical ocean. This water vapor is then converted into heat energy in the eyewall clouds and the rainbands that spiral around the center. In the most intense stages, a clear region (called the eye) forms within

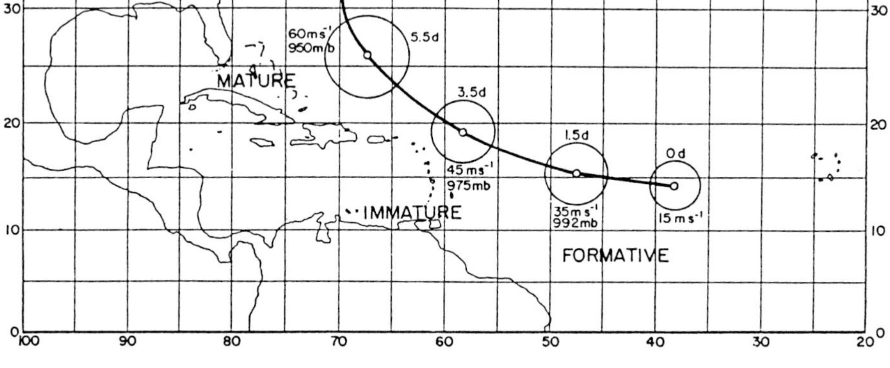

FIGURE 1. Typical changes in size (circles) and intensity in terms of maximum wind speed or minimum sea-level pressure of a recurving Atlantic hurricane (Merrill, R. 1984. *Mon. Wea. Rev.* **112:**1408-1418).

this cloud mass, which then makes it relatively easy to locate the storm center from satellite images. In weaker stages in which this eye is not visible from satellites, the errors in locating the center are much larger.

Storm Motion and Propagation

For a small tropical cyclone, the motion is similar to a cork in a stream. That is, the cyclone moves with the basic atmospheric current in which it is embedded. If this were the only effect, and the high and low pressure regions did not change, the tropical cyclone would simply move around the subtropical high into the cold midlatitude waters or hit land and die. For larger storms, deviations of 2-7 mph have been observed[2] between the typhoon motion and the basic current. When the basic current is weaker than the normal 6 m s^{-1} (14 mph), the addition of this deviation contributes a significant fraction of the total storm motion (Figure 3). This deviation will be referred to as a propagation vector because internal and external influences combine to deflect the storm relative to the basic current, and a combination of the basic current and propagation leads to the total storm motion.

A major effort in a recent Office of Naval Research (ONR) initiative[1] has been devoted to understanding the physical causes of the propagation vector, and how the magnitude and direction of this vector depend on the storm structure and the environmental conditions. As suggested above, the magnitude of the propagation vector increases with size of the storm.[3-5] This effect has been demonstrated in

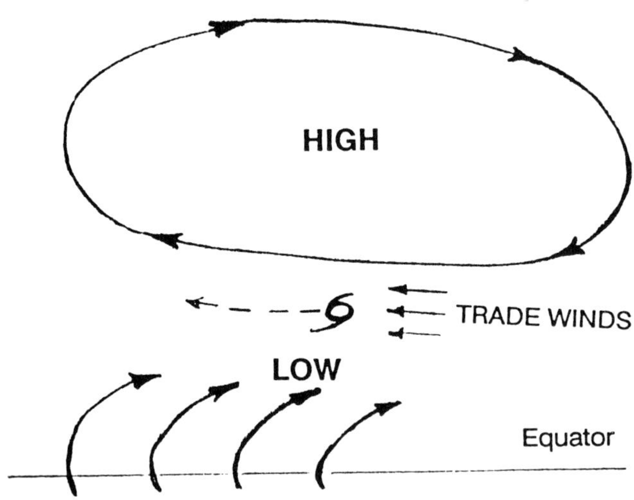

FIGURE 2. Schematic of the environmental features during typhoon (symbol) formation over the western Pacific Ocean. Trade winds between the subtropical high and the low pressure trough near the equator form a basic current that moves most typhoons toward the west-northwest.

numerical models in which no basic current is included. Nevertheless, the cyclone will propagate toward the west and toward the pole simply because the earth is curved and air parcels on the poleward side have more spin around a vertical axis than do air parcels on the equatorward side (Coriolis effect). Movement of these air parcels by the winds in the storm creates a counterclockwise rotating circulation to the west, and a clockwise circulation to the east (Figure 3). These counter-rotating gyres have a nearly uniform flow region between their centers, which typically are separated by 700-935 miles (600-800 km). It is this uniform flow region that propagates the tropical cyclone toward the west and the pole.[5]

Larger tropical cyclones produce stronger gyres that have stronger uniform flow between their centers, which results in larger propagation vectors. Very small tropical cyclones are not as effective in moving air parcels with different rotations about the vertical axis, thus the corresponding gyres are weaker and closer together. Only a weak uniform flow region is produced and the propagation vector is small.

It is the north-south difference in rotation about the vertical axis on a spherical earth that leads to the gyres in Figure 3. Such differences in rotation can also be

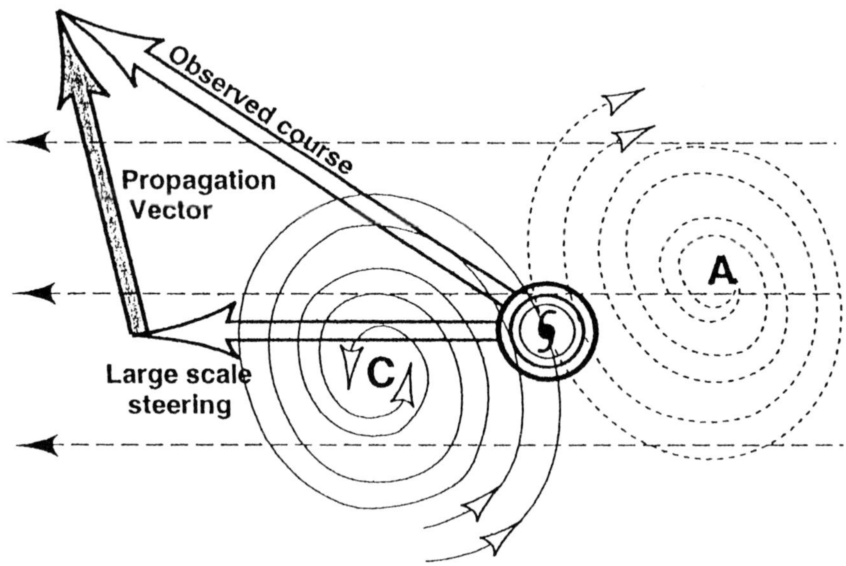

FIGURE 3. Schematic of the relationship between the large-scale steering flow (long dashed) toward the west, symmetric flow around the tropical cyclone (solid circular lines around tropical cyclone symbol), and the cyclonic (light solid) and anticyclonic (short dashed) gyres associated with the propagation vector toward the north-northwest. The net tropical cyclone motion (observed course vector) is the sum of the steering and propagation vectors (Provided by David Titley).

created by the environmental flow. For example, clockwise rotation occurs in the subtropical high pressure cell in Figure 2, and counterclockwise rotation is found around the low pressure trough. This difference in rotation effect tends to offset the spherical earth effect. Farther to the north, the two effects act in the same sense. For the same size tropical cyclone, the westward and poleward propagation would be reduced south of the subtropical high and enhanced on the north side.[6]

Analytical models have been developed[7-10] that predict the amplitude of the counter-rotating gyres in Figure 3, and the magnitude and direction of the uniform flow between the gyre centers. It has been demonstrated[11] that a numerical track prediction model that does not include these counter-rotating gyres in the initial conditions will not have the proper propagation vector until the gyres are generated in 24-48 hours. Thus, the predicted storm motion initially is too slow. If the analytical model[7] is used to insert the counter-rotating gyres in the initial conditions, the propagation vector is immediately present and the correct storm motion results. This basic research result is now being introduced into operational track prediction models,[11] and should result in improved guidance to the tropical cyclone forecaster.

Time-dependent Basic Currents

The tropical atmosphere is more complicated than the simple schematic in Figure 2. Presence of other weather disturbances that have similar horizontal scales as the tropical cyclone makes the determination of the basic current more complex. These disturbances may be confined to the lowest 5 km (18,000 ft) of the atmosphere, or only exist in a layer between 6-15 km (20,000 - 50,000 feet). As the tropical cyclone moves, these disturbances may become part of the environment, so that the basic current varies in time. The sparse meterological observation network over the tropical oceans often prevents an accurate specification of the basic current.

Each of these adjacent weather disturbances may interact with the tropical cyclone. In the western North Pacific, the adjacent disturbance may be another tropical cyclone because the region is so fertile for tropical cyclone formation that multiple cyclones frequently exist. When the separation distance between the two tropical cyclones is less than some critical value around 700 miles (650 km), they begin to rotate about a point between them in a counterclockwise orbit. This rotation is similar to a classical fluid mechanics experiment called the Fujiwhara effect. A re-examination[12] of many tropical cyclone interactions has shown that the two cyclones may suddenly begin to orbit, only later to escape. The classical Fujiwhara merger of the two circulations does not occur very often. Rather, the horizontal wind variations of the larger storm appear to destroy the smaller storm by tearing it apart. Some new theoretical treatments[13,14] of the cyclone interaction problem have explored the capture, escape and merger phases. The rapid escape seems to occur when one of the orbiting cyclones comes under the influence of a neighboring circulation, and is captured by that circulation. The nonlinear interactions between more than two circulations can result in chaotic situations. Thus, chaos may provide guidance as to whether such cases are inherently not predictable. In such situations, the forecaster may be provided guidance that a critical interaction situation is approaching, but the outcome in terms of the possible tracks may be given only in a probabilistic sense.

One of the most critical track forecast decisions is the time when the tropical cyclone moves northward around the subtropical high in Figure 2 and begins to move eastward rather than westward. The time when a cyclone begins to track eastward is called the recurvature time. The forecast is difficult because the cyclone may begin to move toward recurvature and then resume westward motion. Thus, the recurvature might be considered a time of low predictability, perhaps due to multiple interactions as described above.

At other times, large-scale circulations as in Figure 2 appear to determine whether the tropical cyclone will be a recurver or a straight-mover that will continue to move westward until it strikes land and dies. The large-scale circulation changes slowly in amplitude and relative location on time scales of more than 10 days.[10] While the circulation is locked in one mode, a sequence of tropical cyclones will all have the same track type, e.g., recurvers. After the large-scale circulation transitions into the other mode, the opposite track type will occur for some time. Thus, it may be possible to classify the track type by knowing the large-scale circulation at the time the tropical cyclone is first named (maximum winds exceed 17 m s^{-1} or 40 mph). The Naval Postgraduate School track-type forecast technique[15] is less reliable during transitions between modes of the large-scale circulation, or when multiple cyclones exist.

In addition to the intraseasonal variations on time scales of 10-20 d or 30-60 d, the tropical circulation also varies on interannual time scales. The best known of these is the El Niño or Southern Oscillation, which is associated with development of high sea-surface temperatures in the central and eastern equatorial Pacific regions. Gray[16] has demonstrated that this pool of warm water decreases the number of Atlantic hurricanes that normally form from the tropical waves moving from Africa to the Caribbean. The shift in the high sea-surface temperature region associated with El Niño also affects the western North Pacific tropical cyclone formation area. Although the total number of named tropical storms does not appear to be affected as in the Atlantic, the monsoon low in Figure 2 is moved eastward. Thus, the tropical cyclone formation area is also farther east. Accompanying changes in the basic currents lead to more tracks that recurve into the middle latitudes rather than having landfall on East Asia and thus dying over land.

In summary, the basic currents that advect the tropical cyclones have time variability due to adjacent circulations, to intraseasonal oscillations and to interannual changes. These effects on the tropical wind fields must be observed, analyzed and predicted accurately to improve tropical cyclone track prediction.

OBSERVING SYSTEMS

Tropical cyclones (and weather in general) are widely accepted to be a global problem. One goal of the World Meteorological Organization (WMO) has been to improve the observational network. The basic instrument for measuring winds, temperatures and humidity is the rawinsonde, which is carried aloft by a large balloon. Unfortunately, these instruments are too expensive for many developing nations in the tropics, and the number of rawinsondes available for tropical analysis has diminished in recent years. Ocean weather ships that used to be maintained along aircraft routes for search and rescue purposes have been eliminated due to high cost.

Weather satellites have become the primary method of detecting, tracking and estimating the intensity of tropical cyclones. Polar-orbiting satellites with high resolution (1 km) visible imagery and 4 km infrared imagery are very effective in positioning a tropical cyclone during daylight or at night if a clear eye is present. The disadvantage of the polar-orbiting satellite is that the satellite typically views the same area only twice a day. Geostationary meteorological satellites are placed into an orbit so that they view the same area on the earth at all times. Because visible or infrared imagery from these satellites is normally available each hour, the geostationary satellite has become the primary tool for positioning tropical cyclones. Animation of this imagery (as in commonly shown on public television weather broadcasts) is particularly effective in tracking the past motion of the cyclone.

Another critical satellite product is the estimate of the winds at low levels from the tracing of small cumulus clouds, and at around 12 km (40,000 feet) from the tracing of cirrus cloud features that are assumed to move with the wind (Figure 4). These satellite-derived cloud-track winds are a primary data source for tropical ocean analysis because of the scarcity of rawinsondes. However, these winds are generally available at only two levels, and thus do not completely define the basic current that advects the tropical cyclone. Thick cirrus clouds will obscure cumulus clouds at lower levels, so that no low-level, cloud-track winds will be possible. Furthermore, a cloud tracer has to be present, so no winds can be derived in the subsiding cloud-free regions of the environment around tropical cyclones.

Imagery in a water vapor channel (6.7 micron) reveals dry regions in the upper troposphere. Animation of this imagery from some geostationary satellites allows wind estimates to be extracted by tracing moisture inhomogeneities. These wind estimates are generally in the gaps between the cirrus cloud-track winds in the upper troposphere, and thus improve the data coverage in all quadrants of the storm.

Microwave channels onboard the polar-orbiting satellites permit temperature profiling through the cirrus clouds. Thus, the warm core in the upper troposphere of a mature tropical cyclone could be detected if the horizontal resolution of these microwave instruments can be improved, which requires larger microwave antennas. Direct measurements of the warm core would improve tropical cyclone intensity estimates. These microwave channels are sensitive to large raindrops, so that the associated rain rates may be estimated, which again may be helpful for intensity estimates. Other microwave channels are sensitive to ice crystals, so that the curved rainbands may be detected under the cloud shield that often covers the center during tropical depression, tropical storm and weak typhoon stages. Thus, a more accurate center position may be available. However, the narrow swath widths and infrequent repeat orbits for the polar-orbiting satellites that presently carry microwave instruments do not ensure frequent tropical cyclone fixes necessary for accurate initial storm motion vectors.

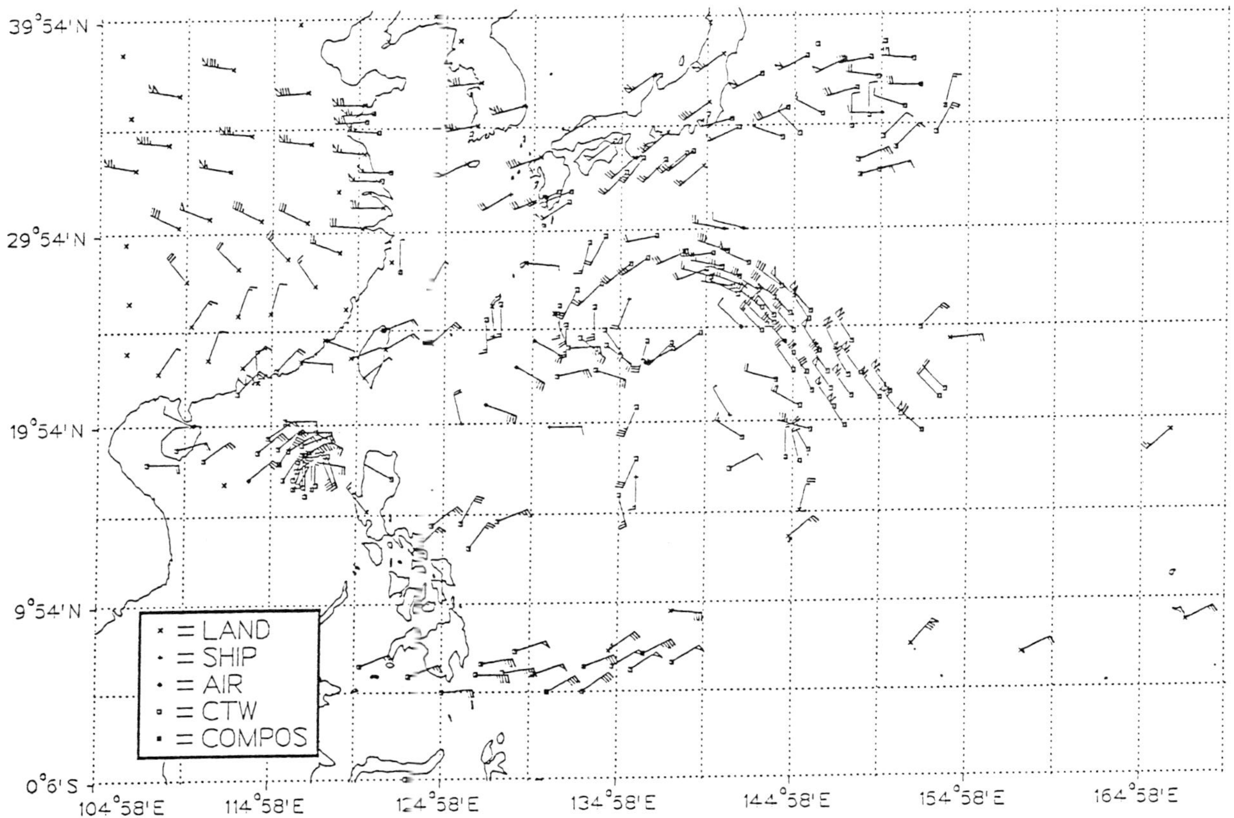

FIGURE 4. Wind vectors at upper levels (200 mb or about 12 km) at 12 UTC 15 September 1990 from a combination of observation platforms (see inset). Most of the open-ocean winds are from a special effort to extract cloud-track winds (CTW) from the animation of geostationary satellite infrared imagery.

Finally, microwave channels are sensitive to the foam associated with whitecaps on the ocean surface. Empirical relationships between the surface wind speed and whitecap coverage allow an inference of the wind speed (no direction). It may be possible to estimate the radius of gale-force winds with this technique. However, the heavy rain rates associated with the central core and the rainbands obscures the surface wind signal, and the higher wind speeds in these regions can not be estimated.

The most reliable and accurate measurement system is the reconnaissance aircraft. An aircraft can be used to search the inner region to locate the center, measure the maximum winds and describe the storm structure in more detail than is possible from satellites. Thus, the reconnaissance aircraft is an essential component of the tropical cyclone warning system of the United States, including rare cases when a hurricane approaches the Hawaiian Islands. However, the rest of the world has no aircraft reconnaissance and must rely on satellite imagery or occasional ship reports to fix the positions of tropical cyclones more than about 300 km from landfall.

Most nations have a coastal network of radars to detect the approach of tropical cyclones within 300 km of landfall. Well-defined eyewall clouds or spiraling rainbands are necessary for accurate positioning. The radar allows fixes of the tropical cyclone center with a typical frequency of 15-30 minutes. However, radar fixes include oscillatory motion that makes it difficult to make consistent and representative estimates of the overall storm motion. New radars with Doppler capability to estimate the motion of hydrometeors are being installed along the U.S. coast lines. These radars will provide wind estimates within about 120 km of the radar site. Thus, the tropical cyclone will have to be quite close to the radar before winds may be estimated.

Recently developed clear-air Doppler radar wind profilers that continuously measure the wind speed and direction directly above the site have great potential for tropical cyclone applications. A 405 MHz radar wind profiler was deployed to Kadena AFB, Okinawa during August-September 1990. Three typhoons passed close enough to Okinawa to generate high winds. For example, high vertical resolution winds (Figure 5) were measured each hour as Supertyphoon (winds exceeding 65 m s^{-1} or 150 mph) Flo passed 100 km (70 miles) to the east of Okinawa. Correlations between the radar wind profiles and the hourly surface winds have been calculated for a larger range of speeds than achieved previously. These observations demonstrate how the surface wind gusts increase more rapidly than does the mean wind speed during the approach of the typhoon.

A future source of observations in tropical cyclones may be Unmanned Aerial Vehicles (UAV) that are equipped with meteorological instruments. The larger of the UAVs (Persus B by Aurora Flight Services) is proposed to fly in the lower stratosphere (about 20 km) above the commercial aircraft levels.[17] Lightweight instrument packages will be deployed automatically along the flight track to measure and transmit the winds, temperatures and humidities. Extremely accurate positioning of the package as it falls will be achieved with the Global Positioning Satellite (GPS) signals. In addition to giving the elevation above the ground as a function of time for plotting vertical profiles, the horizontal displacements with time may be used to estimate the horizontal wind components.

A smaller (wingspan of 3 m) version UAV equipped only with temperature and humidity sensors has also been proposed.[18] This system is called the Aerosonde because it would continually climb upward to perhaps 17 km and then spiral (unpowered) downward to the surface while transmitting the meteorological measurements similar to a rawinsonde. The GPS system would be used for positioning the Aerosonde and for horizontal wind estimates. If the Aerosonde could execute 10 such ascent/descent cycles over approximately 2 day missions, and 10 such missions could be achieved on the average with each UAV, the Aerosonde would be competitive in cost with the rawinsonde. With a satellite communication network in place, the Aerosonde could operate over the oceans where tropical cyclones form and move.

In summary, advances are being made in observation systems for tropical cyclones, even as the number of rawinsondes in the tropics is decreasing. However,

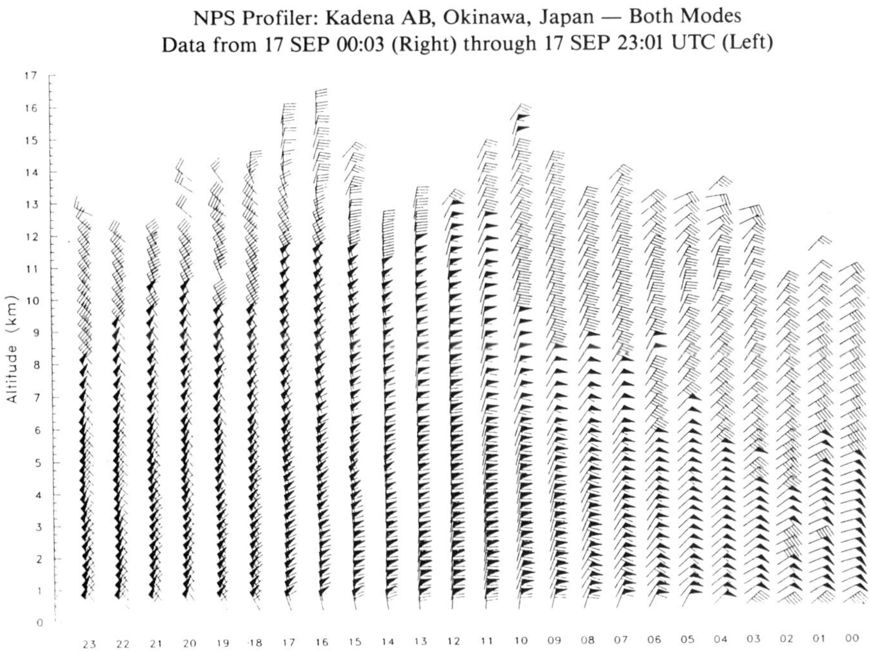

FIGURE 5. Wind speed (flag = 50 knots, long barb = 10 knots) and directions (wind from the north would be arrow pointing from top to bottom) measured by Naval Postgraduate School radar wind profiler at Kadena Air Base, Okinawa from 00 UTC 17 September (right side) to 23 UTC September (left). The profiler provided observations each 360 meters (1180 feet) from 600 meters (1970 feet) above the ground to as high as 16,000 meters (52,000 feet) each hour as Supertyphoon Flo passed within 60 n mi to the east of Okinawa. The maximum wind speeds were greater than 85 knots (100 mph).

many of these proposed systems are expensive. A combination of observational solutions is needed to ensure improved winds for tropical cyclone prediction.

DISCUSSION

Advances in understanding and in observing the tropical cyclone will contribute to improvement in predictions of these dangerous weather systems. However, advances in meteorology are only one component in achieving better and more effective warnings and for avoiding loss of life and unnecessary damage. An effective infrastructure is required to achieve these benefits.

Communication facilities are critical for warning the public. A graduated system of information provision is used to alert the public in hopes of achieving the proper response for the specific threat. In the advanced technology nations, extensive television and radio communication is available to first set a "watch" as the typhoon is still 36-48 h away, and then issue the warning to take shelter 12-24 h prior to onset of damaging winds. In less developed countries, television and radio are not reliable means to alert the public, and a system of warning flags and word-of-mouth is all that is available.

Many factors are involved in the human response to storm warnings. One natural response is to do nothing while awaiting further information or evidence that the warning is justified. Many people are willing to accept risk to their life in order to protect their property. Unfortunately, ignoring the warning can lead to people being trapped when the storm winds, precipitation and flooding are already upon them. The responsibility of the meteorologist is to issue the most accurate warnings that are possible. Because of errors in the track forecast and uncertainties in the size and intensity, the forecaster must warn a larger area than will actually be struck. "Crying wolf" or a perception that the warning may be a false alarm will give an excuse for not responding to the next storm, which could be an entirely different situation.

Disaster preparedness must begin long before the storm strikes. One of the most important components is in education of the public to the threat. This education should start in the schools as part of the science curriculum or in personal safety training. A clear understanding of the nature of the threat and instruction on how to prepare for the threat are necessary. Public awareness is also achieved at the beginning of each hurricane season via television and radio programs. Meteorologists are available to speak at Civic Clubs or other functions to explain the threat.

Local government agencies must have a plan for responding as the tropical cyclone approaches. Regular exercises are required to test the communications among agencies and with the public. Evacuation routes must be planned and tested to move people from areas likely to be flooded or to suffer wind damage. Large migrations of people to coastal areas to live, to retire or for recreation have added greatly to the risk. Many of these people will not be aware of the threat, so that education is essential. In addition, plans must have been made for moving into the area struck by a tropical cyclone to provide food and shelter, repair damage and re-establish services (power, telephone, etc.)

On a longer time scale, disaster preparedness requires adequate zoning and building regulations to withstand a reasonable threat. Building design, choice of materials, and construction practices can minimize damage. For example, the island of Guam in the Marianas Islands was directly struck by three typhoons and had two other typhoons pass within 100 km (60 miles) during 1992. Because of adequate warnings and construction (e.g., most houses are built of reinforced concrete blocks), no loss of life and minimal damage occurred.

No modification techniques are available to reduce the intensity of tropical cyclones, or to re-direct their path. Thus, the meteorologist must strive to make the best possible forecast. In addition, the community must develop an adequate warning system and disaster preparedness plans to address the threat if the potential disaster is to be avoided.

ACKNOWLEDGMENTS

Much of the research described in this chapter has resulted from a five-year Accelerated Research Initiative of the Office of Naval Research Maritime Meteorology Program. My colleagues in this research have shared early versions of manuscripts and provided comments. Mrs. Penny Jones prepared the manuscript.

REFERENCES

1. Elsberry, R.L. and R.F. Abbey, Jr. 1991. Recent Advances in Understanding Tropical Cyclone Motion. Technical Report NPS-MR-91-003, Naval Postgraduate School, Monterey, CA 93943, 92 pp.
2. Carr, L.E., III, and R.L. Elsberry. 1990. Observational Evidence for Predictions of Tropical Cyclone Propagation Relative to Steering. *J. Atmos. Sci.* **47**:542-546.
3. Chan, J. C.-L. and R.T. Williams. 1987. Analytical and Numerical Studies of the Beta-Effect in Tropical Cyclone Motion. Part I. Zero Mean Flow. *J. Atmos. Sci.* **44**:1257-1264.
4. DeMaria, M. 1985. Tropical Cyclone Motion in a Non-divergent Barotropic Model. *Mon. Wea. Rev.* **113**:1199-1210.
5. Fiorino, M. and R.L. Elsberry. 1989: Some Aspects of Vortex Structure in Tropical Cyclone Motion. *J. Atmos. Sci.* **46**:979-990.
6. Evans, J.L., G.J. Holland, and R.L. Elsberry. 1991. Interactions between a Barotropic Vortex and an Idealized Subtropical Ridge. I. Vortex Motion. *J. Atmos. Sci.* **48**:301-314.
7. Carr, L.E., III. 1989. Barotropic Vortex Adjustment to Asymmetric Forcing with Application to Tropical Cyclones. Ph.D. dissertation, Naval Postgraduate School, Monterey, CA 93943. 143 pp.
8. Willoughby, H.E. 1988. Linear Motion of a Shallow-water Barotropic Vortex. *J. Atmos. Sci.* **45**:1906-1928.

9. Peng, M.S. and R.T. Williams. 1990. Dynamics of Vortex Asymmetries and their Influence on Vortex Motion on a Beta-plane. *J. Atmos. Sci.* **47:**1987-2003.
10. Smith, R.K. and W. Ulrich. 1990. An Analytical Theory of Tropical Cyclone Motion using a Barotropic Model. *J. Atmos. Sci.* **47:**1973-1986.
11. Carr, L.E., III and R.L. Elsberry. 1992. Analytical Tropical Cyclone Asymmetric Circulation for Barotropic Model Initial Conditions. *Mon. Wea. Rev.* **120:**644-652.
12. Lander, M. and G.J. Holland. 1993. On the Interaction of Tropical-cyclone Scale Vortices. Part I. Observations. *Quart. J. Roy. Meteor. Soc.* **119:**1347-1361.
13. Holland, G.J. and M. Lander. 1993. The Meandering Nature of Tropical Cyclone Tracks. *J. Atmos. Sci.* **50:**1254-1266.
14. Ritchie, E.A. and G.J. Holland. 1993. On the Interaction of Tropical-cyclone Scale Vortices. II. Discrete Vortex Patches. *Quart. J. Roy. Meteor. Soc.* **119:**1363-1379.
15. Harr, P.A. and R.L. Elsberry. 1991. Tropical Cyclone Track Characteristics as a Function of Large-scale Circulation Anomalies. *Mon. Wea. Rev.* **119:**1448-1468.
16. Gray, W.M. and C.W. Landsea. 1992. African Rainfall as a Precursor of Hurricane-related Destruction on the U.S. East Coast. *Bull. Amer. Meteor. Soc.* **73:**1352-1364.
17. Langford, J.S. and K.A. Emanuel. 1993. An Unmanned Aircraft for Dropwindsonde Deployment and Hurricane Reconnaissance. *Bull. Amer. Meteor. Soc.* **74:**367-376.
18. Holland, G.J., T. McGeer and H. Youngren. 1992. Autonomous Aerosondes for Economic Soundings Anywhere on the Globe. *Bull. Amer. Meteor. Soc.* **73:**1987-1998.

The Oceans: Physical-Chemical Dynamics and Human Impact. Edited by S.K. Majumdar, E.W. Miller, G.S. Forbes, R.F. Schmalz and Assad A. Panah.

Chapter Seventeen

THE MARGINAL ICE ZONE AND OCEAN WAVES

VERNON A. SQUIRE
Department of Mathematics and Statistics
University of Otago
P.O. Box 56
Dunedin, New Zealand

INTRODUCTION

In September, the month of maximum extent, 20 million square kilometers of the Southern Hemisphere are covered by sea ice, whereas at its minimum extent in February the sea ice occupies a mere 4 million km^2.[1] In the Northern Hemisphere the maximum coverage of 14.8 million km^2 occurs in March, while the minimum of 7.8 million km^2 occurs in September.[2] The extreme seasonal variability exhibited by the ice of the Southern Ocean—as opposed to that of the Arctic and sub-Arctic seas combined—is very much due to differences in local geography. In the South, the sea ice circumscribes the Antarctic land mass and offers a free boundary to the north; the majority of the sea ice in the Northern Hemisphere on the other hand grows within the Arctic Basin, a mediterranean sea almost entirely fringed (and therefore sheltered) by land. Concomitantly Antarctic sea ice is predominantly first-year ice whereas its northern partner is on average older and thicker.

Sea ice is a dominant feature of global climate and weather. Important effects are due to altered fluxes of heat, moisture, salt, and momentum, to albedo change, and to the ability of sea ice to reject brine into the upper ocean when growing or to inject fresh water while melting. It is therefore crucial that the morphology of the sea ice canopies of the Arctic and Antarctic be parameterized effectively in general circulation models. It is equally crucial that we understand how these morphologies come about, what the principal factors are in determining them, and how these factors operate.

Because of the intense seas to the north, the sea ice of the Southern Ocean is subjected to the unremitting action of ocean waves.[3] As such, a large fraction has been designated marginal ice zone (MIZ); a region on either side of the ice edge which is significantly affected by open ocean processes. In the Northern Hemisphere the Arctic Basin sea ice is protected by land, but its more southern neighbors, e.g., the Greenland, Bering, Labrador and Barents Seas, all possess striking MIZs. Whereas the sea ice of the Arctic interior takes the form of a continuous sheet, with its relief broken only by the occasional pressure ridge, lead, or polynya, an MIZ is composed of a field of discrete pack ice floes coexisting with open water (Figure 1). The fraction of sea ice in relation to the open water present, i.e., the concentration, the floe size, thickness and roughness distributions, together with the physical character of the sea ice floes themselves, determine MIZ morphology. In the MIZ ocean waves are the overriding determinant of sea ice morphology; they alter the floe size distribution by breaking up the sea ice into smaller ice floes—sometimes even pulverizing it into a slurry, they cause rafting and ridging,[4] they inhibit growth and enhance lateral melting,[5] they drive the ice edge via radiation stress[6,7,8] and they may lead to local ice edge phenomena such as bands and streamers.[9] But the action of ocean waves is not limited to the MIZs alone. Waves have been detected far into the Arctic Basin,[10,11,12,13] some travelling huge distances beneath the ice canopy from their North Atlantic genesis, others created more locally by the action of wind on the underlying ice. Waves may also propagate beneath shore fast sea ice—the sea ice that skirts the various landmasses of the polar regions—causing it to bend rhythmically with their passing.[14,15,16] In this case break-up can again occur suggesting that in some regions, e.g., Newfoundland and the Labrador coast, fast ice may seed pack ice with a distribution of floes of particular size and thickness.[17,18,19]

FIGURE 1. Aerial photograph of the Greenland Sea MIZ.

WAVES IN THE MIZ

The individual ice floes making up a typical MIZ vary in diameter from a few meters near the ice edge to hundreds of meters across within the interior. They may be present in diffuse concentrations ranging from a few percent up to a compact ice field of 100% concentration. A distribution of thicknesses is usually present, certainly reflecting the growth history of the ice but also to some extent relating to past action by winds, ocean currents and waves. When ocean waves enter such a region they are scattered by the floes with the result that energy is gradually lost from the penetrating wave spectrum as it progresses further into the ice interior. The manner in which the energy is lost has been measured on several occasions using a variety of sensors.[20,21,22,23,4,24,25,26,27,28] In all cases, the attenuation of waves with distance into the pack ice takes a negative exponential form, with an attenuation coefficient which decreases with increasing wave period over most of the spectral range. That is, the ice field low-pass filters the waves. The effect of the Bering Sea MIZ on an incident wave spectrum can be seen in Figure 2.[4] The attenuation between stations located near the ice edge and those far into the interior is clear in the acceleration time series and the integrated spectral densities, as is the low-pass filtering effect of the MIZ which tends to make the time series appear "less noisy" as penetration increases. The energy attenuation coefficients at various wave periods have been extracted from the Bering Sea data and are shown in Table 1.

While the expected functional dependence of the attenuation coefficient on period is clear in Table 1 it is unfortunate that the range of periods during the experiment was quite limited. The compilation of several years results presented in[25] gives a more complete picture, which extends to both shorter and longer periods and different ice conditions. Although not occurring in the Bering Sea data reported above,[4] there is some experimental evidence to suggest, for example, that the decay rate may decrease again at very short periods, though it is difficult to see why this should be so on physical grounds.

A series of experiments generalizing the one-dimensional picture described above to fully directional seas is also reported in.[24] The work took place in the Greenland Sea MIZ and used a UK Institute of Oceanographic Sciences pitch-roll buoy in the open water together with a package of similar sensors located on ice floes. The aim

TABLE 1

Energy attenuation coefficients for ocean waves passing through the Bering Sea MIZ. Reproduced from [4].

Period / s	Attenuation coefficient $m^{-1} \times 10^{-4}$
12.2	0.272 ± 0.054
9.4	0.438 ± 0.036
7.6	0.855 ± 0.049
6.4	1.087 ± 0.037
5.5	1.214 ± 0.192

was to consider the progression of a directional sea as it penetrated and travelled further into the pack ice. It was found, using the analysis of[29] that the directional

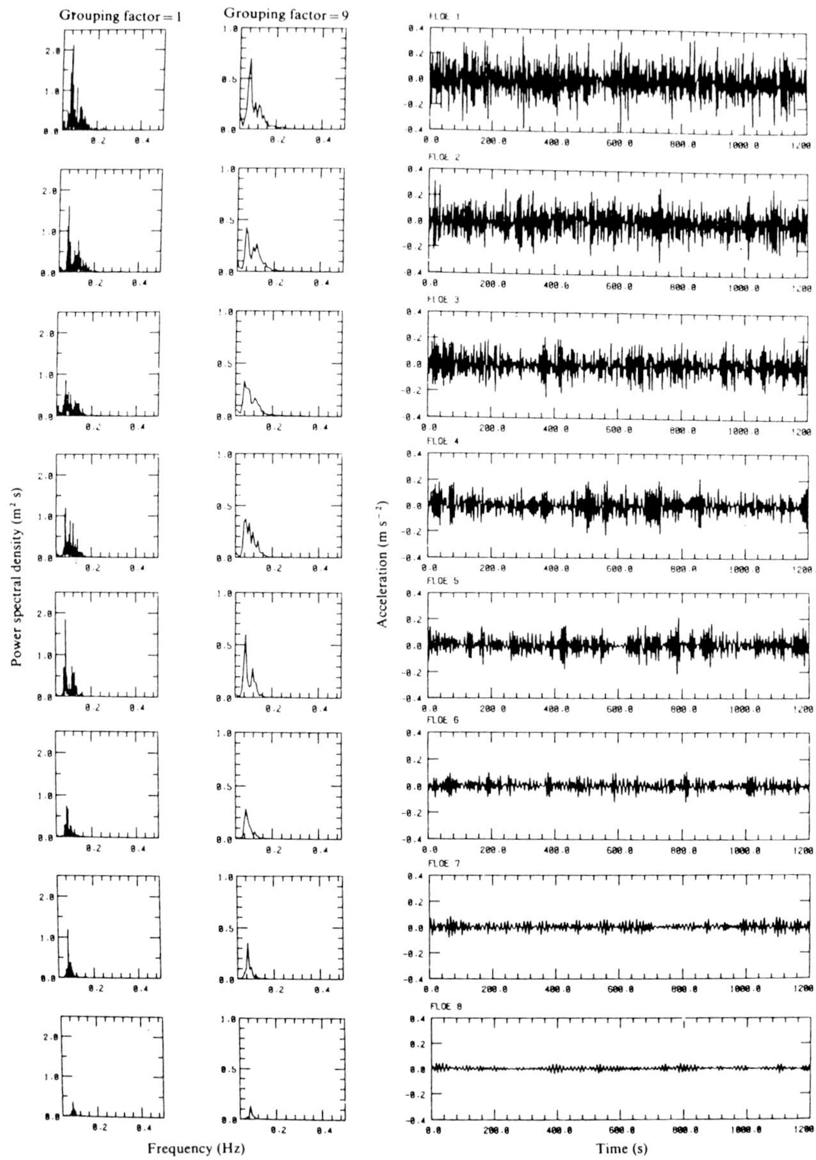

FIGURE 2. Time series and spectral densities for waves penetrating the Bering Sea MIZ. The vertical acceleration time series from the top on the right were obtained using a floe-mounted accelerometer located at penetrations of 0, 0.7, 2.2, 5.3, 10.3, 18.9, 38.6, 65.1 km measured from the ice edge. Raw and smoothed (to 18 degrees of freedom) power spectra are also shown. Reproduced from[4].

spectrum inside the ice field widens at a rate which depends on period until it becomes essential isotropic. At the edge some energy is reflected but the reflexion coefficient is very small even under compact ice conditions. Recently further evidence, obtained from synthetic aperture radar imagery, has emerged which suggests that a critical angle may exist at a compact ice edge, and that refraction can be observed as the waves penetrate the ice cover.[30] We shall see later that this is predicted by theory.

There are currently two approaches to modelling wave propagation into and through an MIZ. The first assumes that each ice floe present scatters a proportion of the incident wave enegy, and that the effect of the entire ice field can be found by incoherently integrating all these individual contributions.[31] The second method, which has more mathematical appeal but which is undoubtedly closer to a parameterization and further from reality, assumes that the MIZ can be represented as a continuum of some type.[26,27,28,32,33]

The Scattering Approach

While the scattering approach is more pragmatic, it rests on our ability to compute the reflexion and transmission characteristics of single floes accurately, a problem which has only just been solved precisely using a Green's function approach, and then only for the two-dimensional case.[34,35] For floes of diameter d each with an energy reflexion coefficient $|R|^2$ occupying a fraction p of the sea surface, the wave energy will decay in the scattering model according to:

$$\frac{\partial E}{\partial x} = 2\alpha E, \tag{1}$$

where

$$\alpha \approx \frac{p|R|^2}{2d}, \tag{2}$$

or for a distribution of floe sizes

$$\alpha \approx \frac{1}{2}\sum_i \frac{p_i|R_i|^2}{d_i}, \tag{3}$$

where the subscript i refers to a particular class of floes within the floe size distribution.[31] As shown in Figure 3b the reflexion coefficient for a single floe can be quite complicated due to the occurrence of resonances within the typical range of periods of an ocean wave spectrum. When a distribution of floes of different sizes and thicknesses is input into equation (3) this effect would be smeared out in the integration.

Because of its importance to the evolution of MIZ morphology Figure 3a plots the maximum strain experienced by a single ice floe as a function of wave period, normalized to a 1 m amplitude wave. The strains experienced are certainly sufficient to break the ice floe which can be expected to fracture when strains at least an order of magnitude less are attained, based on experimental data collected in the Greenland Sea[36] and Bering Sea.[37] Although a monochromatic wave analysis is inadequate these results do suggest that wave break-up may be important, and that it is worthwhile considering the effect of a wave spectrum on a distribution of floe sizes. This is done in Figure 4.[38]

The strain spectral densities of Figure 4 show how an incoming wave spectrum—in this case of Pierson-Moskowitz type though any spectrum may be used—affects ice floes of various lengths at a fixed thickness of 1 m. Ignoring any fine structure in the curves for the moment, the induced strain response increases with length

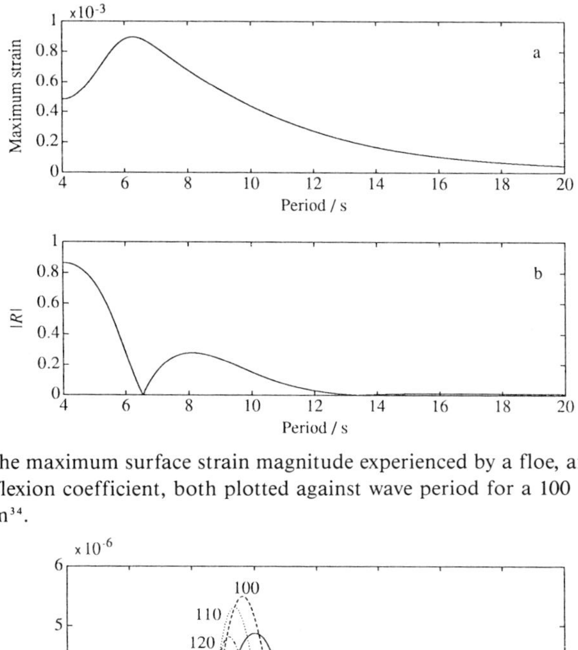

FIGURE 3. a) The maximum surface strain magnitude experienced by a floe, and b) The magnitude of the reflexion coefficient, both plotted against wave period for a 100 × 1 m ice floe. Reproduced from[34].

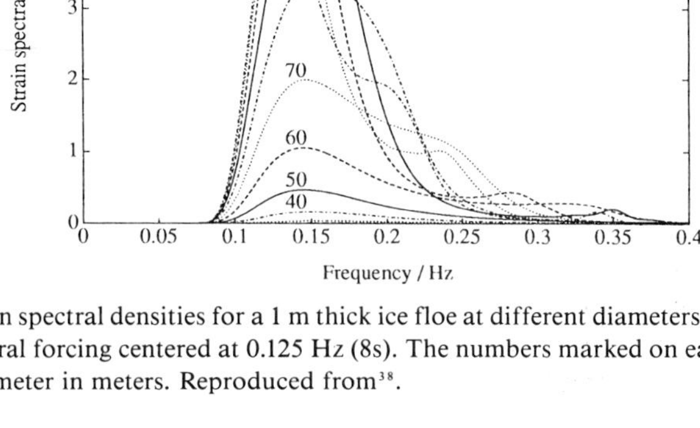

FIGURE 4. Strain spectral densities for a 1 m thick ice floe at different diameters due to Pierson-Moskowitz spectral forcing centered at 0.125 Hz (8s). The numbers marked on each curve represent the floe diameter in meters. Reproduced from[38].

until the floe reaches a certain size (about 100 m). The smallest floes and cakes hardly bend, but as floe length is increased so the waves with significant energy in the open water spectrum begin to affect the floe to a greater extent. Maximum strain densities are reached for floes of about 102 m across, when the energy in the open sea is present at wavelengths which are most favorable to the bending of the floe. For longer floes the peaks begin to decrease, but multiple-cycle tuning can occur so that equivalently large strain density maxima can again occur for longer floes. Fine structure in the curves is generated by resonances at frequencies where the energy is less in the open sea.

The results presented in Figure 4 have important implications to MIZ morphology, since they imply that all floes above a certain size will be broken up into smaller floes. Strain spectral density maxima never drop below the curve for 80 m once that length is exceeded. Thus we expect all floes above a particular size to be destroyed, and taking into account the attenuation offered by the ice floes as the waves penetrate the MIZ, we would expect this floe size to increase with distance from the ice edge. This is precisely what is observed in the MIZ where a zonal floe size distribution is often induced by the waves;[4] the aerial photograph of Figure 1 coincidentally shows many floes roughly 80 m across.

The Continuum Approach

The second method of modelling the MIZ aims to represent its behavior with respect to ocean waves by some deterministic mathematical form, rather like electromagnetic waves in the ionosphere where details of the propagation are controlled by a refractive index which may vary in space. The two models currently in use—the so called "mass-loading model"[39,40,41,42] and the elastic plate model[43,18,44] —are essentially the same model, the latter tending to the former in the limit of zero rigidity.[32] Both derive directly from pioneering work published in 1887.[45] In general it is probably true to say that the mass loading model has been unsuccessful at replicating observations in MIZs but there is some indication that it may work well for wave propagation through frazil ice.[46] Recently a mass-loading model which allows a zonal structure to the MIZ has also been developed,[33] but again this does not reproduce observations well; it appears that a real ice field behaves as though it has some rigidity. A zonally structured, elastic plate parameterization of an ice field is currently under development.

Although originally not intended to represent an MIZ, the elastic plate model described in[43,18,44] for shore fast sea ice has been applied with some success to various MIZs with damping in some form introduced a posteriori. This is of course a parameterization. Using a variational technique it is now possible to compute the normal reflexion and transmission coefficients at the ice edge precisely, and hence to find the deflexion and strain profiles within the sea ice medium.[43,18] Oblique incidence is considered in[44] which generalizes the earlier works. The problem is non-trivial because of the requirement to include additional, rather esoteric wave potentials both within the ice cover and in the open seas off the ice edge; earlier solutions neglected these potentials and were at best approximate. Because of the change of principal wavelength as the waves enter the ice-covered region, some refraction occurs. At zero rigidity, the wavelength is less within the ice; at rigidities typical of sea ice, the principal wavelength is greater. Thus at zero rigidity waves bend towards normal incidence, whereas in the more realistic non-zero rigidity case waves will bend away from the normal and the possibility of a critical angle exists beyond which no penetration of the principal wave into the ice cover is allowed.

In Figure 5 we have plotted critical angle curves for different ice thicknesses. Each curve distinguishes between incidence angles for which penetration into the ice can occur (to the right of the curve) from the angles at which total reflexion occurs at the ice edge (to the left of the curve). As noted earlier, synthetic aperture radar imagery for the Labrador Sea has confirmed that a critical angle effect can occur, at least in compact MIZs, adding confidence to our assertion that some rigidity must be included in the parameterization of the ice cover. A complete picture of the nature of the reflexion/transmission processes is provided in Figure 6 for a 1 m ice floe.

The elastic plate parameterization for an ice field produces no damping of the penetrating wave train and this must be added to the model for it to fit observations; the alternative, namely including a velocity dependent damping term in the plate equation is another approach which has been employed with some success for the case of shore fast sea ice.[47] The most successful form of damping incorporated to date is based on the concept of eddy viscosity,[26,27] which gives an energy attenuation rate α of

$$\alpha = \frac{k\sqrt{\nu\omega/2}}{c_g(1 + k\rho' h/\rho)}, \qquad (4)$$

where ν is the eddy viscosity, k is the wave number, ω is the frequency, $c_g = d\omega/dk$, the group velocity, is found using the elastic model, ρ and ρ' are the water and ice densities respectively, and h is ice thickness. Equation (4) does appear to fit data reported in [25] and [27] well, and even reproduces the unexpected observed decrease in α seen at low periods. However, like all eddy viscosities ν has to be chosen to fit the data and shows wide variation with ice conditions. The recent work described in [28]

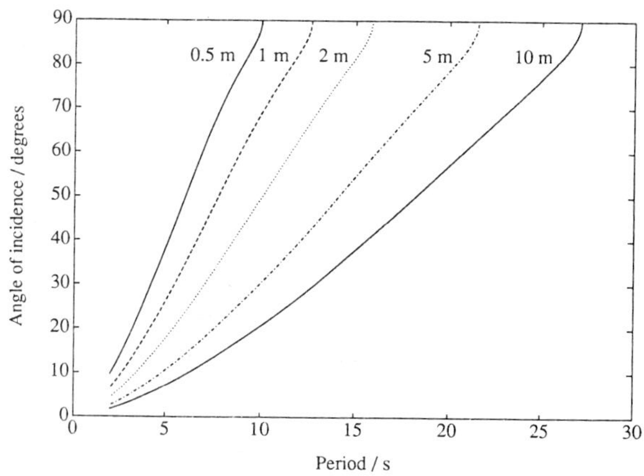

FIGURE 5. Deep water critical angle curves for various ice thicknesses, plotted as a function of wave period. Reproduced from[44].

to some extent overcomes this criticism by establishing an empirical relationship between ν and ice conditions based on dimensional analysis. Specifically it proposed that

$$\nu \propto \frac{AHh}{T} \tag{5}$$

where A is concentration, H is significant wave height, and T is wave period. This simple law to determine ν does appear to fit the data quite well, though more data are really needed to establish or reject the relationship.

DISCUSSION AND CONCLUSIONS

Anecdotal evidence of ocean waves in fields of pack ice dates back many centuries, and is often found in the log books of whaling and sealing ships, or in Inuit tales. Theoretically the problem was looked at over a 100 years ago.[45] The real thrust to understand wave-ice-interaction in pack ice came in the early sixties,[20] however, and has continued to this day. It is fair to say that a great deal of progress has been made on the processes which directly relate waves and pack ice but very little attention has been given to the next step, namely the manner in which wave-induced mutation of the ice pack feeds into the atmospheric and oceanic boundary layers and thence to climate. There are still, of course, outstanding problems associated with the interaction process, mostly concerned with the implementation of a fully three-dimensional model and the effective parameterization of a distribution of floes of different size, shape, thickness and physical composition. Such a model would require as input a measured floe size and thickness distribution, and a directional open water spectrum. It would follow changes in the spectrum as it progressed through the ice field, together with the morphological evolution of the pack ice—particularly due to wave-induced

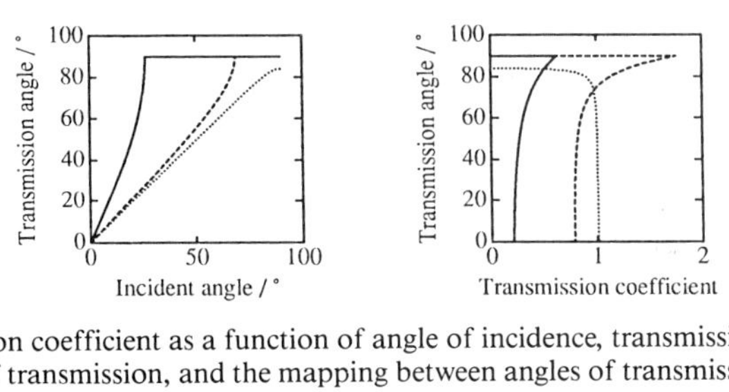

FIGURE 6. Reflexion coefficient as a function of angle of incidence, transmission coefficient as a function of angle of transmission, and the mapping between angles of transmission and incidence, for 1 m thick sea ice. Reproduced from.[44]

fracture but also to highly active (and therefore destructive) wave processes near the ice margin. In an incremental sense one can envisage a series of steps at stationary wave forcing as follows: 1) the wave spectrum enters the MIZ, is attenuated and is made more isotropic with penetration; 2) the waves fracture certain ice floes and therefore modify the MIZ's appearance; 3) because the ice field is now different the wave spectrum at any location is altered; 4) with a different spectrum wave-induced break-up can again occur and concomitantly change the ice field; 5) processes continue until a steady-state is reached. There have already been some attempts to implement such a model but only in the two-dimensional sense.

We summarize here, with references, where we stand with regard to research into wave-ice-interaction in MIZs.

Observations:

- Waves propagating through an MIZ are attenuated preferentially following an exponential decay law; the energy attenuation coefficient α generally decreases as period increases. At very short periods there appears to be some observational evidence of a decrease in α.[4,20-28]
- The directional structure of the sea becomes more isotropic as the waves propagate through a field of discrete ice floes.[24]
- Some reflexion is observed to occur as ocean waves enter an open ice field. For compact pack ice obliquely incident waves may be totally externally reflected due to a critical angle phenomenon. Refraction sometimes also occurs on entry, again depending on ice concentration.[24,30]
- Ocean waves offer a significant contribution to the force balance which controls the dynamics of the ice edge.[6,8]
- Ocean waves create bands and other features at the ice edge.[9]
- Waves are seen well inside the Arctic Basin, some created locally but some travelling great distances from open seas to the south.[10-13]
- Propagation of waves in frazil ice slicks.[46]
- Wave-induced melting of ice.[5]

Theory:

- The normal or oblique penetration of ocean waves into an elastic or visco-elastic plate.[18,43,44,47]
- Strains and displacements induced in ice sheets.[18,44]
- Simple and multiple incoherent scattering theory.[31]
- Ocean wave induced radiation stress affects ice edge position.[6-8]
- Break-up of large ice floes and shore fast ice.[14-19]
- The critical angle and refraction at the ice margin.[16,44]
- Seakeeping and flexural response, and reflexion and transmission charateristics of single two-dimensional ice floes in waves.[34,35,38]
- Wave propagation in an MIZ with varying concentration.[33]
- The mass-loading model.[39-42,46]

Outstanding:

- The three-dimensional response of a single ice floe in waves, including its flexural behavior and break-up.
- Either a fully defensible continuum parameterization of an ice field applicable at any concentration, thickness and floe size distribution, or a scattering theory using realistic ice floes; both to be three-dimensional.
- The morphological evolution of a real ice field due to ocean waves. Both the ice and wave properties would be obtained using remote sensing, probably synthetic aperture radar.
- The relationship between MIZ morphology and climate on local and global scales.

ACKNOWLEDGEMENTS

I am indebted to the University of Otago, and the New Zealand Foundation for Research, Science and Technology for their continued support.

REFERENCES

1. Zwally, H.J. J.C. Comiso, C.L. Parkinson, W.J. Campbell, F.D. Carsey and P. Gloersen. 1983. *Antarctic Sea Ice 1973-1976: Satellite Passive-Microwave Observations.* Report SP-459, NASA Scientific and Technical Information Branch, Washington, DC, pp. 206.
2. Parkinson, C.L., J.C. Comiso, H.J. Zwally, D.J. Cavalieri, P. Gloersen and W.J. Campbell. 1987. *Arctic Sea Ice 1973-1976: Satellite Passive-Microwave Observations.* Report SP-489, NASA Scientific and Technical Information Branch, Washington, DC, pp. 296.
3. Wadhams, P., M.A. Lange and S.F. Ackley. 1987. The ice thickness distribution across the Atlantic Sector of the Antarctic Ocean in midwinter. *J. Geophys. Res.* 92(C13):14535-14552.
4. Squire, V.A. and S.C. Moore. 1980. Direct measurement of the attenuation of ocean waves by pack ice. *Nature, Lond.* 283(5745):365-368.
5. Wadhams, P., A.E. Gill and P.F. Linden. 1979. Transects by submarine off the East Greenland polar front. *Deep-Sea Res.* 26:1311-1328.
6. Squire, V.A. 1989. Super-critical reflection of ocean waves: a new factor in ice edge dynamics? *Annals of Glaciol.* 12:157-161.
7. Squire, V.A. and C. Fox. 1991. The role of incoming waves in ice edge dynamics. *Annals Glaciol.* 15:96-100.
8. Liu, A.K., S. Hakkinen and C.Y. Peng. 1993. Wave effects in ocean-ice interaction in the marginal ice zone. *J. Geophys. Res.* 98(C6):10025-10036.
9. Wadhams, P. 1983. A mechanism for the formation of ice edge bands. *J. Geophys. Res.* 88(C5):2813-2818.

10. Hunkins, K. 1962. Waves on the Arctic Ocean. *J. Geophys. Res.* 67(6):2477-2489.
11. Williams, R.G., N.R. Davis and S.C. Moore. 1990. The Coordinated Eastern Arctic Experiment: SPRI sea-ice studies. *Polar Rec.* 26(158):203-210.
12. DiMarco, R.L., J.P. Dugan and W.W. Martin. 1991. Ice motions forced by boundary layer turbulence. *J. Geophys. Res.* 96(C6):10617-10624.
13. Dugan, J.P., R.L. DiMarco and W.W. Martin. 1992. Low frequency vibrational motion of arctic pack ice. *J. Geophys. Res.* 97(C4):5381-5388.
14. Squire, V.A. 1984. Sea ice. *Sci. Prog., Oxf.* 69:19-43.
15. Squire, V.A. 1984. A theoretical, laboratory and field study of ice-coupled waves. *J. Geophys. Res.* 89(C5):8069-8079.
16. Squire, V.A. 1984. On the critical angle for waves entering shore fast ice. *Cold Reg. Sci. & Tech.* 10(1):59-68.
17. Squire, V.A. 1984. How waves break up inshore fast ice. *Polar Rec.* 22(138):281-285.
18. Fox, C. and V.A. Squire. 1991. Strain in shore fast ice due to incoming ocean waves and swell. *J. Geophys. Res.* 96(C3):4531-4547.
19. Squire, V.A. 1993. The breakup of shore fast sea ice. *Cold Reg. Sci. & Tech.* 21(3):211-218.
20. Robin, G. de Q. 1963. Wave propagation through fields of pack ice. *Phil. Trans. R. Soc. A*255:313-339.
21. Wadhams, P. 1972. Measurement of wave attenuation in pack ice by inverted echo sounding, pp. 255-260. In: T. Karlsson (Ed.) *Sea Ice* Nat. Res. Counc. of Iceland, Reykjavik.
22. Wadhams, P. 1985. Airborne laser profiling of swell in an open ice field. *J. Geophys. Res.* 80(33):4520-4528.
23. Wadhams, P. 1978. Wave decay in the marginal ice zone measured from a submarine. *Deep Sea Res.* 25:23-40.
24. Wadhams, P., V.A. Squire, J.A. Ewing and R.W. Pascal. 1986. The effect of the marginal ice zone on the directional wave spectrum of the ocean. *J. Phys. Oceanogr.* 16:358-376.
25. Wadhams, P., V.A. Squire, D.J. Goodman, A.M. Cowan and S.C. Moore. 1988. The attenuation rates of ocean waves in the marginal ice zone. *J. Geophys. Res.* 93(C6):6799-6818.
26. Liu, A.K. and E. Mollo-Christensen. 1988. Wave Propagation in a solid ice pack. *J. Phys. Oceanogr.* 18:1702-1712.
27. Liu, A.K., B. Holt and P.W. Vachon. 1991. Wave Propagation in the marginal ice zone: model predictions and comparisons with buoy and synthetic aperture radar data. *J. Geophys. Res.* 96(C3):4605-4621.
28. Liu, A.K., P.W. Vachon, C.Y. Peng and A.S. Bhogal. 1992. Wave attenuation in the marginal ice zone during LIMEX. *Atmos. Ocean* 30(2):192-206.
29. Long, R.B. and K. Hasselmann. 1979. A variational technique for extracting directional spectra from multi-modal wave data. *J. Phys. Oceanogr.* 9(2):373-381.
30. Liu, A.K., P.W. Vachon, C.Y. Peng. 1991. Observation of wave refraction at an ice edge by SAR. *J. Geophys. Res.* 96(C3):4803-4808.
31. Wadhams, P., 1986. The seasonal ice zone, pp. 825-991. In: N. Untersteiner (Ed.) *The Geophysics of Sea Ice.* Plenum Publ., N.Y., pp. 1196.

32. Squire, V.A. 1993. A comparison of the mass-loading and elastic-plate models of an ice field. *Cold Reg. Sci. & Tech.* 21(3):219-229.
33. Squire, V.A. 1993. Ocean wave propagation in an ice field with varying concentration. *Proc. 3rd Int. Offshore and Polar Engin. Conf.* 2:724-729, Singapore.
34. Meylan, M. and V.A. Squire. 1993. A model for the motion and bending of an ice floe in ocean waves. *Proc. 3rd Int. Offshore and Polar Engin. Conf.* 2:718-723, Singapore.
35. Meylan, M. and V.A. Squire. 1993. Finite-floe wave reflection and transmission coefficients from a semi-infinite model. *J. Geophys. Res.* 98(C7): 12,537-12,542.
36. Goodman, D.J., P. Wadhams and V.A. Squire. 1980. The flexural response of a tabular ice island to ocean swell. *Ann. Glaciol.* 1:23-27.
37. Squire, V.A. and S. Martin. 1980. A field study of the physical properties, response to swell, and subsequent fracture of a single ice floe in the winter Bering Sea. Univ. of Washington, Sci. Rep. 18, Depts. of Atmos. Sci. and Oceanog., Univ. of Washington, Seattle, USA, 50 pp.
38. Meylan, M. and V.A. Squire. 1993. The response of ice floes to ocean waves. *J. Geophys. Res.* 99(C1):891-900.
39. Weitz, M. and J.B. Keller. 1950. Reflection of water waves from floating ice in water of finite depth. *Commun. Pure Appl. Math.* 3:305-318.
40. Peters, A.S. 1950. The effect of a floating mat on water waves. *Commun. Pure Appl. Math.* 3:319-354.
41. Keller, J.B. and M. Weitz. 1953. Reflection and transmission coefficients for waves entering or leaving an icefield. *Commun. Pure Appl. Math.* 6:415-417.
42. Shapiro, A. and I.S. Simpson. 1953. The effect of a broken ice field on water waves. *Eos Trans. A.G.U.* 34:36-42.
43. Fox, C. and V.A. Squire. 1990. Reflection and transmission characteristics at the edge of shore fast sea ice. *J. Geophys. Res.* 95(C7):11629-11639.
44. Fox, C. and V.A. Squire. 1993. On the oblique reflexion and transmission of ocean waves from shore fast sea ice. *Phil. Trans. R. Soc. A.* In the press.
45. Greenhill, A.G. 1887. Wave motion in hydrodynamics. *Am. J. Math.* 9:62-112.
46. Wadhams, P. and B. Holt. 1991. Waves in frazil and pancake ice and their detection in Seasat synthetic aperture radar imagery. *J. Geophys. Res.* 96(C5):8835-8852.
47. Squire, V.A. and C. Fox. 1992. On ice-coupled waves: a comparison of data and theory, pp. 270-279. In: T.K.S. Murthy, W.M. Sackinger and P. Wadhams (Ed.) *Advances in Ice Technology.* Comp. Mech. Publ., Southampton, UK, pp. 365.

The Oceans: Physical-Chemical Dynamics and Human Impact. Edited by S.K. Majumdar, E.W. Miller, G.S. Forbes, R.F. Schmalz and Assad A. Panah. © 1994, The Pennsylvania Academy of Science.

Chapter Eighteen

OCEANS: THE ORIGIN OF LIFE

LEE M. GRENCI[1] and **JAMES F. KASTING**[2]

[1]Instructor of Meteorology and
Member of the Penn State Weather Communications Group
College of Earth and Mineral Sciences
The Pennsylvania State University
University Park, PA 16802
and
[2]Professor of Geosciences and Meteorology
College of Earth and Mineral Sciences
The Pennsylvania State University
University Park, PA 16802

I am the daughter of Earth and Water,
And the Nursling of the sky;
I pass through the pores of oceans and shores;
I change, but I cannot die.

Percy Bysshe Shelly

INTRODUCTION

The plasma of human blood bears a striking resemblance to sea water. Both fluids have a similar density. They contain nearly the same concentrations of dissolved salts. Both fluids are slightly basic, their alkalinity being almost identical.

This remarkable link between sea water and human physiology is further solidified by the aqueous humor covering the human eye, a lens solution that has a saline similarity very close to sea water. In essence, one could argue that humanity looks at the worlds through lenses coated with ocean water.

Though evidence of the bond between the sea and humanity (as well as the animal kingdom) might be circumstantial, the "coincidence" is too compelling to ignore. Indeed, life on Planet Earth may have evolved from the oceans.

Assuming that the ocean was the first stepping stone in the evolution and emergence of life onto land, the first species to brave a world without sea water was probably already prepared for the challenging migration, capable of carrying part of the sea with them. To survive, organs would need to be bathed in a water-salt solution. Beings with such internal circulatory systems might also be able to retain and regulate fixed volumes of water and salts, another distinct advantage for survival away from the sea. Though many family trees can be traced back hundreds of years, there are none that list the sea as a progenitor. Perhaps they should.

Geological evidence suggests that life has existed on Earth for about three and one-half billion years. Evidence also suggests that life initially consisted of rather primitive organisms. How might the oceans have played a role in the evolution of life, given that early Earth was probably a very hazardous place for any land-dwelling or shallow-sea organisms? One haven could have been the hydrothermal vents, hot oases of life several miles deep on the cold desert floor near the mid-oceanic ridges.

HYDROTHERMAL VENTS

"The ocean is a desert with its life underground, and a perfect disguise above." These lyrics from the song *A Horse With No Name* by America likely refer to vast areas of the Earth's subtropical oceans that receive low rainfall, creating ocean climates as dry as some of the great deserts of Earth's land masses. Moreover, not only can the ocean's surface be considered a desert, the ocean's bottom also qualifies (where waters are sufficiently deep).

Sunlight does not penetrate to great depths in the ocean. Perhaps the faintest of light from the sun can reach down to approximately 200 meters (average depth of the bottom of the photic zone). Thus, the environment on the deep ocean bottom is not hospitable to plants, which require sunlight to make food during the process of photosynthesis. Plants, particularly tiny phytoplankton which serve as the base of the oceanic food chain, must live in the neighborhood of the sea surface.

In 1977 and 1979, expeditions from Woods Hole Oceanographic Institution and other venerable oceanic research groups, armed with a small, submersible vessel named Alvin, arrived over the Galapagos Rift (approximately 8,000 feet deep and part of the Earth's system of mid-oceanic ridges) off the coast of Ecuador. Scientists had returned to this region to explore plumes of hot water rising from hydrothermal vents along the rift that had been discovered during an earlier expedition in 1972 (Average water temperatures coming from the vents were about 17°C compared to surrounding waters of 2°C).

To their surprise, they found large communities of animals—giant tube worms up to 10 feet long with red flesh and red blood being one of the most impressive species. The great number of large animals at such great depths was perplexing, given that the fall of organic matter from the ocean's surface would be much too inadequate to support such vast enclaves of life. There must be another answer—a local answer. Dense clouds of bacteria were discovered to be the base of the food chain for the communities. What plankton can do with photosynthesis in the presence of sunlight, these bacteria can do in the dark using a process called chemosynthesis. Essentially, the bacteria like to eat sulfur, which they obtain from the hydrogen sulfide and sulfur dioxide that spews forth from the hydrothermal vents along the rift.

If organisms are classified on the basis of similarities in their genetic make-up, principally their ribosomal RNA, then the most "primitive' ' organisms—that is, those organisms that appear to have diverged earliest in evolutionary history, are sulfur-metabolizing thermophiles (Pace et al., 1986; Lake, 1988). This implies that all modern organisms are descended from a single ancestor which lived in a high-temperature environment. Perhaps early Earth was simply warm. Then evolution could have taken place anywhere on its surface. But modern organisms could also have descended from bacteria living in hydrothermal vents or hot springs on an otherwise cold early Earth. Indeed, the latter environments are somewhat favored because they both contain abundant elemental sulfur, which is considered gourmet food for primitive organisms. This does not clinch the argument for the vents, however, because elemental sulfur may have been produced globally from photolysis of volcanically-emitted SO_2 and H_2S (Kasting et al., 1989).

Obviously, more information is required in order to decide among the various alternatives. Fortunately, an entirely different line of evidence is available from a rather unlikely source—the Moon.

THE LUNAR CRATERING RECORD

One motivation for studying the Moon is that it preserves an excellent record of the impact environment of the early solar system. As illustrated by Figure 1, the Moon has been bombarded throughout its lifetime by thousands upon thousands of bodies, ranging in size from microscopic interplanetary dust particles to asteroid-sized objects. Mercury, Mars, and even Venus, it now appears based on early returns from the Magellan spacecraft, also show evidence for extensive bombardment. Earth must evidently have experienced a similar impact history; however the geologic record of impacts is much poorer on Earth because of extensive resurfacing of the planet by plate tectonics and by weathering.

The lunar impact record contains two types of information that may be relevant to the origin of life on Earth. The first contains the magnitude of the impact flux as a function of time. American Apollo and Soviet Luna spacecraft brought back Moon rocks from landing sites near several large lunar craters. Radiometric dating of these rocks, in combination with photographic studies of how various craters overlie each other, allows one to estimate when the craters were formed. Such studies indicate that most of the craters, including all the really large ones, were formed between 4.44 billion years ago (the age of solidification of the bulk of the lunar crust) and 3.8 billion years ago. This time interval, which is the same as the interval during which life originated on Earth, is referred to as the period of 'heavy bombardment'. An earlier interpretation of the cratering record, which suggested an isolated episode of 'late heavy bombardment' around 3.8 b.y. ago, has given way to the view that the impact flux declined more or less exponentially with time throughout this interval. Many of the craters that formed during the early part of this period were presumably obliterated by later impacts.

The other type of information concerns the size distribution of the impacting bodies. Laboratory studies of small impacts can be used in combination with scaling relationships to estimate the amount of energy required to produce a crater of a given size. If one has some idea of the velocity and density of the impacting body, this energy can be converted to an estimate of impactor size. For rocky material hitting the moon's surface at a velocity of about 13 km/s, a rough rule-of-thumb is that the diameter of the impacting body is about one-tenth the diameter of the resulting crater. Thus the 930-km Orientale crater (Figure 1) was probably formed by the impact of a 90- to 100-km diameter planetesimal. The largest clearly identifiable impact crater on the Moon, Imbrium, is 1200 km in diameter and would have required a slightly larger or higher-velocity impactor. Both Imbrium and Orientale are thought to be about 3.8 billion years old, indicating that stray 100-km diameter planetesimals were still present in the inner solar system at this relatively late date.

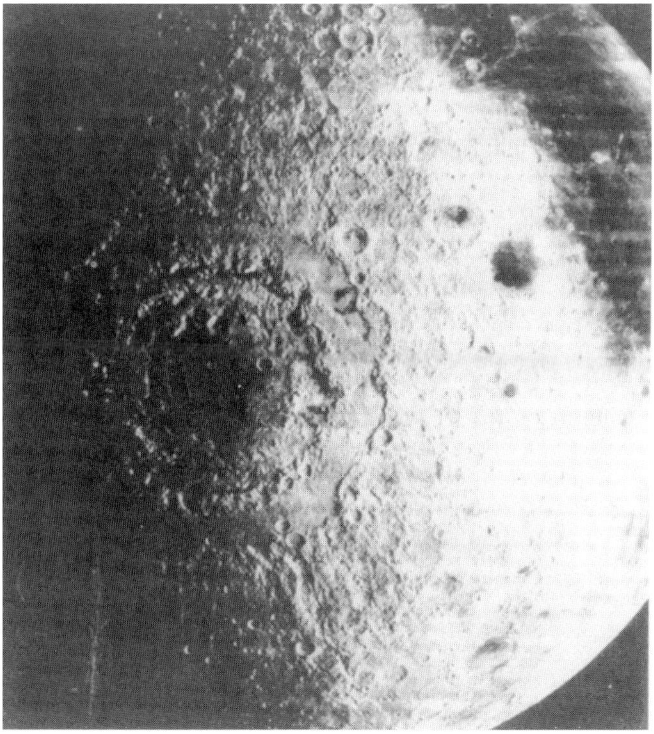

FIGURE 1. Orientale, the second largest impact crater on the Moon's surface (following Imbrium). The crater lies just beyond the western limb of the nearside and is 930 km in diameter. The innermost of the three observable rings is thought to have formed by thermal contraction; this permits an independent estimate of the energy required to form it (about 1.2×10^{26} Joules). The impactor that formed the crater was probably about 95 km in diameter.

IMPACTS ON EARLY EARTH

The planetesimals hitting the Earth during the pre-3.8-billion-year time period were probably even larger and more energetic. Earth's larger mass and higher gravity implies that typical impact velocities should have been at least 17 km/s. Since the kinetic energy of a body is proportional to its velocity squared, this means that impactors of a given size would have released about 70 percent more energy on Earth than on the Moon. The Earth also has fifteen times the surface area of the Moon and about twenty-four times the effective collecting area when its higher gravity is taken into account. Thus, if the planetesimal population included bodies significantly larger than 100 km in diameter, which seems likely, the chances are that most of these bodies should have hit the Earth rather than the Moon. Or to put it another way, the fact that the Moon was not hit by anything larger than 100 km across does *not* imply that the same was true for early Earth.

What would have been the consequences of large terrestrial impacts for life and for the atmosphere? The most immediate consequence for life is the threat of direct thermal annihilation. Numerical studies of giant impact using theoretical computer models indicate that about half the energy of the impact goes into vaporizing rock. The other half is deeply buried and goes into melting mantle material. An Imbrium-sized impact should produce enough rock vapor to increase the total atmospheric pressure by several bars. This rock vapor cools by radiating approximately half its energy to space and half to Earth's surface, both of which are much colder than the 2000 K rock vapor. Since the Earth's surface is 70 percent water-covered, and since there may have been even less land area in the past, most of the downward radiation goes into evaporating sea water. Thus, after the rock vapor had cooled and condensed out, the Earth should have been left with a dense steam atmosphere.

Larger impactors may have been capable of vaporizing the entire ocean. One can calculate thermodynamically that the energy required to accomplish this is about 5×10^{27} Joules. The impact itself must have released four times this much energy, since half the energy is buried and half is radiated away to space. For a 17-km/s rocky impactor, this requires a 440-km-diameter body—roughly four times the size of the object that formed Imbrium. Although nothing this big appears to have hit the Moon, there are two present-day asteroids, Vesta and Pallas, that are about this size, and one, Ceres, that is more than twice as big. If the size distribution of modern asteroids is at all representative of that of the late-arriving planetesimals, it is conceivable—even likely—that the Earth was hit by one or more of these 'ocean-vaporizing' bodies between 4.5 and 3.8 billion years ago. It is even more likely that the Earth was hit by smaller 190-km diameter objects that were capable of evaporating the entire photic zone of the oceans (the uppermost 200 meters). For reasonable planetesimal-size distributions, photic-zone evaporating events should have been commonplace until about 3.8 billion years ago.

IMPLICATIONS FOR EARLY LIFE

The occurrence of large impacts has obvious implications for early life. Because of the production of hot rock vapor, even relatively small impacts would have been lethal to organisms living on land, if there were any such organisms that could survive the expected high solar ultraviolet flux. Benthic (bottom-dwelling) organisms living on continental shelves or on the slopes of volcanic islands would have experienced nearly the same risk. Any impact that evaporated the photic zone would have left them directly exposed to the radiative flux from the rock vapor. Planktonic organisms living in the surface ocean might have been somewhat better off, since they could, in principle, have continued to migrate down to lower levels in the ocean. However, any upwelling current that brought them to the surface would expose them to a layer of boiling water at the top. Since the rock vapor would have persisted for weeks to months, the steam atmosphere for tens to hundreds of years, it seems unlikely that any planktonic organisms would survive a photic-zone-evaporating impact.

It may be illuminating to compare such an impact to the one that is thought to have occurred at the end of the Cretaceous period, 65 million years ago. The K-T (Cretaceous-Tertiary) impactor was probably about 10 km in diameter, based on the amount of iridium it left behind, and should have been sufficiently energetic to evaporate about 20 cm of sea water. Even though this is minuscule by comparison with the events discussed above, the K-T impact apparently resulted in extinctions of phytoplankton and may have killed off the dinosaurs as well. A photic-zone evaporator would release 1000 times as much energy and could be expected to be proportionately more lethal.

Where could organisms have lived and evolved during these hazardous times? One obvious possibility is the mid-ocean ridge hydrothermal vents. These vents today are covered by about 2.5 km of sea water; thus a near ocean-vaporizing impact would be needed to expose them. We don't know how deep the oceans were during the Earth's early history, but there is no good reason to suspect they were significantly shallower than today. A post-impact ocean, being heated from above, should be stable against convection; thus the thermal pulses caused by moderate-sized impacts, even photic-zone-evaporating ones, would not penetrate to the ridge crests. Unless an impacting alien body landed directly on top of it, a vent organism would have been aware of its arrival only because of a transient pressure increase caused by the passage of the initial shock wave.

Putting these observations together with the evidence cited above for the ubiquity of thermophilia and sulfur metabolism among primitive organisms leads to a rather remarkable hypothesis: modern organisms may all be descended from some bacterium that dwelt in the mid-ocean ridge hydrothermal vents. This does not necessarily imply that life originated in the vents; indeed some prebiotic chemists would argue that it is difficult to assemble large complex organic molecules in this type of high-temperature environment. Life could, however, have evolved at the surface at temperatures ranging from rather warm to rather cold, and then colonized the mid-ocean ridges at a later date. The next photic-zone evaporator annihilated all life at the surface, leaving only the thermophylic vent organisms to restart the evolutionary process. This scenario is by no means proven; however, it is consistent with the lessons that can be drawn from molecular phylogeny and from planetary physics.

CONCLUSIONS

The problem of life's origin is far from being solved. But we do have some good clues, and these clues are beginning to fit together into a self-consistent pattern. High temperatures are evidently associated with the early environment of life, if not with life's actual origin. The mid-oceanic ridges were a likely abode for early life because of their warmth, the availability of sulfur, and because they were shielded from impacts. The impacts themselves must have been a dominant feature of the early environment and should be factored into our physical models of the early atmosphere and oceans.

REFERENCES

Kasting, J.F., K.J. Zahnle, J.P. Pinto and A.T. Young. 1989. Sulfur, Ultraviolet Radiation and the Early Evolution of Life. Origins of Life, Volume 19, pp. 95-108.

Kasting, J.F. 1990. Impacts and the Origin of Life. Earth and Mineral Sciences, The Pennsylvania State University, College of Earth and Mineral Sciences, Volume 59, Number 4, pp. 37-42.

Lake, J.A. 1988. Origin of the Eukaryotic Nucleus Determined by Rate-invariant Analysis of rRNA Sequences. Nature, Volume 331, pp. 184-186.

Pace, N.R., G.J. Olsen, C.R. Woese. 1986. Ribosomal RNA phylogeny and the Primary Lines of Evolutionary Descent. Cell, Volume 45, pp. 325-326.

Sleep, N.H., K.J. Zahnle, J.F. Kasting, H.J. Morowitz. 1989. Annihilation of Ecosystems by Large Asteroid Impacts on the Early Earth. Nature, Volume 342, pp. 139-142.

The Oceans: Physical-Chemical Dynamics and Human Impact. Edited by S.K. Majumdar, E.W. Miller, G.S. Forbes, R.F. Schmalz and Assad A. Panah. © 1994, The Pennsylvania Academy of Science.

Chapter Nineteen

THE EVIDENCE FOR LARGE BOLIDE IMPACTS IN THE PRESENT AND PALEO OCEAN BASINS

ROBERT F. FUDALI

Department of Mineral Sciences
U.S. National Museum of Natural History
Smithsonian Institution
Washington, DC 20560

INTRODUCTION

During the past 25 years, planetary probes have convincingly shown that meteorite and cometary bombardment has played a major role in shaping the surfaces of all the terrestrial planets and satellites of the solar system, from Mercury to the moons of Uranus. Concomitantly, terrestrial studies have demonstrated that the bombardment was not confined to the early history of the solar system but has been on-going throughout the entire Phanerozoic. There can be no doubt that this bombardment has repeatedly stressed the Earth's biosphere on a regional scale. And there is good evidence that the high energy end of the impact spectrum has been responsible for at least one of the global mass extinctions that punctuate the paleontological record. It is in oceanic sediments that we find the most precise timing of these mass extinctions preserved. Further, although almost all known terrestrial impact craters are on land, much of the compelling evidence of the most energetic impacts is found in oceanic sediments. In this chapter I briefly review the impact evidence and the possible extinction linkages.

HISTORICAL PERSPECTIVE

In 1953, Kelly and Dachille published (privately) a book[1] wherein they flatly stated that the Earth has undergone repeated, catastrophic collisions with giant meteorites throughout its entire history and will inevitably undergo future collisions, barring man's interference. Among many other consequences, they speculated that some of these collisions were probably responsible for mass extinctions in the biosphere and subsequent radiation by new species into the niches vacated. And so cosmic collisions have been an important factor in driving evolution. As an example they discussed the demise of the dinosaurs and the rise of mammals at the end of the Cretaceous Period. Many of their postulated catastrophic effects were attributed to the production of gigantic waves produced by impacts into the oceans.

Of course, in those days all such catastrophic ideas were considered preposterous. And it didn't help that most of what they wrote was obviously and demonstrably wrong, such that their book has been relegated to a quaint historic footnote.

To be sure, the face of the Moon is pockmarked with huge craters that, as early as 1893, the eminent geologist, G.K. Gilbert, had declared were impact craters after studying their morphologies.[2] And I think it's fair to say that many earth scientists, in 1953, agreed with Gilbert's assessment. But the conventional wisdom was that these collisions occurred very early in the history of the solar system and that no large lunar impact craters were younger than 2-3 billion years old. Certainly we knew of no large impact craters on the Earth. Of the proven handful of terrestrial craters, the largest was Meteor Crater, Arizona (diameter of 1.2 km). It would take a bolide with a kinetic energy of 2-3 megatonnes of TNT equivalent to excavate Meteor Crater. That is a respectable nuclear device but hardly earth shattering. And so matters stood—but not for long.

Over the next two decades the idea of species being snuffed out with a sudden bang rather than a lingering whimper gained a few adherents. De Laubenfels[3] in 1956 and Urey[4] in 1973 independently suggested a giant impact event finished the dinosaurs at the end of the Cretaceous. And Urey went on to suggest "that the geological periods were terminated by such collisions." In between, Opik, in 1958, discussed the catastrophic effects of cosmic collisions[5] and McLaren, in 1970, postulated the ocean impact of a large asterioid as the cause of a specific, late Devonian mass extinction.[6] None of these authors bothered much with details and consequently, unlike Kelly and Dachille, they are routinely referenced today and considered prescient. Perhaps there is a lesson here.

Meanwhile, a wide range of studies, both observational and experimental, had demonstrated that the shock wave generated by a large bolide impact imposed profound macroscope and microscope changes on the target rocks.[7] These so-called shock metamorphic effects could be used to identify impact structures that not only have no associated metoritic signature but may have also lost their original crater morphology to erosion. By 1979 it had become clear that there must be a large number of impact structures > 10 km in diameter on the Earth. Today we count 62 such structures (See Appendix 1) and the tally is far from complete. All but four of these are Phanerozoic in age. Also by 1979, in addition to the known Earth-crossing comets

APPENDIX 1

*Known Terrestrial Impact Craters > 10 Km in Diameter**

Name	Lat.	Long.	Diameter, Km	Age, myr**
Acraman	32° 1′ S	135° 27′ E	90	> 570
Ames	36° 15′ N	98° 10′ W	16	470 ± 30
Araguainha Dome	16° 15′ S	52° 59′ W	40	250 ± 20
Avak	71° 15′ N	156° 38′ W	12	100 ± 5
Azuara	41° 10′ N	0° 55′ W	30	< 130
Beaverhead	49° 36′ N	113° 0′ W	60	~600
Boltysh	48° 45′ N	32° 10′ E	24	88 ± 3
Bosumptwi	6° 32′ N	1° 25′ W	10.5	1.03 ± .02
Carswell	58° 27′ N	109° 30′ W	39	115 ± 10
Charlevoix	47° 32′ N	70° 18′ W	54	357 ± 15
Chicxulub	21° 21′ N	89° 30′ W	180	65 ± 1
Clearwater East	56° 5′ N	74° 7′ W	22	290 ± 2
Clearwater West	56° 13′ N	74° 30′ W	32	290 ± 20
Deep Bay	56° 24′ N	102° 59′ W	13	100 ± 50
Dellen	61° 55′ N	16° 39′ E	15	110 ± 1
Eagle Butte	49° 42′ N	110° 35′ W	19	< 65
El' Gygytgyn	67° 30′ N	172° 5′ E	18	3.5 ± .5
Gosss Bluff	23° 50′ S	132° 19′ E	22	142.5 ± .5
Haughton	75° 22′ N	89° 41′ W	24	23 ± 1
Janisjarvi	61° 58′ N	30° 55′ E	14	700 ± 25
Kaluga	54° 30′ N	36° 15′ E	15	380 ± 10
Kamensk	48° 20′ N	40° 15′ E	25	65 ± 2
Kara	69° 5′ N	64° 18′ E	65	73 ± 3
Kara-Kul	38° 57′ N	73° 24′ E	45	< 225
Karla	54° 54′ N	48° 0′ E	12	10 ± ?
Kelly West	19° 56′ S	133° 57′ E	10	> 550
Kentland	40° 45′ N	87° 24′ W	13	< 300
Lappajarvi	63° 9′ N	23° 42′ E	23	77.3 ± .4
Lawn Hill	18° 40′ S	138° 39′ E	18	> 515
Logancha	63° 30′ N	95° 48′ E	20	< 50
Logoisk	54° 12′ N	27° 48′ E	17	40 ± 5
Manicouagan	51° 23′ N	68° 42′ W	100	212 ± 2
Manson	42° 35′ N	94° 31′ W	35	66 ± 1
Marquez	31° 17′ N	96° 18′ W	22	58 ± 2
Mistastin	55° 53′ N	63° 18′ W	28	38 ± 4
***Montagnais	42° 53′ N	64° 13′ W	45	50.5 ± .8
Nicholson Lake	62° 40′ N	102° 41′ W	12.5	< 400
Oasis	24° 35′ N	24° 24′ E	11.5	< 120
Obolon	49° 30′ N	32° 55′ E	15	215 ± 25
Popigai	71° 30′ N	111° 0′ E	100	35 ± 5
Presqu'ile	49° 43′ N	78° 48′ W	12	< 500
Puchezh-Katunki	57° 6′ N	43° 35′ E	80	220 ± 10
Ries	48° 53′ N	10° 37′ E	24	14.8 ± .7
Rochechouart	45° 50′ N	0° 56′ E	23	186 ± 8
Saint Martin	51° 47′ N	98° 32′ W	40	< 252
Serra da Cangalha	8° 5′ S	46° 52′ W	12	< 300

Appendix continued on next page

APPENDIX 1 *continued*

Sierra Madera	30° 36′ N	102° 55′ W	13	< 100
Siljan	61° 2′ N	14° 52′ E	55	368 ± 1
Slate Island	48° 40′ N	87° 0′ W	30	< 350
Spider	16° 44′ S	126° 5′ E	13	> 570
Steen River	59° 31′ N	111° 37′ W	25	95 ± 7
Strangways	15° 12′ S	133° 35′ E	25	< 420
Sudbury	46° 36′ N	81° 11′ W	200	1850 ± 5
Teague	25° 52′ S	120° 53′ E	30	1685 ± 5
Ternovka	48° 1′ N	33° 5′ E	12	28 ± 10
Tookoonooka	27° 0′ S	143° 0′ E	55	128 ± 5
Upheaval Dome	38° 26′ N	109° 54′ W	10	< 65
Ust-Kara	69° 18′ N	65° 18′ E	25	73 ± 3
Vargeao Dome	26° 50′ S	52° 7′ W	12	< 70
Vredefort	27° 0′ S	27° 30′ E	140	1970 ± 100
Wells Creek	36° 23′ N	87° 40′ W	12	< 300
Zhamanshin	48° 24′ N	60° 58′ E	13.5	0.9 ± 0.1

* Modified after an unpublished list complied by R.A.F. Grieve (Canadian Geological Survey) as of 10/13/92.

** Most ages given are radiometric determinations on impact melt rocks. Upper limits are ages of the target rocks. Lower limits are enigmatic.

*** The only impact crater, so far identified, under water (the continental shelf off Nova Scotia).

(i.e. comets in potentially Earth-colliding orbits), a very modest effort had found 35 Earth-crossing asteroids,[8] estimated to represent about 5% of the total number of Earth-crossers. Clearly, catastrophic encounters with these Earth-crossers was not only possible but inevitable, given enough time.

So the stage was set for a favorable reception when, in 1979, Napier and Clube published a detailed theory of mass extinctions caused by cosmic bolides.[9] They postulated that biotic stresses would be caused both directly by a blast wave and indirectly by the injection of large amounts of nitric oxide and dust into the stratosphere, destroying the ozone layer, blocking sunlight and disrupting the food chain. They further predicted that such events would leave distinctive fingerprints—excess concentrations of platinum group elements at biostratigraphic horizons, worldwide, since these elements are 1000 to 10,000 times more abundant in meteorites than they are in the Earth's crust and upper mantle. Very shortly thereafter, in the same year, Alvarez et al.[10] reported a strong, sharp iridium concentration spike in a marine Cretaceous-Tertiary (hereafter K/T) boundary layer near Gubbio, Italy. And seven months later they added K/T boundary sites showing Ir spikes in Denmark and New Zealand to the Gubbio site.[11] They embraced the "dust in the stratosphere" idea of Napier and Clube (without, unhappily, proper accreditation) as the killing mechanism. A catastrophic theory had again become respectable, 150 years after catastrophism had been banished from geology and replaced by the gradualist uniformitarianism of Charles Lyell. Perhaps there is a lesson here, too—what goes around comes around.

THE FLUX OF LARGE ASTEROIDS AND COMETS TO THE EARTH'S SURFACE

Grieve et al.[12] estimated the flux of large bolides to the earth's surface by counting the impact craters, > 20 km in diameter and < 120 million years old, found on the North American and European cratons. These cratons have little sedimentary cover to hide a 20 km crater and 20 km structures are unlikely to otherwise disappear in 120 myr, given the low erosional (low relief cratons) environment. The assumption then is that all craters > 20 km in diameter and < 120 myr old on these cratons have been found and are the original population. The resultant estimate for the total terrestrial production of impact craters > 20 km in diameter is $(5.4 \pm 2.7) \times 10^{-15}$ per km^2 per year, over the last 120 myr. The result is hardly a robust one, based, as it is on only seven craters found on 3% of the Earth's surface (10% of the land surface). Nevertheless, it does illustrate how woefully incomplete the cratering record is, elsewhere on the Earth. Only 15 craters > 20 km in diameter and definitely < 120 myr old are known on the earth's land surface (Appendix 1). As noted, seven of these occur on the North American and European cratons so only eight are known on the remaining 90% of the land surface—a number that must be too low by almost a factor of 10.

Shoemaker et al.[13] have taken the opposite approach of trying to characterize the possible impactor population at the present time. If the population and size distribution of Earth-crossing bolides is known, and if there is some knowledge of representative orbits, then impact probabilities can be calculated. As of the end of 1989, Shoemaker et al. reported 95 known Earth-crossing asteroids, most of them between 1 and 8 km in diameter and estimated to represent ~6% of the actual Earth-crossing population. That estimate, and the conversion of asteroid brightness (magnitude) to asteroid diameter are not for the faint of heart. In addition to the Earth-crossing asteroids, there are about 400 known Earth-crossing comets with well-characterized orbits. The ratio of the known population of these Earth-crossers to the true population is probably close to a factor of 2, rather than the factor of 20 estimated for Earth-crossing asteroids. But estimates of comet nuclei diameters and densities are fraught with uncertainty. Still, Shoemaker et al.'s estimate of the present cratering rate for craters > 20 km in diameter, by asteroids and comets both, is $(4.9 \pm 2.9) \times 10^{-15}$ per km^2 per year, in decent agreement with Grieve et al.

Using a very different method to calculate the flux of long period and short period comets crossing the Earth's orbit, and adding a large contribution from sporadic comet showers (generated by a numerical simulation), Weissman[14] calculated a present-day crater production rate only about 20% higher than that of Shoemaker et al.'s rate (who considered only long period comets). So, as Weissman commented: "What we have then is a consistent picture, though with large uncertainties."

Cratering rates in terms of impacts per km^2 per year are not terribly enlightening. Expanding the area to the Earth's surface (5.1×10^8 km^2) and the time to 10^8 years, Shoemaker et al.'s estimate, for example, predicts 250 craters > 20 km in diameter have been produced somewhere on the Earth in the last 100 million years. And since Shoemaker et al. have a decent population and size distribution to work with, they can further estimate that, of these 250 craters, 82 were > 30 km, 16 were > 50 km, 8 were > 60 km, 2 were > 100 km and 1 was > 150 km in diameter. Even given the large uncertainties attendant upon these estimates, they clearly show the potential for considerable and repeated mischief in the geosphere and biosphere by large bolide impacts. Of course, ~70% of these impact events will be oceanic impacts.

EVIDENCE IN OCEANIC SEDIMENTS FOR LARGE BOLIDE IMPACTS: TEKTITES AND MICROTEKTITES

Tektites are small, enigmatic, natural glass objects that have received scientific scrutiny since the time of Charles Darwin. The published literature on them is voluminous. For those interested in maximum information with minimal effort, a succinct review article has recently been published.[15]

Tektites were originally found in distinct regions of the Earth's land surface, known as strewnfields: the Czechoslovakian, the North American, the Ivory Coast and the Australasian strewnfields. They have the following distinctive characteristics that, together, absolutely preclude an igneous origin:

1. They have an extremely low water content—always $<0.02\%$ by weight.
2. As a corollary to no. 1, they are depleted in all trace elements that are known to be most volatile at elevated temperatures.
3. They are strongly reduced; the ratio of trivalent to divalent iron is normally less than 0.1, in marked contrast to igneous rocks. This indicates either a very low oxygen partial pressure or a very high temperature during their formation—or both.
4. They commonly contain blebs and stringers of lechetelierite (pure silica glass) that can only reasonably be explained by the melting of pre-existing quartz grains without reaction with the other components of the melt. This requires temperatures considerably in excess of 1700°C (well beyond established igneous temperatures) and a very short interval between melting and quenching (to prevent melt homogenization).
5. For their high silica content (65-85 wt.%) they have too much iron and magnesium and too little potassium and sodium to match known igneous compositions. Their major element chemistry is most nearly matched by sedimentary rocks such as sandy shales or graywackes.
6. Tektites do not contain primary crystals (i.e. crystals grown from the melt) but a subclass of the Australasian tektites contain rare, tiny, relict mineral grains. So far, quartz, zircon, corundum (most likely a breakdown product of an Al_2SiO_5 precursor mineral), rutile, chromite and monazite have been identified.[16] This assembly is only compatible with a variety of igneous and metamorphic rocks and so again indicates a sedimentary deposit fed by these various sources.
7. Coesite is also found in the same Australasian subclass.[17,18] Coesite is a high-pressure polymorph of silica that is very seldom found in normal terrestrial rocks.

8. The Australasian tektites are the youngest and their morphologies least modified by subsequent processes. These morphologies include the so-called splashforms—spheres, discs, rods, dumbbells, and teardrops. Collectively, these morphologies are quite unlike anything produced in a volcanic eruption. Further, tektites found in Australia have undergone severe ablation caused by high speed passage through the atmosphere as solid, rather than liquid bodies. This ablation has strongly modified their original splashforms. The implication is that these australites were ejected from the Earth's surface on suborbital trajectories, became solidified, and reentered the atmosphere downrange at very high velocities (kilometers per second).

Although many details concerning the formation and distribution of tektites remain unknown, there is virtually unanimous agreement on one point: tektites must be produced by the impact of large bolides on the Earth's surface and represent thoroughly melted target rocks. Thus, tektites are proof of large impact events, even when no corresponding impact structures are known. And, since they can be radiometrically dated, they can pinpoint the timing of these events.

In the 1960's, sedimentary cores from oceanic sites, produced by the Deep Sea Drilling Program, were found to contain tiny tektites ($<$ 1 mm in diameter), subsequently termed microtektites.[19] Since then many DSDP cores have greatly extended three of the four original tektite strewnfields (Figure 1).

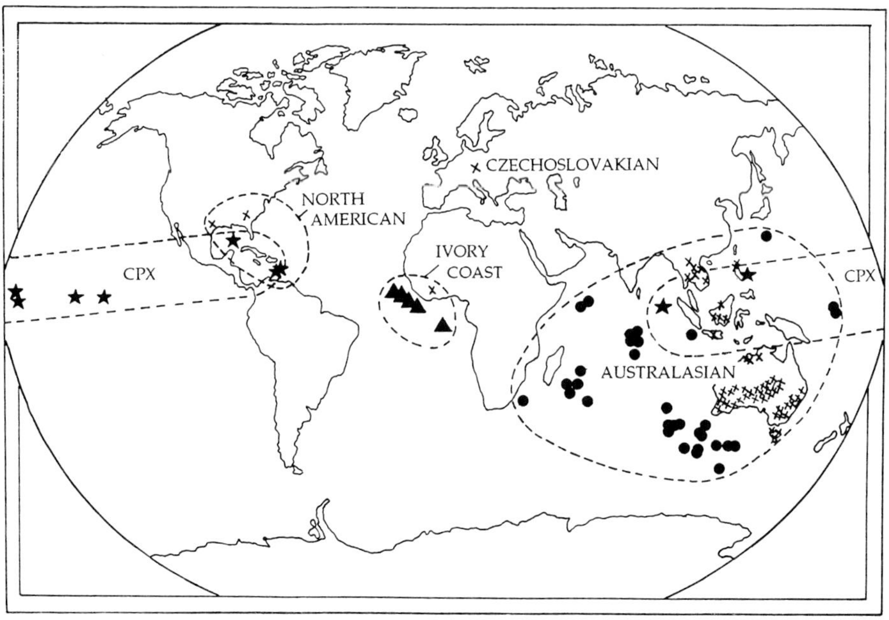

FIGURE 1. Map showing the Czechoslovakian, Ivory Coast, Australasian and North American microtektite strewnfields and the cpx strewnfield. After Glass.[15] x's mark land occurrences. Solid symbols mark DSDP cores that contain microtektites or cpx microspherules.

Microtektites exhibit all the splashform morphologies and physical and chemical characteristics of their larger brethren. And they are datable to their respective strewnfields, not only radiometrically but also by their biostratigraphic relations and their proximity to magnetic reversals, both recorded in oceanic sediments. They do extend the range in tektite chemistry to lower silica, high magnesium compositions.

The Czechoslovakian tektite strewnfield has no known microtektite extension. Based solely on age, this strewnfield is associated with the 14 myr old Ries Impact Structure in Germany.

The Ivory Coast tektites and microtektites show a good chemical match to glassy material found at the (10.5 km diameter) Bosumptwi Crater in Ghana. The radiometric ages of all these glasses are also equivalent at ~1.0×10^6 years.[20] The microtektite horizon is also close to the Jaramillo magnetic reversal, which has been bracketed at 0.97×10^6 years.[21] There are no extinctions or Ir anomaly associated with this event.

The Australasian tektites have been recently re-dated at 770,000 years.[22] The Australasian microtektites are found very close to the Brunhes—Matuyama magnetic reversal which is bracketed at 780-790,000 years.[23,24] This strewnfield covers 10% of the Earth's surface and contains an estimated 100 million metric tonnes of glass.[25] Again, no extinctions or Ir anomaly are associated with this strewnfield, but the cratering event must have had severe regional effects. Given the extent of the strewnfield compared to the Ivory Coast strewnfield, the parent crater must be several times as large as the 10.5 km diameter Bosumptwi Crater. It was most likely an ocean impact because such a large, young crater would be conspicuous on satellite imagery if it were on land. The concentration of microtektites increases, and coesite and shocked debris first appear, in DSDP cores as Indochina is approached. Consequently the best current guess for such a parent crater puts it on the continental shelf in the vicinity of Indochina.

The North America strewnfield was originally thought to be a single field extending from the Caribbean across the equatorial Pacific and into the Indian Ocean. Subsequent work has, however, demonstrated that there are two and perhaps three distinct, somewhat overlapping strewnfields in this region.[26,27,28]

The youngest of these strewnfields, dated at ~35 million years[29] is composed of tektites in Texas and Georgia and microtektites in the Caribbean, Gulf of Mexico and Western North Atlantic. Not only microtektites but tektites, tektite fragments, coesite and shocked quartz and rock fragments have been recovered from DSDP site 612, on the continental slope off New Jersey,[30] and the parent crater is presumed to be nearby. Poag et al.[31] have reported a remarkable 60 meter thick marine boulder bed containing traces of shocked minerals and tektite glass that "is distributed over > 15,000 km^2 beneath Chesapeake Bay and the surrounding Middle Atlantic Coastal Plain and inner continental shelf." They attributed this bed to a giant tsunami, generated by the cratering event responsible for the North American strewnfield. There is no Ir anomaly associated with this strewnfield.

Partially overlapping the North American strewnfield is a microspherule strewnfield that extends from the Gulf of Mexico across the equatorial Pacific. This strewnfield does contain true microtektites, i.e. entirely glass splashforms. But they are subordinate to glassy spherules that contain primary crystallites of clinopyroxene (hereafter cpx). This cpx microspherule strewnfield is a few thousand to a few tens of thousands years older than the microtektite strewnfield and is associated with both a small Ir spike and the extinction of five radiolarian species. No candidate parent crater is known.

There may be a third cpx microspherule strewnfield, older yet, that overlaps the equatorial Pacific field and extends into the Indian Ocean. This needs confirmation.

In addition to the radiolarian extinctions there were major population reductions of foraminiferal species (that did not go extinct) associated with both the cpx microspherules and the microtektites. Further, during the terminal Eocene there were four, closely spaced, stepwise extinctions, involving planktonic foraminifera, culminating at the Eocene/Oligocene boundary.[32] The most recent date for this boundary makes it 1 $\pm$ 0.5 myr younger than the microtektite strewnfield.[33] Although there is no obvious link between these stepwise extinctions and the strewnfields, there is at least a general concurrence in time and the possibility exists that two or perhaps three impact events may have somehow contributed to these biotic changes in the last 1-2 myr of the Eocene.

The one major mass extinction that tektites and microtektites are definitely associated with is the mass extinction across the K/T boundary. Tektites and microtektites have been found in the K/T boundary layer at eight sites in North America and around the Gulf of Mexico.[34,35,36] At six of these sites the glasses have been entirely altered to a variety of secondary minerals but retail their original morphologies and internal textural features. At two sites, Beloc, Haiti and Arroyo el Mimbral, Mexico, some incompletely altered tektites and microtektites still have glass cores. At these sites, splashforms range up to 8 mm in maximum dimension.

The chemistry of these glass cores is a good match for the target rock chemistry of the prime parent suspect—the buried, 180 km diameter Chicxulub Impact Structure on the Yucatan coast. And the $^{40}Ar/^{39}Ar$ ages of the tektites and the melt rocks at Chicxulub are identical at 64.5-65 myr.[37,38]

Smaller microspherules, altered to a different suite of secondary minerals are found at most other K/T sites world-wide; except that they do not overlap the K/T tektite-microtektite field, suggesting a different mode of origin. These microspherules appear to have originally been similar to the cpx microspherules associated with the North American strewnfield. It has been suggested that they represent condensates from a vapor cloud and incorporate some of the impactor, as opposed to the true K/T microtektites that went from solid target to melt directly, with little impactor contribution to their chemistry.[36]

Finally there are presently four other Phanerozoic microtektite occurrences known in the geologic record, all in upper Devonian marine sediments. These are in South China,[39] Western Yukon Territory, Canada,[40] and two sites in Belgium.[41,42] At all these localities the microtektite glasses have only experienced minor, surficial alteration, which is quite remarkable given their 365 myr age. It is tempting to assign these four occurrences to the same strewnfield and to relate this strewnfield to the major extinction in the late Devonian marked by the Frasnian Stage-Famennian Stage boundary. That would be, to put the best face on it, premature. The Yukon occurrence is not yet stratigraphically well-constrained. The South China occurrence is said to lie three conodont zones above the F/F boundary. One of the Belgium occurrences could be associated with the F/F boundary; the other appears to be well above this boundary. The Belgian microtektites are characterized by five chemical groups of bizarre and wildly different compositions. The Yukon microtektites are remarkably uniform chemically, have *very* high calcium contents and do not match any of the Belgian groups. The South China microtektites show more variability than the Yukon microtektites but far less than the Belgian groups and are, again, a poor match for any of the Belgian microtektites. It is true that microtektite chemical compositions have never been a reliable guide to what constitutes a single strewnfield. Still, the chemical differences are so extreme in these upper Devonian microtektite occurrences that one is hard put to entertain a common origin. Clearly, additional work needs to be done on this problem.

THE CRETACEOUS-TERTIARY BOUNDARY EVENT

In addition to the tektites and microtektites at the K/T boundary, which are alone sufficient to brand the boundary event as a large bolide impact or impacts, there is a great deal of additional evidence supporting this conclusion. First and foremost are the strong iridium spikes found at the boundary. Currently more than 105 K/T boundary sites with Ir anomalies are known (W. Alvarez, unpublished data reported in ref. [36]). At sites where other platinum group elements have recently been measured, the Ir/pge's are close to the ratios found in chondritic meteorites (e.g. 43). Shocked quartz (displaying multiple sets of planar deformation features) is found in K/T boundary clays, worldwide.[44] Stishovite (the highest pressure polymorph of silica, which has never been found in a normal terrestrial setting) has been reported from the K/T boundary layer in the Raton Basin of New Mexico.[45] Nanometer-size diamonds have been found in three K/T boundary layers.[46,47] Carbon far above background levels, in the form of fluffy soot, has been found at K/T sites, worldwide. The only reasonable sources so far suggested are global wildfires associated with the boundary event.[48,49] And, within the uncertainties of radiometric dating, two large impact craters are known that are indistinguishable in age from the K/T boundary. These are the previously mentioned Chicxulub Crater in Yucatan, linked to the K/T tektites and microtektites, and the buried, 35 km diameter Manson Crater in Iowa.[50] The Manson Crater is generally believed to be too small to account for the entire array of features at the K/T boundary, worldwide, but could be responsible for some K/T boundary effects.

OTHER MAJOR BIOEVENT HORIZONS

Donovan[51] has defined a mass extinction as one ". . . where there has been a (geologically) rapid reduction in biological abundance and diversity on a global scale." By this definition there have been five mass extinctions in the Phanerozoic, at the K/T, the F/F, the Ordovician/Silurian, the Permian/Triassic and the

Triassic/Jurassic boundaries. It is beyond the scope of this chapter to describe the nature of these mass extinctions and the interested reader is referred to the detailed descriptions to be found in [51]. The point here is that, if the estimates of the flux of large bolides to the Earth's surface are not wildly inaccurate, large bolide impact events could easily account, not only for the K/T mass extinction, but the other four as well.

Unhappily for this idea, and based on more than 3,000 iridium determinations across all these bioevent horizons plus a number of other geologic boundaries from the Precambrian/Cambrian to the K/T boundary, Orth et al. report that the K/T Ir spike appears to be unique.[43] They did find a number of small Ir anomalies but in all cases they could be attributed to one or another of several terrestrial concentration mechanisms. For example, an Ir anomaly near the F/F boundary in the Canning Basin of Australia turned out, on close examination, to be perfectly correlated with a species of stromatolitic bacteria that must have preferentially concentrated the Ir from seawater.

So, the only two Ir anomalies found so far in the Phanerozoic, that can be attributed to bolide impacts and are associated with bioevent horizons are the K/T worldwide spike and the much smaller anomaly associated with a microspherule bed near the Eocene/Oligocene boundary (though only one of these is, by definition, a mass extinction).

And, aside from Ir absences, the other K/T bolide fingerprints have not yet been found at the other mass extinction boundaries. Possible exceptions are the enigmatic microtektites in the upper Devonian and a single report of probable shocked quartz at the Triassic/Jurassic boundary.[52] Both the microtektites and the shocked quartz, if truly associated with mass extinctions, should be found world-wide and should be so searched for.

So, although it would be neat to be able to associate all the Phanerozoic mass extinctions with evidence of large bolide impacts, such is not the case to date. Life is not simple—nor perhaps are mass extinctions thereof.

DISCUSSION

Many times throughout the 1980's, and as recently as 1990 (e.g. [53,54]), it has been argued that volcanism—specifically the massive outpourings of the Deccan flood basalts of India which appear to be coincident in time with the latest Cretaceous is a much more probable kill mechanism for the K/T extinctions than a large bolide impact. Usually cited as evidence are the gradual and/or stepwise extinctions among several groups of organisms leading up to the boundary. Also, at the K/T sites studied in the most detail, the strong Ir spike has complex shoulders on either side where Ir remains well above background values for somewhere between one to several vertical meters of section. If taken at face value as representing an original rather than a secondary distribution, this implies a distribution period spanning up to several hundred thousand years. This would be incompatible with a single instantaneous event and so episodic Deccan volcanism is invoked.

Even in 1990 there were serious objections to the volcanic theory. First it must simply be assumed that Deccan volcanism produced enough iridium, as an aerosol, to account for the K/T anomaly. There is no direct evidence in support of this assumption. A bolide impact does not have this problem. Second, it is highly improbable, if not impossible, that the non-explosive, effusive Deccan eruptions could loft enormous amounts of aerosols, not to mention quartz fragments, microspherules and soot, into the stratosphere for world-wide distribution. Again a bolide impact would do just that for the quartz and microspherules and the soot need not be injected into the stratosphere if the wildfires are global. Then there is the matter of the shocked quartz. Quartz does not crystallize from basalts nor do volcanoes of any kind produce *shocked* quartz. The imaginative suggestion has been made that the Manson Crater (too small to be responsible for the mass extinction) was responsible for the shocked quartz precisely coinciding with the largest Deccan eruption at the K/T boundary[53] but this really seems to be stretching things a bit.

Since 1990, nanodiamonds, stishovite and tektites and microtektites have been found at the K/T boundary. And the 180 km diameter Chicxulub Crater has been identified as an impact crater, tied to the tektites and found to be indistinguishable in age from the K/T boundary. In the past, the volcanic argument was a useful stimulus to on-going K/T studies. Today it is not only indefensible but counter-productive. There are enough problems and complexities in the impact scenario to work through without throwing in a volcanic red herring.

So, what about the complexities in the Ir spike and the stepwise decreases in biodiversity leading up to the K/T boundary? Well, it's probably premature to conclude the shoulders on the Ir spike are primary features. Further, the detailed structure and extent of these shoulders is uncertain. For example, three different laboratories have measured the Ir above and below the main spike at Gubbio, with significantly different results.[55,56,57] And while it is true that some groups of organisms were in decline approaching the K/T boundary, it is equally true that several groups of marine organisms showed no signs of decline right up to the boundary, where they were decimated.[58] The question is, were these declines truly stepwise, implying a suddenly applied stress, or do they only appear stepwise perhaps because of some gaps in the sedimentary record? If the latter, then the declines could have had a number of long-term causes and have no bearing on the sudden K/T mass extinction. If the stepwise extinction pattern is real, and correlated with the Ir complexities, then the only possible extraterrestrial mechanism so far suggested involves a comet shower injected into the inner solar system from the Oort Cloud*, such that 8-10 separate large impact events could take place over 0.5-1 myr before the shower decays. Most of these would be ocean impacts, so it would not be surprising that most of the craters have not been found.

Obviously there is much work still to be done on these problems. Equally obvious, for either one of the above scenarios, the Chicxulub event alone could have

*The Oort Cloud is a theoretical but apparently necessary construct to explain the steady flux of long period comets into the inner solar system. It is assumed to contain billions of comets in a cloud outside the orbit of Pluto and extending halfway to the nearest star.

been responsible for the mass extinction at the K/T boundary. With an estimated impact energy equal to 100 million megatonnes of TNT equivalent, which is 3-4 orders of magnitude larger than the combined nuclear arsenal of the U.S.A. and the former USSR, the problem is not how to kill the biosphere but how to save part of it. Probable kill mechanisms abound, both regional and global; direct blast wave; global wildfires; darkness and cooling (from atmospheric dust and soot); global acid rain and ozone destruction (from enormous amounts of nitric oxide produced by the interaction of the atmosphere with the incoming impactor and the outgoing ejecta plume). And there is ample evidence of chaotic deposits at the K/T boundary in Mexico, the southern U.S. and several sites in the Western North Atlantic and the Caribbean. These have been interpreted as products of a tsunami,[35,60,61] which makes sense because reconstructions of latest Cretaceous sea level[62,63] place the Chicxulub target under 100-300 meters of water when it was impacted. Added to that, the upper part of the Chicxulub target was Cretaceous limestone. So enormous amounts of both water vapor and CO_2 must have been injected into the stratosphere. H_2O and CO_2 are both greenhouse gases and, after the initial darkness and cooling, there should have followed a period of substantial global warming. All in all, not a pretty picture for survival.

Concerning the lack of a geochemical anomaly at the four other mass extinction boundaries, it should be remembered that there is no Ir spike associated with three microtektite strewnfields. Clearly, impact events need not leave a geochemical signature. Why this is so is not presently understood. We have little knowledge about the bulk chemistry of comets so it is safe to suggest some or all comets do not have the platinum group elemental abundances measured in meteorites and therefore do not leave the characteristic meteorite signature. Of course, it's difficult to take this position and, at the same time, advocate a comet shower for the K/T boundary events, as some have done. Numerical simulations[64] suggest that, if the impact is sufficiently energetic, none of the impactor is retained by the Earth—it is vaporized and ejected at escape velocities. But, even if correct, this mechanism cannot be applied to the relatively lower energy responsible for the microtektite strewnfields. Perhaps there are other possibilities.

Alternatively, perhaps there *is* a signature at these boundaries and we just haven't found it yet. It has been remarked: "The Cretaceous-Tertiary boundary seems to be the only major boundary in the stratigraphic record which does not become diffuse when studied in detail."[65] And ". . . each horizon should be examined at numerous sites because preservation of fallout layers is extremely tenous and subject to erosion and mixing processes, especially those from the older geologic periods."[43] So we need to look carefully, at many sites across these mass extinction boundaries, not only for Ir but for shocked minerals, microtektites et al. before we can conclude there really is *no* evidence of a bolide impact. In this regard, microtektites may well be the best bet. They are absolutely diagnostic of impact, the glass is remarkably stable and long-lived, and even when altered, they have been recognized in the early Archean (3.2 to 3.5 billion years ago) in South Africa and Western Australia.[66]

There is another possibility. Catastrophic mechanisms attract megathinkers*. Thus, we have the suggestion that an asteroid impact not only caused the Triassic-Jurassic mass extinction but created "the Bahama Nexxus . . . or mantle plume/hot spot/triple junction [that] . . . triggered the opening of the North Atlantic."[67] And "Impacts of asteroids and comets may have initiated the breakup of Gondwanaland and other continental land masses."[68] And "If the true kill curve is within the range depicted in Figure 2, no killing mechanism other than bolide impact is necessary to explain the Phanerozoic [extinction] record."[69] These are true megathoughts but the truth may be more mundane. Maybe, just maybe, there are causes other than bolide impacts for Phanerozoic mass extinctions, the K/T event represents the only bolide-related mass extinction and we must be badly overestimating the flux of large bolides to the Earth's surface.

This is a very exciting, fast-moving field that has attracted a large number of able scientists. It should not be long before we have good answers to at least some of the uncertainties expressed herein.

Meanwhile, even if we are presently overestimating the large bolide flux by a factor of five, which would be consistent with the K/T event being unique, that still leaves 300 cratering events in the Phanerozoic that were energetic enough to produce craters > 20 km in diameter on land. The 70% of these impacts into ocean targets must have generated enormous tsunami-like waves. Earthquake-generated tsunamis are limited to maximum wave heights of 30-40 meters yet can be tremendously destructive to coastal lowlands. Impact-generated tsunamis can have wave heights equal to the water depth and would not be confined to coastal lowlands. There can be no doubt that such tsunamis must be responsible for many enigmatic erosional surfaces and chaotic deposits whose locations are not readily ascribable to uplift or sea level changes. This has yet to be fully appreciated by the geologic community.

Finally, I return to the question that perhaps initially stimulated the early studies of this field—whatever did happen to the dinosaurs? There has been a decade of heated debate over this question. There is general agreement that the dinosaurs were declining over millions of years prior to the K/T boundary. And some dinosaur species are said to have crossed the boundary and lived another million years into the Tertiary.[70] All this has been taken as evidence by some that no catastrophic event is necessary to explain the dinosaurs' extinction. The terrestrial paleontological record is far harder to read than the marine record and we may never know where the truth lies therein. But with the discovery of the Chicxulub Crater, the issue has become moot. Suffice it to say that if the dinosaurs were not terminated by the Chicxulub impact event, they certainly should have been.

---

*Those who ignore small details and minor inconsistencies as they sweep forward to the grand fallacy.

REFERENCES CITED

1. Kelly, A.O. and F. Dachille. 1953. *Target: Earth. The Role of Large Meteors in Earth Science.* Box 335, Carlsbad, California, 263 pp.
2. Gilbert, G.K. 1893. The Moon's face: a study of the origin of its features. *Bulletin of the Philosophical Society of Washington 12,* pp. 242-292.
3. De Laubenfels, M.W. 1956. Dinosaur extinction: one more hypothesis. *Journal of Paleontology 30,* pp. 207-218.
4. Urey, H.C. 1973. Cometary collisions and geological periods. *Nature 242,* pp. 32-33.
5. Opik, E.J. 1958. On the catastrophic effects of collisions with celestial bodies. *Irish Astronomical Journal 5,* pp. 34-36.
6. McLaren, D.J. 1970. Time, life and boundaries. Journal of Paleontology 44, pp. 801-815.
7. French, B.M. and N.M. Short, Eds. 1968. *Shock Metamorphism of Natural Materials,* Mono Book Corp., Baltimore, 644 pages.
8. Shoemaker, E.M. and three others. 1979. Earth-crossing asteroids: orbital classes, collision rates with Earth, and origin. in *Asteroids,* T. Gehrels, Ed., University of Arizona Press, Tucson, Az., pp. 253-282.
9. Napier, W.M. and S.V.M. Club. 1979. A theory of terrestrial catastrophism. *Nature 282,* pp. 455-459.
10. Alvarez, W. and three others. 1979. Anomalous iridium levels at the Cretaceous-Tertiary boundary at Gubbio, Italy: Negative results of a test for a supernova origin. *Geological Society of American Abstracts with Programs 11,* p. 378.
11. Alvarez, L.W. and three others. 1980. Extraterrestrial cause for the Cretaceous-Tertiary extinction. *Science 208,* pp. 1095-1108.
12. Grieve, R.A.F. 1984. The impact cratering rate in recent time. *Journal of Geophysical Research 89, Supplement,* pp. B403-B408.
13. Shoemaker, E.M., R.F. Wolfe and C.S. Shoemaker. 1990. Asteroid and comet flux in the neighborhood of the Earth. *Geological Society of America Special Paper 247,* pp. 171-180.
14. Weissman, P.R. 1990. The cometary impactor flux at the Earth. *Geological Society of America Special Paper 247,* pp. 171-180.
15. Glass, B.P. 1990. Tektites and microtektites: key facts and inferences. *Tectonophysics 171,* pp. 393-404.
16. Glass, B.P. and R.A. Barlow. 1979. Mineral inclusions in Muong Nong-type Indochinites: implications concerning parent material and process of formation. *Meteoritics 14,* pp. 55-67.
17. Walter, L.S. 1965. Coesite discovered in tektites. *Science 147,* pp. 1029-1032.
18. Glass, B.P. D.W. Muenow and K.E. Aggrey. 1986. Further evidence for the impact origin of tektites. *Meteoritics 21,* pp. 369-370.
19. Cassidy, W.A., B. Glass and B.C. Heezen. 1969. Physical and chemical properties of Australasian microtektites, *Journal of Geophysical Research 74,* pp. 1008-1025.
20. Glass, B.P. and three others. 1991. Ivory Coast microtektite strewnfield; description and relation to the Jaramillo geomagnetic event. *Earth and Planetary*

Science Letters 107, pp. 182-196.

21. Mankinen, E.A. and three others. 1980. Paleomagnetism of the Clear Lake volcanics and new limits on the age of the Jaramillo normal-polarity event. *U.S. Geological Survey Professional Paper 1141,* pp. 67-82.
22. Izett, G.A. and J.D. Obradovich. 1992. Laser-fusion^{40} Ar-^{39}Ar ages of Australasian tektite cores and flanges. *American Geophysical Union Fall Meeting Program and Abstracts,* p. 328.
23. Baksi, A.J., M.O. McWilliams and E. Farrer. 1991. ^{40}Ar/^{39}AR dating of the Brunhes-Matuyama geomagnetic field reversal; implications of the Milankovitch forcing frequency observed in deep-sea cores. *Geological Society of America Abstracts with Programs 23,* p. A91.
24. Izett, G.A. and J.D. Obradovich. 1991. Dating of the Matuyama-Brunhes boundary based on ^{40}Ar-^{39}Ar ages of the Bishop Tuff and Cerro San Luis Rhyolite. *Geological Society of America Abstracts with Programs 23,* p. A106.
25. Glass, B.P., M.B. Swinciki and P.A. Zwart. 1979. Australasian, Ivory Coast and North American tektite strewnfields: size, mass and correlation with geomagnetic reversals and other earth events. *Proceedings Lunar and Planetary Science Conference 10th,* pp. 2535-2545.
26. Keller, G., S. D'Hondt and T.L. Vallier. 1983. Multiple microtektite horizons in upper Eocene marine sediments: no evidence for mass extinctions. *Science 221,* pp. 150-152.
27. Keller, G. and six others. 1987. Late Eocene impact microspherules: stratigraphy, age and geochemistry, *Meteoritics 22,* pp. 25-60.
28. Glass, B.P. and C.A. Burns. 1987. Late Eocene crystal-bearing spherules: two layers or one? *Meteoritics 22,* pp. 265-270.
29. Obradovitch, J.D., L.W. Snee and G.A. Izett. 1989. Is there more than one glassy layer in the late Eocene? *Geological Society of America Abstracts with Programs 21,* p. A134.
30. Glass, B.P. 1989. North American tektite debris and impact eject from DSDP site 612. *Meteoritics 24,* pp. 209-218.
31. Poag, C.W. and six others. 1982. Deep sea drilling project site 612 bolide event: new evidence for a late Eocene impact-wave deposit and a possible impact site, U.S. east coast. *Geology,* pp. 771-774.
32. Keller, G. 1986. Stepwise mass extinctions and impact events: late Eocene to early Oligocene. *Marine Micropaleontology 10,* pp. 262-293.
33. Montanori, A. 1990. Geochronology of the terminal Eocene impacts: an update. *Geological Society of America Special paper 247,* pp. 607-616.
34. Kring, D.A. and W.V. Boynton. 1991. Altered spherules of impact melt and associated relict glass from the K/T boundary sediments in Haiti. *Geochimica et Cosmochimica Acta 55,* pp. 1737-1742.
35. Smit, J. and seven others. 1992. Tektite-bearing, deep-water clastic unit at the Cretaceous-Tertiary boundary in northeastern Mexico, *Geology 20,* pp. 99-103.
36. Smit, J. and six others. 1992. "Tektites" and microkrystites at the Cretaceous-Tertiary boundary: two strewnfields, one crater? *Proceedings of Lunar and Planetary Science 22,* pp. 87-100.
37. Izett, G.A., G.B. Dalrymple and L.W. Snee. 1991. ^{40}Ar/^{39}Ar age of Cretaceous-Tertiary boundary tektites from Haiti, *Science 252,* pp. 1539-1542.

Tertiary boundary tektites from Haiti, *Science 252,* pp. 1539-1542.

38. Swisher, C.C. and twelve others. 1992. Coeval $^{40}Ar/^{39}Ar$ Ages of 65.0 million years ago from Chicxulub Crater melt rock and Cretaceous-Tertiary boundary tektites. *Science 257,* pp. 954-958.
39. Wang, K. 1992. Glassy microspherules (microtektites) from an upper Devonian limestone, *Science 256,* pp. 1547-1550.
40. Boundy-Sanders, S.Q. 1992. Highly calcic, oddly shaped upper Devonian microtektites from Western Yukon Territory. *American Geophysical Union Fall Meeting Program and Abstracts,* p. 328.
41. Claeys, P., J-G. Casier and S.V. Margolis. 1992. Microtektites and mass extinctions: evidence for a late Devonian asteroid impact. *Science 257,* pp. 1102-1104.
42. Claeys, P., J-G. Casier and S.V. Margolis. 1992. A link between microtektites and late Devonian mass extinctions. *American Geophysical Union Fall Meeting Program and Abstracts,* p. 328.
43. Orth, C.J., M. Attrep and L.R. Quintana. 1990. Iridium abundance patterns across bio-event horizons in the fossil record. *Geological Society of America Special Paper 247,* pp. 45-59.
44. Bohor, B.F., P.J. Modreski and E.E. Foord. 1987. Shocked quartz in the Cretaceous-Tertiary boundary clays; evidence for a global distribution. *Science 236,* pp. 705-709.
45. McHone, J.F. and three others. 1989. Stishovite at the Cretaceous-Tertiary Boundary, Raton, New Mexico. *Science 243,* pp. 1182-1184.
46. Carlisle, D.B. and D.R. Braman. 1991. Nanometre-size diamonds in the Cretaceous-Tertiary boundary clay of Alberta. *Nature 352,* pp. 708-709.
47. Gilmour, I. and five others. 1992. Terrestrial carbon and nitrogen isotopic ratios from Cretaceous-Tertiary boundary and nanodiamonds. *Science 258,* pp. 1624-1626.
48. Wolbach, W.S., R.S. Lewis and E. Anders. 1985. Cretaceous extinctions: evidence for wildfires and search for meteoritic material. *Science 230,* pp. 167-170.
49. Wolbach, W.S. and four others. 1988. A global fire at the Cretaceous-Tertiary boundary. *Nature 334,* pp. 665-669.
50. Kunk, M.J. and three others. 1989. ^{40}Ar-^{39}Ar dating of the Manson Impact Structure: A Cretaceous-Tertiary boundary crater candidate, Science 249, pp. 1565-1568.
51. Donovan, S.K., Ed. 1989. *Mass Extinctions: Processes and evidence.* Columbia University Press, New York, 266 pp.
52. Bice, D.M. and four others. 1992. Shocked quartz at the Triassic-Jurassic boundary in Italy. *Science 255,* pp. 443-446.
53. McCartney, K. A.R. Huffman and M. Tredoux. 1990. A paradigm for endogenous causation of mass extinctions. *Geological Society of America Special Paper 247,* pp. 125-138.
54. Courtillot, V. and four others. 1990. Deccan volcanism at the Cretaceous/Tertiary boundary; data and interferences. *Geological Society of America Special Paper* 247, pp. 401-409.

55. Crocket, J.H., C.B. Officer, F.C. Wezal and G.D. Johnson. 1988. Distribution of noble metals across the Cretaceous/Tertiary boundary at Gubbio, Italy: Iridium variation as a constraint on the duration and nature of Cretaceous/Tertiary boundary events. *Geology 16,* pp. 77-70.
56. Rocchia, R. and seven others. 1990. The Cretaceous-Tertiary boundary at Gubbio revisited: vertical extent of the Ir anomaly. *Earth and Planetary Science Letters 99,* pp. 206-219.
57. Alvarez, W., F. Asaro and A. Montanori. 1990. Iridium profile for 10 million years across the Cretaceous-Tertiary boundary at Gubbio (Italy). *Science 250,* pp. 1700-1702.
58. Alvarez, W. and five others. 1984. Impact theory of mass extinctions and the invertebrate fossil record. *Science 223,* pp. 1135-1140.
59. Kauffman, E.F. 1991. Stepwise extinction and environmental chaos: the persistent pattern of mass extinction intervals, including the K/T boundary. *Geological Society of America Annual Meeting, Abstracts with Programs,* p. A184.
60. Bourgeois, J. and three others. 1988. A tsunami deposit at the Cretaceous-Tertiary boundary in Texas. *Science 241,* pp. 562-570.
61. Hildebrande, A.R. and W.V. Boynton. 1988. Impact wave deposits provide new constraints on the location of the K/T boundary impact. *Global catastrophes in Earth History Conference Abstracts, Lunar and Planetary Contribution 673,* pp. 76-77.
62. Bond, G. 1978. Speculation on real sea-level changes and vertical motions of continents at selected times in the Cretaceous and Tertiary periods. *Geology 6,* pp. 247-250.
63. Pitman, W.C. 1978. Relationship between eustasy and stratigraphic sequences of passive margins. *Geological Society of America Bulletin 89,* pp. 1389-1483.
64. Vickery, A.M. and H.J. Melosh. 1990. Atmospheric erosion and impactor retention in large impacts, with application to mass extinctions. *Geological Society of America Special Paper 247,* pp. 289-300.
65. Smit, J. and J. Hertogen. 1980. An extraterrestrial event at the Cretaceous-Tertiary boundary. *Nature 285,* pp. 198-200.
66. Lowe, D.R. and G.R. Byerly. 1986. Early Archean silicate spherules of probable impact origin, South Africa and Western Australia. *Geology 14,* pp. 83-86.
67. Dietz, R.S. 1986. Triassic-Jurassic extinction event, Newark basalts and impact-generated Bahama nexxus. *Meteoritics 21,* pp. 355-356.
68. Oberbeck, V.R., J.R. Marshall and H. Aggarwal. 1993. Impacts, Tillites and the breakup of Gondwanaland. *Journal of Geology 101,* pp. 1-19.
69. Raup, D.M. 1990. Impact as a general cause of extinction: a feasibility test. *Geological Society of America Special Paper 247,* pp. 27-32.
70. Sloan, R.E. 1988. Biostratigraphic case studies of six major extinctions. *Global Catastrophes in Earth History Conference Abstracts, Lunar and Planetary Institute Contribution 673,* pp. 180-181.

The Oceans: Physical-Chemical Dynamics and Human Impact. Edited by S.K. Majumdar, E.W. Miller, G.S. Forbes, R.F. Schmalz and Assad A. Panah. © 1994, The Pennsylvania Academy of Science.

Chapter Twenty

COMMENTARY ON GLOBAL WARMING: The Biogeochemical Cycle of Carbon and the Oceanic Sink of Anthropogenic Carbon Dioxide

FRED T. MACKENZIE[1] and CHRISTOPHER L. SABINE[2]

[1]Department of Oceanography
School of Ocean and Earth Science and Technology
University of Hawaii
Honolulu, Hawaii 96822
and
[2]Program in Atmospheric and Oceanic Sciences
Princeton University
Princeton, NJ 08544

INTRODUCTION

As in the solid Earth sciences, a new subdiscipline of oceanography has developed in recent years. Environmental science, environmental geology and environmental geochemistry have all arisen as autonomous subdisciplines because of the effect of human activities on natural systems. In like fashion, environmental oceanography deals with human forcings on the global ocean and the effect of human activities on physical and biogeochemical processes in the ocean. The problem of an enhanced greenhouse effect with its potential for global warming is an especially active area of oceanographic research today. The potential of future global warming derives from the fact that human activities are leading to the release of gases to the atmosphere that are radiatively active. These gases, like CO_2, CH_4, N_2O and the synthetic chlorofluorocarbons (CFCs), are presently accumulating in the atmosphere. Their accumulation and potential for absorbing long-wave infrared energy reradiated from the planetary surface can lead to a warming of the Earth's climate. The oceans play a major role in the biogeochemical cycling of the climatologically important gases and are a major sink for CO_2, the most talked about greenhouse gas. Thus, the interest of oceanographers in the problem of global warming.

Scientific considerations of global warming and in particular the role of oceans and the terrestrial biosphere as sinks/sources of anthropogenic CO_2 remind us somewhat of the Indian fable of The Six Blind Men and The Elephant. Six blind men went for a walk and stumbled upon an elephant. The lead man touched the side of the elephant and exclaimed, "We have reached a wall." A second man put his hand on the elephant's trunk and said, "No, it is not a wall, the object is round like a snake." The third man felt the elephant's tusk and argued, "It is sharp, so it's a spear." The fourth touched the leg and decided the object was tall and therefore a tree. A fifth felt the ear and argued the object was wide like a fan. The sixth touched the tail and explained the object was thin and thus was a rope. They had a strong argument about what the elephant was. Soon the Maharajah came along and said to the group that the object was a big animal. He told the men, "To find the truth, you must put all the parts together and look at the whole." The six blind men then went off to debate the animal's identity from a holistic point of view.

In this paper, we discuss three aspects of the global carbon cycle that are relevant to considerations of global climate change: (1) global warming, the biogeochemical cycle of carbon and the oceanic sink of CO_2 (2) CO_2 and $CaCO_3$ interactions in the oceans and (3) the future of atmospheric CO_2 change. These topics all require a holistic point of view of the planetary system, and application of the new Earth System Science. It is now necessary for oceanographers, terrestrial biologists, geochemists, climatologists, etc. to work together to solve problems of global environmental change.

THE PROBLEM OF GLOBAL WARMING

Global warming is a topic of current scientific controversy and worldwide concern. This concern was demonstrated by the United Nations Conference on Environment and Development held in Rio de Janeiro in the summer of 1992. Representatives from 178 nations and a whole host of non-governmental organizations attended this meeting. The topic of global warming has even reached the ultimate of expression in the form of the comic pages and in political cartoons. It is a problem that has global dimensions and may be the major coupled environmental and human infrastructure problem of the next century.

The Earth's surface is in a state of dynamic change. Change is more the rule than constancy. This natural change makes it very difficult to isolate the causes affecting the temperature record of the planet during the past several centuries of potential human impact on this record. This is the reason for the debate between Jim Hansen, a climatologist at NASA's Goddard Space Institute, his scientific and political supporters, like Vice-president Al Gore, and other scientists and politicians, like John Sununu, former White House chief of staff. Hansen testified before Congress in 1988 that the buildup of greenhouse gases in the atmosphere over the last century is responsible for the half of a degree centigrade observed increase in global mean temperature over this time period. His viewpoint was strongly disagreed with, for example, by individuals of the George C. Marshall Institute, who argued that without greenhouse gases to warm things up, the world would probably get cooler in the next century. This debate continues today.

The idea of global warming is not new. In 1896, Svante Arrhenius, a distinguished Swedish chemist, recognized how the Earth's atmosphere could maintain a relatively equable climate when the major gases of nitrogen and oxygen do not absorb any of the infrared radiation escaping the surface of the planet. He discovered that small amounts of carbon dioxide in the atmosphere could absorb large amounts of heat. Without the natural greenhouse gases of water vapor, carbon dioxide, methane and nitrous oxide in the atmosphere, the planet would be 35 degrees cooler than its present mean temperature of about 15°C, or -20°C. Arrhenius also reasoned that the burning of the fossil fuels, coal, oil and natural gas, could eventually release sufficient carbon dioxide to the atmosphere to warm the planet. Interestingly, and somewhat prophetically, he calculated the amount of warming expected from a doubling of the carbon dioxide content of the atmosphere, and arrived at a value of + 5°C. This is within the range of results obtained from global circulation models (GCMs) today.

What does the record of the past tell us? The trace gas composition of the atmosphere and temperature have changed substantially over geologic time. Atmospheric carbon dioxide concentrations have been ten or more times greater than present during the last 600 million years of Earth history. Global temperatures have been ten or more degrees warmer at times. In more recent glacial-interglacial time, mean global temperature has varied by several degrees centigrade, and atmospheric carbon dioxide and methane contents by a factor of about 1.5 and 2, respectively. Since the last glacial maximum of 18000 years ago, atmospheric carbon dioxide has climbed from about 180 to 280 ppmv, and methane from about 350 to 750 ppbv in the late 18th century; sea level has risen 150 meters and mean global temperature has risen about 5°C.

How have the fossil fuel burning, deforestation, agricultural, industrial and transportation activities of humankind modified the chemical composition of the atmosphere? Because of these activities, atmospheric carbon dioxide has risen from 280 to nearly 360 ppmv from the late 18th century to present, methane has increased from 750 to about 1700 ppbv, nitrous oxide from 285 to 310 ppbv and concentrations of synthetic chlorofluorocarbons have reached levels of several hundreds of parts per trillion over this same time period. These atmospheric trace constituents are all greenhouse gases and by absorbing reradiated long wave, infrared radiation from Earth's surface can lead to global warming of the planet. An enhanced greenhouse is the potential result.

It is still a subject of controversy among scientists whether or not the erratic observed increase in mean global temperature of 0.5°C since 1870 is due to increased greenhouse gas concentrations in the atmosphere. The potential direct climate forcing owing to accumulated greenhouse gases over this period of time is small, only about 2 watts per square meter. Thus, the temperature signal owing to an enhanced greenhouse is difficult to isolate at this time from the observed record of temperature change. The future accumulation of greenhouse trace gases in the atmosphere because of human activities depends on many factors, including population growth, per capita energy consumption, energy sources and the behavior of the coupled ecosphere-climate system of the Earth. For example, one scenario predicts a doubling of the carbon dioxide content of the atmosphere by the middle of the 21st century over the 18th century value of 280 ppmv.

What are some consequences of this human-induced atmospheric compositional change? Increased greenhouse gases in the atmosphere have the potential to modify the world's climate. A super-interglaciation may occur. Because enhanced concentrations of these atmospheric gases will absorb more reradiated Earth energy, the distribution of the net radiant energy flux and hence global temperature and its distribution may change. One "business as usual" scenario obtained from a general circulation model predicts for the end of the next century a 2 to 5°C increase in mean global temperature, with temperature changes most dramatic in the high polar latitudes of both hemispheres. The direct climate forcing for such a temperature change is only 4 watts per square meter. However, this change in climate forcing and resultant change in global temperature and its distribution can lead to reorganization of the atmosphere-ocean circulation system. One potential outcome of such a reorganization could be increased hurricane frequency and severity and storm surge. With a potential sea level rise of 30-100 centimeters by the end of the 21st century because of partial melting of continental ice sheets owing to global warming and other factors, such increases could be potentially devastating for low-lying countries like Bangladesh and the atoll nations of the Pacific. Because of warming, the hydrologic cycle may change, leading to increased precipitation in some regions and decreased precipitation and soil moisture in other areas. Drying out of large continental regions, like the midwest "breadbasket" of the U.S., could take place because of global warming. Because of warming and change in soil moisture in soils, the distribution of insect pests and soil microbiota could be modified, leading to changes in the natural rates of exchange of trace gases between the land and the atmosphere. Global warming could also modify natural gas exchange between the ocean and atmosphere. Obviously, these physical and biological changes could give rise to a variety of social, economic and political problems. The poorer nations of the world which are already on the threshold of starvation or political disruption and are least capable of dealing with global warming and its consequences will be vulnerable to economic and political upset.

It should be emphasized that the whole topic of global warming is still a matter of scientific controversy. Some scientists argue that the positive temperature change of the past 100 years is a sign that human civilization is experiencing an enhanced greenhouse brought about by releases of greenhouse trace gases to the atmosphere from human activities. Other scientists contend the temperature record can be explained by other causes. The consequences of a global warming are even more debatable. The jury is still out on global warming.

TEMPERATURE DURING THE PAST 100 YEARS

Since the late 1700s, atmospheric CO_2 levels have increased by nearly 30% from 280 to 360 ppmv. Equivalent CO_2 levels have risen 40% from 310 ppmv one hundred years ago to 430 ppmv in 1992. With this increase in equivalent CO_2, the global circulation models predict several changes in climate over this 100 year period. These include globally a warming of at least 1°C with a reduction in the range of diurnal temperature, an increase in precipitation and cloud cover, a slight decrease in stratospheric temperatures, a small decrease in the area of coverage of snow, glacial ice and sea ice, and a slight rise in sea level. To some degree, all these changes are present in the observational record of climate and sea level.

The model predictions, however, for the past 100 years do not agree in detail with the observational record. In particular, the amount of warming of the planet globally is less than that predicted for the equivalent CO_2 increase of 120 ppmv. Also, much of the warming took place prior to the 1940s, before the substantial increase in atmospheric trace gas concentrations. Between 1940 and the mid- to late- 1970s, the observational record actually shows a very small decrease in temperature. This trend is followed by increasing global temperatures with eight of the warmest years on record in the 1980s and 1990s. Furthermore, in contrast to most model calculations, the Southern Hemisphere has warmed more than the Northern Hemisphere, particularly since 1950. These observations suggest that the temperature record of the past 100 years is due in part to climatic forcings other than that of increasing trace gas concentrations. Several forcings have been postulated, including variations in the solar constant, volcanic aerosol emissions and anthropogenic (and natural?) emissions of gaseous sulfur compounds.

One interpretation of the temperature history of the earth since the late 1800s is that the early part of the record still represents in part a recovery from the Little Ice Age of the 15th through mid-19th century. Temperatures during this cold period at times were 0.5°C cooler than a global average of 15°C. It should be recalled that recovery from the last glacial stage of the Wisconsin has not been continuous and unidirectional. The recovery has involved a series of temperature fluctuations, including the cool periods of the Younger Dryas and Little Ice Age and the warm intervals of the Holocene Climatic Optimum and the Medieval Warm Period. Of interest here is the observation that during the Little Ice Age when temperatures fell, atmospheric CO_2 levels as recorded in air bubbles in ice cores also appear to have fallen. Furthermore, it appears that the fall in temperature may have preceded somewhat the fall in atmospheric CO_2, although the reality of this time lag is an open question. This observation, however, is in accord with the record from the Vostok ice core in which during cooling phases, changes in atmospheric CO_2 and CH_4 concentrations lag temperature changes. The point is that some of the erratic rise in global mean temperature between the late 1800s and 1940 of about 0.34°C may be a result of recovery of the climate system from the Little Ice Age. If this is true, then we would not expect a strong relationship between the buildup of anthropogenic CO_2 and the temperature record of at least the early part of this century.

Now we come to the global mean temperature record between 1940 and the mid-1970s. For this period of time, a slight erratic cooling of about 0.1°C is observed in the record. The cause of this cooling is debatable. Several forcing mechanisms have been suggested including decreased solar activity, increased stratospheric volcanic dust and sulfate aerosol loadings and most recently increased tropospheric sulfate aerosol, particularly in the Northern Hemisphere, derived from anthropogenic SO_2 emissions. The latter mechanism is most appealing because of the increased global rate of emission of anthropogenic SO_2 that has occurred in the post-W.W.II years, primarily in the Northern Hemisphere. This SO_2 forms sulfate aerosol in the troposphere which can act to cool the earth by stimulating the formation of brigher clouds that reflect solar radiation back to space and by backscattering of incoming solar radiation because of the presence of sulfate particles in cloud-free air. Whatever the actual mechanism, it is obvious that the greenhouse forcing on temperature was modulated for the period 1940 to mid-1970.

Since mid-1970, global mean temperature has risen erratically at a linear rate of 0.02°C/year or 0.32°C between 1974 and 1990. For comparison, the linear rate between 1910 and 1939 was 0.014°C/year or 0.43°C for this period of time. Thus, the rate of rise of temperature between mid-1970 and 1990 is about 1.4 times faster than that of the rise prior to the mid-century slight cooling beginning in 1940.

THE ROLE OF THE OCEAN IN GLOBAL WARMING

Most of the carbon released from fossil fuel burning and land use activities will eventually end up in the oceans. Its fate there will be determined by circulation and biogeochemical processes operating in that reservoir. Fossil fuel combustion and land use practices released 7.0 +/– 1.2 Gt Cy^{-1} (gigatons of carbon per year) to the atmosphere in the 1980s. The amount of carbon retained in the atmosphere, 3.2 Gt Gy^{-1}, is a direct measurement obtained from Moana Loa, Hawaii and other observational sites. Its uncertainty is small, only +/– 0.1 Gt y^{-1}. The ocean uptake of 2.0 +/– 0.8 Gt Cy^{-1} is derived from ocean circulation models that have been calibrated by radiocarbon measurements. The budget does not balance. The sum of the ocean and atmosphere sinks is smaller by 1.8 +/– 1.4 Gt Cy^{-1} than the sources. This imbalance is referred to as the "missing sink".

The problem of evaluation of the contemporary carbon cycle and its imbalance has been with us for at least 15 years. Broecker, et al. in 1979 recognized the imbalance in the global carbon budget, the "missing" carbon, the difference between the amount of carbon released to the atmosphere by the burning of fossil fuels and the amount accumulating in the atmosphere and invading the ocean. It cannot be overemphasized that the former sink is a measurable quantity in the carbon budget, whereas the strength of the latter sink involves models of the ocean-atmosphere system. It should also be recognized that during most of the 1970s, terrestrial ecosystems were thought to be a global sink for fossil fuel carbon. It is likely that this perception by the majority of the scientific community was an outgrowth of the definition and use of the biotic growth factor, β. This term expresses the effect of carbon dioxide fertilization on carbon storage in long-lived terrestrial plants. It is the factor that relates the net annual carbon uptake of green plants (NPP) to elevated atmospheric carbon dioxide concentrations. However, the findings in the late 1970s and 1980s that important amounts of carbon are released to the atmosphere from deforestation and other changes in land use threw a wrench in the argument that terrestrial ecosystems are net sinks of fossil fuel carbon dioxide, and also increased the size of the "missing" carbon flux. Now, we are faced with balancing five flux terms in the global carbon cycle: fossil fuel emissions, deforestation and land use emissions, accumulation in the atmosphere, uptake in the ocean and uptake in terrestrial ecosystems, including living plants, litter and soil organic carbon.

The jury is still out on the "missing" carbon problem. However, data on the difference of carbon dioxide partial pressures across the air-sea interface and the observed latitudinal gradients of atmospheric carbon dioxide concentrations and carbon isotopic composition require a significant, Northern Hemisphere, mid-latitude sink for anthropogenic carbon dioxide. This requirement has been satisfied by some members of the scientific community by asserting that this sink is terrestrial. Indeed, if one accepts the lower estimates of the deforestation and land use flux and the higher estimates of the strength of the ocean sink, it can be concluded that for the 1980s, the global carbon cycle could be balanced. However, the necessary accumulation of large amounts of carbon in northern temperate and boreal forests has not been observed. Thus, if there is a mid-latitude terrestrial sink for anthropogenic carbon dioxide, it is likely to be small in living plants, and if accumulating in soils, will be difficult to measure, because of the large standing stock of carbon in soils. It seems most likely, then, that the carbon cycle balance will be most efficiently closed by global atmospheric and oceanic observational programs already underway.

Despite the present uncertainties involving carbon fluxes between the three reservoirs of atmosphere, oceans and terrestrial biosphere, the lack of strong biotic feedbacks in the ocean portend that for the future warming of the next several decades, the ocean's capacity for absorbing carbon dioxide will lessen. Thus, it is imperative that the role of terrestrial ecosystems in the carbon cycle now and during the warming transient be resolved. The potential for release of substantial amounts of carbon during warming is high because of the temperature effect on total ecosystem respiration and the continuing deforestation and land use activities of human society. The feedback is such that warming begets warming!

BIOLOGICAL FEEDBACKS IN THE OCEAN

The oceans are currently an important sink for anthropogenic carbon dioxide, on the order of 2 +/− 0.8 Gt C y^{-1}. On the basis of a great deal of experience and theoretical arguments for the physical and chemical behavior of the oceans, the warming of the planet is likely to decrease the capacity of the ocean to store carbon, as the carbonate ion of oceanic water is titrated by invading anthropogenic carbon dioxide, and as the surface waters of the oceans warm and their capacity for holding dissolved CO_2 is further reduced. Both processes are positive feedbacks.

Negative feedbacks on the global warming of the next several decades appear to be few. A potentially significant negative feedback is the "CLAW" hypothesis that suggests a complex connection between phytoplankton community composition and production, zooplankton grazing and bacterial decomposition, and the flux of dimethylsulfide (DMS) from the sea surface. Its potential as a feedback is uncertain and controversial. Resolution of the uncertainty will come about as the scientific community gains a fuller understanding of dimethylsulfide production in the ocean and its atmospheric interactions.

The oceanic production of trace gases other than DMS (OCS, N_2O, CO, NH_3) which are involved in climate forcing may also be proportional to net ecosystem production. In a warmer world, it is likely, as with DMS, that the production of these gases will be enhanced, creating the potential for feedbacks on global warming. However, this conclusion is tenuous, and the magnitude of the feedbacks on the time sale of the warming is small.

Changes in net ecosystem metabolism in the ocean can constitute a potential feedback on atmospheric carbon dioxide and hence global warming. However, on the time scale of decades, changes in oceanic metabolism are unlikely to be primarily related to additions of inorganic nutrients and organic carbon from the land. Changes in rates of nutrient upwelling cannot significantly affect the net metabolism of the ocean because upwelling waters deliver dissolved carbon, nitrogen and phosphorus in the Redfield ratio of production. The direction of change in the trophic status of the ocean owing to global warming is difficult to assess. Indeed, it is conceivable that with warming, competing processes may lead to no or little change in the net metabolism of the ocean. On the one hand, excess inorganic nutrient loading, particularly of coastal margins, is likely to result in enhanced eutrophication and burial of organic carbon—a negative feedback on atmospheric carbon dioxide accumulation and hence global warming. On the other hand, respiration of enhanced organic carbon loading of the marine ecosystem could lead to an increased flux of carbon dioxide to the atmosphere—a positive feedback. Whatever the case, changes in ocean metabolism on the time scale of decades of the warming will not constitute a significant feedback.

There are several other potential biotic feedbacks on global warming involving the ocean. Increases in UV-B flux owing to stratopheric ozone depletion may affect net ocean metabolism, particularly in the Antarctic. The composition and distribution of biotic species and iron, molybdenum and other bioessential trace metal fluxes to the ocean may change and alter ocean metabolism. Decomposition of the large reservoir of dissolved organic carbon in the ocean may occur. The biological processes associated with these possibilities can change the partial pressure of carbon dioxide in surface waters and hence the strength of the ocean sink. However, there is little experiential or theoretical information to substantiate the argument that these potential feedbacks are quantitatively significant on the decadal time scale. Thus, because of physical and chemical processes, the capacity of the oceans to store carbon during warming and accumulation of carbon dioxide in the atmosphere will decrease. In the case of ocean, on the time scale of the warming of the next several decades, the warming will feed the warming!

GEOCHEMICAL MODELS

In a series of papers (Smith and Mackenzie, 1987; Wollast and Mackenzie, 1989; Sabine and Mackenzie, 1991; Morse and Mackenzie, 1990), we developed the reservoir relationships and process arguments for the long-term (geological), biogeochemical cycle of carbon and the human-induced peturbations to that cycle shown in Figure 1. The details of Figure 1 are thoroughly discussed in the above references; here, we emphasize a few points germane to ocean-atmosphere CO_2 transfer.

Carbon is deposited in marine sediments as reduced carbon (C°) in organic matter and as oxidized carbon (C^{4+}) in calcium- and magnesium-bearing carbonates. Prior to human activities, particularly fossil fuel burning and land use, the ocean was a net heterotrophic system; that is, the ocean consumed more organic carbon than it produced. The excess heterotrophy probably occurred mainly in the coastal

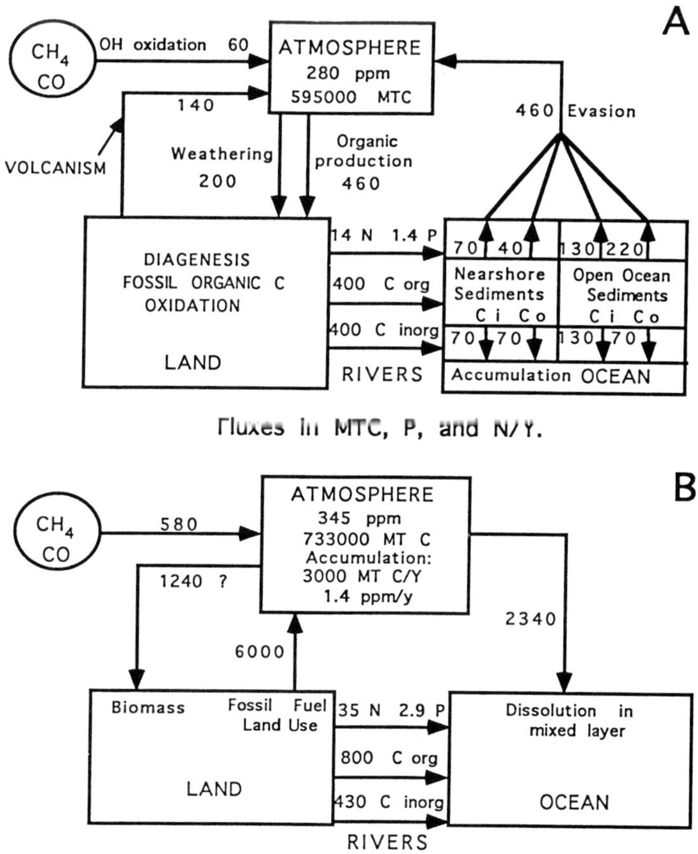

FIGURE 1. A: Long-term geological carbon cycle. B: Anthropogenic perturbation of the carbon cycle. Fluxes in million tons of carbon, phosphorus, and nitrogen per year. Modified from Wollast and Mackenzie (1989).

margin. This net ocean-atmosphere flux, in part, fueled organic production on the land. Any enhancement of organic carbon fluxes to the seafloor over the past century could represent a sink of anthropogenic CO_2 (Sabine and Mackenzie, 1991).

Calcium carbonate accumulation on the time scale of human generations does not represent a sink of anthropogenic CO_2 (as recently suggested by some investigators, e.g., Kinsey and Hopley, in press) because in the reaction

$$Ca^{2+} + 2HCO_3^- = CaCO_3 + CO_2 + H_2O, \tag{1}$$

CO_2 is released to the ocean-atmosphere system. The two carbon atoms in bicarbonate are partitioned during deposition of $CaCO_3$ into one carbon in the mineral and one carbon released as CO_2. This CO_2 enters the ocean-atmosphere system but eventually is used in the weathering of $CaCO_3$ on land, resulting in production of two HCO_3^-.

Because of the pH of the ocean and its carbonate buffer system, some of the CO_2 released during precipitation of $CaCO_3$ will not leave the oceans quickly. Smith (1985) estimates that about 40% of the CO_2 released in reaction (1), or based on today's fluxes, about 6.6×10^{12} moles CY^{-1} (79.2 $MTCY^{-1}$), will be involved in organic metabolism and consequent production of organic matter. This carbon eventually must be returned to the atmosphere or in $61{,}100 \times 10^{12}$ moles C (atmospheric mass)/6.6×10^{12} moles CY^{-1} = 9250 years, all the CO_2 in the atmosphere would be consumed. This residence time calculation provides a feeling for the time scale of the feedback mechanisms involving return of this CO_2 to the atmosphere. Oxidation of this organic carbon and return of CO_2 by vertical mixing processes to the ocean-atmosphere system occur as the organic matter sinks through the oceanic water column. This process accounts in part for the increase of dissolved inorganic carbon with depth in the ocean. Also, any organic carbon that escapes the ocean to sediments is metabolized during early diagenetic processes involving aerobic and anaerobic bacterial reactions or undergoes oxidative decay upon uplift of the sediments and exposure to weathering processes (Garrels, Lerman, and Mackenzie, 1976). Both early diagenetic and weathering processes involving organic matter return CO_2 to the atmosphere.

We emphasize that on a longer time scale, both organic and inorganic carbon accumulation can be a sink of atmospheric CO_2. This is the case, because in the geochemical cycle of carbon, some of the dissolved inorganic carbon in rivers comes from the weathering of silicate minerals according to the schematic reaction

$$CaSiO_3 + 2CO_2 + 3H_2O = Ca^{2+} + 2HCO_3^- + H_4SiO_4^0 \tag{2}$$

This carbon is precipitated in the ocean as calcium carbonate (eq. 1), releasing one carbon back to the atmosphere and sequestering one carbon in $CaCO_3$. Today, about 12% of the Ca^{2+} in rivers is derived from weathering of silicate minerals (Mackenzie, 1992); this represents a silicate-mineral weathering flux of about 5.9×10^{12} moles $CO_2\ Y^{-1}$ (67.2 $MTCY^{-1}$). Thus, 2.95×10^{12} moles of carbon as $CaCO_3$ are sequestered annually on the seafloor from the weathering of Ca-silicate minerals. If there were no return mechanism for this CO_2, the atmosphere would be depleted of CO_2 in $61{,}100 \times 10^{12}/2.95 \times 10^{12}$ = 20,700 years. The return mechanism is related to the volcanic and metamorphic CO_2 flux (e.g. Berner, et al., 1983).

Human interference in the biogeochemical cycle of carbon has reversed the role of the ocean as a CO_2 source; it is now a net sink of atmospheric CO_2. The strength of this sink is debatable (e.g. Broecker, et al., 1979; Sarmiento and Siegenthaler, in press). Table 1 summarizes the anthropogenic CO_2 balance. Several of the flux estimates given in Table 1 are highly controversial. The deforestation (land use) flux has a range of at least 600-2600 $MTCY^{-1}$ and the net ocean flux is in the range 1000 - 2400 $MTCY^{-1}$. Net carbon fixation in the terrestrial biosphere is very uncertain. During the last decade, estimates of this flux have ranged from a CO_2 source (c.f. Houghton, et al. 1983; Houghton, 1990) to a net terrestrial uptake in the range 200 $\pm$ 1700 $MTCY^{-1}$. This is a good example of the parable of The Blind Men and The Elephant. On the one hand, terrestrial ecologists in general see the damage that deforestation and other land use activities have done to forests, particularly in the sub-tropical and tropical rainforests. These activites release CO_2 to the atmosphere. They see no good evidence from boreal and temperate forests and soils that regrowth or CO_2 stimulation are increasing carbon storage in these ecosystems. They favor the ocean as the net sink for anthropogenic CO_2. On the other hand, many oceanographers see difficulties in the ocean as the only large reservoir for the uptake of anthropogenic CO_2 and tend to favor a net sink in the terrestrial biosphere. Whatever the case, one must not forget to take into account the minor oceanic CO_2 sinks, other than the synoptic air-to-sea gas invasion, of Table 1. These are mainly related to the riverine fluxes of inorganic and organic carbon to the oceans and to excess nutrient input to the oceans via the rivers and atmosphere. These sinks could account for about 600 $MTCY^{-1}$ (Wollast and Mackenzie, 1989; Sabine and Mackenzie, 1991). Inclusion of these sinks in the anthropogenic CO_2 budget, and recalling that the steady state carbon system prior to human interference required a sea-air flux of 460 $MTCY^{-1}$, help to bring the budget into line. However,

TABLE 1

Estimated sources and sinks for anthropogenic CO_2.

CARBON DIOXIDE SOURCES	MTC Y^{-2}
Fossil fuel burning†	5890
Deforestation	1210
Total	**7100**
CARBON DIOXIDE SINKS	MTC Y^{-1}
Atmosphere	3000
Invasion of CO_2 gas into ocean	1000-2340
Increased stream derived POC and DOC to the oceans	400
Nutrient fertilization in the oceans and increased organic carbon burial	280
Increased dissolution of $CaCO_3$ in the oceans	50
Increased DIC flux to the oceans from increased weathering	30
Increased carbon fixation in terrestrial biosphere	?
Total	**4760-6100**

†from Marland et al., 1989; remaining values from Sabine and Mackenzie, 1991

balancing the carbon cycle may still require a net terrestrial sink, if we accept recent land use carbon emission values, even if the oceans have a sink strength on the order of 2000 $MTCY^{-1}$. In line with the fable of The Blind Men and The Elephant, the terrestrial and ocean scientists need to get together and look at the problem from a holistic standpoint.

PENETRATION OF ANTHROPOGENIC CO_2 AND REACTION WITH $CaCO_3$

It is unlikely that anthropogenic CO_2 has penetrated sufficiently deep in the Atlantic and Indian Oceans to dissolve significant quantitites of $CaCO_3$ minerals. However, it is probable that even in these oceans, small quantities of the very soluble magnesian calcites ($Ca_xMg_{1-x}CO_3$) produced by benthic marine organisms may be dissolving because of anthropogenic CO_2 additions. It has been demonstrated that bank-derived, suspended benthic carbonate particles or these particles deposited on periplatform slopes in the Bahamas region and in the nearshore areas of the Hawaiian Archipelago may be dissolving at depths shallower than the aragonite lysoclines in these regions. This dissolution may involve magnesian calcites and anthropogenic CO_2.

In the North Pacific, anthropogenic CO_2 has reached a depth of 1500 meters (Chen et al., 1987). We demonstrate here that at the Hawaii Ocean Time-Series Station ALOHA in the North Pacific gyre, located 100 km north of Oahu, Hawaii, anthropogenic CO_2 has reached a depth where it probably is dissolving carbonate minerals, including pelagic aragonite.

Figure 2 shows the carbonate ion concentration as a function of depth at Station ALOHA. Lines for various aragonite and calcite saturation depths are also shown on the figure. The depth at which a particular mineral saturation line crosses the observed CO_3^{2-} concentration in seawater represents the depth below which the waters are undersaturated with respect to that mineral. The depth of cross-over of the 50% saturation lines with the CO_3^{2-} concentration curve indicates that the waters below that depth are less than 50% saturated with respect to the mineral of interest. The CO_3^{2-} concentration crosses the equilibrium value for aragonite at about 500 m. At 700 m and greater depths, the waters are at 60%, or less, saturation with respect to aragonite. Sabine (1992) has shown from this observation and other considerations that dissolution of aragonite suspended in the water column or sedimented on the slopes of the Hawaiian Archipelago can begin at depths of 500-700 m in the central Pacific Ocean.

Following the method of Chen (1982), Sabine calculated the depth of maximum penetration of anthropogenic CO_2 at station ALOHA. The calculated depth of penetration is 700 m. Thus, it appears that anthropogenic CO_2 molecules may have penetrated deep enough in parts of the central Pacific to dissolve pelagic and benthic aragonite particles suspended in the water column or sedimented on bank slopes at depths below 500 m. Benthically-produced magnesian calcite phases containing more than 8-12 mole percent Mg may be dissolved at even shallower depths, because these phases have solubilities greater than that of aragonite.

The dissolution of $CaCO_3$ and the production of alkalinity in the ocean is the mineral sink for anthropogenic CO_2, not the precipitation of carbonate phases (eq. 1). This carbon sink for anthropogenic CO_2 can be permanent, because unless the carbonate production of the ocean is altered, the carbon will remain in the ocean as alkalinity. In the Pacific alone, there are 2.6 x 10^{10} MT of carbon in aragonite available for dissolution between the depth range of 200 to 1000 m.

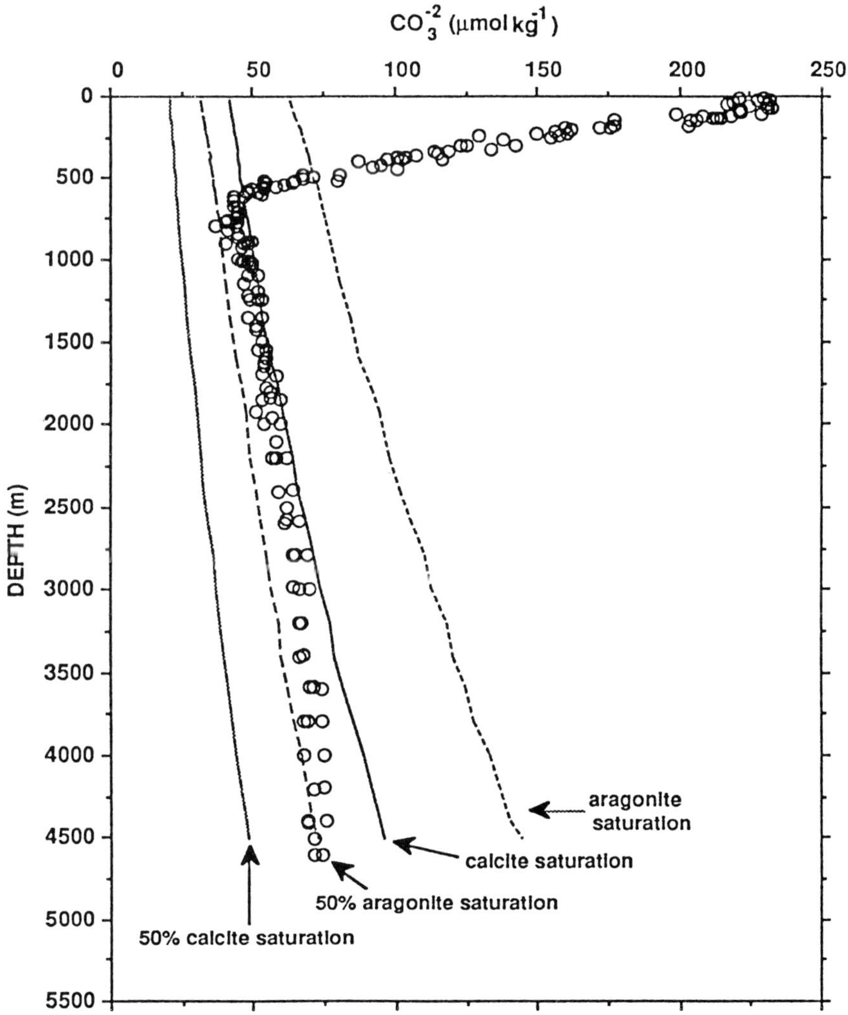

FIGURE 2. Carbonate ion concentration as a function of depth for HOT-3 through HOT-15 at Station ALOHA. Lines indicate carbonate ion concentrations at 50 and 100% saturation with respect to aragonite and calcite.

FUTURE ATMOSPHERIC CO_2 CHANGE

Introduction to Model

Any future scenario of atmospheric CO_2 change necessitates a reasonably complete understanding of the behavior of the complex biogeochemical cycle of carbon. Most global models employed today to develop scenarios of future atmospheric CO_2 concentrations resulting from fossil fuel burning and land use activities consist of four major reservoirs: atmosphere, non-living terrestrial biosphere, living terrestrial biosphere and ocean. The oceanic sink usually involves only the process of dissolution of CO_2 gas in the upper ocean. Enhanced biological production and sequestration of organic carbon in the ocean are generally not considered a quantitatively important sink of anthropogenic CO_2, because enhanced production would require additional nutrients for the ocean's photic zone.

The Revelle-Munk (RM) (1977) model is a four-box model involving terrestrial organic detritus, terrestrial photosynthesizing biomass, ocean, and atmosphere. It attempts to account for the observation that only 56% of the fossil fuel CO_2 production has remained in the atmosphere. Revelle and Munk hypothesize that the additional CO_2 is sequestered by the oceans and terrestrial biomass. The RM model is based on the assumptions that (1) the carbon cycle was in steady state prior to 1860, (2) the increase in atmospheric CO_2 observed over the last century has caused an increase in photosynthetic production such that the net production of the terrestrial biomass is autotrophic and (3) the invasion of CO_2 into the ocean is limited by the buffer mechanism of seawater (the "Revelle factor") and the slow exchange

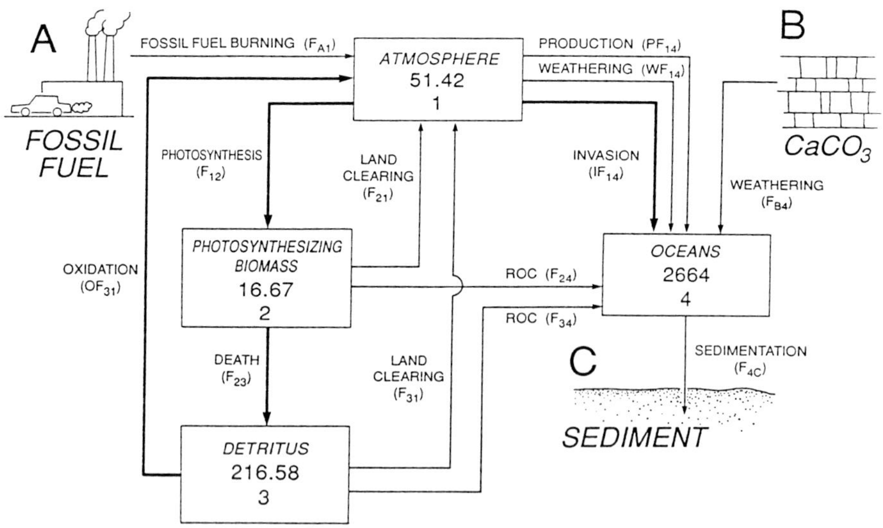

FIGURE 3. Schematic diagram of the modified Revelle-Munk (1977) model of atmospheric CO_2. Reservoir masses are in units of 10^{15} moles C.

of the surface and deep waters of the oceans. By considering only the oceanic sink of "invasion", which can also return CO_2 to the atmosphere when atmospheric concentrations drop, the conclusion from the model was that the terrestrial biosphere must be the ultimate sink for *all* of the anthropogenic CO_2.

The purpose of this modeling exercise is to demonstrate the potential importance of minor ocean sinks (Mackenzie, 1981; Sabine and Mackenzie, 1991) in modulating future atmospheric CO_2 use stemming from fossil fuel and land use CO_2 emissions to the atmosphere. A portion of the CO_2 taken up by these additional sinks remains in the oceans, thus making both the terrestrial biomass and the oceans ultimate sinks for some of the anthropogenic CO_2. The modeling is done by modifying the original Revelle and Munk (RM) model of the modern CO_2 cycle, using the STELLA™ computer program. Details of the original model are given in Revelle and Munk (1977) and of the modified model in Mackenzie et al. (1993). A schematic diagram of the modified RM model is shown in Figure 3.

Results and Discussion

The perturbation to the modified RM model starts in 1860 when the carbon system is assumed to be in quasi-steady state. Obviously, this condition may not have existed for the pre-1860 state of the system. The system is then perturbed by

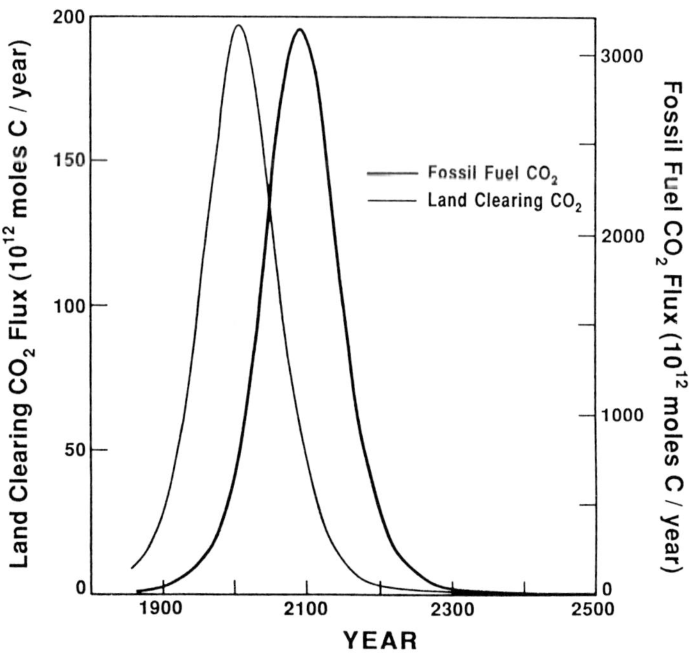

FIGURE 4. Anthropogenic CO_2 fluxes to the atmosphere used in the modified Revelle-Munk model.

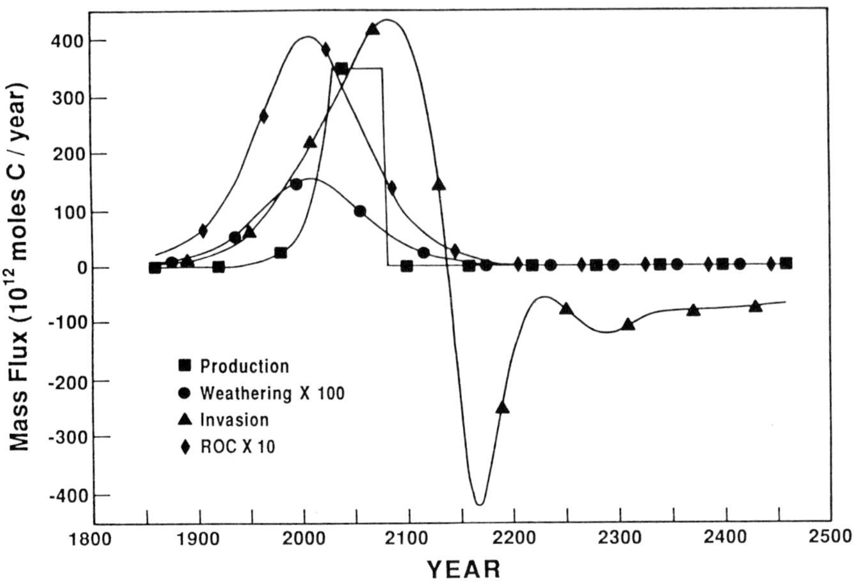

FIGURE 5. Carbon fluxes into the oceans calculated from the modified Revelle-Munk model.

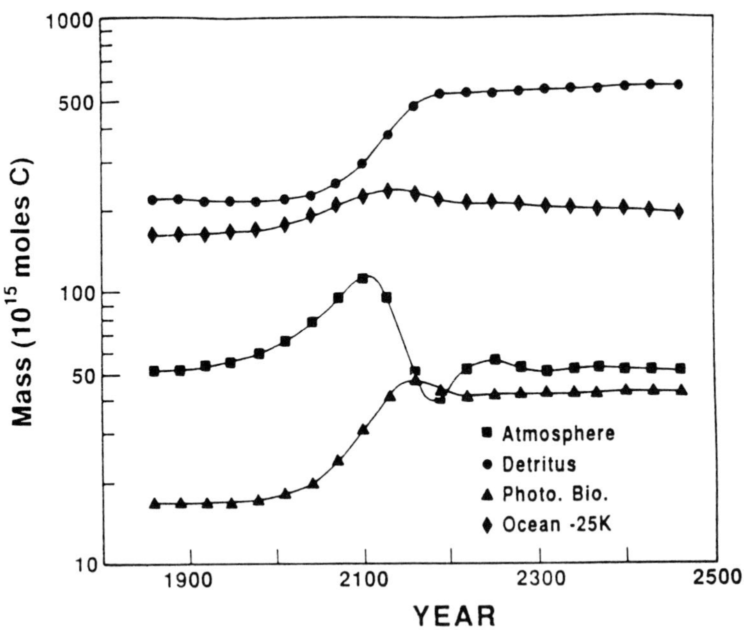

FIGURE 6. Partitioning of carbon among the reservoirs of atmosphere, detritus, photosynthesizing biomass, and oceans based on calculations using the modified Revelle-Munk model.

introducing the CO_2 fluxes owing to fossil fuel burning and land clearing into the system. The temporal changes in the two anthropogenic fluxes from 1860-2460 A.D. are shown in Figure 4.

Figures 5 and 6 show the results of the calculations using the modified RM model. Comparison of the original RM model with that developed here leads to several conclusions. First, the additional organic and inorganic carbon fluxes do not greatly affect the future course of atmospheric CO_2, as calculated in the RM model. The mass of CO_2 in the atmosphere at its maximum is not significantly different between the two models, but in the modified RM model the time of attainment of this maximum is delayed by ten years. This delay is a direct result of sequestration of carbon in the minor oceanic sinks. Second, in the modified RM model, the terrestrial biomass is not the *ultimate* reservoir for all of the anthropogenic CO_2, as was the case in the original RM model. 10.4 x 10^{15} moles of carbon accumulate in the sediment reservoir, or about 2.5% of the estimated 416.7 x 10^{15} moles of carbon that could eventually enter the atmosphere from fossil fuel burning and land use sources.

The additional fluxes in the modified RM model also result in more carbon storage in the ocean. Most of the carbon introduced into the ocean comes from the living and non-living biomass reservoirs via the rivers. Thus, both the photosynthesizing biomass and detritus reservoirs are smaller in the modified RM model than in the original model at any given time. The largest difference in mass between the two models for the ocean reservoir is 16.9 x 10^{15} moles. The modified model also yields, on average, a photosynthesizing biomass reservoir which is 6.4% smaller than that in the original model and a detritus reservoir which is 3.5% smaller. The differences in the reservoir sizes do not make much of a difference in the long run, because it still takes about 2600 years for the system to return to a steady state.

Although the modifications to the RM model do not have substantial effects on future atmospheric CO_2 levels as calculated by Revelle and Munk (1977), the addition of the minor fluxes to the original model enhances its capability. Also, these additions provide the basis for a more quantititative model that attempts to describe fully the behavior of the carbon cycle on the time scale of human civilization. One additional sink that we are presently incorporating into the modified RM model is the dissolution of the carbonate minerals calcite and aragonite in the deep ocean and of highly soluble magnesian calcites in the shallow ocean. Both processes will increase the *ultimate* storage capacity of the ocean for anthropogenic CO_2.

ACKNOWLEDGEMENTS

This reasearch was supported by NSF grant EAR89-15536, EAR93-16133 and NOAA.

REFERENCES

Berner, R.A., A.C. Lasaga and R.M. Garrels. 1983. The carbonate silicate geochemical cycle and its effect on atmospheric carbon dioxide over the past 100 million years. Amer. J. Sci., *283*, 641-683.

Broecker, W.S., T. Takahashi, H.J. Simpson and T.H. Peng. 1979. Fate of fossil fuel carbon dioxide and the global carbon budget. Science, *206*, 409-418.

Chen, C.T. 1982. Oceanic penetration of excess CO_2 in a cross section between Alaska and Hawaii. Geophys. Res. Lett., *9*, 117-119.

Chen, C.T., M.R. Rodmen, C.L. Wei, E.J. Olson, R.A. Feely and G.F. Gendron. 1987. Carbonate chemistry of the North Pacific Ocean. NDP-029. Oak Ridge National Laboratory, Oak Ridge Tennessee, Carbon Dioxide Information Analysis Center.

Garrels, R.M., A. Lerman and F.T. Mackenzie. 1976. Controls of atmospheric O_2 and CO_2: past, present and future. Amer. Scientist, *64*, 306-315.

Houghton, R.A., J.E. Hobbie, J.M. Melillo, B. Moore, B.J. Peterson, G.R. Shaver and G.M. Woodwell. 1983. Changes in the carbon content of terrestrial biota and soils between 1860-1980: A net release of CO_2 to the atmosphere. Ecological Monograph, *53*, 235-262.

Houghton, R.A. and G.M. Woodwell. 1989. Global climate change. Scientific American, *260*, 36-44.

Houghton, R.A. 1990. The global effects of tropical deforestation. Environ. Sci. Technol., *24*, 414-422.

Kinsey, D. and D. Hopley. (in press). The significance of coral reefs as global carbon sinks - response to greenhouse. Global Warming and Planetary Change.

Mackenzie, F.T. 1981. Global carbon cycle: Some minor sinks for CO_2. *In:* Flux of Organic Carbon by Rivers to the Sea, G.E. Likens, F.T. Mackenzie, J.E. Richey, J.R. Sedwell, and K.K. Turekian, eds., U.S.D.O.E. Conf. - 8009140, NTIS U.S. Dept. Commerce, Springfield, VA, 360-384.

Mackenzie, F.T. 1992. Chemical mass balance between rivers and oceans. *In*: Encyclopedia of Earth Science, Vol. 1, Academic Press, N.Y., 431-445.

Mackenzie, F.T., L.M. Ver, C. Sabine, M. Lane and A. Lerman. 1983. C, N, P, and S global biogeochemical cycles and modeling of global change. *In*: Interactions of C, N, P, and S Biogeochemical Cycles and Global Change, R. Wollast, F.T. Mackenzie and L. Chou, eds., Springer-Verlag, Berlin, 1-61.

Marland, G.T., A. Boden, R.C. Griffin, S.F. Huang, P. Kanciruk and T.R. Nelson. 1989. Estimates of CO_2 Emissions from Fossil Fuel Burning and Cement Manufacturing, based on the United Nations Energy Statistics adn the U.S. Bureau of Mines Cement Manufacturing Data. ORNL/CDIAC-25, NDP-030, Oak Ridge Natl. Lab., Oak Ridge, TN.

Morse, J.W. and F.T. Mackenzie. 1990. Geochemistry of Sedimentary Carbonates. Amsterdam, Elsevier, N.Y., 707 pp.

Revelle, R. and W. Munk. 1977. The global carbon cycle and the biosphere. *In:* Energy and Climate, Natl. Res. Coun., Geophys. Study Comm., eds., Natl. Acad. Press, Washington, D.C., 140-158.

Sabine, C. and F.T. Mackenzie. 1991. Oceanic sinks for anthropogenic CO_2: International Journal of Energy • Environment • Economics, *1*, 119-127.
Sabine, C.L. 1992. Geochemistry of Particulate and Dissolved Inorganic Carbon in the Central North Pacific. Univ. Hawaii Ph.D. thesis, 249 pp.
Sarmiento, J.L. and U. Siegenthaler. (in press). New production and the global carbon cycle. J. Geophys. Res.
Smith, S.V. 1985. Physical, chemical and biological characteristics of CO_2 gas flux across the air-water interface. Plant, Cell and Environ., *8*, 387-398.
Smith, S.V. and F.T. Mackenzie. 1987. The ocean as a net heterotrophic system: Implications from the carbon biogeochemical cycle. Glob. Biogeochem. Cycles, *1*, 187-198.
Tans, P.P., I.Y. Fung and T. Takahashi. 1990. Observational constraints on the global atmospheric carbon dioxide budget. Science, *247*, 1431-1438.
Wollast, R. and F.T. Mackenzie. 1989. Global biogeochemical cycles and climate. *In*: Climate and Geo-Sciences, Kluwer Academic Publishers, Dordrecht, 453-473.

The Oceans: Physical-Chemical Dynamics and Human Impact. Edited by S.K. Majumdar, E.W. Miller, G.S. Forbes, R.F. Schmalz and Assad A. Panah. © 1994, The Pennsylvania Academy of Science.

Chapter Twenty-One

OCEAN MINING: Problems and Prospects

FILLMORE C.F. EARNEY

Department of Geography
Northern Michigan University
Marquette, Michigan 49855

INTRODUCTION

A first encounter with the topic of ocean mining might prompt one to think of the world's many continental shelf petroleum production areas. Indeed, petroleum does account for a major share of both the offshore mineral tonnage produced and capital invested, but many other non-metallic and metallic minerals have long been extracted in the oceans. And, today, visionaries in the mining industry and in government believe the deep oceans may also provide many of our future mineral needs.

WHY MINE THE SEABED?

Why should anyone want to mine the seabed—continental shelf or deep oceans? Advantages include: 1) opportunities to find usable mineral deposits, as onshore sources become depleted or too inaccessible; 2) water provides a relatively cheap transport mode for mined products and needed logistical materiel; 3) transshipment nodes (ports) are already in place; 4) environmental pollution constraints may not be as rigorous as those applied to onshore operations; 5) greater flexibility is available for siting ore processing establishments; 6) states may gain greater mineral resource autarky. Disadvantages are: 1) a need for very large capital investments in exploration and development; 2) source areas may be great distances from markets; 3) weather-related downtime can be a greater problem than onshore;

4) unknown environmental problems and special extraction difficulties may occur; 5) some potential oceanic mineral commodities are already in oversupply from onshore sources; 6) numerous political and economic problems may be encountered in obtaining mining rights and maintaining control of proprietary technologies (Earney, 1990, p. 2). Not all these topics or all minerals produced in the oceans (as those from seawater) can be considered here, but a selective examination should provide some understanding of the problems and prospects associated with ocean mining. The main focus is on continental shelf and deepwater hydrocarbons and on deep seabed hard minerals.

CONTINENTAL SHELF HARD MINERALS

Using land-based tunnels, the ancient Greeks mined zinc and lead from beneath the Aegean Sea, and Cornishmen extracted tin off their Cornwall Peninsula. Mercury, scheelite (a tungsten ore), copper, nickel, iron ore, tin, gold, coral, barite, aragonite, shells, sulphur, coal, and sand and gravel have been or are won from the seabed (Earney, 1980, p. 11; Earney, 1986, pp. 307-319).

Sand and Gravel

Two of the most important of the hard minerals produced by dredging in the offshore are sand and gravel building aggregates and industrial silica sands used in glass and foundry molds. Japan is the world's largest producer of offshore sand and gravel aggregates; output comes mainly from the Inland Sea region and from along the western shore of Kyushu. The United Kingdom (UK) follows Japan in total annual output; its industry dates from the 1500s.

Producers in both Japan and the UK—and others as in the United States (US)—have problems with other coastal-zone activities such as fishing and recreation, because of sediment plumes, beach erosion, and deficits in sand replacement. As onshore and gravel reserves continue to deplete, especially near urban regions, coastal states are likely to become more dependent on inshore continental shelf areas. US Bureau of Mines specialists in aggregate resources estimate that the US's demand for construction-grade gravel and sand should be approximately 163 million metric tons by the year 2000 (Davis and Tepordei, p. 695).

Coal

The UK's Firth of Forth seabed (off Scotland) has provided coal by land-based tunnels since at least the 1600s. Many other states continue to depend on offshore coal production for part of their needs, including Japan, Taiwan, Chile, Turkey, New Zealand, Australia, and Canada. One of Canada's mines (off Cape Breton Island) recently suffered a major explosion, and in the mid-1980s, one of Japan's establishments experienced a severe fire; both accidents took lives. Employment in both the UK and Japan's undersea coal mines has declined during the last decade. Given the competition of large-scale onshore open-pit coal mining, the future of offshore producers seems problematical.

Sulphur

Numerous industries require sulphur in their production processes, as in steel manufacture, in paper milling, and in paint fabrication. The US has an important sulphur-producing site in the offshore of Louisiana. Producers have long extracted this sulphur, which is associated with salt domes (Ackerman, pp. 222-226). The sulphur, resulting from complex bacterial and chemical reactions, is obtained via the Frasch process, whereby superheated water (325°F) is injected into the sulphur deposits. Pumping systems move the liquid sulphur onshore where it is loaded onto barges that transfer it to a deepwater port on the Mississippi River and, ultimately, onto ocean carriers for transport elsewhere in the US or for export. Operators have recently discovered additional large sulphur reserves, which will likely be developed in the future, despite strong competition from onshore production and from byproduct sulphur captured by pollution control programs in industry.

CONTINENTAL SHELF AND DEEPWATER HYDROCARBONS

Some oil industry specialists in the early 1980s predicted that by the year 2000 the offshore would supply 50% of the world's annual production of crude oil. This will not happen. Nevertheless, offshore crude oil production in the early 1990s, accounted for approximately 30% of the world's annual output (British Petroleum Company, p. 5; Mid-year Report, p. 42). Most offshore petroleum output comes from relatively shallow water areas. In 1991, seven states produced more than one million barrels per day (bbl/d) of offshore crude. Mexico led the world, followed by Norway, Saudi Arabia, UK, Nigeria, Indonesia, United Arab Emirates, United States (US), Venezuela, and India (Table I). The US in 1991 produced more than twice the offshore natural gas of its nearest competitor, the UK. Eight countries produced more than one billion cubic feet per day (bcft/d) (Table II).

Producers now pump petroleum from nearly all of the world's oceans and major seas, with the exception of Antarctica's offshore. In the early 1990s, offshore waters receiving the greatest exploration attention were Africa's Guinea Coast, the Middle East's Red Sea and Persian/Arabian Gulf, Asia's South China Sea, Europe's North Atlantic, Russia's Arctic Ocean and Okhotsk Sea, and Brazil's South Atlantic coast. The latter two areas exemplify two important problems facing petroleum producers seeking black gold from beneath the oceans, i.e., the increasingly difficult constraints of severe climatic conditions and ice and of water depth.

Deepwater Frontiers

As near-shore petroleum sources become depleted, more and more producers are seriously considering deepwater areas, those in excess of 1000 feet (ft). Geologists have learned that these deepwater areas, both close to the continents and far offshore, provide many more opportunities than previously thought possible.

TABLE I

Top-Ten Offshore Crude Oil Producers, 1991

Country	Production (thousands bbl/d)
Mexico	1,945
Norway	1,862
Saudi Arabia	1,780
United Kingdom	1,770
Nigeria	1,400
Indonesia	1,374
United Arab Emirates	1,101
United States	891
Venezuela	828
India	750

Source: Mid-year Report: World Offshore Oil Production. 1992. *Offshore* 52, No. 6:42.

TABLE II

Top-Ten World Offshore Gas Producers, 1991

Country	Production (millions of cft/d)
United States	12,780
United Kingdom	5,010
Norway	3,173
Australia	2,102
Malaysia	1,860
Netherlands	1,750
Mexico	1,200
Russia	1,190
Indonesia	730
United Arab Emirates	706

Source: Mid-year Offshore Gas Production. 1992. *Offshore* 52, No, 6: 42.

Brazil:

Petrobrás, the national petroleum company of Brazil, is the world's leader in what its personnel characterize as "a great leap downward." During the last decade, Petrobrás engineers have pushed Brazil's production patch ever deeper into the Atlantic. Petrobrás expects to drill at water depths of 6500 ft by the year 2000 (Political Pressure..., p. 8). Petrobrás is certainly pushing toward this goal; it already produces petroleum at a water depth of 2,500 ft (Petrobrás..., pp. 32, 34).

In early 1992, Petrobrás decided to develop the Campos Basin's Barracuda field, lying 60 miles (mi) offshore from the state of Rio de Janeiro and at water depths of 2000 to 3300 ft. Geologists put the field's proven reserves at 1.1 billion bbl of crude; production should total approximately 200,000 bbl/d. Petrobrás' managers expect to have a pilot system operating by the end of 1993; it will cost $218 million and have an anticipated payback time of two years and eight months, assuming oil prices remain at early 1992 levels (Barracuda..., p. 8).

By late 1993, Petrobrás expects to have installed two identical special subsea manifold systems that can operate at a depth of 3300 ft. This unit will operate on

Brazil's second-largest and super-giant Albacora field. Gravity alone holds the unit on the seabed, which eliminates the need for piles during installation and allows repairs topside, both major savings. A regular drilling rig assisted by a remotely operated vehicle can install the manifolds, all without assistance from a special vessel. The new manifold units can work in tandem; while one is under repair or shut down, the other will continue producing normally (Manifold..., p. 7). Petrobrás in 1991-1992, also, installed the world's two largest monobuoys; these units permit the simultaneous mooring of two 120,000 dwt ships. One ship loads petroleum onto tankers and the other transfers crude to an onshore terminal. An exceptional characteristic of these monobuoys is their 1300-ft seating depth, the deepest of any in the world (Largest Monobuoys..., p. 8).

The expertise and equipment of Petrobrás is in great demand throughout the world, especially in areas with exceptionally deep deepwater fields. Petrobrás is working in foreign oil exploration production in offshore Angola, Norway, and the UK.

United States:

Because of federal and state moratoria on continental shelf drilling along much of the US's offshore coastal zones, many firms are looking to deeper waters. For example, a consortium of 11 major oil companies is presently considering the feasibility of developing an economically viable and low-risk system in the Gulf of Mexico that uses shallow-water (600-800 ft) host platforms to extract oil via production lines from wells as distant as 60 miles offshore and at water depths of as much as 6000 ft. (Wheeler, p. 26).

A fundamental problem that discourages exploration in the US's deepwater areas is the US Mineral Management Service's reluctance to approve prolonged periods of gas flaring, needed to prove the viability of new discoveries. Without these long-assessment periods for reserve determination, producers hesitate to go ahead with development. As of late 1992, producers considered a recoverable reserve of 150-200 million bbl as the minimum discovery needed to make a field's development economic, given a market price of $20/bbl and an $8/bbl lifting cost. With the present world market price for crude at about $18/bbl and a current lifing cost of $10/bbl, the Gulf-of-Mexico deepwater fields will remain uneconomic (Is It Time..., p. 54).

Presently, few financial incentives exist for producers to explore and develop deepwater areas. In late 1992, however, the US Congress considered—as part of a National Energy Strategy Bill—having offshore operators pay a reduced royalty for production in shallow waters (less than 656 ft). Another late-1992 Congressional bill, the OCS Deep Water Production Incentives Act, seeks royalty abatement for petroleum developments seaward of the 656-ft depth. Under the act, until oil prices exceed $33/bbl for two consecutive quarters, royalty payments would be suspended. The Congress may act on this bill in 1993 (Crowden, p. 13).

Although activity in the US's Gulf of Mexico has recently shown signs of reviving, overall offshore drilling is inadequate to maintain the US's present gas reserves. At present drilling rates, US gas reserves will deplete by the year 2000 and oil a few years later (Dobson and LeBlanc, p. 23). Because of the very unfavorable economic and political climate in the US's offshore, a drastic reduction in the number of active oil firms, workers, and drilling rigs has occurred in recent years. Firms have either folded or moved to more politically friendly and economically attractive areas, as in Asia, Africa, and Russia.

Russia's Arctic Frontier

Arctic regions are becoming increasingly attractive to petroleum producers. Russia's continental shelves—perhaps the most exciting of the high-latitude frontiers—also, may be exceptionally frustrating. Drillers have discovered large hydrocarbon reserves in the Okhotsk, Barents, and Kara Seas (Figure 1). Many oilmen believe these waters will be the focus of the world's next major offshore bonanza (Russian Federation, p. 48). They could be right. Personnel with Minneftegasprom, Russia's national oil corporation, contend that the country's continental shelves may ultimately prove to hold a total of some 550 billion bbl of hydrocarbons (gas and oil). Minneftegasprom claims producers could see profits of 15-20 percent (Russian shelf..., p. 8). Both the reserve and profits estimates may be too optimistic, especially the latter, given the internal chaos and lack of established supply and tranportation infrastructures in Russia. But major petroleum markets

FIGURE 1. Russia's Barents and Kara Seas Frontier.
Source: After George, Dev. 1993. Shtokmanovskoye to Rosshelf: Cernomyrdin to Kremlin. *Offshore* 53, No. 1: 6. With permission.

could become available to foreign-based firms within Russia, because production by domestic producers has tumbled during the last two years.

One major field alone, the Rusanovskaya in the Kara Sea (offshore of the famous onshore Urengoi field), is estimated to contain 282 trillion cubic feet (cft) of gas. The field lies in only 165 ft of water, but presents severe ice conditions, having only a two- to three-month period annually when icbreakers are unneeded (Knott, p. 31). Another field, the Shtokmanovskoye, in the Barents Sea to the west of Novaya Zemlya, holds an estimated 141 trillion cft of gas and gas condensate (Russian Federation, p. 48). This area has occasional periods when icebergs are present. Iceberg scouring in arctic regions poses a major threat to completion and pipeline systems, requiring specially designed units encased within shielding and recessed into the seabed (Earney, 1980, p. 159).

These finds are encouraging Russians to talk about Murmansk's becoming the Houston of the high Arctic or the Aberdeen of the Barents and Kara Seas (Major Hydrocarbon..., p. 34). On the other hand, cost estimates indicate that to develop the Shtokmanovskoye field alone will require $20 billion (Johansen, p. 31). Although Minneftegasprom claims producers could see profits of 15-20%, western firms may have difficulty sharing in these. For example, Russia's President, Boris Yeltsin, in late 1992 gave in to nationalistic pressure when he handed over exclusive development rights for the Shtokmanovskoye field to a large 19-member industrial military organization, the Russian Shelf Development Company (Rosshelf). This firm is a complete neophyte in petroleum exploration and development. Yeltsin's decision, in effect, excluded from participation the consortium Arkitcheskaya Zvezda (Arctic Star)—made up of Conoco, Norsk Hydro, Neste, and Metra—an experienced group which had in progress a feasibility study scheduled for completion in December 1992 (George, 1993, No. 1, p. 6). Such actions by the Russian government cannot but discourage potential investors.

Although the Russians have done only limited exploration in the country's Laptev and western East Siberian Seas, they believe these waters contain large hydrocarbon reserves. During the summer of 1992, a US firm, Fairfield Industries, entered into a joint offshore venture with the Sakha-Yakut Autonomous Republic in northeastern Russia (George, 1992, p. 56). Estimates put reserves here at between 25-60 billion bbl of oil and gas. Again, sea-ice conditions are extremely bad in this region.

To the south, in the Okhotsk Sea, near Sakhalin, production (with the help of Japan) has been under way for several years. This region, too, has not been fully explored and should become a truly significant offshore oil region in the near future, especially given the significant petroleum requirements of Japan. A consortium of three oil firms (Marathon, McDermott, Mitsui) has a $6-$10 billion appraisal- and development-project concession from Russia, but because of poor service and transport infrastructures on the island and the mainland and a dispute over whether other partners should be allowed into the partnership, no work has begun (George, 1992, p. 56). A major liability of this region is its very severe winters and associated ice conditions (ice ridges sometimes reach a depth of 120 ft), not to mention its location within a very active seismic zone (Designing..., p. 33).

Political Frontiers

If deepwater and arctic frontiers present difficulties, political frontiers pose even more intractable situations. Boundary disputes occur in nearly all known or potentially exploitable hydrocarbon-rich offshore regions. During the past decade, numerous states have settled marine boundary or resource-sharing disputes; but, many have not.

North Atlantic:

A recent example of a dispute settlement concerns Canada's claims to 15,000 square miles of waters surrounding France's St. Pierre & Miquelon Islands, off Newfoundland. The parties put the issue before an international court of arbitration. The court granted France a 200-mi-long and 10.5-mi-wide corridor from the islands to the Atlantic's high seas. Both states may soon make exploration licenses available within the disputed areas. The International Court of Justice in 1984 settled a dispute between Canada and the US over their international boundary in the Gulf of Maine, lying between Nova Scotia and the Massachusetts-Maine coasts. The settlement made neither side happy, but both accepted the Court's compromise.

Still another, but less formal, example of a pacific settlement concerns an agreement between Norway and the United Kingdom over rights to petroleum reserves in the Statfjord field within the North Sea, a case where the field straddles the median line, as do several others. The "consultant's agreement" increased Norway's share of the field's total reserves by 1% (from 84% to 85%). Norway also received a retroactive return of oil from UK producers, valued at $430 million. Norway's producers, in turn, compensated their UK counterparts for $72 million of the field's development costs (Norwegian . . ., p. 7).

South China Sea:

One of the most volatile contemporary regions of dispute focuses on southeast Asia's waters, especially among littoral states adjacent to the South China Sea. Players in this drama include the People's Republic of China (PRC), Taiwan, Vietnam, Thailand, Malaysia, Indonesia, Brunei, and the Philippines (Figure 2). The PRC has long produced modest amounts of petroleum within its interior provinces and more recently within its Bohai Gulf, part of the Yellow Sea. Because of the PRC's growing population, need for hard currencies, diversifying industrial base, and depleting onshore petroleum reserves, the government is pursuing a vigorous offshore exploration and development program in its East China and South China Seas. Here lies the dilemma. Which waters belong to the PRC and which to its neighbors?

Because the PRC does not recognize Taiwan as an independent state, it considers waters surrounding Taiwan as the PRC's domain. Indeed, based on historical evidence of commercial domination and occupance, the PRC lays claim to all islands and their surrounding waters within the South China Sea. This claim includes all the Spratly Islands, which sprawl over an area of nearly 70,000 square mi, and it places the PRC's boundary claim within 20 mi of Malaysia's Sarawak and Sabah coasts (Earney, 1981, p. 123). To bolster its claim, in late 1992, the PRC awarded an exploration license to Crestone Energy for exploration work in an area near the Spratly Islands (Vietnam-China . . ., p. 15). The PRC's vehemence in its claim to all South China Sea waters may become even more intransigent, because: 1) few discoveries have been made in the South China Sea near the Chinese mainland and 2) waters adjacent to the coasts of Vietnam, Malaysia, Brunei, and the Philippines' Palawan Island have proven exceptionally petroleum rich.

Two decades ago, the PRC proved its willingness to use military force to gain what it wants. In January 1974—during what has been called the "100-minute-war"—Beijing militarily occupied the Paracel Islands, then held by South Vietnam. After the fall of South Vietnam in 1975, the Hanoi government occupied six islands in the Spratly Group, an act loudly protested by the PRC. Adding fuel to the fiery disputes are: 1) conflicts between the PRC and the Philippines over the latter's claim to and use of several islands in the Spratly Group; 2) Taiwan's occupation of Itu Aba—another island of the Spratly Group—; and 3) Japan and Taiwan's claim to the Senkakus, lying in the southernmost reaches of the East China Sea, not to mention boundary issues in the Gulf of Tonkin, situated between the PRC's Hainan Island and the Vietnam mainland. In addition, South Korea and the PRC do not agree upon a proper division of the Yellow Sea. Numerous concessionaires have been caught in the middle of these disputes, having to withdraw from the area and suffering financial losses and exposing workers to danger. Indeed, in the past, shots by naval vessels have been exchanged between Vietnam and the PRC in the Gulf of Tonkin. Consequently, some observers are calling upon the United Nations or the International Court of Justice to help settle these claims (LeBlanc, p. 8).

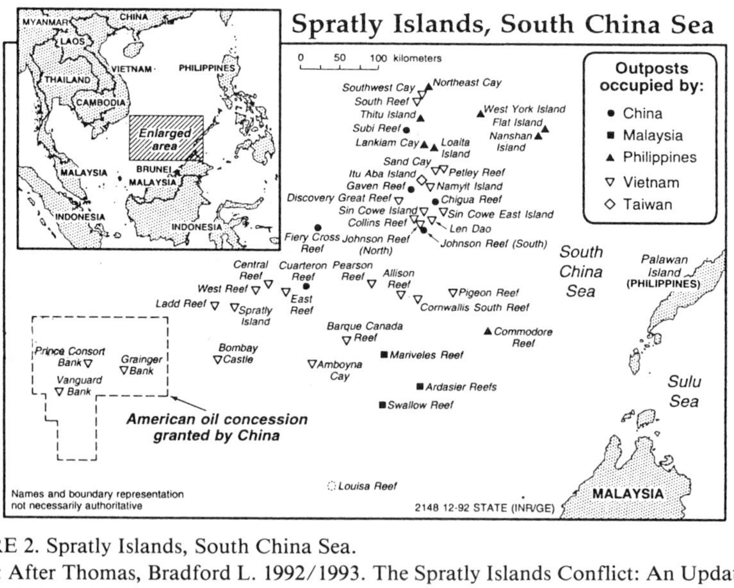

FIGURE 2. Spratly Islands, South China Sea.
Source: After Thomas, Bradford L. 1992/1993. The Spratly Islands Conflict: An Update. *Geographical Notes* 2, No. 4:6.

Other Hot Spots:

Other areas with petroleum potential and past and continued volatility include long-standing conflicts between Argentina and the United Kingdom in the Falkland Islands/Malvinas region and between Greece and Turkey in the eastern Aegean Sea. More recently, boundary disputes have occurred between Cambodia and Thailand in the Gulf of Siam. Confrontations, also, have emerged in the Middle East between Egypt and Sudan and between Saudi Arabia and Yemen in the Red Sea. In the Persian/Arabian Gulf, Saudia Arabia and Qatar, Bahrain and Qatar, and Iran and the United Arab Emirates have been disputing offshore waters and islands (George, 1993, No. 3, pp. 31-32). And a potential problem looms between Zaïre and Angola in the latter's Cabinda region (offshore and onshore). Because Zaïre's Mobutu Sese Seko is supporting a secessionist movement in Cabinda, oil producers are fearful that Seko's ambitions, as well as recently renewed fighting between political factions in Angola proper, will rob them of new investments in offshore Cabinda (Africa . . ., p. 10).

Offshore disputes are not limited to international conflicts. Bitter controversies over control of the offshore and associated oil revenues have occurred between Canada's provincial government in Newfoundland and the national government in Ottawa. Similar offshore jurisdictional confrontations between the US's federal and state governments (California, Texas, Louisiana) also have vividly illustrated intranational petropolitics.

Cooperation:

If on the one hand there exist numerous conflicts associated with offshore oil regions, there are, on the other, at least a few examples of cooperation. In 1992, Thailand and Malaysia signed an agreement to allow their national oil companies to explore waters disputed by the two states (Gulf of Thailand . . ., p. 10). Norway and Iceland have an arrangement to work together in the exploration of waters lying between Norway's Jan Mayen Island and Iceland; Norway and Russia pursue joint seismic exploration programs in the Barents Sea; and Japan and South Korea have a Joint Development Zone in waters between South Korea's Cheju Do Island and Japan's Ryukyu Islands. Neither the Norway-Iceland nor the Japan-South Korea cooperative program has resulted in petroleum finds, but they do demonstrate that states can cooperate in offshore petroleum ventures.

DEEP SEABED

For more than three decades, marine geologists and mining engineers have been evaluating prospects for mining the deep seabed, i.e., those oceanic regions lying seaward of the continental shelves. Although many mineral markets have long been glutted, entrepreneurial consortia have invested hundreds of millions of dollars in developing prototype seabed mineral mining, lifting, and processing equipment. Despite tremendous investments in money and time, glutted mineral markets and unresolved political issues debated for more than two decades under the auspices of the Third United Nations Law of the Sea Conference (UNCLOS III) have, in part, precluded deep seabed mining. Someday, however, this oceanic storehouse will be tapped, because it holds a surprising array of metallic minerals contained in nodules, crusts, and muds.

Glutted Markets

In the 1980s and early 1990s, of those minerals produced onshore that have been, are, or may be produced in the oceans, mining firms had a production overcapacity (relative to demand); in the mid-1980s these overcapacities ranged from 55% for bromine (produced from onshore well brines and from seawater) to 9% for gold (produced from hard-rock mines and from onshore and offshore coastal placers). Other mineral industries with an overcapacity included cobalt (33%), nickel (32%), manganese (32%), zinc (22%), copper (21%), and lead (19%), all available from the deep seabed (US Bureau of Mines, various pages). The problem of overcapacity will likely continue throughout the 1990s. Thus, in the near future, offshore production of a given mineral will occur only if: 1) this source can compete technologically and economically with onshore producers; 2) substitutes and recyclable materials do not become available; and 3) states' strategic concerns demand it.

The International Political Arena

From ancient times to the present, kings and statesmen have concerned themselves with controlling the oceans as resource reservoirs and as modes of transport and communication. Perceptions have varied through time about what specific marine resources should be used and to what distance offshore control should be exerted by coastal states.

As technological capabilities have increased, the capacity of industrial enterprises to exploit ever-deeper oceanic areas has improved. Thus, some observers have questioned whether the capability to exploit should be the sole criterion for laying claim to oceanic resources, such as seabed minerals. One of the first to raise this issue was Arvid Pardo (Malta's ambassador to the UN); in 1967 he called upon the world community of states to reserve as a "common heritage of mankind" all nonterritorial seabed resources and to establish some form of international control in seabed use. Subsequently, the UN began a series of meetings (UNCLOS III), where some 150 states debate how ocean resources and space should be allocated and administered.

UNCLOS III

A primary focus of UNCLOS III was how to achieve an equitable use of seabed minerals and to assure that economically underdeveloped and/or landlocked states could share in this bounty. The developed states with the capital resources and technological capabilities to exploit deep seabed minerals and those that did not (what came to be called the Group of 77) were antagonists in the debates.

On 10 December 1982, after nearly ten years of debate and compromise within

UNCLOS III, the majority of the participants adopted (as a package) a Law of the Sea Draft Convention (Treaty), scheduled to come into force one year after the sixtieth state's ratification. On 16 November 1993, Guyana became the sixtieth state to ratify. The US, Turkey, Venezuela and the Federal Republic of Germany in 1982 voted against adoption of the Treaty, and to date, neither these four states nor any other with the know-how to exploit deep seabed resources has ratified the Treaty. Why not?

Their reasons for not ratifying vary. The US's concern focuses mainly on problems associated with the administration of seabed mineral resources in a treaty-designated region termed "the Area," i.e., oceanic waters lying beyond what is called the 200-mile Exclusive Economic Zone (EEZ), within which a coastal state controls all resource exploitation. An administrative body, the International Seabed Authority (ISA)—composed of an Assembly, a Council, a Secretariat, and an Enterprise—has responsibility for administering all resource exploitation within the Area (United Nations, pp. 52-64). The main controversies between the Group of 77 and the developed states focused on voting-majority issues, on membership in the Council, and on prerogatives of the Enterprise—which will mine seabed minerals for the ISA and/or in association with developing states party to the Convention (Earney, 1985, pp. 105-107). Industry was (is) concerned that the ISA can: 1) insist that, with compensation (at fair-market prices), proprietary mining and processing technologies must be transferred from mining consortia to the Enterprise; 2) cancel permits to mine; and 3) set annual production ceilings on nickel, cobalt, copper, and manganese, all strategic minerals. Many third world states consider the ceilings issue very important. For example, Zaïre and Zambia depend heavily on cobalt exports and do not want to see overproduction and depressed prices on the world market (Cobalt . . ., p. 15).

Ferromanganese Nodules and Crusts

Ferromanganese concretions found in lakes have been known to northern Europeans for centuries; the marine variety was not discovered until 1869, when an expedition led by A.E. Nordenskiöld (sailing on the *Sophia*) hauled some nodules aboard in the Kara Sea. A century later, John L. Mero, who had done intensive investigations on manganese nodules in the 1950s, published a book titled *The Mineral Resources of the Sea* (Mero, 1965). This volume caught the attention of industry and scientists who began to give serious consideration to manganese nodules as a source of cobalt, nickel, copper, and manganese, in that order of importance.

Investigators have found manganese nodules in all the world's oceans, although they are uncommon in the Arctic Ocean (Figure 3). Nodules form best when sedimentation rates are low, as in the mid-North Pacific, where few major rivers debouche into the ocean. Among others, additionally important factors include having materials available to act as nuclei around which the nodules can accrete, water depths of 10,000 to 20,000 ft. and strong bottom currents and benthic activity that keep the nodules from becoming buried (Earney, 1990, pp. 73-79).

The world's prime manganese nodule region lies south and southeast of the

Hawaiian Islands, and extends eastward toward Mexico, especially in an area lying between the Clarion and Clipperton Fracture Zones. Several international consortia have applied for and received approval to mine pre-selected sites in this region, once the LOS Treaty goes into force. Even though the Treaty will go into force on 16 November 1994, it is doubtful that mining will begin before well into the twenty-first century. Mineral markets are weak, needed mining and processing technologies require more refinement, and many political uncertainties still exist. The various mining consortia have, however, already done much research in developing usable mining systems, including ship-suspended hydraulic, pneumatic, and continuous-line bucket dredges. The latter system has major problems, whereby the dredge buckets become filled with sediments (rather than nodules) and the cables become tangled.

Recently, the seabed mining consortia have shown more interest in manganese crusts than in nodules. The desired mineral content (especially cobalt) is higher, and the crusts occur in shallower waters (1500 to 8000 ft) than do nodules; these crusts occur in greatest abundance on the slopes of Pacific Ocean islands and seamounts. Some crusts, south of the US's Johnston Atoll are as much as 6 inches thick, although most are less. Crusts will require different mining systems than do nodules. Thus, mining consortia hoping to mine them will need to develop new mining equipment and techniques.

Polymetallic Sulfides

Plate tectonics is a key to understanding the presence and characteristics of the polymetallic sulfide minerals. Where earth's crust is forming through the divergence of oceanic-plate spreading centers, ideal conditions occur for the genesis of these minerals. When upward-moving magma from the asthenosphere comes into contact with lithospheric rocks containing downward-percolating seawater, adjacent

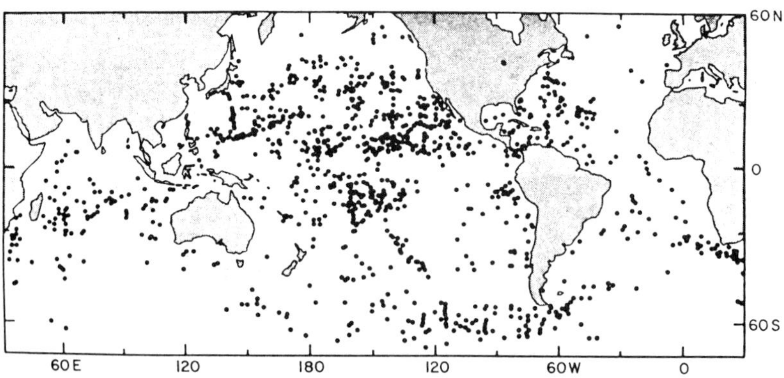

FIGURE 3. World Distribution of Manganese Nodules.
Source: Heath, G. Ross. 1982. Manganese Nodules: Unanswered Questions. *Oceanus* 25, No. 3:37. With permission.

to spreading centers, metal-rich solutions develop. As these solutions emerge from the spreading centers, they come into contact with the cold seawater, forming precipitates that build into platforms and chimney-like vents.

It is likely that the world's first deep sea mining operations will begin neither on the abyssal plain of the North Pacific nor on the side of some seamount or island, but rather within a deep of the Red Sea, a relatively young spreading zone that separates Saudi Arabia and Sudan. Scientists in the 1950s discovered salinity and temperature anomalies here. Subsequent investigations in the 1960s demonstrated vast concentrations of hydrothermally formed polymetallic muds trapped in an area now called the Atlantis II Deep (7,000 ft). These precipitated sediments (3-mi wide, 8-mi long, and up to 115-ft thick) contain an estimated 32 million tons of metals (Cruickshank, M., et al., p. 24)—iron, zinc, copper, lead, silver, and gold (Rona, pp. 130-131).

In 1976 a German firm, Preussag AG, received a contract from the Saudi Arabian-Sudanese Red Sea Commission (formed by the two countries in 1974 for joint research) to develop an environmentally safe mining and beneficiating system (Amman, p. 103). Preussag has tested and carefully mapped Atlantis II's polymetallic muds (which are the consistency of shoe polish), and has developed prototype mining and processing technologies. Although Pruessag seems ready to begin work, no Red Sea metals have yet been mined for commercial use, probably because of low metal-market prices.

CONCLUSIONS

Although miners produce numerous metallic and nonmetallic minerals on the world's continental shelves, petroleum will continue to dominate offshore mineral extraction efforts. Indeed, as onshore petroleum resources continue to deplete, the oceans will become an increasingly important source for crude oil and natural gas. Now, after 50 years of production, many shallow offshore areas are also beginning to deplete. In coming years, deepwater regions (as off the coasts of Brazil and the US) will become more attractive, especially if crude oil and gas prices increase and if lifting costs can be reduced. Oilmen now view arctic regions more positively (as in Russia's Barents and Kara Seas), despite the hazards posed by severe climatic and ice conditions. Both deepwater and arctic regions will experience intensified exploration and development in coming years.

Deep seabeds harbor a great variety of metals, all contained within ferromanganese nodules and crusts and also within polymetallic sulfides associated with spreading zones. But, the great depths at which these resources occur, the hazards of mining on the high seas, the large capital investments required to explore and to develop abyssal-plain and spreading-zone minerals, the present weakness of many metal markets, and the still unsettled political situation surrounding mining under the now-ratified UNCLOS III Treaty suggest that deep seabed mining is not imminent. Indeed, deep seabed minerals may remain for many years in Davy Jones's locker.

REFERENCES

Ackerman, John M. 1992. Main Pass—Frasch Sulphur Mine Development. *Mining Engineering* 44, No. 3:222-226.

Africa: Cabinda in Question. 1993. *Offshore* 53, No. 1:10.

Amann, Hans. 1985. Development of Ocean Mining in the Red Sea. *Marine Mining* 5, No. 2:103-116.

Barracuda: Another Super Field. 1992. *Petrobrás News* No. 138:8.

Bergeron, Harry A. et al. 1992. Alba Field First to Use Horizontal Wells for Primary Development. *Offshore* 52, No. 8:36, 38, 43, 88, 90.

British Petroleum Company p.l.c. 1992. *BP Statistical Review of World Energy.* Corporate Communications Services, BP, London, p. 42.

Cobalt: Zaire and Zambia Reaffirm Price Ceiling. 1987. *Minerals and Materials: A Bimonthly Survey* (April/May):15-16.

Crowden, Michael. 1992. Can Deepwater Economics Be Improved Without Higher Prices? *Offshore* 52, No. 9:13.

Cruickshank, Michael et al. 1987. *Marine Mining on the Outer Continental Shelf.* Environmental Effects Overview, OCS Report 87-0035. Minerals Management Service, US Department of the Interior, Washington, D.C., pp. 66.

Davis, L.L. and V.V. Tepordei. 1985. Sand and Gravel, pp. 689-703. In United States Bureau of Mines. *Mineral Facts and Problems, 1985.* USBM Bulletin 675, USGPO, Washington, D.C., pp. 956.

Designing Production Structures for the Russian Arctic. 1992. *Offshore* 52, No. 10:33, 53.

Dodson, James K. and Leonard LeBlanc. 1993. US Gulf Activity Reviving, But Drilling Insufficient to Halt Declining Reserves. *Offshore* 53, No. 1:23-25.

Earney, Fillmore C.F. 1980. *Petroleum and Hard Minerals from the Sea.* Edward Arnold, London, pp. 281.

__________. 1981. China's Offshore Petroleum Frontiers: Confrontation? Conflict? Cooperation? *Resources Policy* 7:118-128.

__________. 1985. Law of the Sea, Resource Use, and International Understanding. *Journal of Geography* 84:105-110.

__________. 1986. Seashells and Cement in Iceland. *Marine Mining* 5, No. 3:307-319.

__________. 1990. *Marine Mineral Resources.* Routledge, London, pp. 387.

East China Sea. 1992. *Offshore* 52, No. 9:10.

FLEC Reflects Mobutu's Designs on Cabinda. 1992. *Offshore,* No. 8:10.

George, Dev. 1992. Operators, Service Contractors Settle Down to New World Balance. *Offshore* 52, No. 5:38, 40, 45-46, 48, 51, 54, 56, 59, 62, 128.

__________. 1993. Shtokmanovskoye to Rosshelf: Chernomyrdin to Kremlin. *Offshore* 53, No. 1:6.

__________. 1993. Mideast Continues as Dependable Producer, Makes Up for Drop in US, Russian Production. *Offshore* 53, No. 3:31-32, 45.

Greenland Round. 1992. *Offshore* 52, No. 8:12.

Gulf of Thailand Agreement Signed. 1992. *Offshore* 52, No. 7:10.

Heath, G. Ross. 1982. Manganese Nodules: Unanswered Questions. *Oceanus* 25, No. 3:37-41.

Howes, John. 1992. Unmanned Structures Set to Grow as Recoverable Resources Shrink. *Offshore* 52, No. 9:82, 85.
Is It Time for a Collective Effort in Deepwater Field Development? 1992. *Offshore* 52, No. 8:52, 54, 56, 59, 87,
Japan-Vietnam Oil Deal Going Through. 1992. *Offshore* 52, No. 3:13.
Johansen, Bård. 1992. Many Challenges Ahead for Exploration, Development in Norwegian-Russian Arctic. *Offshore* 52, No. 10:31-32.
Knott, David. 1992. Natural Gas Dominates European Development Plans for North Sea. *Offshore* 52, No. 8:33-34.
__________. 1992. Western Groups Making Progress in CIS Partnership Pacts. *Offshore* 52, No. 4:29-30.
Largest Monobuoys in the World. 1990. *Petrobrás News,* No. 133:8.
LeBlanc, Leonard. 1992. No-man's Seabed. *Offshore* 52, No. 10:8.
Major Hydrocarbon Structure in Barents Sea, Kara Sea Regions. 1993. *Offshore* 53, No. 1:34-35.
Major Philippine Oil Find Will Help Economy. 1992. *Offshore* 52, No. 3:13.
Manifold Designed for Record Depth. 1991. *Petrobrás News,* No. 137:7.
Mid-year Report. 1992. *Offshore* 52, No. 6:42.
North America: French-Canadian Dispute. 1992. *Offshore* 52, No. 8:12.
Norwegian Information Service. 1991. Norway Gets 1 Per Cent More of Oil Field. *Norinform,* No. 23:7.
Petrobras Preparing for Next Step in Deepwater Production. 1992. *Offshore* 52, No. 9:32, 34.
Political Pressure, Posturing, Intrigue at Rio Oil Show. 1992. *Offshore* , No. 12:8.
Rona, Peter A. 1985. Hydrothermal Mineralization at Slow-spreading Centers: Red Sea, Atlantic Ocean, and Indian Ocean. *Marine Mining* 5, No. 2:117-145.
Royalty Relief. 1992. *Offshore* 52, No. 10:40.
Russia-Yakutia Polar Research. 1992. *Offshore* 52, No. 5:12.
Russian Federation. 1992. *Offshore* 52, No. 5:48.
Russian Shelf Harbors Huge Potential. 1992. *Offshore* 52, No. 3:8.
Saga's Tordis Field Development Pushing the Boundaries of Subsea Technology. 1992. *Offshore* 52, No. 8:44, 46.
Status of the United Nations Convention on the Law of the Sea. 1993. *Law of the Sea Bulletin.* No. 24:2.
Thomas, Bradford L. 1992/1993. The Spratly Islands Conflict: An Update. *Geographical Notes* 2, No. 4:2-6.
United Nations. 1983. *United Nations Convention on the Law of the Sea.* UN, New York, pp. 224.
United States Bureau of Mines. 1985. *Mineral Facts and Problems.* USBM Bulletin 675. USGPO, Washington, D.C., pp. 956.
Vietnam-China Dispute Discussed. 192. *Offshore* 52, No. 12:15.
Vietnam Getting Down to Business. 1992. *Offshore* 52, No. 5:10.
Wheeler, S.A. et al. 1993. Low-risk Approach to Deepwater Development Proposed. *Offshore* 53, No. 1:26-29, 44.

The Oceans: Physical-Chemical Dynamics and Human Impact. Edited by S.K. Majumdar, E.W. Miller, G.S. Forbes, R.F. Schmalz and Assad A. Panah.

Chapter Twenty-Two

COASTAL ZONE AND OCEAN RECREATION

EDGAR M. HOPKINS
Department of Geology and Environmental Science
University of Pittsburgh at Bradford
Bradford, PA 16701

INTRODUCTION

In the coastal zone, two great realms of the earth, land and sea, interact. Humans may find here a sense of renewal in experiencing the physical nature and rich life forms of the oceanic realm. The various environments enjoyed are produced by climatological, oceanographic, and geological processes.

The temperate and tropical zone coasts have the widest appeal due to their comfortable climates. Geologic settings include barrier island systems, estuaries, rocky shorelines, reefed coasts and deltaic plains. Different environments support many types of recreation including beach activities, fishing, shellfishing, migratory waterfowl hunting, bird watching, hiking, camping, shell collecting, skin diving, whale watching, and boating.

Worldwide, rise in human activity in the coastal zone is raising complex environmental and social issues. Desire, time and money for recreation is burgeoning, particularly in the developed countries (West 1990).

Increasingly, private ownership of recreational coastal lands is prohibiting free access to many shores. Government must respect private rights while assuring expanded public access to the (non-owned) "underwater lands", which are held in the public trust in most countries.

Rising spending on new technology provides additional economic justification for recreational use of the coastal zone environment (See West 1990). Body boards and sail boards are examples of new technology which increases markets and enjoyment levels without environmental impact. However increasing use of off-road vehicles and personal water craft "deepens" the "user footprint" and enlarges user-impacted zones.

The ocean environment is very dynamic. Storms create hazards to life and property. High user populations incur large costs of reconstruction. Stabilization structures on retrograding coasts are costly and environmentally degrading.

Oceanic cruising is increasing as more find the time and money to travel. It has fewer users and infinitely more space than coastal zone recreation. Nevertheless, cruise related tourism has environmental and social impacts on coastal destinations. Also, rising ecotourism (cruising to wild, unspoiled islands) poses new threats to the rare, fragile ecosystems it targets. Introduction of non-native species is one threat. (Edwards 1992).

Coastal zones are particularly endangered because they are geographically linear, hence limited in global abundance. Coastal systems most attractive to visitors/vacationers tend to evolve quickly to an "anthropogenic" setting. Many governments recognize a new public *ethos* regarding the environment and have set a holistic goal of zero-net-loss of remaining natural coastal environments to protect areal viability of ecosystems. Several new management models are being "tested" in the 1990's.

Recreation requires a healthy resource base. Government and private conservation groups have protected some threatened coasts. However, conflicting public and private demands for use, and governmental bureaucracy and fractionation into single-purpose agencies retard conservation efforts.

CLIMATE AND RECREATION

Climatic Factors

Rain has a negative influence on outdoor recreation. Lack of "sun days" along coasts may be a function of humid ocean air moving over the land.

Water temperature is critical to in-water activities. Cold water conditions limit in-water activity and define seasonal recreation limits.

The oceans moderate coastal climate: water stores heat and gives it off when the air above is colder; evaporation cools the air and provides the bulk of moisture for continental rainfall. Surface water temperatures change as oceans circulate horizontally and vertically, to warm or chill a given coast.

Coastal Climate Types

Continental coasts draw recreationists fleeing inland summer heat to seek cool ocean breezes. Hundreds of tropical and subtropical islands draw winter tourists, particularly in the western Pacific Ocean and the Caribbean.

Because an island's height, as well as its latitude, controls rainfall amounts, many islands have distinctive climates, fostering a unique pioneer fauna and flora. The ecotourism industry takes advantage of the distinctive character of uninhabited oceanic islands (Edwards 1992).

Bailey (1976) explains differences in coastal climates in different oceans, different sides of oceans, different hemispheres, and different latitudes. Of 14 distinct coastal climates defined by Bailey (and equated to Axelrod's [1960] coastal vegetation types), 10 are warm enough for significant recreation; Of these, 4 are limited by raininess (marine, tropical, subtropical, and temperate), 3 are ideal with only slight or seasonal humidity (subhumid tropical, summer-dry subtropical and wet-winter temperate), and 3 are limited by dryness (warm semiarid, warm arid, hyperarid).

Tropical Coasts

Coral reefs and mangrove swamps are warm temperature ecosystems which influence coastal recreational potential in the tropics.

Coasts within 20 degrees of the equator fall into either a wet, hot tropical zone, or a more favorable subhumid and year-round warm outer tropical zone. A northward shift of the less desirable inner tropical zone in the "landlocked" Indian Ocean spans all its northern coasts.

The Caribbean and the Gulf of Mexico are in the more favorable outer zone. Although the Gulf lies north of 20 degrees, it is influenced by the warm Florida current. Across the equator, similar poleward expansions of the outer tropical zone on the west sides of the oceans warms eastern South America and eastern Australia (Bogdanov 1963).

Subtropical/Temperate Coasts

The greatest east-west differentiation of climate is in the subtropical zones. The vast subtropical surface current gyres transport warm water poleward on the western sides of oceans and transport cold waters equatorward on the eastern sides of oceans. This warms lower latitudes on the west sides of the oceans (eastern U.S.A, China and Japan). It also warms the downstream, current impinged eastern sides in the *northern* hemisphere only (western North America and Britain/northern Europe). These poleward western coasts have *maritime climates,* much warmer (and wetter) in the winter than inland. Maritime climates allow out-of-water coastal zone recreation to continue through winter, despite higher latitudes.

Contrasting with the foggy, rainy maritimes north of central California, U.S.A. is the southern California coast, with year-round (mostly) dry, sunny weather, excellent for all types of coastal recreation. Such sub-tropical western coasts lie further along the cool, returning part of the subtropical current gyres. Still further south, the still cool offshore waters take moisture from the air, producing the coastal desert climate of Baja California, and the Sonoran desert of northwest Mexico (See Belt 1989).

The larger ocean area under the *southern* hemisphere westerlies causes stronger subtropical current gyres, which brings more cold water northward along the western coasts. This causes the true desert coasts, the Atacama of western South America and the Namib desert of southwestern Africa (Bailey 1976). Ecotourism in the Skeleton Coastal Park of the Great Dune Sea along the coast of Namibia is described by Hodgson (1982), Hamilton (1983) and Bartlett and Bartlett (1992).

On the east sides of continents in the subtropics, including the eastern U.S.A. coast, the situation is nearly reversed from that on western coasts. As mentioned, warm currents flow away from the equator here, and rainy subtropical coasts persist to 35 degrees away from the equator. Southern species of fish are caught as far north as Cape Hatteras, North Carolina due to the influence of the warm Gulf Stream (See Cobb 1987).

Rainy subtropical climates are replaced further poleward by rainy temperate climates (Bailey 1976). In the northeastern U.S.A., typical of eastern temperate coasts, only waterfowl hunters brave the foreboding early winter weather. Yet in the summer it is very hot just a few miles inland. Seasonal temperature extremes of the northern hemisphere subtropical to temperate east coasts reflect the passage of westerly winds across the wide northern continents.

COASTAL GEOMORPHOLOGY AND RECREATION

Coastal Classification

Coastal geomorphic forms and substrate material affect coastal population levels, accessibility and, by fixing environments, potential recreational activities. Factors of coastal genesis include (tectonic) land movement, subaerial processes causing original landforms, and coastal and landform modification by marine processes after drowning.

The energy level of physical processes in marine environments may vary from nil to colossal. Based on the energy levels and geologic time of action, the form of coasts may either reflect previous (nonmarine) processes (primary coasts) or specific modern marine processes (secondary coasts) (Shepard 1976). The material make-up of the original landforms also affects how easily landforms are modified by marine forces.

Regional Distribution of Coastal Types

Coastal relief and material make-up are linked to each other according to type of tectonism, either raising of the land (tectonic activism) or slow subsidence (tectonic passivism). These correspond to regional settings of global tectonic-plate interaction: where plates are colliding, uplift and lithified rocks are typical; where they are diverging, plates cool and subside, and sediments (granular rock fragments or chemical precipitates) accumulate to great thickness over rock surfaces. The two plate settings produce "emergent" and "submergent" coasts, respectively. (However, local areas of uplift or subsidence may exist in either plate setting).

World maps showing plate tectonic settings of coasts are given by Inman and Nordstrom (1971) and Davies (1973). Generally, the Pacific Ocean is dominated by orogenic or uplifting coasts. (Major exceptions are volcanic islands that have sunk to accumulate sediment and reefrock). The Atlantic and Indian Oceans are dominated by subsiding coasts, except in the Caribbean and Mediterranean regions where small tectonic plates are converging.

Regional Marine Energy Variations

The wind-wave climate and the range between high and low tide level are the main factors of physical process variation along coasts. Winds produce waves, which erode rock by impact. Wave-generated currents remove or deposit sediment. Wind may also move sediment subaerially.

Tides move the impact point of waves, by changing sea level, and generate their own underwater currents which can remove or deposit sediment around narrow straits and inlets or around shallow shoals and banks.

The size of waves breaking on beaches is a major influence on recreation. Large breakers are favored by body surfers and board surfers. Others may prefer calmer surf where children can safely play, boats can be launched, etc.

During storms, when winds throw large chaotic waves on all shores, most recreation ceases. During fair weather, when beaches may be crowded, the greatest differences in wave heights between beaches may be seen.

Large swells are not common in small ocean basins or along coastal segments facing short fetches. Neither are they common opposite a wide offshore shelf, which causes traversing swells to lose energy (and height).

"High energy" beaches usually involve large swells, as well as numerous storms. Northern New Jersey is a good example along the eastern U.S.A. barrier island coast. Large swells also characterize many beaches along the western U.S.A. rocky coast, with its very large fetch and active-plate-margin, narrow shelf. "High energy" beaches are generally excellent "surfing" beaches.

"Low energy" beaches along the eastern U.S.A. rocky coast of Maine are caused by the great width of the shelf across the Gulf of Maine. Examples of "low energy" beaches along a barrier island coast stretch from south of Cape Hatteras (central Outer Banks of North Carolina) to the barrier beaches and "Sea Isles" of South Carolina and Georgia. Most of the Gulf of Mexico has low energy beaches as well. (The "Big Bend" coast of northwestern peninsular Florida, which has very small beaches dotting extensive open-Gulf tidal marshes, has even been described as a "zero energy" coast).

Tidal range varies according to ocean basin geometry and coastal configuration. Typically, tidal ranges are higher in large embayments such as semicircular "bights" and funnel-shaped estuaries or bays. A greater tidal range "fluctuates" wave breaking point, modifying beach form. If sufficiently high, beaches are eliminated, resulting in coasts with wide tidal flats backed by ocean-front marshes or mangals (mangrove swamps).

On rocky coasts with low waves and moderate tidal range, such as the southern Maine, U.S.A. coast mentioned above, high tidal ranges produces low energy pocket beaches which are intermediate between (sandy) beaches and (muddy) tidal flats. Such marginal recreation areas are about all that is left for the public, as burgeoning private home ownership has preempted most of the rocky shorelines in southern Maine (Duscik 1974, p. 49).

Recreation potential is limited along coasts with a very high tidal range due to the lack of a beach and extensive tidal-cycle changes in shoreline position. In some cases dangerous currents are generated. Shellfishing is a popular activity, done with a keen eye toward the tide change.

COASTAL SYSTEMS

The endless variety of world coasts offers a great range of recreational opportunities. The overall location of major coastal types is outlined below, and recreational aspects are discussed in terms of one or two examples from the U.S.A. Other areas are mentioned where relevant.

Barrier Islands

Barrier islands have excellent recreational attributes, including the barrier-front sand beach, the cooling oceanic microclimate (barriers may lie many miles to sea), and the backbarrier lagoons, offering protected water recreation. Sunbathing, swimming, beachcombing, jogging, fishing, surfing, windsurfing, sailing, canoeing and kayaking are all popular beach activities. Swimming, fishing, crabbing, shellfishing, migratory waterfowl hunting, bird watching and boating, including waterskiing and parasailing are pursued in the backbarrier lagoons. Recreational activities away from the barrier shores vary from camping and hiking on "primitive" barriers, to jogging, biking, rollerblading, golf, tennis, etc. on developed barriers.

Extensive barrier island chains in moderate to warm climates occur in the following regions: in the *Atlantic Ocean,* eastern U.S.A., northern Europe, southeastern Brazil, and along the Gulf of Guinea in west-central Africa; in the *Indian Ocean*, southern and southeastern India and western and southern Australia; in the *Pacific Ocean,* eastern Japan, northern Mexico, Guatemala and El Salvador, Ecuador and part of Peru. Finally, the northern *Gulf of Mexico* and parts of the northern *Mediterranean* Sea as well as numerous inland seas, have extensive barrier chains. (See Hayes 1979 and Bird 1984). More locally, deltaic barrier island chains develop.

Barrier islands are massive linear sand bodies created over thousands of years by wind-driven waves. They form along low open coasts having abundant supplies of terrigenous sandy sediment, hence are most common on trailing (passive) continental margins. Barrier island formation is antithetic to limestone deposition, hence is more common away from the tropics. The formation of barrier islands is incompletely understood. Origins include: drowning of a mainland dune ridge above a wave cut scarp; sand build up above a submarine break in slope, and; inlet cut-off of a spit grown laterally away from an erosional promontory. Lateral growth may be most important with inherited topographic relief (See Halsey 1979).

Barriers differ in size and shape according to differences in the ratio of wave forces to tidal forces (Hayes 1979). The best development of barriers worldwide is in microtidal settings, where wave dominated barriers tend to be long and narrow with only occasional narrow inlets, allowing human development with minimal bridge construction.

Slight structural uplift or the growth of a large delta may form coastal promontories. Due to heightened wave attack and diminished tidal range on such coasts, wave dominated barriers are more common here.

Tide-dominated barriers tend to be short with wide, relatively permanent inlets. The strong ebb tidal action at the inlets builds large seaward-extending sand shoals which promote deposition at the ends of the islands, particularly the downdrift side, producing the characteristic "drumstick" aerial configuration (Hayes 1979). Such barriers are more common in large coastal embayments, which damp waves but enhance tidal range. Many bridges are necessary to fully access all the short barriers.

Barrier vegetation varies with latitude. It affects physiography through dune formation and changes cultural aspects (Godfrey et al. 1979). Beachfront dunes are a reserve of sand protecting the barrier from severe erosion by large storms, analogous to the floodplains of rivers, i.e.), a "contingency plan of nature" for handling rare, high energy events.

Erosion and Retrogradation of Barriers

Barrier islands are among the most dynamic of coastal systems. The prevalence of barrier islands in the modern world may be due to a slowing of sea level rise 6,000 years ago. A renewed rise of 10 to 15 centimeters per century is driving many barriers landward at an accelerated rate, threatening their survival (Williams et al. 1990, Peck and Williams 1992).

Comparison of historical with modern maps of barrier islands show that some barriers fronting the eastern Mississippi Delta plain have retreated more than 10 meters per year, averaged over the last 100 years (Williams et al. 1992). Such barriers, formed by wave reworking of subsiding, abandoned lobes may submerge with time (Penland et al. 1988). Many large barriers along the non-deltaic coasts of the eastern U.S.A. are also severely eroding (Williams et al. 1990). Some have retreated entirely past older *landward* limits, exposing, on the modern beach face, peat layers and rooted tree stumps which formed behind the island. However, not all modern barriers are retreating. With abundant sediment supply or land uplift some are growing upward or accreting seaward.

Both tropical summer cyclones and extratropical winter storms track across subtropical barriers. Water/wind damage to human structures and loss of life can be catastrophic during such storms. Evacuation, across a single narrow causeway, may be prevented by high water or traffic jams.

Dunes backing many barrier beaches are, as mentioned, a major bulwark against storm erosion, and are now protected by statute from human disturbance. Installation of "snow fences" and dune-grass planting may initiate or restore damaged dunes (Carter 1988). "Growing" of dunes may not be entirely beneficial, however. Many wave-dominated barrier systems, such as those of the North Carolina U.S.A. Outer Banks, are in a state of quasi-equilibrium as they slowly migrate landward by storm "overwash" of sand. Artificial dune construction begun in the 1930's may now be heightening storm erosion effects by stabilization-induced beach steepening (Dolan 1973). What once seemed like a human solution to erosion of barrier islands is now seen to be merely a delaying tactic.

"Overwash" also feeds ocean waters into lagoons behind (inlet-poor) microtidal barriers, preventing stagnation by "overfreshening". Engineered dune enhancement on the Outer Banks has been blamed for loss of aquatic vegetation and degradation of fish and waterfowl habitats (Madsen 1984).

New knowledge of surf-zone-current hydrodynamics allows mapping of sediment transport systematics. Engineered structures have also improved greatly over the last decade. Reversal of short-term beach erosion is often possible. However, a new question is being asked. Even with knowledge of coastal currents and short term future beach position within the natural system, is it wise and affordable to interfere? Not all engineering projects have succeeded. Some result in unexpected and deleterious changes to adjacent beaches, such as in the Ocean City, Maryland area, where a jetty built to stabilize an inlet resulted in sand starvation to the immediately downdrift barrier, luckily undeveloped. A several mile long stretch of this barrier retreated landward as much as a half mile in 60 years (Williams et al. 1990).

"Soft engineering" methods refer to beach nourishment by the hydraulic pumping of sand, as opposed to "hard" structures such as jetties and sea walls. This method has worked on many beaches but it is expensive. The cost of rebuilding Miami beach to protect threatened large hotels was 60 million dollars, but the new beach has been successfully maintained for more than 10 years (Carter 1988 and Williams 1993). (A hard-engineered, "dune ridge substitute" was also constructed here).

Funding limitations and aesthetic conditions are also halting some projects. In recent years, many geologists and planners have advocated a graceful retreat from eroding barrier shorelines by moving threatened buildings, or demolition (See Dean et al. 1988).

Coastal "set-back" regulations for new vacation home construction became a widely used method to decrease hazards to structures in the 1980's. However, restrictions against rebuilding (after Hurricane Hugo, of beach-front houses that had been "grandfathered" into a "setback" law) were recently overturned in court. This decision reaffirms the landowner's prerogative to use his/her land as seen fit, regardless of costs for disaster relief, etc. It brings into question the legality of mandatory "setbacks" (Williams 1993 and Leeworthy 1993).

Human Development on Barrier Systems

Over the last half century, many barrier islands have been converted from pristine to highly "anthropogenic" by burgeoning recreational use. Many barriers close to large cities are themselves essentially urbanized and plagued with overcrowding problems (West 1990). With easy travel, most barriers not specifically protected are subject to rapid development, as sights are set on potential new tourism meccas (See Fabbri 1990).

The evolutionary steps in development of a barrier from pristine to anthropogenic are universal. The change is from nature-dependent to artificial forms of recreation, along with destruction of the natural base that supported the earlier recreation. What is most troubling is the accelerating rate at which such development is happening. Discussion exceeds space limitation here. The interested reader is referred to West (1990), Carter (1988), Marx (1975), Duscik (1974), and Ketchum (1972).

Preservation of Barrier Island Ecosystems

To counter the loss of natural environments associated with barrier systems, the U.S.A. government, in 1982, declared a moratorium on providing federal funds for causeways, roads and related infrastructures in support of development interests (Williams 1993 and Leeworthy 1993).

Unfortunately, the preservation of barriers (and nature-based recreation) has arrived rather late in the day for the U.S.A. Whereas protection of backbarrier areas began in the 1930's as part of the National Wildlife Refuge system, the National Seashore Program did not begin until 1962. As of 1982, there were only 10 National Seashores versus over 100 National Wildlife Refuges along the U.S.A. coast. National Seashore beaches support a wide array of recreational activities, generally for a small fee. State beach parks and recreation areas comprise additional smaller sites.

The Nature Conservancy and other private environmental preservation groups have purchased additional pristine barriers for protection. The Nature Conservancy purchased a unique stretch of 14 small undeveloped barriers along the Delmarva penninsula of the east central U.S.A. coast, just south of the Assateague National Seashore. Very limited recreation by boaters, compatible with the main goal of protection, is allowed.

Rocky Coasts

The Pacific Ocean, surrounded by active plate margins, has the greatest global abundance of rocky coasts. The northern Mediterranean, northern Atlantic, and eastern Indian Ocean also have extensive rocky coasts. Finally, many oceanic islands have (volcanic) rocky coasts.

It should be stressed that there is great range in geologic character of rocky coasts. Unique appearance leads to the emphasis on geology being an important consideration in coastal protection plans of countries with rocky shorelines such as Heritage Coasts of the United Kingdom (Williams 1992). Rocky coasts may also exhibit much greater local variety than lowland coasts (Bird 1984). This is credited with being one of the principal attractions of California's coast (Heiman 1986).

Rocky coasts have unique recreational potential, relating to their scenic beauty. Hikers, bikers or drivers along cliff-top coastal highways enjoy panoramic views of the sea and coast from their elevated position. Campers may hike coastal trails beyond roads, and kayakers, such as the "Tsunami Rangers" of the U.S.A. explore inaccessible rugged cliffs and sea caves. Kayakers in the northeastern and western U.S.A. number 100,000.

Problems of shoreline access, as well as access and safe harbor from the sea, plague many rocky coasts. Some beaches, difficult to climb down to, at least remain relatively uncrowded. Some cliffed shorelines, such as portions of the White Cliffs of Dover, England, defy access by land.

Large stretches of some rocky coasts are fronted by excellent beaches. The presence and character of beaches on uplifted coasts depends upon lithology (rock type), rock structure and relief. Erosion of headlands may supply sand for beaches, or it may issue from rivers draining the uplands.

Much of the western U.S.A. coast is characterized by regularly spaced rocky, cliffed headlands, formed originally as interstream divides. Waves form a "low tide" rocky platform below the major cliffed headlands, but others are fronted by a narrow beach, with offshore rock formations (sea stacks and arches) adding to the scenery. Between widely spaced headlands are extensive mainland "pocket beaches". Estuarine waters may be ponded at the mouths of seasonal streams by sand or gravel spits which form bars analogous to barrier islands, but of much smaller scale (See Arden 1984).

Rocky shorelines have unique, rocky-bottom marine fauna supporting surf fishing from low rock promontories, and SCUBA diving sites for spearfishing and shellfishing. Kelp forests (the Giant Kelp [*Macrocystis*] in U.S.A.) support many subtropical to subpolar rocky bottoms ecosystems.

The 650 miles from central California to northern Oregon is the most isolated coast in the U.S.A. The water is too cold for swimming. Fishing, camping, and hiking are primary activities. Sunny days occur less than half the year. The area is foggy and cool in the summer. In the winter it may be sunny, or battered by intense North Pacific storms. Beachcombing, hiking, and full wetsuit SCUBA diving peak in autumn, delayed by summer fogs and cloudy water (Heiman 1986, Morgan and Morgan 1977 and Miller 1979).

Erosion problems along rocky coasts are less severe than along barrier islands because the rocky points act like giant groins or breakwaters. Beach erosion is generally without as much consequence to human culture, because no building is done below backshore cliffs. However, cliff rock falls and slope failure by slump or debris flow are severe problems in some areas, especially if vacation homes, etc. are built on uplands too close to the edge of slopes (Griggs et al. 1991).

Coral Reef Coasts

"Rocky" coasts of limestone reefs occur in warm clear water. Fish are abundant, and the coral reefs and palm-covered isles or mainland beaches draw SCUBA divers and winter-fleeing tourists from both hemispheres (Anderson 1979).

The global reef ecosystem forms a tropics-wide belt. *Major* reefs form where lime sediments accumulate off carbonate-lowlands in the absence of a source of terrigenous sediment. Rapid cementation forms a hard substrate which allows coral colonization. Such reefs occur throughout the Caribbean including South Florida, off Brazil, in the western Indian Ocean, Red Sea and Persian Gulf, and off the north half of Australia.

Oceanic volcanic islands in tropical waters, most abundant in Indonesia, Micronesia, and Polynesia, are also surrounded by (coral) reefs. Some volcanos sink through tectonic plate motion while coral grows upward, forming a circular atoll reef surrounding a deep water lagoon.

Coral reefs are one of the most diverse ecosystems on land or sea. For example, the Great Barrier Reef, now partially protected by the Australian government, supports 300 coral species, 1,500 species of fish and more than 4,000 molluscan species (Beatley 1991). It is surprising that mankind could influence such a vast biological system, spread across thousands of islands and 5 continents. However, reefs are very sensitive systems, and as tourism reaches coral islands, impact follows. Florida's reefs and the Great Barrier Reef exhibit degradation by ship groundings, anchorages, harvest by collectors, oil spills, sewage outfall, etc. (See Christie and Johnson 1991 and Hildreth 1992). Much of the Australian reef and the entire Florida reef tract, lagoon, and keys were recently made into federal Sanctuaries.

Estuaries

Much recreational activity occurs from small boats in estuaries, including fishing, hunting, shellfishing, cruising, and water skiing. Hunting, shellfishing, fishing, and bird watching are also carried out from shore.

Abundant modern estuaries relate to rising sea level caused by melting of the latest Pleistocene ice sheets. One may define an estuary as an elongate body of brackish to saline water, open at one end to the sea. The two major categories are drowned glacially modified river valleys (fjords) and drowned river valleys (ria's). Ria's are the most important to recreation because they dominate the tropical and temperate zones.

Fjords

Fjords are river valleys deepened by glaciers and later flooded by the sea. They are often deep and may exhibit somewhat oceanic conditions. Most have rocky bottoms, with some glacial sediment cover. Those with recreational potential are located in temperate zones, including northern Europe, southern New Zealand, and northwestern and northeastern North America. Open water fishing, whale watching, SCUBA diving, camping on isolated islands and hiking along fjord-head trails are major activities.

Spanning the western U.S.A./Canada border is a world class example of a rocky, fjord coast (Queen Charlotte and Johnstone Straits, Strait of Georgia, Strait of Juan De Fuca, and Puget Sound). Vancouver Island and the Olympic Peninsula, with its huge Rain Forest, separate the area from the Pacific. Flooding of tectonic lands and glaciated river valleys has formed a huge inland waterway. A quarter of a million boats ply the waters of the Strait of Georgia during the summer, but winter is preferred by SCUBA divers, due to better underwater visibility (Kohl 1980).

Fishing, shellfishing, waterfowl hunting, hiking, camping, Killer Whale watching, and eagle watching is done from kayaks, canoes and larger craft. Protection Island is a rookery for a multitude of (oceanic) seabirds.

There are pollution problems in Puget Sound from timbering runoff and industry and sewage effluent from the large urban area around Seattle, Washington and Vancouver, B.C. (Richards 1984). Puget sound still has more than 100,000 acres of wetland, but historical loss due to human activity is about 70 percent (Edgerton and Parrish 1987).

Ria's

Ria's are typified by a drowned, tributary pattern. Each tributary has a central channel through the ancient river floodplains, which form shallow flats surrounded by extensive tidal flats and marshes. The sedimentary substrate includes sandy and muddy bottoms. The marshes, swamps, sloughs, bogs, wet meadows, and shallow ponds skirting estuaries are collectively known as wetlands. Abundant phytoplankton, aquatic bottom vegetation, and especially marsh grasses, render ria estuaries as extremely biologically productive ecosystems (See Hitchcock 1972).

A "barred" estuary is one that is partially closed off from the sea by barrier and/or spit systems. Under these conditions, shore-parallel backbarrier lagoons may be combined with the drowned river system to multiply the area of shallow water, tidal flats, and marshes. On the eastern coast of the U.S.A., the "Sounds" separating the Outer Banks of North Carolina from the mainland provide an excellent example of such an estuary. Presently, fishing tends to be good only near the mouth of this estuary, because overfreshening (mentioned earlier) and pollution (as seen in Curituck Sound, the northern lagoon of this system) tend to be endemic to barred estuarine systems due to longer "flushing-times."

Large estuaries may influence the development pattern of coastal human settlement and growth. For example, the "Eastern Shore" of the Chesapeake Bay, isolated by the upper reaches of the Bay, has escaped most of the "Megopolian" urban sprawl of the east-central U.S.A. seaboard. The Chesapeake Bay Bridge opened this area to easy access from Baltimore and Washington, and the Bay Bridge-Tunnel across the Bay mouth now forms a through-route along the Atlantic coast, but long isolation of the Southern Delmara Peninsula barriers allowed survival in a natural state.

Isolation of southern New Jersey along the Delaware Bay had a similar effect on early development, but the closer distance to New York and Philadelphia has resulted in fairly intensive barrier development, despite the lack of a bridge across the wide mouth of the Bay.

Urbanized Estuarine Recreation

Two thirds of the largest cities in the world lie on or near estuaries (see Nordstrom 1992). "Anthropogenesis" of estuaries often springs from the excellent potential for port facilities. The central channel requires little deepening for ship traffic, and so a large city may grow near the head of the flooded portion of the river. The growth of additional port facilities expands along the shore, and the low, marshy coasts are "reclaimed" by dredge and fill operations to build ship channels and dry land areas. The U.S.A. has lost over half of its coastal wetlands to infill operations since European colonization.

Commercial/industrial landowners in urbanized estuaries have typically ignored the recreationist or saw him/her as a possible obstruction to operations. Port facilities refused public access to the water to avoid amenity costs and legal responsibilities (Walsh 1987).

A recent development in California, U.S.A. has been proactive development of recreation facilities in urbanized estuaries. Container ships freed terminal space in the 1960's. Governmental management of the coastal zone became holistic in the face of rising recreational demands. The Harbor Department of Los Angeles spent $300 million to develop recreational facilities on empty terminal land, including a theme village, maritime museum, marine museum, cruise center, hotels, marina and fishermans' wharf. By 1987, over 16 percent of the Port's revenue came from marine tourism and recreation. The development spared open areas along the heavily stressed coast of southern California (Walsh 1987).

The estuarine beach environment, previously ignored, is now also recognized as an important recreational resource (Nordstrom, 1992). The dynamics of such beachers differs from open coast mainland and barrier beaches, leading to differences in optimal use, and management. One aspect is the greater incidence of water contamination.

Pollution of estuaries is most critical to recreation insofar as the declining health of estuaries and injury to and loss of their associated wetlands not only removes a direct area of recreation, for fishing, shellfishing, hunting, bird watching, and perhaps the type of activities in ports discussed above, but it is also highly injurious to open coast and ocean recreation fishing, because a high percentage of the fished-for species critically depend on the estuaries and related protected waters.

The Chesapeake Bay

Fishing, sailing, shellfishing, waterfowl hunting and birdwatching head the list of recreation in Chesapeake Bay. The fight to save Chesapeake Bay is one of the most publicized of environmental struggles in the U.S.A. The Bay's criticality to the history and commerce of the region is well documented (Fisher 1980). The U.S.A. Environmental Protection Agency (EPA) completed a 7 year, $27 milion study of the Bay in the early 1980's (Bateman 1984).

Chesapeake Bay is still in grave environmental danger, perhaps not as much from the inhabitants of the immediate region, as from the pollutants coming down the rivers that drain into the Bay. The watershed of the Chesapeake is over 40 million acres, 25 times as extensive as the estuary itself (Horton 1993). One quarter of the basin is farmed, and fertilizer runoff is one of the major causes of over-nutrification in the Bay. It has been more resistant to remediation than sewage effluents, the other main cause of over-nutrification (Horton 1993).

Industrial pollutants also come down the Susquehanna River, the main river feeding the bay. The Environmental Protection Agency (EPA) also identified over 30 pollution point sources in cities on smaller tributaries of the Bay (Horton 1993). Influx of toxic chemicals, such as a Kepone spill on the James River, constitutes a major environmental problem to the Bay (Fisher 1980 and Bateman 1984).

Another environmental problem in the Bay has been loss of habitat to housing developments, marinas, power plants and industries. Maryland has declared a 1,000-foot buffer zone along the bayshore, but suburban sprawl is eating up waterfront in the Virginia part of the western bay (Horton 1993). Remaining saltwater marshes are concentrated in the east-central part of the bay. Many are protected as National Wildlife Refuges. Aquatic vegetation is most abundant off these marshes. It is critical to the marine habitat, and a major waterfowl food. Loss of aquatic vegetation was one of three serious problems identified by the EPA study.

Recreational fisheries for bluefish, striper, shad and croaker have shown severe declines. Oystering, long famous in the upper two thirds of the bay has all but died out. Commercial Blue Claw crabbing has supplanted the failed commercial oyster industry, providing nearly half of the national catch. The crabs may become overfished at this level (Horton 1993). A recent book sponsored by the Chesapeake Bay Foundation details the problems and some of their solutions (Horton and Eichbaum 1993).

Tropical Estuaries

Many tropical coasts have mangals (mangrove forests) along estuaries and lowlands instead of salt marshes. Major mangal coasts include Angola and Mozambique in southern Africa, northern Australia, western India, western Indonesia, northwestern Africa, and northern South America to the southern Gulf of Mexico.

Puerto Rico professionals build vacation homes on stilts in the (public) mangrove swamps (Fiske 1992), but generally such areas are resistant to the type of development seen for other coasts because of accessibility only by boat. Also, mosquitos and biting flies or midges are often legion.

Recreational significance relates primarily to fisheries productivity. Mangroves replace marsh grasses in providing very high primary production. In the south Florida area, 75 percent of game fish and 90 percent of commercial fish depend on the region's mangal (Gore 1977).

Mangals are often destroyed ("reclaimed") by dredging and landfill. In southwestern Florida, where part of the Ten Thousand Islands mangrove swamp was threatened, the Army Corps of Engineers denied permission for a large housing development. The mangrove swamps of south Florida were valued in the late 70's at $4,000 dollars per acre per year to fishing, tourism, and recreation. This mangrove ecosystem has suffered, however, by channelization in the Everglades, which causes too rapid freshwater runoff. A decline in fishing has been blamed on this (Gore 1977).

Mangrove forests are being destroyed in the Philippines, Thailand and peninsular Malasia, largely by conversion to agriculture and by timbering (Beatley 1991). India reportedly had one half of its extensive mangrove forests cut down between 1963 and 1977.

Delta Plain Coasts

Delta geomorphology reflects river/wave/tide interaction, leading to a tripartite classification (Wright and Coleman 1973). Wave dominated and tide dominated deltas have recreation similar to mainland beaches and estuaries, respectively, and are not discussed further. Deltas contain many environments, some found in other settings already discussed. Rich soil, easy water transport, and rich marine faunas support large human populations, despite disastrous flooding and ground subsidence.

Freshwater swamps and marshes of upper delta plains are replaced in lower delta plains by salt marshes in temperate climates and mangals in tropical climates. Much recreation takes place on the lower delta plain, a wide zone of river and sea influence. Extensive bays and wetlands support large wildlife populations and thus hunting and fishing. Small barrier inlands, also popular for recreation, form at the seaward edge of lobes of the delta abandoned by the main river flow.

Mississippi Delta

The impact of geological events on recreation and settlement on the Mississippi Delta in the northern Gulf of Mexico, U.S.A. in historic times is documented by Williams et al. (1992). The geologically ephemeral nature of barriers along this delta-rim has resulted in the loss of many resort and fishing villages to storms and erosion.

Marsh tracts are leased for waterfowl hunting, individually or by hunting clubs. On *cheniers* (small sand ridges) on the western part of the delta plain, clubs build sizable road accessed camps. On the lower southeastern part of the delta, individual hunter camps are accessible only by boat (Davis and Detro 1975). Five million waterfowl overwinter in the coastal marsh supporting an $80 million industry (Davis 1987).

Fishing brings in $50 million annually. Recreation varies seasonally. Some activities peaking twice a year (Davis and Detro 1975). Boating, shrimping, crabbing, crawfishing, and trapping are also popular activities.

TYPES OF COASTAL ZONE AND OCEAN RECREATION

A brief discussion of fishing and hunting affords an analysis of the resource base from an environmental and ecological perspective. Because healthy natural ecosystems are of import to general (nonconsumptive) coastal recreation as well as commercial fisheries (Ketchum 1972), this seems the best course given limited space. For other interests, shell collectors are referred to Tucker (1991) and Zahl (1969), skin divers are referred to "Skin Diver", boaters to "Cruising World", "Sail", and "Yachting", whale watchers to Corrigan (1991), and ecotourists to Daniel (1993).

Marine Recreational Fishing

Fishing satisfies myriad human needs (Voiland 1984). Marine recreational fishing is economically important, being the number one activity from boats (West 1990). The U.S. Fish and Wildlife Service found participation in the U.S.A. to have grown by 2.7 times between 1955 and 1980, while fishing-related expenditures went up by seven times (Schmied 1984). Anglers spent over $7.2 billion on marine recreational fishing in the U.S.A. in 1985 (See Ross 1991).

Surveys of the National Marine Fisheries Service of NOAA show growth to have slowed in 1975, a trend which continued to 1991. Catches also appear to have declined slightly. The conclusion of NOAA is that the U.S.A has reached the limits of the resource.

The big game fish, marlin, sailfish, tuna, etc. range the "Blue" waters of open oceans. They are fewer in number and are the most costly to catch, requiring charter of a boat, captain and mate, or boat ownership. Satellites can now guide anglers to offshore water fronts along which game fish such as billfish and tuna may gather (Boreman, 1984).

Some traditional big game fishing sites have declined, particularly off the U.S.A. coast (Leeworthy 1993). Commercial fishing, often by foreign boats, may be the culprit, although heavy sport fishing may share blame. Many follow a voluntary catch and release program, but charter boat captains may ask to sell fish onshore (Leeworthy 1993).

The small game (pelagic) fishery involves some chumming from headboat and charterboat, but is done primarily from private, medium-sized sea-going boats, especially for Salmon on the west U.S.A. coast and Bluefish on the east.

The economical one day offshore fishing trip is likely to be on a headboat with from 25 to 75 anglers. Rocky bottoms have the most demersal fish species. "Fishing grounds" also exist on sandy shelves, or headboats travel to shipwrecks that draw rocky bottom species. Fishing success around wrecks and drilling platforms has spurred fisheries agencies to increase siting of "artificial reefs". Such structures may actually "grow" fish, increasing the stock available to the recreational angler (Boreman 1984). Catches ten times better than away from artificial reefs are reported (See West 1990). Reef siting factors are presented by Radonski et al. (1985). Most are offshore, but nearshore reefs are also placed for SCUBA spearfisherman (West 1990).

Onshore, "driftfishing" in economically-rented boats in inlets is popular. Some headboats fish a short distance off the beach for "doormat" flounder and large specimens of other fish caught from shore.

Fishing in lagoons and estuaries is primarily from small boats, over bottom features such as drop-offs, rocks, or banks. For example, in Laguna Madre, in south Texas, the slightly hypersaline shallow flats near the mainland support an excellent recreational seatrout fishery.

Shore anglers comprised 40 percent of the entire marine recreational fishermen in 1976 (West 1990). Surf fishing is very popular. Costs are low. Many species might be caught at any given time, and seasonal variation is typical. Both demersal and pelagic fish may be caught, although the former are caught more consistantly, with brief "runs" of the latter. On sandy shores, jetties or fishing piers extending beyond the surf zone are popular with many.

Full justice cannot be done to the complex issues of marine fishing in a short space. The Proceedings of the Marine Recreational Fisheries Symposium annually addresses current issues in the U.S.A. (See for example Stroud 1984). Extensive involvement of the National Marine Fisheries Service of NOAA and the U.S. Fish and Wildlife Service includes research, surveys, monitoring programs, enforcement, etc. Private groups include; International Game Fish Association, National Coalition for Marine Conservation, and Sport Fishing Institute. Angling guides for the U.S.A. include Ross (1991) (New England), Freeman (1976) (Atlantic coast), Squire (1977) (Pacific coast), and Bauer 1962 (circum-U.S.A.).

In 1950, the U.S.A. Federal Aid Sport Fish Restoration Act imposed a 10 percent excise tax on importers/manufacturers of fishing equipment. States received funding based on a formula. (Sigler and Sigler 1990).

In 1986, passage of the Wallop-Breux Amendment rewrote this Act, assuring additional money for boating safety/use and also for wetland trust protection. Funding for wetlands projects are based on federal applications, in compliance with the U.S. National Environmental Protection Act (NEPA) (Sigler and Sigler 1990 and Leeworthy 1993). Funds were basically frozen by the Reagan administration but resumed thereafter.

The 1976 passage of the Magnuson Fishery Conservation Act (FCMA) marked the official recognition that both commercial *and* recreational fishing were viable industries, important to employment, economy and health within the U.S.A. (Schmied 1984). This act provided 7 national standards to guide preparation of fisheries management plans, but *none* addressed habitat or ecosystem factors (Hair 1984 and Schmied 1984).

Fishing has declined markedly in U.S.A. lagoons and estuaries due to chemical pollution or habitat degradation or, more rarely, overfishing. Offshore fisheries are affected; for example, greater than 50 percent of the fish/shellfish recreational catch (and 96 percent of the commercial catch) in the southeastern U.S.A. depends on the estuarine environment (Hair 1984). Similar estimates are made around the U.S.A. and in other countries.

Many fish breed at an estuary mouth, lay eggs offshore, and juvenile fish swim back into the estuary to grow before returning to sea (Saenger 1988).

Chemical pollution also affects fish *on* shelves along urbanized coasts. NOAA's Northeast Monitoring Program (NEMP) found disease and mutations in several fish species in the New York Bight, PCB's in fish (generally below the 5 PPM hazard level to humans), (reduced) oxygen levels fatal to fish along the New Jersey coast during the summer of 1976, and high faecal bacterial levels in many estuaries *and* along the open coast, including some species harmful to fish and humans (Boreman 1984).

Health concerns among anglers is growing. Twice as many ex-fishermen gave up fishing in one area due to perceived pollution than due to not catching enough fish (Hair 1984). If a certain species is announced to contain toxic chemicals, fishing for it will drop markedly, but will partially regrow as the memory of the event fades (Hair 1984).

The problem of health endangerment to specific groups due to heavy recreational fishing in polluted areas, once limited to the "Cajun" community on the lower Mississippi Delta, now includes urban U.S.A. immigrants from Southest Asian and the Carribbean (West 1990).

The conflicts between commercial and recreational fisheries have increased as competition for many of the same species has focused on smaller populations (Denisof and Rote 1991). Recreational anglers are gaining strength through numbers, organization and (economic) recognition. Hair (1984) points out that conflicts between the fisheries over declining species can cloud issues of habitat degradation, so often the real culprit.

The immediate challenge to fisheries management is to balance the available resource to needs of all users, a difficult task in times of rising demand and falling supply! Within the last decade, U.S.A. federal agencies appear to have adopted a new *ethos* stressing scientific and holistic management of resources. Fisheries, as with most ocean resources, have global linkages of users. Ultimately the problems must be addressed on our coasts and on all the other coasts of the world.

Shellfishing

Space does not permit a full discussion of shellfishing. Some include both crustaceans and molluscs under this heading. Most popular recreationally harvested molluscs are oysters, clams (several) scallops, and mussels. Crustaceans include shrimp, blueclaw crabs, and lobster (several). A separate survey on recreational shellfishing in the 1985 survey of U.S.A. recreational fisheries suggests that shellfishing accounted for about 24 percent of total marine recreational fishing expenditures (NOAA, USFWS 1991).

As with fin-fishing, commercial *and* recreational advantage is taken of the resource. As with fin-fishing, pollution and loss of habitat is critical, even more so because so many of the shellfish sought live their entire lives within the estuaries and other coastal water bodies. Shellfishing around the U.S.A. coast has seriously declined. NOAA now keeps track of 122 estuaries and sub-estuaries. The historical loss of a large percentage of estuarine environments has been discussed elsewhere. The loss continues. There was a 6 percent decline in area of shellfish waters between 1985 and 1990; Out of 18.7 million remaining acres, 1.2 million acres were closed to shellfishing (Leeworthy 1993).

Health hazards associated with consumption of molluscan shellfish from polluted waters are in the news. Raw shellfish (including those eaten in restaurants) are a risk when commercially harvested! Nearly all U.S.A. oysters come from the Mississippi delta region, and the harvester can fish closed waters without getting caught. There is no federal inspection of fish in the U.S.A. The best bet is to harvest your own shellfish!

Migratory Waterfowl Hunting and Birdwatching

Coastal bird hunting involves an appreciation of the outdoors, particularly the marshlands where so many birds congregate. Birdwatching may seem an antithetic pursuit, but differs primarily in being passive and nonconsumptive, whereas hunting is active and consumptive. Recreationists of both activities have preserved habitats of shorebirds and waterfowl. Ducks Unlimited and the Audubon Society are the two lead private groups in U.S.A. bird conservation.

Waterfowl (*Anatidae*), including swans, geese, and ducks, have approximately 225 species worldwide of which some 60 are native to North America (Kortwright 1942). Most of these birds are migratory, breeding during summer in the subpolar prairies and tundra, and overwintering on or near coasts, mostly in the subtropics. Because the wintering grounds are so compressed compared to the breeding grounds, the total population depends more on these environments, although breeding ground droughts cause wild fluctuations in population (Johnsgard 1968). There are 3 Flyways funneling birds from the breeding grounds to wintering areas, the Eastern to the east coast, the Central to the Gulf of Mexico, and the Western to the west coast.

Waterfowl hunting in the eastern U.S.A. has a history intertwined with human development of the coasts and the decline of marsh environments in backbarrier lagoons and estuaries. To paraphrase from Holland's (1975) discussion of the Back Bay National Wildlife Refuge of Virginia, U.S.A.:

> Commercial "market hunters" reduced populations by the late 1800's. Northeastern U.S. industrialists saw excellent hunting to the south and bought up marshland between New Jersey and South Carolina, establishing exclusive huntclubs which used wardens to keep the "locals" off their shooting grounds (beginning, perhaps, an ongoing conflict between "natives" and "tourists" on some east coast barrier islands). The clubs carefully managed the resource and marsh. In 1918, market hunting was stopped. In 1929, National Wildlife Refuges were authorized by the Migratory Bird Conservation Act. In 1934, the "Duck Stamp Act" banned the Spring hunting season and began raising money from hunters for management. Since waterfowl populations depend on sufficient water on mid-continent breeding grounds as much as on hunting pressure and coastal marsh health, two great declines of waterfowl populations in the U.S.A. coincide with the major droughts of the 1930's and 1960's. In response to declines in the 30's, the Back Bay National Wildlife Refuge was established by land purchase and condemnation of over 4,000 acres belonging to two hunt clubs. Soon, the original marsh managers, who also managed the barriers *de facto*, were gone. The next wave to hit the east coast barriers, after the hunt clubs began to sell off land, was the vacation home business. Marsh was ditched for mosquito control, drained and filled for housing, and lagoons were subjected to increasing sewage effluent. The barrier building boom began and hasn't stopped yet.

While more than 70 *coastal* National Refuges were established around the lower 48 states between 1929 and 1982. (The wetland-rich east coast has nearly twice as many refuges as the Gulf of Mexico plus the West Coast), barriers and non-refuge marsh continued to decline.

The northern Gulf, particularly Texas and the Mississippi Delta has many waterfowl wintering grounds servicing the Central Flyway, the most waterfowl-rich of the three.

More than 24 species of ducks and geese and the Pacific Brant are important game species on the West Coast. They utilize more than 400 separate estuaries and related waterfowl havens comprising approximately 400,000 acres (Edgerton and Parrish, 1987). The largest are Puget Sound, Columbia River estuary and San Francisco Bay.

The loss of wetlands on the West Coast has been severe; exceeding 70 percent of two largest estuaries, Puget Sound and San Francisco Bay (Edgerton and Parrish, 1987). Small duck hunting clubs still own and manage, for sustainable yields, a large number of wetlands. However, higher taxes and rising costs, including purchase of water for flooding at certain times, threatens to allow agricultural or other uses to take over.

The crowding of ducks onto smaller wetlands can cause outbreaks of deadly waterfowl diseases (Edgerton and Parrish, 1987). Also, pesticide, herbicide, and metal residues from farming operations along rivers draining to the lagoons and estuaries have increased in the last several decades, causing poisoning of waterfowl (Madsen 1984).

Most government refuges and management areas in the west, more than one million acres, provide some public hunting. Private management holdings (most being duck hunting clubs) surpass one million acres (Addy, 1987).

Several Federal Acts passed in the 1960's have fostered protection of the remaining coastal wetlands. State projects have also been undertaken, and the private sector, most notably National Audubon Society, Nature Conservancy and Ducks Unlimited, manage many wetlands for waterfowl and other wildlife.

The future of waterfowl conservation in North America depends partly on recreational users (hunters and watchers). Geese have generally retained healthy populations, but North American duck populations have continued to diminish, from 80 million in 1984 (Madsen) to 44 million in 1992 (Mitchell).

Habitat loss is the greatest problem, both in the breeding grounds and the (coastal) wintering grounds. Duck stamp purchases, ($300 million sold between 1934 and 1987), and "Pittman-Robertson" Federal Aid in Wildlife Restoration Program-funds (federal excise taxes on arms and ammunition, with matching state funds from hunting licenses), as well as private money from Ducks Unlimited dues and contributions to preservation groups can buy habitat, the key element in conservation (Madsen 1984 and Whitesell 1987).

COASTAL ZONE PLANNING AND MANAGEMENT

The coastal zone is the most lived in "environment" on this planet (Marx 1975). The urbanization around estuaries and suburbanization of barriers and other beach-fronted coasts is rapidly destroying habitats essential to biodiversity and the resource base (Fabbri 1991 and Beatley 1991). Yet, multiplicity and intensity of economic and political interest continue to hamper planning and management toward maintainable resource allocation (Ketchum 1972 and Fischer 1989).

Coastal zone programs all state concern for the environment, but in *negative* programs, need must be shown before development is allowed, whereas development is encouraged in *positive* systems (Carter 1988). The negative coastal zone management (CZM) model of the U.S.A. is mandated by high private ownership and advanced habitat loss (Carter 1988).

Previous to the 1970's, the U.S.A governmental structure for coastal zone planning and management included many agencies, each dedicated to a single marine resource, but without binding decision-making powers. There was little consensus between agencies (Fischer 1989).

The 1972 passage of the U.S.A. Coastal Zone Management Act (USACZMA) reflected public outcry over media-reported coastal ocean environmental decline and catastrophies (Knecht 1991). Four stated goals were to protect environments, diminish human hazards, find access for the public, and coordinate governance (Carter 1988). The most influential aspect of the Act has been to elevate the level of governmental decision making, stressing regional or national interests, while retaining state and local flexibility (Hershman 1975 and Owens 1992). In 1990, the CZMA, with improvements, was reauthorized for five years. Owens (1992) presents five recommendations for reform to make the CZMA more effective. Cicin-Sain (1992) calls for further adoption of *ocean management* versus traditional *marine resources management*, i.e., interagency cooperation to minimize conflicts between coastal zone users *while* ensuring continued existence of the resources. Knecht (1991) advocates expanding the area-based, multiple purpose format to the "American Ocean" as defined by the 12-mile limit and 200 mile Exclusive Economic Zone (EEZ). (Worldwide, some dozen National Marine sanctuaries have adopted this strategy).

The eliminiation, where possible, of arbitrary political boundaries allows more holistic planning. Cicin-sain (1992) advocates integration by spatial orientation (towards ocean management), i.e., everything in an area must be considered. However, one might add that the area should encompass a *natural* space, based on containment of process, fauna, etc.

The concept of planning zones reflecting natural spatial systems of the earth started with the river basin. Extension of the river basin model to the coastal zone has become a reality in the managed restoration of the Chesapeake Bay. The watershed "sources" not only the fresh water of the Bay, but many of the pollutants reaching the Bay. The Cheasapeake Bay Foundation and governments of several states cooperate to manage the entire watershed, e.g., Pennsylvania now monitors agricultural manure introduction throughout the Susquehanna River basin as part of the clean-up effort. Local planning entities are also being involved.

Where humans impact a coastal zone, they become an important part of that ecosystem. This is reflected in the management of the Florida Keys as a National Marine Sanctuary encompassing the modern reef trend, the back reef lagoon and the adjacent (anthropogenic) keys. The system can thus be monitored, e.g., against sewage pollution by "live-aboards", 78,000 permanent residents and 2.5 million annual visitors (Leeworthy 1993).

A last example of natural spatial units is the recognition by Taylor (1991) of the difference between small and large estuary systems in California. Because of a lack of oceanic pelagic predator fish in the small estuaries, certain small-fish species survive there. The recognition of the small estuaries as distinct ecosystems helps insure their preservation.

Recreational Planning and Management

U.S.A. participation in outdoor recreation is estimated at 89 percent (Leeworthy 1993). Over 50 percent vacation on coasts (Ketchum 1972). According to NOAA, outdoor recreation is the number two source (behind human health) of human benefit from the coastal zone.

With regard to recreation, management is being asked to seek a balance between "pristine-unused" and "anthropogenic-fully used" which meets both user demand (conflict resolution function) and human health needs (public trust function) (See Cicin-Sain 1992).

Coastal environments have variable abilities to absorb human pressure (Carter 1988, Ch. 10). Set-asides of relatively unimpacted areas allow scientific study. Often such areas prove to be so rare and so sensitive that a rationale develops for continued protection. Reserves improve (nearby) recreation in many indirect ways, e.g. birdwatching and sustainable hunting around them, improved fishing along the coast, etc.

In addition to exclusive recreational use, as illustrated above, uses may be multiple, or displaceable (Ketchum 1972). Multiple use may allow recreation to develop in certain anthropogenized areas which also serve other functions (e.g., shipping, etc.). Displacement of uses not dependent on coastal resources, e.g. a coastal military base being moved inland, allows conversion of this land to recreational use.

Traditionally, recreational economics followed a strict cost analysis (See Freeman 1992). This approach, called "utilitarian anthropocentrism" by Douglas (1991), states that if the cost analysis of the present use of the resource is not the most profitable, then it should be converted to the more profitable use. (The USACZMA states that recreation or other uses must yield in favor of conservation when an ecosystem is threatened.)

One aspect of worth is the rarity of a resource relative to its demand. This has been addressed in the U.S.A. and elsewhere by national inventories of coastal environments by the federal government. In 1962, the Outdoor Recreation Resources Review Commission reported that only 6 percent of the U.S.A. coastal lands suitable for recreation were held by the federal government. Public and private recreational facilities inventoried by NOAA (1988) and NOAA and the U.S. Forest Service (1992), respectively, constituted a very small percent of the U.S.A coasts.

Recreational value of healthy coastal environments compares favorably to all other uses (Leeworthy 1990). One troublesome aspect to evaluating recreational environments is the difficulty in setting economic limits on permissible degrees of damage. Although the public can act without economic justification (e.g., through the Endangered Species Act and the Clean Water Act), such is required for cleanup of low-level (non-point source) pollution; a Reagan administration executive order decreed that, if a regulation had high cost, the benefits of remediation had to be quantified. Investigating recreational values (as opposed to some alternate use) of a non-pristine site thus requires completion of (costly) baseline studies in other areas (Leeworthy 1993).

The public trust cannot be served unless the resource base is maintained. It would seem to matter little whether an environment is displaced by industry, such as the Hackensack Meadows of the Hudson River estuary (See Squires 1992) or by intense, artificialized recreation, such as the marshlands behind the Atlantic City gambling casinos.

The wide range of recreation (from wilderness recreation to full-blown tourism) makes evaluation of the environmental compatibility of coastal recreation complex. Qualification is necessary to define the rarity of a site, e.g., natural environments may vary from wild to hardened, to facilitated, to enhanced, depending on cultural impact (Kenchington 1993).

Also, unlike natural systems which may be homeostatic, social systems are Markovian, where one change predetermines the next, Kenchington (1993) finds "anthropogenesis" irrevocable through time because of feedback between environmental change and recreation choice. This "activity creep" from nature-based to synthetic activities is especially driven by real estate and development pressures (Kenchington, 1993). Management must limit activity creep or steer it so as to preserve those systems which have so far escaped major impacts.

Sociologic status of the user is important. Klausner divides this aspect into man's relationship to nature, his relationships to fellow man, and his relationship to himself (See Smith 1983). Different types of users factor into these relationships, from the individualistic "wilderness" user, to the ecotouristic "pilgrim", to the "interested" user, to the "distracted" user, to the user needing environmental "background", to one for whom the natural environment is "irrelevant". All must be accommodated *and* survival of the environment and its ecosystems must be assured.

Other studies concentrate on the economic status of the user (See West 1990). The economically advantaged recreationist can avoid the increasing crowds by travelling to distant sites, or by seeking out the open water in his/her own boat. Others are not so fortunate. The public trust function of government must steer the development of coasts so as to prevent the displacement/nonaccess of segments of the public by economic forces. The native population has been forced off Hilton Head Island, N.C. by increasing land taxes associated with high-income tourism (Blockson 1987). Developers should be required to set aside sectors of equal recreational quality for the non-land owning public to have access to their tidelands.

CONCLUSION

Recreation in the coastal zone is diverse, popular, and economically viable. However, it has contributed to the deterioration of coastal zones, especially in the developed countries of temperate (northern) climes, which have depleted many of their coastal resources. Negative management programs have arrived late in the day. Nevertheless, in the developed countries coastal zones are becoming places of concern to citizens (Knecht 1991), and places where governments are attempting multiple purpose, area-based management practices.

Despite its problems, the USACZMA and its amendment legislation has had much practical success. Updating and integration of state coastal management through a council structure or a designated lead agency was important (Carter 1988). Private agencies in the U.S.A. system are largely in a watchdog role, but can often move faster than government to avoid major loss of habitat by intermediate purchase, etc.

Developing countries, beginning industrial expansion and anxious for (recreational) tourist dollars from the developed countries, have adopted positive plans (Knecht 1992); these must be long range to minimize pervasive environmental degradation. Ecotourism and nature-based tourism promise to be more profitable in the long run, considering the value placed on unspoiled settings by tourists

(Kenchington 1993).

The international effort is in its infancy, but has appeared in the Mediterranean and Caribbean under the Regional Seas Program of the United Nations (Knecht 1991). As the world shrinks, realization of travel/recreation interconnectedness across the larger oceans (as for whaling, fishing, trade, etc.) should follow.

The separation of beach recreationists into two groups, urban and suburban middle income is noted by West (1990). The increasing demand for urbanized recreation is spreading intense development along the coastal zone. An alternative, floating artificial islands, complete with recreational facilities is discussed by Droege (1992). Such "techno-redemption" could allow preservation of natural coasts into the future (See Marx 1975 and Droege 1992).

Ultimately, the public commitment to saving its coastal zones may influence the level of anthropogenic modification of the coastal and open marine environments of the planet. Only by finding ways to achieve a balance between economic development and protection will natural coastal and oceanic environments continue for the common good of mankind.

ACKNOWLEDGEMENTS

Thanks are given to Drs. V.R. Leeworthy of NOAA and S.J. Williams of the USGS for providing the author with time and materials during his visit to Washington, D.C. Any misinterpretations of information from these individuals or utilized references are the sole responsibility of this writer. Thanks also go to the University of Pittsburgh at Bradford for a Faculty Development Grant which assisted the author in completion of this chapter.

REFERENCES

Addy, C.E. 1987. The Atlantic coastal Area: How it Was. pp. 210-217. In: *Fireside waterfowler: Fundamentals of Duck and Goose Ecology.* D.E. Wesley and W.G. Leitch (Eds.) Ducks Unlimited Inc., Stackpole. Harrisburg, PA, pp. 352.

Anderson, A. Jr. 1979. *The Blue Reef.* Alfred A. Knopf, NY, pp. 259.

Arden, H. 1984. East of Eden: California's Mid-coast. *National Geographic,* 165:4:424-461.

Axelrod, D.I. 1960. Coastal Vegetation of the World, pp. 43-58. In: W.C. Putnam et al. (Eds.) *National Coastal Environments of the World.* U. Calif. Press.

Bailey, H. 1976. Coastal Climates in Global Perspective, pp. 65-72. In: S.J. Walker (Ed.) *Geoscience and Man. Vol. XIV - Coastal Research.* School of Geoscience Louisiana State University Publication, pp. 153. Baton Rouge, LA, pp. 153.

Bartlett, D. and J. Bartlett 1992. Africa's Skeleton Coast. *National Geographic,* 181:1:54-85.

Bateman, H.H. 1984. Congress Views the Chesapeake Bay, pp. 5-8. In: Stroud, R.H. (Ed.) *Marine Recreational Fisheries 9-Marine recreational Fisheries Development.*

Proc. 9th Ann. Mar. Rec. Fisheries Symposium. National Coalition for Marine Conservation, Inc. Publication, Savannah, GA, pp. 218.

Bauer, E.A. 1962. *The Salt-Water Fisherman's Bible.* Doubleday and Co. Inc. Garden City, NY, pp. 192.

Beatley, T. 1991. Protecting Biodiversity in Coastal Environments: Introduction and Overview. *Coastal Management*, 19:1-19.

Belt, D. 1989. Baja California: Mexico's Land Apart. *National Geographic,* 176:6:714-745.

Bird, E.C.F. 1984. Coasts: An Introduction to Coastal Geomorphology. Blackwell Publ. Inc., NY, pp. 320.

Blockson, C.L. 1987. Sea Change in the Sea Islands. *National Geographic,* 172:6:735-763.

Bogdanov, D.V. 1963. Map of the Natural Zones of the Ocean. *Deep Sea Research,* 10:520-23.

Boreman, J., M.P. Sissenwine, M.C. Ingham, and W.G. Smith. 1984. Marine Recreational Fisheries Opportunities. pp. 145-156. In: Stroud, R. H. (Ed.) *Marine Recreational Fisheries 9-Marine recreational Fisheries Development.* Proc. 9th Ann. Mar. Rec. Fisheries Symposium. National Coalition for Marine Conservation, Inc. Publication, Savannah, GA, pp. 218.

Carter, R.W.G. 1988. *Coastal Environments.* Academic Press, NY, pp. 609.

Cobb, C.E., Jr. 1987. North Carolina's Outer Banks: Awash in Change. *National Geographic,* 172:4:485-513.

Corrigan, P. 1991. *Where the Whales Are: Your Guide to Whale Watching Trips in North America.* Globe Pequot Press, Old Saybrook, CT.

Christie, D.R. and P.J. Johnson. 1990. State Ocean Policy Initiatives in Florida. *Coastal Management,* 18:283-296.

Cicin-Sain, B. 1991. California and Ocean Management - Problems and Opportunities pp. 5-11. In.: Domurat, G.W. and Wakeman, T.H. (Eds.) *The California Coastal Experience.* Amer. Soc. Civil Engineers, NY, pp. 311.

Daniel, J.E. (Ed.). 1993. *The Buzzworm Magazine Guide to Ecotravel.* Buzzworm Books.

Davies, J.L. 1973. *Geographical Variation in Coastal Development.* Hafner Pub, NY.

Davis, D.W. and R.A. Detro. 1975. Louisiana's Marsh as a Recreational Resource, pp. 91-98. In: S.J. Walker (Ed.) *Geoscience and Man. Vol. XII-Coastal Research.* School of Geoscience Louisiana State University Publication, Baton Rouge, LA, pp. 115.

Davis, D.W. 1987. The People, Boats, Homes, and Economics of "the Bayou Country", pp. 11-1 to 11-18. In: S. Penland and J.R. Suter (Eds.) Barrier Shoreline Geology, Erosion, and Protection in Louisiana, Coastal sediments '87 Field Trip Guide, Am. Soc. Civil Engineers, NY.

Dean, R., O. Pilkey and J. Houston. 1988. Eroding Shorelines Impose Costly Choices; *Geotimes*, 33:9-14.

Denisoff, C.R. and J.W. Rote. 1991. Future of Commercial Fisheries in California, pp. 61-73. In: Domurat, G.W. and Wakeman, T.H. (eds.) *The California Coastal Experience.* Amer. Soc. Civil Engineers, NY, pp. 311.

Dolan, R. 1973. Barrier Islands: Natural and Controlled, pp. 263-278. In: D.R. Coates (Ed.) *Coastal Gemorphology*. Publ. in Geomorphology, SUNY, Binghamton, NY.
Douglas, P.M. 1991. Ethics, Politics, Art, Accountability Quest for Quality in Coastal Management, pp. 12-31. In: Domurat, G.W. and Wakeman, T.H. (Eds.) *The California Coastal Experience*. Amer. Soc. Civil Engineers, NY, pp. 311.
Droege, P. 1992. To Live in Harmony with the Sea: An Envelopment Strategy for the Next Three Generations. *Coastal Management,* 20:73-88.
Duscik, D.W. 1974. *Shoreline for the Public: A Handbook of Social, Economic and Legal Considerations Regarding Public Recreational Use of the Nation's Coastal Shoreline*. MIT Press, Cambridge, MA, pp. 257.
Edgerton, L.F. and Parrish, J.M. 1987. West Coast Wetlands, pp. 199-204. In: *Fireside waterfowler: Fundamentals of Duck and Goose Ecology*. D.E. Wesley and W.G. Leitch (Eds.) Ducks Unlimited Inc., Stackpole Harrisburg, PA, pp. 352.
Edwards, S.E. 1992. The Demand for Galapagos Vacations: Estimation and Application to Wilderness Preservation. *Coastal Management,* 19:155-169.
Fabbri (Ed.). 1990. *Recreational Uses of Coastal Areas: A Research Project of the Commission on the Coastal Environment, International Geographical Union.* Kluwer Academic Publishers, Boston, MA, pp. 285.
Fischer, D.W. 1989. Governing the U.S. Coastal Zone: An Unresolved Issue. *Ocean and Shoreline Management,* 13:21-34.
Fisher, A.C., Jr. 1980. My Chesapeake-Queen of Bays. *National Geographic,* 158:4:428-467.
Fiske, S.J. 1992. Sociocultural Aspects of Establishing Marine Protected Areas. *Ocean and Coastal Management,* 18:25-46.
Freeman, A.M. III. 1992. *The Economics of Valuing Marine Recreation: A Review of the Emperical Evidence;* draft submitted to U.S. Env. Protection Agency, Off. of Policy Planning and Evaluation, Washington, DC, pp. 37.
Freeman, B.L. 1976. *Anglers Guide to the U.S. Atlantic Coast; Fish, Fishing Grounds, and Fishing Facilities,* NOAA Nat. Fish. Serv., Wasington, DC.
Godfrey, P.J. Leatherman, S.P. and Zaremba, R. 1979. A Geobotanical Approach to Classification of Barrier Beach Systems, pp. 99-126. In: *Barrier Islands From the Gulf of St. Lawrence to the Gulf of Mexico*. Academic Press, NY, pp. 325.
Gore, R. 1977. The Tree Nobody Liked. *National Geographic,* 151:5:668-689.
Griggs, G.B. Pepper, J.E. and Jordan, M.E. 1991. California's Coastal Hazards Policy: A Critique, pp. 89-107. In: Domurat, G.W. and Wakeman, T.H. (Eds.) The California Coastal Experience. Amer. Soc. Civil Engineers, NY, pp. 311.
Hair, J.D. 1984. Confronting Habita Restraints, pp. 117-124. In: Stroud, R.H. (Ed.) *Marine Recreational Fisheries 9-Marine recreational Fisheries Development.* Proc. 9th Ann. Mar. Rec. Fisheries Symposium. National Coalition for Marine Conservation, Inc. Publication, Savannah, GA, pp. 216.
Hamilton, W.J. 1983. The Living Sands of the Namib. *National Geographic.* 164:3:364-377.
Halsey, S.W. 19XX. Nexus: New Model of Barrier Island Development, pp. 185-210. In: *Barrier Islands From the Gulf of St. Lawrence to the Gulf of Mexico.* Academic Press, NY, pp. 325.

Hays, M.O. 1979. Barrier Island Morphology as a Function of Tidal and Wave Regime, pp. 1-28. In: *Barrier Islands From the Gulf of St. Lawrence to the Gulf of Mexico.* Academic Press, NY, pp. 325.

Heiman, M. 1986. *Coastal Recreation in California: Policy, Management, Access.* Inst. of Gov. Studies, U. Calif. Berkeley, CA, pp. 170.

Hershman, M.J. 1975. Coastal Zone Management: Accommodating Competing Coastal Uses, pp. 109-115. In: H.J. Walker (Ed.) *Geoscience and Man. Vol. 12, Coastal Resources.* School of Geoscience Louisiana State University Publication, Baton Rouge, LA, pp. 115.

Hildreth, R.G. 1992. Australian Coastal Management: Some North American Perspectives on Recent Queensland and Other Initiatives. *Coastal Management,* 20:255-265.

Hodgson, B. 1982. Namibia: Nearly a Nation? *National Geographic,* 161:6:755-797.

Holland, D.F. 1975. Back Bay National Wildlife Refuge: Some Parallels in Implementing the Coastal Zone Management Act, pp. 55-60. In: Coastal Plains Center for Marine Development Services. *Report of Conference on Marine Resources of the Coastal Plains States.* Savannah, GA, pp. 88.

Horton, T. 1993. Chesapeake Bay - Hanging in the Balance. *National Geographic,* 183:6:2-35.

Horton, T. and W.M. Eichbaum. 1991. *Turning the Tide: Saving the Chesapeake Bay.* Island Press, Washington, DC, pp. xx.

Hitchcock, S.W. 1972. Fragile Nurseries of the Sea: Can We Save Our Salt Marshes? *National Geographic,* 141:6:728-765.

Inman, D.L. and C.E. Nordstrom. 1971. On the Tectonic and Morphologic Classification of Coasts. *Jour. Geology,* 79:1-21.

Johnsgard, P.A. 1968. *Waterfowl: Their Biology and Natural History.* U. Nebraska Press, Lincoln, NE, pp. 138.

Kenchington, R. 1993. Tourism in Coastal and Marine Environments - A Recreational Perspective. *Ocean and Coastal Management,* 19:1-16.

Ketchum, B.W. (Ed.). 1972. *The Water's Edge: Critical Problems of the Coastal Zone.* Coastal Zone Workshop, M.I.T. Press, Cambridge, MA, pp. 393.

Knecht, R.W. 1991. Changing Perceptions of the 'American' Coastal Ocean. *Ocean and Coastal Management,* 17:318-325.

Kohl, L. 1980. British Columbia's Cold Emerald Sea. *National Geographic* 157:4:526-551.

Kortwright, F.H. 1942. *The Ducks, Geese and Swans of North America.* Amer. Wildlife Inst., Telegraph Press, Harrisburg, PA, pp. 476.

Leeworthy, V.R. 1990. The Environment as an Economic Asset: Three Empirical Examples in Florida. Gulf of Mexico Program's Conference, The Environmental and Economic Status of the Gulf of Mexico, New Orleans, LA.

Leeworthy, Dr. Vernon R. 1993. (personal interview) Strategic Environmental Assessments Division, National Oceanic and Atmospheric Administration, Building #4, Silver Springs, MD.

Madsen, J. 1984. American Waterfowl: Troubles and Triumph. *National Geographic,* 166:5:562-599.

Marx, W. Modern Man and the Coastal Environment, pp. 99-107. In: H.J. Walker (Ed.) *Geoscience and Man. Vol. 12, Coastal Resources.* School of Geoscience Louisiana State University Publication, Baton Rouge, LA, pp. 115.

Miller, M. 1979. Oregon's Lonely, Lonely Coast. *National Geographic,* 156:6:698-823.

Mitchell, J.G. 1992. Our Disappearing Wetlands. *National Geographic,* 182:4:2-45.

Morgan, J. and N. Morgan. 1977. California's North Coast. *National Geographic,* 152:3:330-363.

NOAA. 1988. *National Estuarine Inventory Data Atlas, Vol. 4: Public Recreation Facilities in Coastal Areas.* USGPO, Washington, DC, pp. 123.

NOAA/USFWS. 1991. *Recreational Shellfishing in the United States: Addendum to 1985 National Survey of Fishing, Hunting, and Wildlife-associated Recreation.* U.S. Depts. Commerce and Interior, Washington, DC.

NOAA/U.S. Forest Service. 1992. *Private Commercial Suppliers of Outdoor Recreation Facilities in the Coastal United States: National Coastal Recreation Inventory Project.* U.S. Depts. Commerce and Agriculture, Washington, DC, pp. 31.

Nordstrom, K.F. 1992. *Estuarine Beaches: An Introduction to the Physical and Human Factors Affecting Use and Management of Beaches in Estuaries, Lagoons, Bays, and Fjords.* Elsevier, NY, pp. 215.

Owens, D.W. 1992. National Goals, State Flexibility, and Accountability in Coastal Zone Management. *Coastal Management.* 20:143-165.

Peck, D.L. and S.J. Williams. 1992. Sea Level Rise and Its Implication in Coastal Planning and Management, pp. 57-73. In: P. Fabbri (Eds.) Ocean Management in Global Change, Elsevier Science Pub.

Penland, S., R. Boyd and J.R. Suter. 1988. Transgressive Depositional Systems of the Mississippi Delta Plain: A Model For Barrier Shoreline and Shelf Sand Development. *Jour. Sed. Petrology,* 58:6:932-949.

Radonski, G.C., R.G. Martin and W.P. Dubose, IV. 1985. Artificial Reefs: The Sport Fishing Perspective, pp. 529-536. In: F.M. Ditri (Ed.) *Artificial Reefs - Marine and Freshwater Applications.*

Richards, B. 1984. The Olympic Peninsula. *National Geographic,* 165:5:644-673.

Ross, M.R. 1991. *Recreational Fisheries of Coastal New England.* U. Mass. Press, Amherst, MA, pp. 279.

Saenger, P. 1988. Estuaries and Bays, in: R. Allen (Ed.) *Go Fish Australia.* Austr. Recr. and Sport Fishing Confed. Inc.

Schmied, R.L. 1984. Tools and Development for Fisheries Development, pp. 145-156. In: Stroud, R.H. (Ed.) *Marine Recreational Fisheries 9-Marine Recreational Fisheries Development.* Proc. 9th Ann. Mar. Rec. Fisheries Symposium. National Coalition for Marine Conservation, Inc. Publication, Savannah, GA, pp. 218.

Shepard, F.P. 1976. Coastal Classification and Changing Coastlines, pp. 53-64. In: S.J. Walker (Ed.) *Geoscience and Man. Vol. XIV - Coastal Research.* School of Geoscience Louisiana State University Publication, Baton Rouge, LA, pp. 153.

Sigler, W.R. and Sigler, J.W. 1990. *Recreational Fisheries Management. Theory and Application.* Univeristy of Nevada Press.

Smith, S. 1983. *Recreational Geography.* Longman Inc., NY, pp. 210.

Squire, J.L. 1977. *Anglers Guide to the U.S. Pacific Coast, Marine Fish, Fishing Grounds, and Facilities,* NOAA Nat. Fish. Serv., Washington, DC.

Squires, D.F. 1992. Quantifying Anthropogenic Shoreline Modification of the Hudson River and Estuary from European Contact to Modern Time. *Coastal Management,* 20:4:343-353.

Stroud, R.H. (Ed.) *Marine Recreational Fisheries 9-Marine Recreational Fisheries Development.* Proc. 9th Ann. Mar. Rec. Fisheries Symposium. National Coalition for Marine Conservation, Inc. Publication, Savannah, GA, pp. 218.

Taylor, T.L. 1991. Coastal State Parks Wetlands-Islands of Aquatic Diversity-Challenges of Maintaining Functional Ecosystems, pp. 223-231. In: Domurat, G.W. and Wakeman, T.H. (Eds.) *The California Coastal Experience.* Amer. Soc. Civil Engineers, NY, pp. 311.

Tucker, A.R. 1991. The Shell Collector Threat. Sea Frontiers. 37:5:14-19.

Voiland, M.P., Jr. 1984. Coastal Fishing Access: Needs and New Initiatives, pp. 97-104. In: Stroud, R.H. (Ed.) *Marine Recreational Fisheries 9-Marine Recreational Fisheries Development.* Proc. 9th Ann. Mar. Rec. Fisheries Symposium. National Coalition for Marine Conservation, Inc. Publication, Savannah, GA, pp. 218.

Walsh, D. 1987. Pleasure in the Ports: Can Recreation and Shipping Live Together? pp. 737-742. In: Marine Technology Society and IEEE Ocean Engineering Society Oceans 87-The Ocean, An International Workplace, Conference Proceedings, Halifax, Nova Scotia, CA.

West, N. 1990. Marine Recreation in North America. In: Paolo Fabbri (Ed.) *Recreational Uses of Coastal Areas: A Research Project of the Commission on the Coastal Environment, International Geographical Union.* Kluwer Academic Publishers, pp. 285.

Whitesell, D.E. 1987. The Future, pp. 337-339. In: *Fireside Waterfowler: Fundamentals of Duck and Goose Ecology.* D.E. Wesley and W.G. Leitch (Eds.) Ducks Unlimited Inc., Stackpole Harrisburg, PA, pp. 352.

Williams, A.T. 1992. The Quiet Conservators: Heritage Coasts of England and Wales. *Ocean and Coastal Management,* 17:151-167.

Williams, S.J., K. Dodd and K.K. Gohn. 1990. *Coasts in Crisus.* U.S. Geological Survey Circular 1075, USGPO, pp. 32.

Williams, S.J., S. Penland, and A.H. Sallenger, Jr. (Eds.). 1992. Louisiana Barrier Island Erosion Study-Atlas of Shoreline Changes in Louisiana from 1853 to 1989. U.S. Geol. Svy. Misc. Invs. series 1-2150-A, pp. 103.

Williams, Dr. S.J. 1993. (personal interview) National Coastal Geology Program, Office of Energy and Marine Geology, U.S. Geological Survey. National Center MS 914, Reston, VA.

Wright, L.D. and J.M. Coleman. 1973. Variations in Morphology of Major River Deltas as Functions of Ocean Wave and River Discharge Regimes. *Bull. Amer. Assoc. Petroleum Geologists.* 57:370-398.

Zahl, P.A. 1969. The Magic Lure of Seashells. *National Geographic,* 135:3:386-428.

The Oceans: Physical-Chemical Dynamics and Human Impact. Edited by S.K. Majumdar, E.W. Miller, G.S. Forbes, R.F. Schmalz and Assad A. Panah.

Chapter Twenty-Three

EARTH'S OCEANS: The Rarest Jewel in the Solar System

LEE MICHAEL GRENCI
Instructor of Meteorology and
Member of the Penn State Weather Communications Group
College of Earth and Mineral Sciences
The Pennsylvania State University
University Park, PA 16802

I do not know what I may appear to the world; but to myself I seem to have only been like a boy playing on the seashore, and diverting myself in now and then finding a smooth pebble or a prettier shell than ordinary, whilst the great ocean of truth lay undiscovered before me.

Sir Issac Newton

INTRODUCTION

One of the most striking features in photographs of Earth taken by satellites thousands of miles in space is the overwhelming impression of blue, a majesty of color befitting the fact that seventy percent of the Earth's surface is covered by oceans. Water is the only known substance in the universe that can coexist as a gas, liquid and solid within a relatively small range of temperatures and pressures found only near the surface of Planet Earth. In the atmosphere, this trinity of phases is the basis for the Bergeron-Findeisen process, an in-cloud synergy between water vapor, water and ice that creates much of the precious precipitation that falls to earth. Our nearest solar-system neighbors, Mars and Venus, likely had water running on their surfaces at some time, perhaps even seas and oceans. Venus lost almost all of its water. Mars may have been able to preserve some of its water by locking it away in ice below the planet's surface and in the polar caps. But the Red Planet is devoid of surface water today. Meanwhile, Earth, which received approximately

10^{24} grams of water during its embryonic period, has lost only .2% of the planet's original endowment. Each year, Earth loses about 5 x 10^{11} grams of water to space, equivalent to the amount contained in a small lake (Trefil 1984). As the sun increases in luminosity over the remainder of its ten-billion year lifetime (the sun is already about four and one-half billion years old), the Earth may have difficulty preserving its oceans.

Geological evidence supports the fact that Earth has had water on its surface for at least 3.8 billion years. Large bodies of water will likely exist for at least a few billion more years. No other planet in this solar system has been blessed with oceans or seas of such longevity. Like the porridge of the Baby Bear in the fairy tale of Goldilocks and the Three Bears, Earth is "just right" with regard to temperatures (and pressures) conducive for having and keeping water on its surface. Mars is too cold. Venus is too warm. Thus, the scientific investigation into this watery quandary among the three planets has been christened *The Goldilocks Problem.* The traditional solution for the *Goldilocks Problem* has long been a fine balance between the amount of incident solar energy and increasing or decreasing the greenhouse effect of the planet's atmosphere. It appears to be more complicated.

MARS: TOO COLD

They were the first cosmic close-up photographs in history. The alluring model was Mars. The photographer was Mariner 9, a 1960s-vintage spacecraft that maneuvered skillfully above the Red Planet and snapped away with a photographic frenzy reminiscent of a New York advertisement campaign. After years of peering through terrestrial telescopes at fuzzy, inconclusive images of Mars, scientists were ecstatic. They were also stunned by what appeared to be channels that looked as though they had been cut by liquid water (Figure 1).

The idea that running water was once part of Mars' landscape was not new. In the late 1800s and early 1900s, Percival Lowell, an astronomer with a particular passion for Mars, theorized that the network of unnaturally straight lines he had observed on the Martial surface was a system of irrigation canals built by an advanced civilization of Martians trying to control the flow of water on a desiccated, dying planet. But Lowell's canals were optical illusions, mirages created by the Earth's atmosphere distorting the details of the Martian surface. More significantly, these phantoms swayed Lowell's imaginative mind, a mind so hopeful for life on Mars that it ignored the possibility that the apparitions conceivably could have been refractive tricks played on his eyes.

The icy, white caps covering the planet's poles that Lowell viewed through his telescope were real, however. Knowing that water was trapped in the ice, he imagined that the advanced civilization was using every means at its disposal to tap the precious but scarce resource. Lowell's imaginative theory was airtight by his own standards.

Lowell died in 1916, long before the Viking spacecraft that touched down in the mid 1970s proved that the irrigation systems he had telescopically observed were mere illusions. The ice that Lowell observed covering the poles was not the water nirvana he had imagined. The ice is not steadfast. It waxes and wanes in response to changing Martian seasons. Moreover, the ice is only a few centimeters thick, hardly the significant source of water that Lowell had imagined could have aided the Martian civilization trying to keep pace with planetary dehydration. Finally, all of Mars' polar ice is not water-ice. The southern pole contains frozen carbon dioxide (dry ice), a useless substance for a desiccating planet. Lowell was much too optimistic about the amount of available water on the Martian surface.

The atmosphere of Mars is dry too, though there is enough water vapor in the Martian atmosphere to support water-ice clouds. Clouds made of dry ice (most of the thin Martian atmosphere is composed of carbon dioxide) also populate the planet's pink sky. But, in a relative sense, the atmosphere is essentially bone dry. In December of 1990, measurements of infrared energy from water vapor taken by the Very Large Array radiotelescope near Socorro, New Mexico, revealed the smallest concentration of the gas ever observed in the thin Martian atmosphere (*Science News,* December, 7, 1992). Indeed, if all the water vapor in the atmosphere had condensed onto the planet's surface, an "ocean" just three micrometers deep would cover the planet, a sea not deep enough to cover the thickness of the human hair.

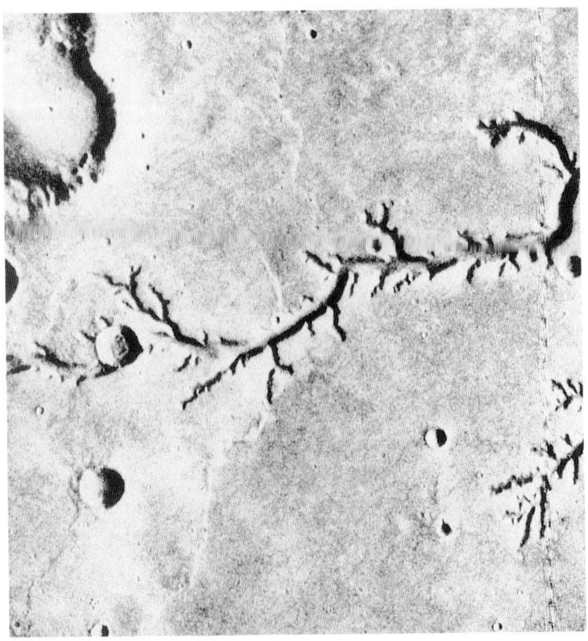

FIGURE 1. In the 1960s, Mariner 9 photographed the surface of Mars, revealing many channels whose presence would suggest the possibility that the planet was once warm enough to support running water.

There were, however, extenuating circumstances. The basis for comparison of these land-based measurements was data gathered by Mariner 9 and the Viking spacecraft. Radiation readings of water vapor from these space missions were gathered during or soon after dust storms, common weather phenomena on a planet whose scenery looks like a extra-arid rendition of the Sahara Desert and the Atacama Desert of South America (where it hasn't rained in some areas for more than 50 years). At any rate, long-lived dust storms likely warm the planet, a warming sufficient to melt ice on the Martian surface. In turn, the melt water would rapidly evaporate, thereby edging up the concentration of atmospheric water vapor slightly above typical levels.

Despite the appearance of Mars being dry, it's not. Though much of Mars' original endowment of water has been forever lost to space, scientists believe that a significant quantity is still locked away in the form of ice beneath the Martian surface. In the region poleward of 30 degrees latitude, enough permafrost may exist to provide enough water to cover the entire planet of Mars to a depth of several hundred meters. Indeed, scientists who theorize about the terra forming of Mars (making Mars habitable) assume a relatively large inventory of available water in the form of subterranean ice (Toon, et al. 1991). Unlike Earth, with water present on its surface for at least 3.8 billion years, Mars has only barren relics of what once were basins for running water.

After the stunning pictures of the Martian run-off channels were beamed back to Earth in the 1960s, Carl Sagan and his colleagues hypothesized that these nearly 4 billion-year-old conduits were cut by running water in a relatively warm environment (Sagan and Muellen, 1972). They suggested an elevated greenhouse warming, enhanced by ubiquitous ammonia. But ammonia is unstable when bombarded by ultraviolet radiation from the sun (Kuhn and Atreya, 1979), its atoms breaking apart like a rack of billiard balls scattering when struck forcefully by a cue ball. That is, ultraviolet radiation converts ammonia to nitrogen and hydrogen, with hydrogen, the featherweight of all gases, escaping to space. Moreover, ammonia is irreversibly converted to nitrogen at a rate much faster than it can naturally be replenished. Sagan's "favorite son", ammonia, soon fell from grace.

Methane was next on the suggested list of greenhouse gases. Methane, like ammonia, is also photochemically unstable (Kuhn and Atreya, 1979). But something else happens when methane is photodissociated. Photochemical smog is made, a fact supported by Saturn's moon Titan, whose atmosphere contains methane. Titan's surface is hidden, shrouded by photochemical fog which gives the moon a hazy, orange appearance. For Mars, it is feasible that just enough photochemical smog could deflect incoming ultraviolet radiation, thereby protecting ammonia from photodissociation. But too much smog would cool the planet and promote sub-freezing temperatures at the surface. The theory is a long shot.

It is interesting to note, that when looking for oceans in our solar system, Titan was once thought to have a global sea of ethane or some other liquid hydrocarbon. By bouncing radar off the surface of Titan, scientists have confirmed that Titan does not possess a global ocean, though they have not dismissed the possibility of ponds or lakes.

Carbon dioxide was also an obvious choice, a choice that stood the test of time of research—until recently. Kasting (1991) showed that in the frigid upper atmosphere of Mars, carbon dioxide-ice clouds can form, clouds that tend to cool the planet's surface and thereby limit or almost eliminate the possibility of liquid water, given the sun's current luminosity. Thus, there are now serious doubts about the role that carbon dioxide played in the evolution of the run-off channels of Mars. Indeed, a scientific scenario based on enhanced CO_2 warming and concomitant rain (or snow) has fallen from favor in the minds of some scientists.

On close inspection of the run-off channels, they bear little resemblance to the river beds of Earth: they are not as branched as Earth's rivers - relatively truncated compared to the nomadic tributaries of our planet. Geomorphologists who have examined photographs of the run-off channels of Mars think that they may have formed from the sapping, or leaching, of ground water (or ice). One possible mechanism would be a discharge of ground water from a release of geothermal heat, energy that was readily available billions of years ago when the Red Planet was young and internally hot. But numerous discharges would have been required to completely carve out the channels, a process that is not accepted by some scientists.

But there are two researchers, Ivett A. Leyva of the California Institute of Technology in Pasadena and Stephen M. Clifford of the Lunar and Planetary Institute of Technology in Houston, who did not give up on the theory, taking a cue from a catastrophic incident on planet Earth. In 1964, the great Alaska earthquake, in addition to destroying towns like Valdez, produced geysers and a bevy of other watery uprisings as far as 400 kilometers from the quake's epicenter. Moreover, changes in levels of water in wells as distant as the town of Perry, Florida, 5500 kilometers from the epicenter, were linked to the great quake. Impressed with such stunning geological evidence, the scientists devised a model for large asteroids slamming into Mars during the period of heavy bombardment around four billion years ago. According to this model, intense shock waves spreading across the planet could have been generated when these giant alien rocks impacted. The shock waves would have exerted tremendous pressure on porous rocks and the water trapped within these rocks. Inevitably, the water pressure would have caused fissures in the Martian surface, rapidly releasing vast quantities of underground water. In turn, the process of erosion would have then carved out the channels.

The answer to running water on Mars is important, not only because of the scientific interest in solving the Red-Planet problem, but because the solution will further help us to understand why Earth was and continues to be blessed with vast oceans, seas that are apparently unique to the solar system. Perhaps the answer to the dilemma of Mars must come from a more cosmic perspective while incorporating planetary-scale considerations. Perhaps the solution emanates from the sun, the source of energy for warming the "porridge" of our solar system.

It is conceivable that the young sun was more energetic than the solar-evolution models (mathematical schemes designed by astronomers that attempt to mimic how stars developed over their lifetimes) *predict* it was four billion years ago. Had the young sun been a mere two percent larger than currently believed, it would have been sufficiently luminous (the luminosity of the sun is a function of its mass) to support liquid water on the Martian surface (Kasting, 1991). If indeed the sun had been slightly larger after the "Big Bang", then it would have had to shed weight via a robust solar wind to attain its current mass, a cosmic crash diet that has not yet been observed from other solar-type stars in our galaxy (Kasting, 1991). Thus, there are uncertainties about the evolution of the sun and its influence on the early Martian climate.

One of the most recent developments in the study of Mars that might explain the run-off conundrum is a new calculation that reveals the chaotic wobbles of the axis of spin of the planet over the past 40 million years (Touma and Wisdom, 1993). Unlike the Earth, whose axis of spin has remained relatively stable over time, tilted at an angle close to 23.5 degrees to the plane of its orbit, Mars may have been a wild "wobbler", oscillating between extreme tilts of 11 and 49 degrees. In the case of the Earth, the gravitational pull of the moon has had a significant stabilizing effect. Mars on the other hand, has relatively tiny satellites, too small to prevent the spin axis of the planet from being erratic. If the obliquity of Mars were high in the past, ice (particularly near the poles) probably became unstable and totally melted (or came close to a total meltdown). If high obliquity (high tilt angles) occurred during a time when a very elliptical orbit brought the planet relatively close to the sun, the melting of ice and the subsequent cutting of aquifers from discharges of water then becomes a viable theory (Kasting, 1991).

That Mars had liquid water on its surface is an intriguing thought. However, in the final analysis, no matter how difficult it is to explain the startling photographs of the run-off channels taken by Mariner 9, Mars **does not** have oceans. Substantial amounts of water are probably locked in ice beneath the Martian surface, but, at least for now, a cosmic key does not exist. Like the porridge for the Mama Bear, Mars is too cold. As the sun increases in luminosity during its main-sequence stage, perhaps some melting could occur in the distant future and the ancient run-off channels might once again turn into babbling brooks. But oceans, in the truest sense of the word, are not likely to form. The atmosphere is too thin and the planet too small to keep hydrogen (hydrogen is a building block of water) from escaping to space. In our search for alien seas of water, the run-off clues on the surface of Mars have, at least temporarily, come to a dead end. We must now turn in another direction - toward Venus. Percival Lowell would be disappointed.

VENUS: TOO HOT

Venus has been Earth's more private neighbor, having long ago shrouded itself with a cloak of mystery comprised of thick clouds laced with sulfuric acid (Kasting et al. 1989). But dauntless spacecraft like Magellan and the Pioneer Venus Orbiter have penetrated the Venusian atmosphere, unlocking some of the mysteries hidden from terrestrial telescopes for centuries. Armed with sophisticated sensors such as radar, Pioneer descended into the Venusian atmosphere on October 8, 1992, burning up soon thereafter and ending its 14-year old mission orbiting around the second planet. Beaming back invaluable data to Earth right up to the final moments,

Pioneer allowed scientists to take a giant step toward solving the quandary of ancient oceans on Venus.

In the past, the idea of oceans once lapping the shores of Venus bordered on the absurd to some scientists. The classical explanation, the runaway greenhouse effect, maintained that Venus was never able to retain a single drop of water on its hot surface (average planetary temperature today is about 850 degrees Fahrenheit). Its proximity to the sun assured sufficient solar energy that any water that appeared on the planet's surface rapidly vaporized.

Scientific pessimism about water on Venus had its roots in a more extreme theory, a belief that Venus was never endowed with much water from the very beginning. This scenario, if correct, would make the debate over Venusian oceans somewhat moot. But Venus likely got what Earth got - a lot of water, enough water, that, if all of the endowment existed as vapor in the atmosphere, it would exert a pressure roughly 265 times the average sea-level pressure on Earth.

At an embryonic stage in the evolution of the two planets, Earth and Venus were massive enough to exchange planetesimals, large chunks of interplanetary rocks that hurdled through space in the aftermath of the "Big Bang", surrendering their precious cargo of elements and minerals upon impact with the growing, accreting planets. If planetesimals brought water to Earth, then they brought water to Venus, given the gravitational give-and-take of planetesimals between the two similar, neighboring planets.

Unlike Earth, Venus' trust fund went rapidly bankrupt. But atmospheric measurements taken by the Pioneer Venus Orbiter have shown that the trust fund was likely not spent all at once. Indeed, there was once probably enough water available to cover the entire surface of Venus with an ocean 25 to 75 feet deep (NASA News Release, Ames Research Center, March 24, 1993).

The task of putting together the pieces of the jigsaw puzzle of the veiled Venusian surface went to Magellan. Magellan was a "scavenger-hunt" craft, a makeshift machine built with spare parts from Galileo, an unused Voyager and other derelict space vehicles. It was launched from the space shuttle Atlantis on May 4, 1989. Sixteen months later, Magellan, using radar as its eyepiece, began the arduous and unprecedented task of mapping the surface of Venus. In just two years, Magellan sent enough data back to Earth to give scientists a composite picture of Venus more complete and more detailed than any map that currently exists of Earth's ocean floor. However, Magellan did not find any run-off channels like those of Mars, not even a small trace that might indicate that water had once flowed on the surface of Venus was discovered.

Among the geological features discovered by Magellan were impact craters, similar to those on the moon that will be discussed in Chapter 18. By counting the number of craters and comparing the result to how many impacts should occur in a given period of time (the moon provides an excellent, comparative record for extraterrestrial bombardment of the early solar system by bodies as large as asteroids), scientists have deduced that the current surface of Venus is roughly a billion years old. Moreover, evidence indicates that Venus likely hasn't had liquid water on its surface for at least a billion years.

With regard to clues leftover from the period prior to a billion years ago, they have been erased, given that the surface of Venus has been resurfaced by volcanism. If anything was to be learned about what happened to all the water, scientists believed that the answers must come from the atmosphere.

Hydrogen is one of the building blocks of water. If hydrogen escapes to space, then, essentially, water is lost as well. Earth is fortunate. Most of the planet's water vapor lies in the lowest two kilometers of the atmosphere. What little water vapor reaches altitudes near the cold tropopause, most of it condenses out as ice crystals or water drops. Because the gradient in concentrations in water vapor below and above the tropopause is small, upward diffusion of water vapor to altitudes where it can be photodissociated by high doses of ultraviolet radiation is also very small. Thus, Earth loses little hydrogen. Moreover, it loses much less deuterium, ("heavy hydrogen", consisting of one proton and one neutron), owing to Earth's gravitational pull and deuterium's larger weight.

Because of the principal of solar abundance, Earth and Venus had, at one time, the same amount of hydrogen and deuterium. That is, the ratio D/H was about the same for both planets. From measurements taken from the Pioneer Venus Orbiter, the ratios are vastly different today, with D/H on Venus 150 times greater than the ratio for Earth (NASA News Release, March 24, 1993). "This means that much more hydrogen had to escape to build up the present high deuterium concentration," said Dr. Thomas Donahue of the University of Michigan, head of the Pioneer Venus science steering group. "A lot more hydrogen lost means a lot more water early on. This also rules out theories of a dry-from-the-beginning Venus, whose present meager supply (total water vapor in the Venus atmosphere exerts about .0038 times the average sea-level pressure on Earth) comes from an occasional comet impact."

Many scientists believe that Venus once had oceans and lost them to evaporation as the sun increased in luminosity. As in the fairly tale of Goldilocks, the "porridge" just got too hot. After evaporation, they theorize that ultraviolet radiation split water vapor molecules apart, allowing hydrogen to escape to space. The escape mechanisms for hydrogen (and deuterium) are complex processes. The most recent breakthrough shedding light on the escape mechanisms came from Donahue and Hartle (1992), who contend that there are large losses of hydrogen on the dark side of Venus during periods of high solar activity. The concentrations of hydrogen ions (H^+) in the nighttime Venusian ionosphere, as measured in the fall of 1992 (solar activity diminishing toward a minimum) by the Pioneer Venus Orbiter, were an order of magnitude less than concentrations obtained by the same instruments during 1978-1980, a time of maximum solar activity.

Their conclusion: "almost all escape occurs during periods of solar maximum." Regardless of the complexity of this and several other proposed escape mechanisms, Venus is just too hot to have oceans. Ironically, though Venus receives 1.9 times more solar radiation than Earth, its thick overcast of sulfuric-acid clouds reflects 80 percent of incoming solar energy. Indeed, Venus absorbs much less of the sun's radiation than Earth. It is the tremendous amount of carbon dioxide (CO_2 in the Venus atmosphere exerts 90 bars of pressure: CO_2 on Earth exerts .00035 bars of pressure) that creates the king of all greenhouse effects in this solar system. But hot is hot, no matter what the reason: Venus has no oceans now. It never will. It is too late. Venus' water is lost forever in space.

EARTH: JUST RIGHT (FOR NOW, AT LEAST)

Liquid water has been on planet Earth for 3.8 billion years, its presence forensically revealed by the sediments in ancient rocks found in Isua, West Greenland. For all of this time, Earth appears to have had its own Goldilocks fairy tale—just the right setting on the planet's climate dials with regard to solar energy and greenhouse effect. From what has been discovered about Mars and Venus, Earth has been astronomically lucky. Had it formed either much closer to or farther from the sun; had the sun been an F-star (the largest and brightest of stars with a relative short lifetime) or an M-star (the smallest and least bright of the stars); or, had the greenhouse been significantly different, Earth's oceans may not have lasted through the millennia, if, in fact, they would have existed at all.

Recent calculations (Kasting, et al. 1992) suggest that in order to have liquid water on the surface of a planet in the solar system today requires the planet to lie between .95 astronomical units and 1.37 astronomical units from the sun, where one astronomical unit (abbreviated A.U.) represents the average distance that the Earth is from the sun. To put these two values into interplanetary perspective, Venus lies at a distance of .75 A.U. from the sun, while Mars resides at a distance of 1.52 A.U.

The narrow zone between .95 A.U. and 1.37 A.U. is called the Habitable Zone, a corridor in which water would likely exist on any planet residing in this zone. The .95 A.U. value is non-negotiable, given the current and increasing luminosity of the sun. The 1.37 A.U. is a conservative estimate, given that some theories suggest that the outward edge of the solar system's Habitable Zone, assuming all planetary factors involved were highly optimal (such as the size of the planet and its concomitant gravitational pull), could possibly extend beyond the orbit of Mars near 1.77 A.U.

The inward value, .95 A.U., should turn the heads of those concerned about future earthlings. As the sun increases in luminosity during its main-sequence time, Earth may have to face the same fate of Venus—losing its oceans, losing its **water**, forever to space.

The Continuously Habitable Zone (CHZ), a theoretical corridor that has remained conducive to habitability over the solar system's lifetime of four and one-half billion years, was a concept originally conceived by Hart (1978). If the Habitable Zone (HZ) is a current snapshot of the solar system, the Continuously Habitable Zone can be considered a full-length movie. Hart's original calculations were stunning, the CHZ defined as the narrow band of space between .95 A.U. and 1.01 A.U. Earth was lucky indeed. According to Hart's calculations, Earth has been uncomfortably close to having its oceans either freeze or boil away.

But Hart's model neglected the strong stabilizing feedbacks on climate by greenhouse gases, CO_2 being the most popular gas in the scientific community (Walker, et al. 1981). Over long time scales (in excess of 500,000 years), atmospheric concentrations of CO_2 are controlled by the carbonate-silicate cycle, a process that accounts for approximately 80 percent of the carbon dioxide exchanged between the Earth's crustal rock reservoir and the atmosphere. During cold epochs, for instance, the process allows carbon dioxide to slowly build up in the atmosphere, thereby gradually warming the planet. But Kasting and Caldeira (1992) argued that oceans can freeze much faster than carbon dioxide can be extracted from rocks via the carbonate-silicate cycle and put back into the atmosphere. Their calculations also suggest that, since the carbonate-silicate cycle is slow on the trigger, CO_2-ice clouds could have formed, cooling the planet and sending Earth into a tailspin toward irreversible, planet-wide glaciation. Nonetheless, they did not abandon the idea that greenhouse warming was somehow very important in maintaining habitability during Earth's early years, when the sun was about 70 percent as bright as its current luminosity.

Given that the sun's luminosity has changed over time, the CHZ is not the same as the HZ. Using greenhouse-gas feedbacks, Kasting calculated that the CHZ lies in the band between .95 A.U. and 1.15 A.U., the latter value once again being a conservative estimate.

Is there extraterrestrial life? If the fundamental requirement for life as we know it is liquid water on the surface of a planet, then the HZ and the CHZ around other star systems is extremely relevant. Broader habitable zones around other star systems probably exist, given the wide range of size and luminosities of other stars. But the crux of the question of extraterrestrial life is the need to know whether other stars do indeed have their own planets.

Most astronomers assert the affirmative, but, so far, there is no proof. Proof has proven to be elusive. But within the next 4 or 5 years, the proof may finally come. It will come from a cosmic eyewitness—the Hubble Space Telescope.

The chances appear to be good. Fogg (1992) pursued the idea of extraterrestrial life, taking into account stellar evolution and Bode's Law (Kasting et al. 1992), an unproved algorithm that predicts the spacing of planets around a star (it is very accurate when applied to our solar system). Assuming an inner edge of .95 A.U. and an outer edge of 2.0 A.U. for our own sun's HZ (a bit optimistic, but reasonable, assuming that there are indeed broader HZ's in the cosmos), Fogg estimates that one out of every 39 star systems possesses a planet with liquid water on its surface. The nearest planet with water on its surface should lie within 14 light years of Earth.

Such a star trek is a long way to go, unless a version of the Enterprise's warp drive is developed in the near future. So NASA scientists have opted for SETI, a program christened the Search for Extraterrestrial Life. The basic philosophy of the program is that if life exists on another planet, perhaps it has evolved into intelligent life, capable of building radio telescopes like those on Earth. Thus, NASA's detection equipment is trained to the heavens, searching for radio waves from alien transmissions. The possibilities seem to be very good. Percival Lowell would have approved.

SUMMARY

Earth's oceans are unique to the solar system. Though Venus and Mars likely had water on their surfaces at one time, water flows no more. Only Earth's seas have stood the test of time. Mars became too cold; Venus too hot. Apparently, Earth has been just right, though the planet's luck could run out in the future as the luminosity of the sun steadily increases. Perhaps there are more planets like Earth in the cosmos, with oceans of water lapping their shores. The search continues.

REFERENCES

Donahue, Thomas M. and Richard E. Hartle. 1992. Solar Cycle Variations in H^+ and D^+ Densities in the Venus Ionosphere: Implications for Escape. *Geophysical Research Letters,* Vol. 19, Number 24:2449-2452.

Fogg, M.J. 1992. An Estimate of the Prevalence of Biocompatible and Habitable Planets. *J. Brit. Interplanet. Soc.* 45:3-12.

Hart, M.H. 1979. Habitable Zones around Main-sequence Stars. *Icarus* 37:351-357.

Kasting, James R. 1991. CO_2 Condensation and the Climate of Early Mars. *Icarus* 94:1-13.

Kasting, James F., Daniel P. Whitmore and Ray T. Reynolds. 1993. *Icarus* 101:108 pp.

Kasting, James F. and Ken Caldeira. 1992. Possible Irreversible Glaciation of Early Earth Caused by CO_2 Clouds. *Nature* 359:226 pp.

Kasting, James F., James B. Pollack and Thomas P. Ackerman. 1984. Response of Earth's Atmosphere to Increases in Solar Flux and Implications for Loss of Water from Venus. *Icarus* 57:335-355.

Kuhn, J.T. and R.E. Dickerson. 1979. Ammonia Photolysis and the Greenhouse Effect in the Primordial Atmosphere of Earth. *Icarus* 37:207-213.

Sagan, C. and G. Mullen. 1972. Earth and Mars: Evolution of Atmospheres and Surface Temperatures. *Science* 177:52-56.

Toon, Owen B., Christopher McKay and James Kasting. 1991. Making Mars Habitable. *Nature* 352:489-496.

Touma, Jihad and Jack Wilson. 1993. The Chaotic Obliquity of Mars. *Science* 259:1294-1297.

Trefil, James. *A Scientist at the Seashore.* Charles Scribner's Sons, New York, New York, pp. 5.

The Oceans: Physical-Chemical Dynamics and Human Impact. Edited by S.K. Majumdar, E.W. Miller, G.S. Forbes, R.F. Schmalz and Assad A. Panah. © 1994, The Pennsylvania Academy of Science.

Chapter Twenty-Four

OIL POLLUTION IN THE OCEANS: Sources, Control and Problems

E. WILLARD MILLER
Professor of Geography
Associate Dean for Resident Instruction, Emeritus
College of Earth and Mineral Sciences
The Pennsylvania State University
University Park, PA 16802

INTRODUCTION

Small quantities of oil have polluted the oceans from underground seepages, but the amounts were so small that there was little environmental damage. Only since oil production has increased in small consuming areas and massive transportation is required to move the oil to the consuming centers has ocean pollution become a major problem. While regulations have gradually evolved to control transport movement, operational pollution and accidents, when hundreds of thousands of barrels of oil are spilled, create environmental and economic havoc in the areas where they occur. While progress is being made to control oceanic oil pollution, enforcement has been slow and remains inadequate.

SOURCES OF OIL POLLUTION

There are four major sources of oceanic oil pollution - natural seepages from underground deposits, pollution from land sources, spills from oil exploitation, and oil pollution from tankers due to operational activities and accidents.

Natural Seepages from Underground Deposits

In many areas of the world, notably the Gulf of Mexico, North Sea, and Persian Gulf, oil penetrates the surface through fissures from oil sands. This oil then escapes into the oceanic waters. This is a natural phenomenon and will continue indefinitely. No reliable figures are available for the amount of oil escaping into the oceanic water each year from fissures, but estimates place the amount between 1,400,000 and 4,200,000 barrels.

Pollution from Land Sources

A major source of oil pollution in the oceans is from the land. Much of the oil spilled on land is ultimately carried by streams to the ocean. With the passage of the Clean Water Act in the 1960s the discharge of oily wastes directly into streams has been lessened. Oily water from plants may be discharged accidentally into river systems, especially during high water stages. The deliberate discharge of oil into a stream is unlawful and penalties are high.

In the past, large quantities of used oil were poured into drains or into the ground during the maintenance of motor vehicles. This practice is now prohibited and the discharge is small. In the operation of vehicles some oil is expelled through the exhaust system which drips onto the roadbed. Much of this oil is ultimately deposited into the oceans. The amount of oil from these sources is unknown, but estimates vary from as low as 7,000,000 barrels to 20,000,000 barrels annually.

Spills from Oil Field Development

In the early days of offshore drilling it was not uncommon for the oil pressure in the producing formation to be greater than expected. Spills occurred when the oil-well casing fractured. Little attention was given to these occurrences until the Santa Barbara, California oil spill, beginning on June 28, 1969 (Easton, 1972). Since then several spills from exploitation have occurred in the North Sea and the Gulf of Mexico. In each of these spills less than 700,000 barrels of oil were spilled, but the public awareness was great and demands to protect the environment increased (Baldwin, 1990).

The equipment and technology to control spills have been greatly improved in recent years. But under the best of conditions some oil will be spilled. It is now estimated that one well in 1,000 will spill small amounts of oil, but only one well in 10,000 will spill sufficient quantities of oil to create an environmental hazard.

Operational Pollution from Tankers

Many of the oil fields of the world are located far from major consuming countries. Thus oil must be transported as either crude or refined products. For most of these areas, such as the Middle East, North Africa, Nigeria, and Venezuela, the only means of shipment is by tankers. As a result of the growing demand for petroleum, the world movement of oil increased from a little less than 3,500 million barrels in 1960 to about 12,250 million barrels in 1973. Since then the amount of oil transported by tankers has increased slowly due to the great rise in oil prices and the development of conservation measures.

The oil discharged into the oceans from tankers comes from four major operational activities - ballasting and deballasting, tank cleaning, discharging bilge water, and transferring oil from one bunker to another. The operational spills are rarely, if ever, spectacular in nature but estimates indicate these sources may spill 10,000,000 barrels of oil into oceanic waters each year.

Ballasting and Deballasting. An oil tanker is a single product vehicle carrying either crude or refined oils. As a consequence the vessel carries oil in one direction only and then returns unloaded to its source area. When the oil is unloaded the tanker floats very high in the water for the weight of the vessel is not sufficient to immerse it deeply into the water. In many unloaded vessels the draught of the bow is only a few feet, and half the propeller may be exposed. In order to remedy this situation a number of the ships' tanks are filled with water. Since the tanks or pumps originally contained oil, the ballast water is contaminated.

Because the ballast water must be removed before a new shipment of oil is received, it must be cleaned at sea for contaminated ballast water cannot be discharged when the vessel is in port. This requires hours of activity in cleaning the tanks and pumps and the excess oily residue that cannot be discharged into the sea is stored in the vessel for later disposal. All vessels are required by international regulations to keep a record of this activity.

Tank Cleaning. Tank cleaning is a major source of oceanic pollution. When oils, particularly those with heavy waxy or asphaltic components are contained in a vessel a small portion of these substances will collect on the tank walls. These layers may be several centimeters thick and commonly consist of 0.3 to 0.5 percent of the oil originally found in the tank. These residues are refinable oils but have long been considered to be waste because of their difficult recovery. In a large vessel they can amount to 5,000 to 7,000 barrels or more in the oil tankers and as much as 2,000 barrels in pipelines and pumps. This provided a potential of nearly 40,000,000 to 63,000,000 barrels of oil to be discharged into the sea annually.

The attempt to control tank cleaning pollution began in the 1920s, but as late as the 1950s it was customary to discharge tank washings and bilge water directly into the ocean. Significant advances were made at the 1954 Oil Pollution Convention when large areas adjacent to ports were defined where discharge of these oily waters was prohibited. Since then with increased movement of oil on the oceans the regulations have been strengthened. The 1969 Intergovernmental Maritime Consultation Organization's regulations have been approved so that no oily substance may be discharged except under the following conditions:

1. The tanker is proceeding enroute.
2. The instantaneous rate of discharge of oil content of any mixture does not exceed 60 liters per minute.
3. The total quantity of oil discharged on a ballast voyage must not exceed 1/15,000 of the total carrying capacity; and

4. The tanker must be more than 50 miles from the nearest land whenever any oily mixture is being discharged.

When these regulations are observed the small amount of oil that enters the ocean is rapidly dispersed and degraded.

Bilge Water Discharge. Small amounts of water, fuel and lubricating oils collect in the bottom of all vessels. After a given time these oily liquids must be pumped out of the vessel. Because these waste liquids are easily observable in the water, their discharge has been prohibited in most ports. If the vessel must remove these oily waters while in port, they are usually provided with a discharge hose from the bilge pump to a cofferdam where the water is stored for later processing.

In order to provide international standards the IMCO 1973 Pollution Convention established strict guidelines for the disposal of bilge water. While these guidelines have not been ratified by some governments, they are slowly being implemented. The International Chamber of Shipping (ICS) has encouraged their acceptance. Under these guidelines bilge water discharge is permitted only when a vessel is enroute and is more than 12 miles from land and only when the oil and water is separated so that the oil content does not exceed 100 ppm. Oil-water separators are a standard feature of most modern-day vessels. Nevertheless, these voluntary regulations are frequently ignored in the open ocean. It is estimated that 2,000,000 barrels of oil are deposited from bilge waters annually in the oceans.

Oil Transfer Operations. There are a number of operations associated with tankers when spills may occur. At times oil must be transferred within the vessel. This transfer is performed by using bunker transfer pumps. In this process, the most common cause of spills is overfilling the tanks for this operation may require several hours and the pumping rate may be misjudged.

Oil may also be spilled when it is transferred from ship to ship or from ship to a barge. This practice has greatly increased in recent years as tankers have grown in size and cannot enter shallow harbors. The oil is thus unloaded from the supertanker in the open sea into smaller vessels that can enter the shallow ports.

Tanker Accidents

Because of the large volume of oil carried and the number of tankers engaged in this activity, occasional accidents occur.

Structural Failure. The most common accident creating an oil spill is due to structural failure of the vessel. This single cause accounts for over one-third of the accidents. To illustrate, on March 16, 1978 the Amoco Cadiz, with a cargo of 1,500,000 barrels of light Arabian crude oil was approaching the English Channel. About 10 miles off shore there was ruder failure and, before a tug could aid the vessel, the weather deteriorated and the tanker grounded on the Brittany coast. Although the crew was rescued, on March 24 the vessel split in two spilling its entire cargo. A large proportion of the oil was carried onto the coast of Brittany.

Grounding the Vessel. The second most important cause of accidents is grounding the vessel. The first supertanker to ground was the Torrey Canyon on Seven Stones Shoal about 21 miles off Cornwall's Land End on March 17, 1967. The tanker spilled 840,000 barrels of Kuwait oil into the sea.

In American waters, the most spectacular grounding was that of the Exxon Valdez on March 24, 1989 on Bligh Reef in Prince William Sound shortly after leaving Valdez, Alaska. Several hundred thousand barrels of crude were spilled into the sound. The vessel was guided by an inexperienced pilot and did not follow the correct channel. This illustrates that many of the groundings are due to human error.

Accidents are also caused by the physical elements of the region. On January 5, 1993 the oil tanker Braer was driven aground by a heavy storm at the southern tip of the Shetland Islands, about 100 miles north of Scotland in the North Sea. By January 12 huge waves driven by hurricane-force winds had shattered the ship's hull, spilling its cargo of 620,000 barrels of crude oil. The accident was caused when power in the vessel failed due to sea water contamination in the fuel lines. On January 12 oil gushed from the hull. Fortunately the vessel was carrying a light grade of crude oil that was quickly thinned out and scattered by the high winds and rough sea. Nevertheless a light sheen of oil spread along 45 miles of Shetland's shoreline. The high winds also blew oil residues inland, fouling buildings and roads and contaminating gardens and sheep pastures. Conservationists estimated 6,000 seabirds were killed. The Scottish government placed a ban on all fishing within 50 miles of the spill site, and many European food wholesalers cancelled their orders from commercial salmon farms elsewhere in the Shetlands. The British government began a public inquiry about the Braer wreck. British transport secretary John MacGregor indicated that the inquiry would include a study to determine whether British laws provided sufficient safeguards against future severe oil spills.

Collision of Vessels. Collision of tankers is the third major cause of accidents causing oil spills. A spectacular collision between two tankers, the Ven Oil and the Ven Pat occurred December 16, 1977 off Port Elizabeth, South Africa. Each vessel was of 325,000 registered gross tons. One vessel was loaded, but the other was in ballast. As a result of the collision about 210,000 barrels of oil were spilled, polluting a considerable area of the South African coast. Collisions are usually due to negligence. If the sophisticated navigation equipment carried by tankers, particularly the supertankers, is used properly, these collisions could be avoided.

Fires and Explosions. Fires and explosions constitute the fourth major hazard causing tanker spills. These hazards frequently occur after another accident has damaged the vessel. For example, the Jacob Maersh struck an underwater obstruction while trying to berth at Oporto, Portugal. This accident triggered an explosion and fire and the entire cargo of 240,000 barrels of crude oil was either burnt or spilled into the ocean.

At times a fire may remain undetermined. On August 6, 1983 a Spanish supertanker laden with about 1,225,000 barrels of Persian Gulf oil burst into flame about 80 miles northeast of Capetown, South Africa. Some eight hours after the fire started the ship broke apart in choppy seas. The Benguela Current swept the oil toward South Africa's Langebaan Lagoon, a 15-mile long coastal sanctuary for marine and bird life. Although a large percentage of the oil burned, oil streamed from the bow and stern spreading to form a 3-mile wide slick. Further, it was feared that if the bow and stern sank close to shore, oil would ooze from the wreckage for weeks. At the height of the fire, black smoke rose 8,000 feet into the sky and could be seen 180 miles away. Farmers along the coast, where wind delivered a choking day-long smoke, reported an oily film that covered grainfields and vineyards. Newly shorn sheep became oil slicked, and black goo accumultated on houses and cars.

Other causes of accidents include breakdowns, ramming of vessels and reasons that cannot be determined.

CONTROL OF OIL SPILLS

Although massive oil spills have been occurring for nearly three decades, there is still no acceptable technology to protect the oceanic environment. No nation has developed plans to manage a major oil spill including such aspects as a computer system to track the spill movement, the location of vulnerable environmental areas, the availability of personnel or an acceptable clean-up technology. Many techniques have been attempted but all have limitations. One of the major problems in the development of the techniques to control oil spills is that they occur in a wide variety of physical and biological environments. Each area thus has unique features that require a very specific approach.

Crude Oil Recovery. After an oil spill an initial response is to attempt to recover as much oil as possible. To illustrate, after the Valdez oil spill there was a "window of response" that lasted about 72 hours during which time removal of oil from the surface was possible. However, oil removal equipment was not available. The initial "window" ended when the oil was emulsified during a stormy day. After the storm a significant portion of the emulsified oil could have been recovered by mechanical pick up. This period also represented the optimal time for preventive booming to protect sensitive beaches (Harrold, Marcus, Wallace, 1989).

During the 10-day-window there was a lack of coordination between industry, state and federal agencies. In this period only about 5 percent of the oil was removed, dispersed or burned. With an optimal response another 10 to 20 percent of the oil could have been prevented from reaching shore. After the first ten days, the oil had covered the beaches, and little could have been done to reduce its impact on the beaches. During severe weather in Prince William Sound no amount of equipment or dispersants could have kept the oil from the beaches.

Coastal Control. A number of techniques have been attempted to remove oil from coastal areas. In the early coastal oil deposits at Santa Barbara and from the Torrey Canyon spills, detergents were utilized. In the Torrey Canyon disaster over 1,000,000 gallons of detergent were sprayed on Cornwall beaches. It was hoped that the detergent would break up not only the crude oil but also the mousse, a water and oil emulsion. The mousse, formed by wave and wind action, is bound together by the asphaltic oil residues. When detergent is poured on the beach, the ensuing tides cause the detergent, oil and mousse to move downward in the sand as much as 5 feet. As a consequence, there is a structural breakdown of the compounds creating a quicksand. Further, the interaction of the oil and detergent causes a caustic effect on the skin as well as a stench. As a result of these early endeavors it was found that detergents are only effective in cleaning oil from rock surfaces (Jordon, Payne, 1980).

There have also been attempts to find materials that will absorb the oil as it reaches the shore. Talc and perlite have been somewhat successful, but straw and sawdust have proved to be most effective. In Prince William Sound after the Valdez accident steam and hot water were utilized to remove oil from the rocks. Many questions have been raised about this technique that used high temperatures and high pressure sprays, repeated 10 to 20 times, to clean the oil from the beach. In the cool waters of the Sound the hot waters have a tendency to kill all life thus destroying the natural ecosystem. Of all the techniques the most effective method to clean rock is sandblasting, but this is time consuming and relatively expensive.

In beach clean-up the process used depends considerably on whether tourism or the saving of the fishing industry is most important. For example, the major fishing industry along the coast of Brittany has been saved by pumping the oil and mousse into scavenging vessels and by the use of stearated chalk to sink the mousse to the ocean floor. The mousse, once deposited on the bottom, does not surface for it has a density close to seawater and remains in place. In a short time sediments solidify the entire mass of material.

ENVIRONMENTAL IMPACTS

The ecological damage to fauna and flora after an oil spill varies greatly depending upon physical conditions of the region. Although a large number of studies have been made on the effects of oil spills, conflicting interpretations do not provide a sound basis for the development of standardized controls.

Birds

As oil reaches a shoreline birds suffer the most evident immediate loss. In the Santa Barbara oil spill it was estimated a minimum of 6,000 and possibly as many as 15,000 birds died as a result of oil contamination. In the Torrey Canyon spill it was reported that at least 30,000 birds were killed on the Cornwall and Brittany coasts. The U.S. Fish and Wildlife Service estimated that in Prince William Sound more than 33,000 birds died from oil contamination. In addition, an observer of aberrant bird behavior indicated that 75 percent of the bald eagles of the area failed to nest in 1989.

In all coastal disasters there have been major attempts to save the birds. However, when the birds are covered with oil, the survival rate is amazingly low, usually between one and 11 percent. The death of the seabirds is due to shock, improper cleaning and handling, and ingestion of oil and toxic chemicals by preening. Many birds die of starvation and exhaustion, being unable to hunt for food.

Birds also die from pneumonia because the insulating ability of the feathers is destroyed by oil accumulation. In the Torrey Canyon spill the Plymouth laboratory reports indicated that man-made pollutants, such as detergents, were far more damaging to birds than the oil from the spill.

Although the initial impact on the bird population is devastating, there is no indication that permanent damage has occurred. Nevertheless an ecological disaster is possible. The coasts of southern England and Brittany are breeding grounds of such seabirds as the guillemots, rajorbills, shags, preffins and the great northern divers. The Santa Barbara area is the breeding ground of the brant. If the birds had migrated to these areas at the height of the oil spills, entire species could have been exposed to oil pollution.

Marine Life

Of the oil spills affecting coastal United States, the Santa Barbara and Prince William Sound are best documented. The effect on marine life varies greatly. In the Santa Barbara spill plankton and surf grass suffered significantly. Some colonies of acorn and goose-necked barnacles were virtually destroyed.

In most of the early oil spills the amount of damage to marine life was not conclusive due partly to lack of records prior to the spill. In the Santa Barbara spill statistical surveys conducted by the state, federal and academic investigators showed no significant damage to the fish population due to the spill. The major problem in determining the effect of the Santa Barbara oil spill on the fish population was that there was little information on the movement of migratory fish. Large numbers of anchovy, mackerel, and bonito entered and left the channel during the spill's early months. Dr. Herbert W. Frey, a senior marine biologist of the California Fish and Game Commission stated, "We have detected no impairment of numbers, reproduction of feeding activities attributed to oil pollution. That's not saying that five years from now we won't find something different."

The Valdez oil spill occurred in a sparsely populated region in which fishing is the major economic activity in the towns of the Prince William Sound. Thus, this region provides the opportunity for a case study of the effects of an oil spill on a major fishing region. In 1988, the year prior to the oil spill, these waters yielded \$131 million worth of salmon, herring, halibut and shellfish. The spill occurred just prior to the herring season. As an immediate response, the herring season was canceled with a loss to local fisherman of about \$12 million. The oil spill also occurred just prior to the scheduled release of millions of salmon fry to the ocean.

The long-term economic impacts on the fishing industry can not be evaluated for a number of years. In 1993 Exxon indicated that there was little environmental damage done to the region. In contrast environmental groups still persist in the viewpoint that ecological damage will be evident for many years.

Mammals

Oil spills have an immediate effect on mammals in the sea. The Santa Barbara oil spill area is noted for the California sea lions. Of the 881 living and dead sea lion pups observed on San Miguel Island by the scientists from the State Fish and

Game Commission, slightly less than half were oily, but about two-thirds of all dead pups were oily. Thus many more pups that were oily died than would have occurred under natural conditions.

Only three dead dolphins were found in the Santa Barbara Channel during the early days of the oil spill. While an autopsy of one did not reveal it died from oil contamination, eyewitnesses indicated its breathing hole was filled with oil and its lungs were ruptured. The dolphins, however, fled the area before great damage was inflicted on their population.

During the oil spill five dead whales were washed ashore on the California coast. Of these, three were grey whales, one was a sperm and one was a pilot whale. As with the other mammals the effect of the oil contamination was not conclusive. It appeared that oil caused the death of one of the whales.

In the Prince William Sound area the devastation of the mammal population was more evident. This area is the home of about 5,000 sea otters. They were especially vulnerable to oil because it reduced their insulation and buoyancy of their fur. In an attempt to survive they can ingest fatal doses of oil when licking it off their coats. Of the dozens of otters that were cleaned of oil in the weeks after the spill, only a few survived. Hundreds died in the first days after the spill on remote oil-fouled shores. By fall 1989 an initial evaluation of damage to the animal life of Prince William Sound was made. It is conservatively estimated that 980 otters, 30 harbor seals, 17 grey whales, and 14 sea lions had died.

ECONOMIC COSTS

The economic cost to Exxon of the Valdez oil spill exceeded $500 million in the first two years after the accident, but the ultimate cost will not be determined for many years. In 1991 Exxon was fined $100 million, but this amount is being contested in the courts as being too low. There is also a court decision that Exxon pays Alaska $1.2 billion for damages. This decision is also being contested. Besides the major costs there are scores of smaller claims that will be adjudicated by the courts in the next decade. In contrast, the large sums of money infused into the fishing community for clean-up costs were generally viewed by the local inhabitants as a boon of virtually unprecedented proportions. It not only cushioned the losses from fishing, but provided additional revenues. The local labor supply was inadequate and more than 4,000 workers were attracted to the region.

SOCIAL CONDITIONS

Because the communities on Prince William Sound are isolated they provide case studies as to how people react under stress. To measure the social impact on a community, the team of Picou, Gill, Dyer, and Curry (1990) made a survey of the people of Cordova and Petersburg as to how the oil spill had changed their lives.

Cordova is dominantly a fishing village on Prince Edward Sound isolated from other settlements by mountains, glaciers and the sea. Cordova fisherman hold 55 percent of the salmon and 44 percent of the herring licenses of the Prince William Sound area. The towns people have a history of subsistence practices stemming directly from a Native-American heritage. About 20 percent of the residents of Cordova are Native-Alaskans. Although the oil spill did not reach the Cordova beaches, it impacted critically on fishing grounds used by local fishermen.

Petersburg was used as the control town for it was not affected by the oil spill. The research design included the collection of data in a stratified random sample of households, an ethnographic sample of native Alaskans, and a random telephone survey of the inhabitants.

Four questions were asked of residents in Cordova and Petersburg. The first question was, have you noticed any changes in the way your family gets along together? In the control area (Petersburg) 91 percent responded no change and 9 percent a change, while in the impact (Cordova) area the response was 61 percent no change and 39 percent a change in family relationships. Of the question, have you made any changes in the plans for the future, the families of the control area responded 86 percent no, and 14 percent yes, but in the impact area 51 percent answered yes, and 49 percent no. To add to the information, the question, have other family members changed their future plans, was asked. The impact area families indicated 70 percent no and 30 percent yes, and in the control area, 83 percent no and only 17 percent yes. Finally the question was asked, have things changed for you at work? The Cordova residents responded 68 percent yes and 32 percent no, while the Petersburg residents indicated 81 percent no change and only 19 percent a work change (Picou et al., 1990).

The results indicate that the spill had a significantly greater social disruption in Cordova than in Petersburg. These disruptions included family relations and future plans of the community members. The four researchers who conducted this study felt that the general uncertainty that characterized the Cordova residents was directly related to the threat posed by the spill for future economic viability in the community.

DEVELOPMENT OF ENVIRONMENTAL LAWS AND REGULATIONS

The first international controls on shipping began in the 1940s when the United Maritime Authority was established (Cycon, 1981). In 1948 the UMA was succeeded by the United Maritime Consultative Council (UMCO). In that year governments were invited to attend a United Nations Maritime Conference in Geneva to consider the establishment of an international body within the United Nations. As a response the Convention created the Intergovernmental Maritime Consultative Organization. It was not, however, until 1958 that the required 21 nations ratified the Convention and IMCO began to operate. In its initial years this body was primarily concerned with development of safety regulations.

IMCO does not have the power to enact laws. As the word "consultative" in its title implies it can pass recommendations, convene conferences, draw up conventions, and facilitate consultation among member states. The resistance to IMCO as a law-making body was based on the grounds that this action would introduce political implications into an area that was fundamentally technical in nature. Although the initial Convention of IMCO did not mention a pollution control function, IMCO extended its activities in this direction at an early date. By 1965 the pollution control aspect had become so important that IMCO established a subcommittee on Oil Pollution under its Maritime Safety Committee.

With the Torrey Canyon oil spill in 1967 IMCO increased its activites to control oil pollution in the oceans. For example, it organized the Conference on Marine Pollution in 1973 and the Tanker Safety and Pollution Prevention Conference in 1978. IMCO established a permanent committee, the Marine Environment Protection Committee, to examine the legal aspects of accidents at sea that create pollution problems. Within IMCO the Maritime Safety and Pollution Control Committees are now of equal importance. The increased involvement in pollution control has given the organization vitality. In the control of oil pollution of the oceans, IMCO has recommended a vast number of regulations. The individual nations have made many of these regulations laws and have extended them to protect their own waters. Regrettably IMCO has a small staff and a very limited budget to carry out an effective program. Most of the regulations developed by the United Nations do not treat the management of oil spills after an accident but rather the problems of oil transportation.

MANAGING ACCIDENTAL OIL SPILLS

The difficulty in preparing for and responding to accidental oil spills stems from the fact that they are extremely rare events (Harrold, 1989). Well over 99 percent of the oil is transported on the oceans safely. Society has not learned how to deal effectively with low probability, high consequence events. The attitude of the public and governmental officials toward these events tends to be polarized. When there is low probability the normal reaction is that the event will never occur and therefore any response plan is wasteful. In contrast, others recognize the consequence of the possible disaster, and would totally prohibit the activity. In the latter situation, the catastrophe develops high initial interest, and if it does not reoccur, interest quickly diminishes. A system has been developed to handle small accidents that occur routinely, but a procedure has not been developed to deal rationally with catastrophic and rare events (Wenk, 1986).

The major oil spills beginning with the Torrey Canyon of the late 1960s and culminating with such recent spills as the Exxon Valdez and the Shetland Islands spill of 1993 demonstrated that these catastrophic accidents cannot be ignored.

An initial step is to reduce the risk of an accident during normal transportation of oil. Most present-day tankers are equipped with modern navigational facilities. Personnel must be trained and regulations must be enforced to ensure the maximum safety in navigation. In addition procedures must include the making of the port facilities and the harbor safer for navigation of the modern, huge tankers. These efforts must include control of the vessel traffic in the harbor and the enforcement of personnel standards.

A longer-term change must require all tankers be built with a double hull. With the traditional tanker when the single sheet metal hull is pierced the oil immediately spills into the sea. With a double hull the possibility of piercing the inside hull is greatly reduced. However, the world's tanker fleet is now adequate to transport the present demand for oil to consuming centers. Because an oil tanker has a life of 30 to 40 years, there will be few new tankers built in the immediate future.

In spite of all precautions, accidents will occur. A system must be developed that can effectively control the environmental and economic damages and keep social disruptions to a minimum. In all of the spills the determination of the ecological impact on the ecology of the area has been hampered by a lack of scientific information. A hodge-podge of scientific studies has been compiled, but no nation has developed a comprehensive analysis.

To correct this deficiency in the United States, the U.S. Environmental Protection Agency in 1990 began a long term project known as EMAP - The Environmental Monitoring and Assessment Program. The program will collect basic scientific data on the nation. When this information is placed on a grid it can provide basic environmental and ecological information for any ocean area around the United States. If, for example, a tanker is grounded in the San Francisco Bay area, scientists could immediately secure information about the environment of the area. EMAP will provide a geographic information system that will be vital for decision making in oceanic disasters.

Beyond these intial scientific and technological endeavors there must be contingency planning in the high risk areas so that an adequate response can be mounted at the time of the accident.

CONCLUSIONS

Considerable progress has been made in controlling oil pollution of the oceans. The Intergovernmental Maritime Consultative Organization has developed a vast body of laws and regulations that has reduced spills in the normal movement of petroleum. Accidental spills, however, remain one of the greatest hazards in the movement of oil. Although accidents are now rare they are frequently devastating to coastal areas when they do occur. While accidents may never be eliminated, the problem of oil pollution in the environment is recognized and better controls to handle spills remain a fundamental problem.

SELECTED REFERENCES

Asadi, G.V. and H. Vaughan. 1990. Dynamics and Structural Damage of Tanker Ships Running Aground. *Journal of Hazardous Materials.* 25(Oct):61-73.

Baldwin, Malcolm F. 1990. The Santa Barbara Oil Spill. *University of California Law Review.* 42(May):33-76.

Boisch, Donald F., Carl H. Hershner and Jerome H. Milgram. 1974. *Oil Spills and Marine Environment.* Cambridge, MA: Ballinger, 114 pp.

Borthwick, A.G.L. and S.A. Joynes. 1992. Laboratory Study of Oil Slick Subjected to Nearshore Circulation. *Journal of Environmental Engineering.* 118(Nov.-Dec.):905-922.

Brown, Joseph E. 1978. *Oil Spills: Danger in the Sea.* New York: Dodd Mead, 123 pp.

Cole, G. Mattney. 1993. *Assessment and Remediation of Petroleum – Contaminated Sites.* Boca Raton, FL: Lewis Publishers.

Conrad, Jon M. 1980. Oil Spills: Policies for Prevention, Recovery and Compensation. *Public Policy.* 28(Spring):143-170.

Cormack, D., et al. 1982. *Oil Map Device for Oil Recovery on the Open Sea.* New York: State Mutual Bank.

Cycon, Dean E. 1981. Calming Troubled Waters: The Developing International Regime to Control Operational Pollution. *Journal of Maritime Law and Commerce.* 13:35-51.

Davidson, A. 1990. Valdez Reflections. *Sierra.* 75(May-June):42-49 + .

Dempsey, Paul Stephen and Lisa L. Hellwig. 1980. Oil Pollution by Ocean Vessels—An Environmental Tragedy: The Legal Regime of Flags of Convenience, Multilateral Conventions, and Coastal States. *Denver Journal of International Law and Policy.* 10(Fall):37-87.

Easton, Robert. 1972. *Black Tide: The Santa Barbara Oil Spill and Its Consequences.* New York: Delacorte Press, 336 pp.

Fairhall, David and Philip Jordan. 1980. *The Wreck of the Amoco Cadiz.* Briarcliff Manor, NY: Stern and Day, 256 pp.

Francis, R.D., et al. 1992. Rapid Disappearance of Oil from a Moderate-sized Spill: Cleanup Effort Coupled with Natural Degradation Processes. *Marine Technology Society Journal.* 26:26-33.

Frey, R.W., et al. 1989. Coastal Sequences, Eastern Buzzards Bay, MA: Negligible Record of an Oil Spill. *Geology.* 17:461-465.

Galt, J.A., et al. 1991. Fate and Transport of the Exxon Valdez Oil Spill. *Environmental Science and Technology.* 25:202-209.

Grambles, B.H. 1990. Federal Oil Spill Legislation in the Wake of the Exxon Valdez. *Marine Technology Society Journal.* 24:5-11.

Harper, J.B. 1992. The Leading Edge of an Oil Slick, Soap Film, or Bubble Stagnant Cap in Stokes Flow. *Journal of Fluid Mechanics.* 237:23-32.

Harrold, John, Henry Marcus and William A. Wallace. 1989. *The Management of a Maritime Crisis: The Integration of Planning, Prevention and Response.* Boulder, CO: The Natural Hazards Research and Information Center, Quick Response Report, #34.

Jackson, J.B.C., et al. 1989. Ecological Effects of a Major Oil Spill on Panamanian Coastal Marine Communities. *Science.* 243:37-44.

Jordon, Randolph E. and James R. Payne. 1980. *Fate and Weathering of Petroleum Spills in the Maritime Environment: A Literature Review and Synopsis.* Ann Arbor, MI: Ann Arbor Science. 170 pp.

Kardyban, E. 1992. The Effect of Waves on the Oil Slick at a Retention Boom. *Journal of Energy Resources Technology.* 114:31-37.

Kennicu, M.C., 2nd. 1990. Oil Spillage in Antarctica [Grounding of Bahia Paraiso].

Environmental Science and Techology. 24:620-626.

Krahn, M.M., et al. 1992. Mass Spectrometric Analysis for Aromatic Compounds in Bile of Fish Sampled after Exxon Valdez Oil Spill. *Environmental Science and Technology.* 26:116-126.

Lynch, B. and J. Smith. 1981. *Dispersants for Oil Spill Clean-Up Operations at Sea, on Coastal Waters and Beaches, 1979.* New York: State Mutual Bank.

Mason, Rachel. 1989. *Community Preparation and Response to the Exxon Oil Spill in Kodiak, Alaska.* Boulder, CO: The Natural Hazards Research and Information Center, Quick Response Report #36.

Ohlendorf, Harry M., Robert W. Risebraugh and K. Vermeer. 1978. *Exposure of Maritime Birds to Environmental Pollution.* Washington, DC: Department of Interior, Fish and Wildlife Service. 40 pp.

Petrow, Richard. 1968. *In the Wake of Torrey Canyon.* New York: D. McKay, 250 pp.

Picou, Steven, Duane A. Gill, Christopher L. Dyer and Evans W. Curry. 1990. *Social Disruption and Psychological Stress in an Alaskan Fishing Community: The Impact of the Exxon Valdez Oil Spill.* Boulder, CO: The National Hazards Research and Information Center, Quick Response Report #35.

Pritchard, P.H. 1991. Bioremediation as a Technology: Experiences with the Exxon Valdez Oil Spill. *Journal of Hazardous Materials.* 28:115-130.

Riser-Roberts, Eve. 1993. *Biomediation of Petroleum Contaminated Sites.* Boca Raton, FL: Lewis Publishers.

Smets, Henri. 1983. The Oil Spill Risk: Economic Assessment and Compensation Limit. *Journal of Maritime Law and Commerce.* 14: 23-43.

Special Report: Troubled Waters. 1989. *Americus Journal* 11 (Summer):10-31. (Five articles assessing the aftermath of the Valzez oil spill).

Swift, M.R., et al. 1992. Diversion Oil Booms in Current. *Journal of Waterway, Port, Coastal and Ocean Engineering.* 118:587-598.

U.S. Congress. House. Committee on Interior and Insular Affairs. Subcommitee on Water, Power and Offshore Energy Resources, 1989-1990. *Investigation of the Exxon Valdez Oil Spill. Prince William Sorel, Alaska: Oversight Hearings: Pts 1-3, May 5- July 28, 1989.* 101st Cong. 1st sess. Washington, DC: GPO, 3 pts.

Valenti, M. 1991. Oil Recovery: Technology that Tames Large Spills. *Mechanical Engineering.* 113:44-49.

Wardley-Smith, J., ed. 1982. *The Control of Oil Pollution on the Sea and Inland Waters.* New York: State Mutual Bank, 264 pp.

Wenk, Edward J. 1986. *Tradeoffs: Imperature of Chance in a High-Tech World.* Baltimore, MD: John Hopkins University Press.

Yaroch, G.N. 1991. The Nestucca Oil Spill: A Christmas Story. *Chemtech,* 21:722-727.

The Oceans: Physical-Chemical Dynamics and Human Impact. Edited by S.K. Majumdar, E.W. Miller, G.S. Forbes, R.F. Schmalz and Assad A. Panah. © 1994, The Pennsylvania Academy of Science.

Chapter Twenty-Five

POLLUTION IN THE MEDITERRANEAN SEA: Problems and Control

E. WILLARD MILLER

Professor of Geography and Associate Dean for Resident Instruction (Emeritus)
College of Earth and Mineral Sciences
The Pennsylvania State University
University Park, PA 16802

INTRODUCTION

The Mediterranean Sea is the major inland water body of the world. The people residing on its shores dating back to prehistory have irreversibly transformed this fragile environment. It is, however, only since the 1960s that it has been recognized that the vast Mediterranean Basin is in danger of being devastated by increasing pollution. This chapter describes and analyses some of the most pressing environmental problems. Because of the deterioration of the ecosystem it is now recognized that an international effort is needed to maintain a viable environment.

PHYSICAL AND HUMAN CHARACTERISTICS

The Mediterranean Sea has a number of physical characteristics that encourage the build up of pollution. It is nearly an inland sea for its only connection to the Atlantic Ocean is through the narrow, shallow neck of the Strait of Gibraltar. It takes almost a century for the Mediterranean water to be renewed. Within the basin the tides are small and the weak currents are regional so that pollution remains largely where the rivers discharge their waste into the sea (Figure 1).

The Mediterranean Basin has long, hot and dry summers and warm moist winters. Evaporation of the water in summer far exceeds the precipitation so that pollution increases during the warm season. This is especially true on the south (African) side of the basin which receives only 15 percent of the region's rainfall.

The population of the Mediterranean Basin has experienced remarkable growth in recent decades. In 1950 the total population was about 212 million. By 1985 it has reached 356 million. By 2025 it may exceed 550 million. Most of this growth will occur in North Africa, Turkey, and Syria. The countries along the southern shore of the Mediterranean have among the highest growth rates in the world. North Algeria's population will increase by 30 percent this decade and Morocco by 25 percent.

Today about 38 percent of the population of the countries around the Mediterranean crowd into 15 percent of the coastal zone. In Algeria 53 percent occupy 3 percent of the land area. In Spain the share of the population along the Mediterranean coast has risen from 12 percent in 1900 to 35 percent presently. It is predicted that the urban population in North Africa will more than triple by 2025.

NEED FOR POLLUTION CONTROL

It became evident in the 1960s that pollution had become a major problem in the Mediterranean Basin. In 1972 Lord Peter Ritchie-Calder stated (Ritchie-Calder, 1972):

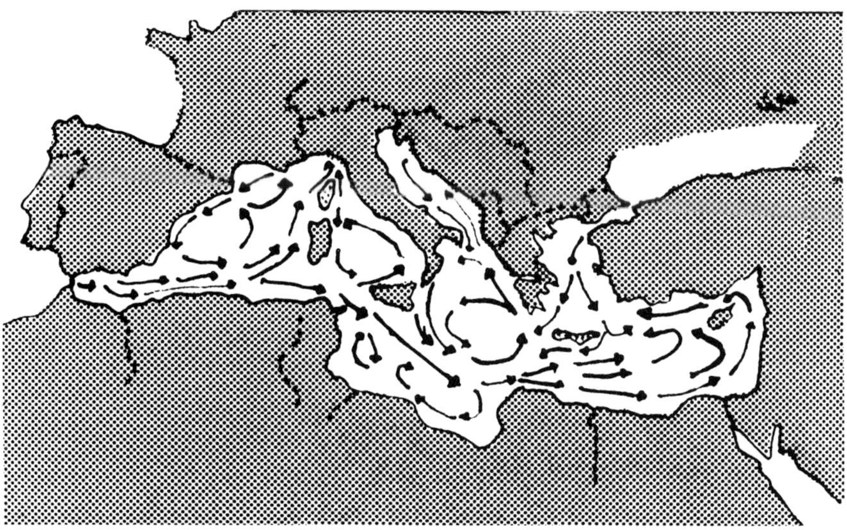

FIGURE 1. General Flow of Surface Currents. Because of the configuration of the Basin and the pattern of islands, currents of the Sea are highly interrupted. As a consequence pollution is highly concentrated due to the type of current movement. The only connection of the Mediterranean Sea with its Atlantic Ocean is at the narrow Strait of Gibraltar. Water of low salinity enters the Sea in a surface current. The highly concentrated saline water of the Sea flows into the Atlantic from a sea bottom current. These two currents flowing in opposite directions maintain a balance between evaporation and maintaining the level of the Sea. Source: Compiled from various sources.

> The Mediterranean Sea is sick. It needs intensive care, day and night nursing. By nature it has always been delicate but its condition has greviously deteriorated in recent years. The short term prognosis is obvious: on present trends, things will get worse because the effects will be multiplied and magnified by an increase in industrial activity. Recreational beaches will be tarred and ordured. The sea will be out of bounds for bathers. Seafood will be a health hazard. In terms of epidemic disease, the Mediterranean can become a biological time bomb. The trees will be dying around the coasts, suffocated or poisoned by polluted sea winds (Haas, 1990).

SOURCES AND LOCATION OF POLLUTION

With over 100 million people living in the coastal areas of the Mediterranean, pollution comes from the emission of untreated or inadequately treated wastes that flow into the sea. There is also oil pollution for the Mediterranean is a major navigation route for tankers coming from the eastern and southern Mediterranean as well as those coming through the Suez Canal. About 40 percent of the world's tanker traffic passes through the Mediterranean. (Table 1).

There are 537 cities of over 10,000 population on the Mediterranean coast. Most of these cities produce a great deal of pollution. Few of them have sufficient sewage treatment facilities. Even when they do exist many do not function effectively. More than two-thirds of Italy's secondary treatment plants work poorly or not at all. All of the sewage of Athens is pumped raw into the Saronic Gulf. Most of the organic, untreated wastes contain microorganisms and bacteria that spread such diseases as typhus, hepatitis, typhoid, cholera, dysentery, polio and others. These microorganisms can be particularly hazardous when they contaminate seafood. In 1973 an outbreak of hepatitis in Naples was traced to polluted shellfish (Brisou, 1976).

TABLE 1

Types and Sources of Mediterranean Pollution

Region	Organic Matter	Nutrients	Specific Organics	Pesticides	Metals
1	40.7	43.1	21.8	30.0	35.8
2	28.1	31.6	19.8	22.3	38.0
3	10.8	6.8	27.5	21.9	6.4
4	3.8	4.7	9.0	10.1	5.2
5	16.6	13.0	19.9	15.7	14.4
Total	100.0	99.2	98.0	100.0	99.8

1. Spain, France, Italy (Northwestern basin and Tyrrhenian Sea)
2. Italy, Yugoslavia, Greece (Adriatic and Ionian Seas)
3. Morocco, Algeria, Tunisia, Libya (Southern coast)
4. Lebanon, Israel, Egypt, Libya (South Levantin Sea)
5. Greece, Turkey, Cyprus, Syria, Lebanon (Central and Aegean Sea)

Source: United Nations Environment Programme, 1984, Annex 2.

Industry since the 1940s has expanded faster in the Mediterranean countries than in the world as a whole. The industry is concentrated in the northern countries near industrial cities. France, Spain, and Italy have about 85 percent of the Basin's manufacturing with pollution concentrated at Genoa, Marseille, and Barcelona. In the Adriatic pollution concentrations are at Venice in Italy and at Split and Rijeka in Yugoslavia. In Greece, pollution pools have grown in the Aegean near Athens and Thessaloniki. Three-fourths of all industrial pollution comes from northern countries (Helmer, 1977).

The principal polluting industries are chemical, iron and steel, petrochemicals and refineries (Bliss-Guest, 1981). These industries generate heavy metals and other inorganic chemical wastes which are often directly toxic to humans and fish (Charbonnier, 1977). These wastes accumulate in the food chain as they are transmitted from plankton up through the large predators. Large fish are likely to have higher concentrations of toxic materials than smaller fish. Radioactive wastes from inland nuclear power plants in Spain, France, Italy, Greece, and Egypt are deposited in the Sea by rivers (El-Sayed, 1979).

Oil spills continue to be a major source of pollution in the Mediterranean Sea. Through standard operations tankers create oil pollution along the coast as well as in the open sea. About 400 million tons of petroleum are unloaded annually in Mediterranean ports. Of the 19 crude oil-loading terminals in the Mediterranean, only 9 have deballasting stations. In the 10 ports without such facilities, tankers discharge their ballast into the sea before entering a harbor. There have been at least 8 major oil spills due to accidents since 1970. The oil clots not only mar the aesthetics of beaches but limit other activities. The oil clogs fishermen's nets and the engines of their boats.

One of the most annoying evidences of oil pollution in the Mediterranean is the continuing presence on the coasts of tar balls. It is estimated that of the 600,000 tons of oil spilled in the sea about 30 percent of it reaches the coast. This type pollution can only be controlled by the strict application of the conventions developed by the International Maritime Organization (IMO).

Agricultural practices produce a runoff of organic and inorganic fungicides, biocides, pesticides and fertilizers which reach the sea and contribute to the buildup of inorganic chemicals such as organochlorines, organophosphates, and carbonates which lead to entrophication and loss of fisheries.

In the open sea, the Mediterranean, more than any other body is polluted by floating debris such as plastics and package wrappings (Morris, 1980). Much of this waste material is not or barely biodegradable. This material comes not only from naval and merchant vessels but also from rivers emptying their waste into the sea. This problem has become increasingly serious within the past few years, particularly for the clogging of fishing nets and cannot be solved without the close surveillance of the coastal rivers. The use of biodegradable packaging, as now required in Italy, will reduce this nuisance in the Mediterranean.

Each year more than 100 million international tourists visit the Mediterranean coast. Pollution of coastal waters through the discharge of waste water from tourists adds to the pollution from permanent sources. Although precise data are not available, many tourists have become ill with typhoid, infectious hepatitis, and other diseases after swimming in the Mediterranean (Evison and Tosti, 1980). In some areas, swimmers had a one-in-seven chance of getting a skin infection. Tourists come mainly in the hot summer months when water is most scarce. As a result in many areas additional untreated sewage flows into the Sea.

ADOPTION OF THE MED ACTION PLAN

In the 1970s the countries of the Mediterranean Basin recognized that the pollution problem could no longer be ignored. As a response, all nations (except Albania) met in Barcelona to develop a unified approach to the pollution problem. In February 1975, the 16 nations adopted the Barcelona Convention, a comprehensive plan of attack (Dobbert, 1980).

The plan, designed by the United Nations Environmental Program, had as its basic provision the framework within which countries could develop projects to control various aspects of pollution. It combined the comprehensive UNEP approach with the existing country initiatives (Figure 2). The Plan also recognized the particular problems of the individual countries.

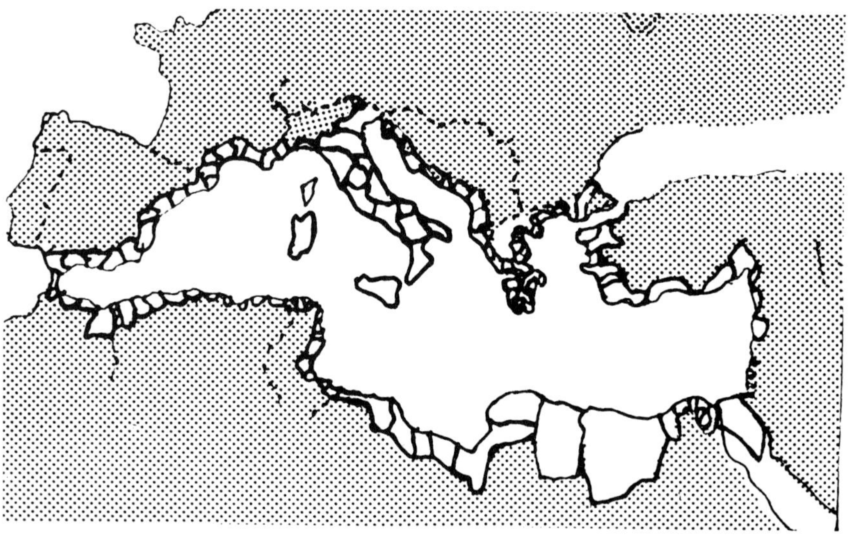

FIGURE 2. Regions of the Mediterranean Plan. The coastal area of the Mediterranean Sea was divided into small units so that each area could develop specific plans to control pollution. For countries such as Italy and Greece these divisions cover most of the country while for other countries such as Spain and Algeria they form a very small area. All of the rivers flowing into the Mediterranean, except the Nile, have a small basin. The planning regions are small with most coinciding with a stream basin.

The Med Plan consists of four components, each is mutually reinforcing (Thatcher, 1977):

1. The integrated planning of the development and management of the resources of the Mediterranean Basin.
2. A coordinated pollution monitoring and research program in the Mediterranean.
3. A framework convention and related protocols with technical annexes for the protection of the Mediterranean environment.
4. Institutional and financial implications of the action plan.

The integrated planning components maintained the basic, holistic viewpoint of UNEP. Its fundamental goal was to relate problems of massive pollution to patterns of land use due to long-term demographic and economic focus. It was envisioned that economic and environmental planning would be coordinated. Because this was a rather new concept to the governments of the Mediterranean, most countries have had difficulty in relating the two objectives. Initially most nations were concerned with the direct cause and effect of pollution. France was the only nation prior to the Med Plan that expressed interest in integrated pollution planning.

The monitoring and research component was included to collect data on the actual state of pollution. It began as a monitoring device but has evolved to include more research on controlling pollution. Because the amount of and extent of pollution was not known by the individual countries on the Mediterranean Basin the assessment component, called Med Pol, received immediate support. The assessment of the Mediterranean pollution was developed in two phases. Med Pol I was carried out between 1976 and 1981 during which time the extent of the problem was defined. Med Pol II between 1982 and the present has as its major objective the reduction of the sources of pollution.

The legal components were intended to support existing laws and regulations under UNEP's supervision. The countries agreed to the need for legal arrangements in order to attain regional pollution controls. For example, the U.N. Food and Agriculture Organization (FAO), has prepared regulations to control agricultural pollution from land sources and the Intergovernmental Maritime Organization (IMCO) has drafted legislation to control pollution of oil by marine dumping.

In the implementation of the Med Plan there was some debate among the countries as to what constituted appropriate supporting institutional programs. The Less Developed Countries (LDCs) desired programs that would develop their marine science capabilities. As an initial step they wanted a regional operation center to coordinate information and resulting planning. The LDCs also supported the development of comprehensive regional arrangements for pollution control which would be legally binding on the participants. In contrast, Spain, France, and Italy, who already possessed substantial marine laboratories, wanted only minimal regional cooperative schemes and preferred a decentralized organization that would only facilitate information exchange.

In the Barcelona Convention development of the organizational structure, stronger monitoring and assessment proposals were approved than the Developed Countries (DC) desired, but the supporting administrative arrangements remained unspecified. UNEP was charged with devising the administrative support for the Plan, and with overseeing and coordinating the projects within each component of the plan.

IMPLEMENTATION OF THE MED PLAN

The Med Plan began in 1976 with seven pilot studies. The first phase consisted of determining the amount and type of pollution in the region and the establishment of a continuing monitoring program supported by governments rather than by an international organization such as UNEP. A coordinating headquarters was established in Athens, with a Mediterranean staff supported by annual contributions by the 16 participating countries. In order to further the development of controls regional centers were also established to coordinate such programs as the preservation of species, integrated development planning, oil spill management, and developmental projects of specific interest to developing states. The 16 governments met biannually to review the programs.

In 1980 the Land-Based Sources Protocols, Med X, were signed but did not become effective until 1983. This is the most important protocol of the Med Plan for it establishes limits on industrial, mineral, and agricultural emissions into the Mediterranean (Figure 3). A Med X study determined that 85 percent of all pollutants in the Mediterranean Sea originated on the land and that 80 to 85 percent was transmitted to the sea by rivers. The Med X study also found that 80 percent of the sewage reaching the Mediterranean was untreated.

This protocol requires long negotiation between the developed and less developed countries of the Mediterranean Basin. Controlling land-based pollution sources directly influences economic development. The ultimate adoption of this protocol, and its subsequent ratification by Algeria and Egypt, who initially opposed it,

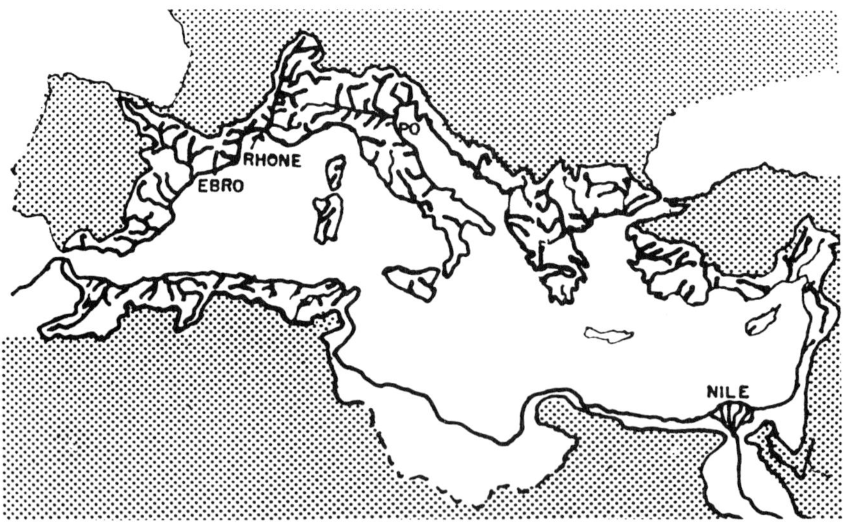

FIGURE 3. The Mediterranean Watershed. The principal rivers of the Mediterranean Basin are the Nile, Po, Rhone, and Ebro. Because the rivers are the major sources of pollution for the Sea, the rivers basin provides the basis for the extent of the Med Plan.

demonstrates a reversal of the position that considered environmental protection and economic growth as incompatible goals.

The Med Plan efforts have expanded over time and the marine pollution controls have gradually become more comprehensive. By 1982 the number of pollutants covered by the protocols had increased greatly. The sources of pollution that were regulated grew from oil spills from tankers to land-based emissions from agriculture, industry, and municipal sewers. Thus, in the period from 1976 to 1983 the control grew from an exploratory framework to a set of regulations over a wide variety of specific substances.

The Land-Based Sources Protocols' "black list" banned the use of the nine most-toxic substances groups, and its "gray list" designates 13 groups of less-toxic substances to be controlled by discharge permit. As a response to these controls, nations are required to implement practices to conform to common pollution standards. Nevertheless, progress has been slow in the implementation of these regulations.

SOME IMPLEMENTATION PROCEDURES

Because there have been political and economic disagreements between certain countries of the Mediterranean Basin, tacit rules of diplomatic behavior have prevented the implementation of the 1975 Convention. Conflict between the Arab States and Israel was averted when it was agreed that the Israel delegation would maintain a very low profile. Both groups have since abided by these ground rules. The PLO was not invited to the Convention. The United States attended the 1975 Conference as an observer, but has not attended any subsequent meetings after being convinced that the Sixth Fleet would remain unaffected by Med Plan activities.

In order to maintain technical cooperation, certain diplomatic behavior prevails. No country is criticized for the amount and type of pollution emitted into the Mediterranean. The attitude prevails that each country has problems and if there is to be a solution, pollution control measures must be developed by all nations.

Specific political problems continue to be recognized. For example, Syria ratified the 1975 Barcelona Convention with the qualification that ratification did not constitute a recognition of Israeli sovereignty. The Turkish delegation commonly submits a letter to the UNEP secretariat protesting the Greek presence on Cyprus. This letter is received and then ignored. The European countries were unwilling to have the Soviet Union as a member of the Med Plan though there is the participation of the Black Sea countries of Bulgaria and Rumania. As a result, the Mediterranean Basin was defined as the area between the Strait of Gibraltar and the southern limits of the Strait of Dardanelles between the Mehmetcik and Kumkale lighthouse in Turkey. These boundaries exclude nonlittoral states with rivers that flow into the Mediterranean, such as Portugal, Switzerland, and the Sudan from compliance with the convention.

COMPLIANCE WITH MED PLAN

As the Med Plan has evolved, two distinct groups of nations are now recognized as having different development strategies. One group consisting of France, Greece, Israel, Algeria, and Egypt have developed strict and broad measures for controlling marine pollution. The second group consisting of the remainder of the Mediterranean countries have weaker domestic regulations and give only moderate support to the Med Plan.

The five countries that give Med Plan the greatest support have well developed marine science establishments. The marine scientific community, allied with ecological scientists, have led their governments toward more comprehensive marine pollution controls.

The development of a more comprehensive program in Algeria illustrates the evolution of pollution control in the Mediterranean. In 1975 the Centre de Recherches Oceanographiques et des Peches (CROP) was established with scientific personnel. Their first task was to study the causes of lowered coastal fish yields. In 1979 and 1980 CROP published its first report describing the industrial and public health dimensions of Algerian coastal pollution. It was recognized that marine pollution was a major problem. As a response the government began to implement plans to control pollution in order to increase fish production.

The gradual involvement of the scientific community was a major factor in the evolution of the Egyptian environmental policies. Scientists were initially isolated from the government. In the late 1960s and early 1970s an 85 percent drop in the fish catch in Lake Marlfret led the National Academy for Scientific Research and Technology to fund studies monitoring the quality of the lake by oceanographers. Other studies followed on marine pollution. From these scientific reports the scientists provided the foundation for the development of governmental policies. In 1982 the government introduced laws to protect the Nile River, and in 1983 ratified the Land-Based Sources Protocol and the Specialty Protected Areas Protocol in order to protect the Nile and coastal areas.

NATIONAL REGULATIONS

Since 1975 national efforts to control marine pollution in the Mediterranean have become more comprehensive. National governments have cooperated with the international Med Plan to develop a comprehensive approach to pollution control. As an example, the Med Plan's Regional Oil Contacting Center (ROCC) was influential in the development of national efforts to control oil pollution.

The degree to which nations have developed environmental policies are not uniform. In general the countries with the broadest pollution control policies are those in which the ecological epistemic community had established areas of national policy making through regulating environmental authorities (Haas, 1990). These countries have also been the leaders in developing the international aspects of the Med Plan.

By 1985 many countries had created environmental agencies or ministries. The national programs have gradually evolved from uncoordinated national policy-making among ministries without clear mandates and insufficient budgets to dedicated environmental ministries with their own budgets and regulatory responsibilities. The regulatory ministries developed specific programs to control pollution. These regulatory institutions are staffed by members of UNEP's epistemic and marine scientists. The Med Plan strengthened the national institution by publicizing Mediterranean pollution problems.

Of the Mediterranean countries, France has the best developed governmental environmental institutions. In 1970 the High Commission for the Environment was established followed in 1971 by the Ministry for the Protection of Nature and the Environment. This Ministry was entrusted with the protection of natural sites and with preventing, controlling, and reducing all forms of environmental pollution. In 1974 the program of the Ministry was enlarged to cover the overall quality of life and renamed the Ministry for the Environment and Quality of Life. In 1981 the name was again changed to the Ministry for the Environment and it was raised to full ministerial rank.

Within the Ministry for the Environment, a number of coordinative interministrial committees have evolved. A fund, known as the Funds d'Intervention pour la Nature et l' Environment (FIANE), was created to subsidize private investment in pollution control equipment. The Interministrial Committee on the Sea was created to control maritime resources, particularly fishing, within a 200-mile maritime zone. At its first meeting in 1979 the committee approved a program to control Mediterranean oil pollution from tankers.

Six River Basin Agencies, the Interministrial Commission on Water and the Ministry of the Environment's National Committee on Water regulate France's river pollution. There is now strict control of river pollution of the metal plating, the pulp and paper, the chemical, and most food processing industries.

Most Mediterranean countries have developed domestic legislation to control marine pollution. The legislation normally covers oil pollution and offshore dumping. France, Greece, Algeria, Egypt, Spain, and Yugoslavia have implemented legislation to control land-based sources of pollution. France, Greece, and Israel require environmental impact assessments for coastal developments. The regulations vary greatly from country to country and are usually not compatible.

The legislation of the developed nations and that of the less developed nations have evolved on different schedules. In the developed countries, general legislation often accompanies or preceded the creation of environmental ministries. For example, most countries of the European Economic Community have established specific marine pollution standards for their coastal waters. In the lesser developed countries, the institutionalization of environmental groups normally occurred before ministries drafted legislation.

Of all types of pollution, the nations have made the greatest efforts to control oil spills. When the Med Plan was established in 1975 initial efforts called for the control of oil pollution from spills and accidents. To implement the procedures the Regional Oil Combating Centre (ROCC) was established in Malta. The centre has an emergency communications network to notify national oil pollution control authorities and to coordinate action. National and multi-national oil spill contingency plans have also been developed. Before the Barcelona Convention of 1976 only four nations (France, Greece, Israel, and Italy) had emergency contingency plans. By 1983, only Albania had failed to begin to prepare contingency plans for an oil spill emergency.

Outside the Med Plan there have been a number of regional efforts to combat pollution. For example, in 1976 the governments of France, Italy, and Monaco (RAMOGE) agreed to coordinate antipollution efforts along the Riviera. The program includes the measurement of pollution loads, construction of industrial and domestic waste water treatment plants, and monitoring to assure the maintenance of common environmental quality standards (Boxer, 1982).

Other countries have prepared agreements to protect the Mediterranean from oil spills. In 1974 Italy and Yugoslavia signed the Agreement on Cooperation for the Protection from Pollution of the Waters of the Adriatic Sea and Coastal Areas. A joint commission of the two states coordinated research and prepared measures to combat pollution. In 1979 the Italo Greek Agreement on the Protection of the Ionian Sea was signed. In 1982 the Union of Greek Shipowners and the Panhellenic Seamen's Federation created the Hellenic Marine Environmental Protection Association (HELMEPA) to monitor oil spills and provide information to shipowners. In 1983 France, Greece, Italy, Lebanon, Monaco, Tunisia, and Yugoslavia became members of the International Maritime Consultative Organization (IMCO).

RESULTS AND PROBLEMS

Since the Mediterranean Action Plan has been in effect since 1976, a basic question now is, has the pollution of the Sea improved? There is no question that progress has certainly been made in recognizing the problem and initial plans developed. Not only has the secretariat of MAP from its headquarters in Athens encouraged member nations to develop programs for action, but in addition to the international efforts the individual nations have established environmental agencies and have developed research staffs in order to develop sound planning.

There is evidence of some success in decreasing the pollution in the Sea (Haas and Zuckman, 1990). For example, swimming is now safe in the Cannes and Nice areas of coastal France. Oil pollution has also decreased due to controls imposed by national efforts and the International Maritime Organization (IMO).

In spite of these modest reductions of pollution in limited areas, there is little evidence of general improvement. Each year 2.5 million tons of heavy metals and fossil fuels and 300,000 tons of fertilizers are still emptied into the Mediterranean. In addition, 80 to 90 percent of the human waste is still deposited untreated into the Sea.

There have been many problems in the implementation of the Barcelona Convention. One of the major deterrents has been the initial act of constructing sewage treatment plants. Estimates run as high as $100 billion. The greatest need for these sewage plants is in the countries on the southern shore of the Mediterranean. Funds could possibly be raised by taxing the tourist industry, but many countries believe that such a tax would destroy the tourist trade and are reluctant to impose such a tax. Funds might also be secured by transferring them from the northern countries to the southern countries of the Basin. It has been difficult, however, to persuade northern countries that it is to their economic advantage to help the countries of the South.

In spite of the international proposals encouraging cooperation among nations, there is still opposition among nations in developing pollution control measures. For example, at the 1990 Conference on Security and Co-operation in Energy, attended by 34 European nations and the United States and Canada, Turkey representing the North African nations, proposed to ban the traffic of toxic wastes from the industrial nations to the developing countries. France opposed the word "ban" and had it replaced by the word "suggest". This weakened the resolution so that it is of little importance. The conflict over this resolution reflects the fundamental economic difference between the industrial nations on the North and the lesser industrialized nations on the southern shore of the Mediterranean. Hentati Adel of the National Tunisian Agency for the Protection of the Environment reflected the frustration of the lesser developed countries when he stated, "Many developing nations would be more willing to adopt Western standards of environmental control, but when they try to implement them, the Western multinationals threat to withdraw their factories because, as they rightly point out, if they were going to abide by the rules of their nations of origin they would have stayed at home. There exists a kind of double morality which is impossible to abide by."

The implementation of pollution control in a particular country has been directly related to the economy of the country. To illustrate, France, followed by Spain, has developed legislation to protect the sea grass off their shores (Amico, 1979). There was strong evidence that the sea grass, Posidonia oceanica, was disappearing. This grass provides food and shelter for some 400 species of algae and several thousand species of fish and mollusks. The sea grass generates so much oxygen that it has been called "the lungs of the seabed." It grows in clean water to depths of 120 feet. The losses of grass have been greatest in the areas of untreated sewage. The untreated sewage encourages the growth of plankton which prevents light from reaching the sea floor causing the sea grass to die. Fishing trawlers also destroy the sea grass by scraping the sea beds. The sea grass, by providing food, is vital to the fishing industry. Research indicates that 400 square meters of sea grass can produce 2,000 tons of fish a year. The presence of sea grass is thus absolutely fundamental to the proper functioning of the Mediterranean ecosystem.

CONCLUSION

It is now recognized that environmental problems of the countries of the Mediterranean Basin must be approached as a unit. In 1990 the European Commission stated, "It is now time to move from the pilot and preparatory stage to the action stage. To further implementation the 1990 Conference developed the Nicosia Charter which stated the environmental goal is "to identify, finance, and implement those actions necessary to bring about a situation in which all significant environmental problems have been eliminated by the year 2025, at the latest. From past experience it is recognized that this will be a Herculean task. The central tenet is that the European Community, together with the World Bank, the European Investment Bank and the United Nations Environmental Program will commit resources, financial and research, to achieve this goal. While the task is great, failure will result in an environmental disaster.

SELECTED REFERENCES

Amico, Vincenzo and others. 1979. Concentrations of PCBs, BHCs and DDTs Residues in Seaweeds of the East Coast of Sicily. *Marine Pollution Bulletin.* 10:177-179.

__________ and others. 1979. Levels of Chlorinated hydrocarbons in Marine Animals from the Central Mediterranean. *Marine Pollution Bulletin.* 10:282-284.

Bellan, G. and J.M. Peres. 1972. La Pollution dans le Bassin Mediterranean in *Marine Pollution and Sea Life.* West Byfleet, England: Fishing News, Ltd.

Bliss-Guest, Patricia A. 1981. The Protocol Against Pollution from Land-Based Sources: A Turning Point in the Rising Tide of Pollution. *Stanford Journal of International Law.* 17.

Boehmer-Christiansen, Sonia. 1984. Marine Pollution Control in Europe Regional Approaches. *Marine Policy.* 8.

Boxer, Baruch. 1982. Mediterranean Pollution: Problem and Response. *Ocean Development and International Law.* 10 (April):Nos. 3-4.

Brisou, J.F. 1976. *An Environmental Sanitation Plan for the Mediterranean Seaboard.* Public Health Papers #62. Geneva: World Health Organization.

Brown, R.A. and H.L. Huffman. 1979. Extractable Organics and Hydrocarbons in the Mediterranean Sea. *Marine Pollution Bulletin.* 10:291-298.

Charbonnier, D. 1977. Prospects for Fisheries in the Mediterranean. *AMBIO.* 6:No. 6.

Dobbert, Jean-Marie. 1980. Protocol to Control Pollution in the Mediterranean. *Environmental Policy and Law.* 6.

El-Habr, Habib and H.L. Golterman. 1987. Input of Nutrients and Suspended Matter into the Golfe du Leon by the River Rhone. *Chemosphere.* 16:417-418.

El-Sayed, Mahmoud Kh. and others. 1979. Mercury Pollution of Mediterranean Sediments around Alexandria, Egypt. *Marine Pollution Bulletin.* 10:84-86.

Evision, Lilian and Elsabetta Tosti. 1980. Bathing Water Quality in the North Sea and the Mediterranean. *Marine Pollution Bulletin.* 11:72-75.

General Fisheries Council for the Mediterranean. 1972. *The State of Marine Pollution in the Mediterranean and Legislation Controls.* Studies and Reviews. No. 51.

Grenon, Michel and Hihel Batisse, eds. 1989. *Futures for the Mediterranean Basin: The Blue Plan.* New York: Oxford University Press, 256 pp.

Haas, Peter M. 1990. *Saving the Mediterranean: The Politics of International Environmental Cooperation.* New York: Columbia University Press. 303 pp.

__________ and J. Zuckman. 1990. The Med is Cleaner. *Oceanus.* 33 (Spring):38-42.

Hager, Wolfgang. 1974. The Mediterranean: A European Mare Nostrum? *Orbis.* 18, No. 1:Spring.

Hardstedt-Romes, M. and F. Laumond. 1980. Zinc, Copper and Cadmium in Zooplankton from the N.W. Mediterranean. *Marine Pollution Bulletin.* 11:133:138.

Helmer, Richard. 1977. Pollutants from Land-Based Sources in the Mediterranean. *AMBIO.* 6:No. 6.

International Registry of Potentially Toxic Chemicals. 1978. *Data Profiles for Chemicals for the Evaluation of Their Hazards in the Environment of the Mediterranean Sea.* Vols. 1 and 2. Geneva: United Nations Environmental Program.

Morgan, Joseph R. 1987. Large Marine Ecosystems: An Emerging Concept of Regional Management. *Environment.* 29 (Dec):No. 10.

Morris, Robert J. 1980. Floating Plastic Debris in the Mediterranean. *Marine Pollution Bulletin.* 11:125.

Osterberg, Charles. 1977. Taking the Pulse of the Mediterranean. *International Energy Agency Bulletin.* 9 (Feb.):No. 1.

__________ and S. Keekes. 1977. The State of Pollution of the Mediterranean Sea. *AMBIO.* 6:No. 6.

Papadopoulou, C. and others. 1978. Zinc Content in Otoliths of Mackerel from the Aegean. *Marine Pollution Bulletin.* 9:106-108.

Ress, Paul Evan. 1986. Mediterranean Sea Becoming Cleaner. *Environmental Conservation.* 13:Nov. 3.

Ritchie-Calder, Peter. 1972. *The Pollution of the Mediterranean Sea.* Bern: Herbert Land.

Saliba, Louis. 1978. Protecting the Mediterranean. *Marine Policy.* 2 (July):No. 3.

Schlain, Avi. 1976. The Community and the Mediterranean Basin. In Kenneth J. Twitchett, ed. *Europe and the World: The External Relations of the Common Market.* London: Europa.

Stansell, John. 1990. A Mediterranean Holiday from Pollution. *New Scientist.* (May):28-29.

Thatcher, Peter S. 1977. The Mediterranean Action Plan. *AMBIO.* 6:No. 6.

Tuncel, G., G. Ramelow and T.I. Balkas. 1980. Mercury in Water, Organisms and Sediments from a Section of the Turkish Mediterranean Coast. *Marine Pollution Bulletin.* 11:18-22.

Vadrot, Claude-Marie. 1977. *Mort de la Mediterranee.* Paris: Actuels Iseuil.

Zsolnay, A. 1987. Spatial and Temporal Variation of Pelagic Tar in the Mediterranean Sea. *Chemosphere.* 16:399-404.

The Oceans: Physical-Chemical Dynamics and Human Impact. Edited by S.K. Majumdar, E.W. Miller, G.S. Forbes, R.F. Schmalz and Assad A. Panah.

Chapter Twenty-Six

LAND-BASED SOURCES OF MARINE POLLUTION IN SRI LANKA

H.H. COSTA
Department of Zoology
University of Kelaniya
Kelaniya, Sri Lanka

INTRODUCTION

Sri Lanka is an island of about 64,000 km^2 situated between 5°55′ and 9°51′ north and 79°41′ and 81°54′ east. The island has a coast line of 1700 km, more than a kilometer to every 40 km^2 of land of which nearly 9/10 consists of sandy beaches. No part of the island is more than 120 km from the sea. In the country's coastal areas are located the major urban centers as well as the multiplicity of human activities vital to the life of the nation.

The coastline is dotted with a large number of lagoons, more so in the eastern and western coast and in the northern extremity than in the southern coast.

A conspicuous feature of the marine environment is the continental shelf which extends for a distance varying from 8-40 km with an average depth of 65 m. It is narrow around the southern part of the island but widens out in the north and merges with the platform that surrounds India. On the outer edge is a comparatively steep shelf falling to 1800 m or more and taking around 20 km to reach the general level of the Indian Ocean. The shelf is sometimes broken by deep submarine valleys where deep waters occur within a few kilometers off the coast near Trincomalee, Kumana, Matara and Panadura (Figure 1).

The economic activities are mainly related to coastal resources and among others include fisheries, mining, tourism, etc.

As in most other developing countries, concern for the environmental problems is a relatively recent phenomenon. Data relevant to marine environment for instance are not being gathered in coordinated manner nor the little available data have been subjected to any detailed analysis. This situation is partly because of the diffused resources management responsibilities. Many authorities are responsible for the collection of data. With data scattered, unpublished or available in unprocessed form analysis becomes difficult, when possible at all.

As defined by UNEP (1985) "Land-based sources" means:

i) Municipal, industrial or agricultural sources both fixed and mobile, on land, discharges of which reach the marine environment, in particular:

   From the coast including from outfalls discharging directly into the marine environment and through run off:

   Through rivers, canals or other water sources and via the atmosphere.

ii) Sources of marine pollution from activities conducted on offshore fixed or mobile facilities within the limits of national jurisdiction.

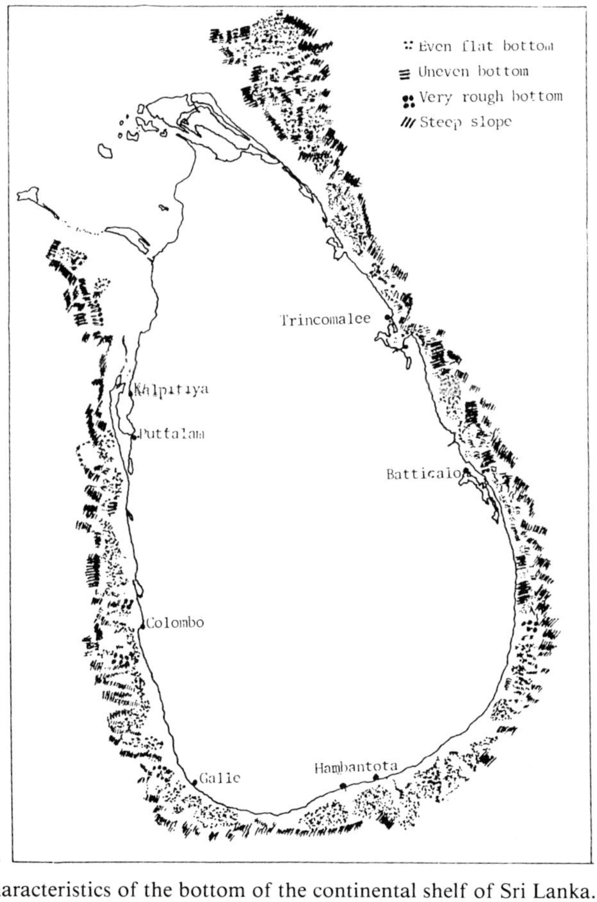

FIGURE 1. Characteristics of the bottom of the continental shelf of Sri Lanka.

MARINE POLLUTION FROM COASTAL OUTFALLS DISCHARGING DIRECTLY AND THROUGH RUNOFF

Pollution from Municipal Wastes and Sewage

The largest amount of organic pollution is due to human settlement (Costa 1980). There are 24 administrative districts in Sri Lanka of which 13 abut the coast line. The total population in the coastal districts is 8,228,298 which is about 55% of the total population. The 1981 census data disclosed that one out of every three houses lacked toilet facilities.

Figure 2 shows the comparison of potential mass of BOD_5 discharged to the marine waters by population of the coastal districts. BOD_5 estimate was based upon a per capita discharge of 55 grammes per habitant per day. The estimate of the value of BOD_5 from domestic sewers entering the marine environment is around 8,000,000 tonnes per annum.

Apart from some small establishments and the Katunayake Investment Promotion Zone, there is no sewage treatment. Sewage from Colombo is discharged untreated to the sea through two ocean outfalls of 2000 m and 1400 m to the north and south respectively.

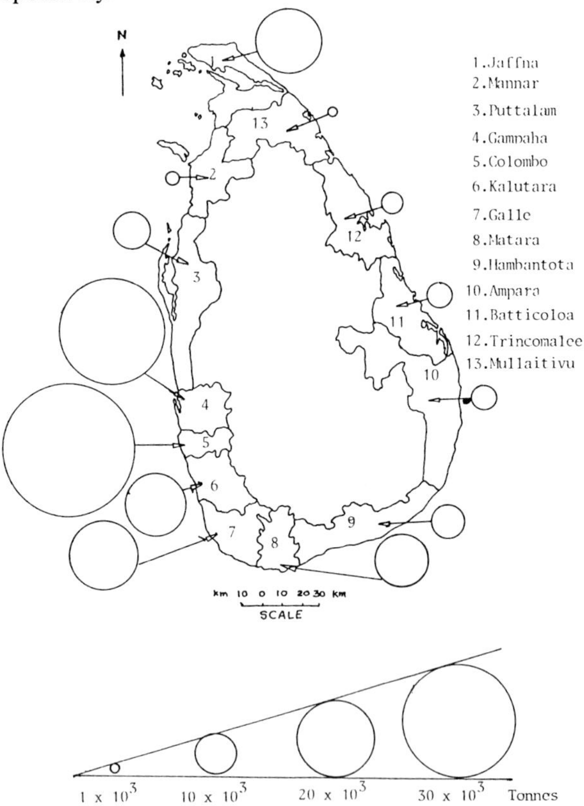

FIGURE 2. Potential mass of BOD_5 discharged by the population of coastal districts of Sri Lanka.

Pollution from Industrial Wastes

Sri Lankan economy is primarily agricultural. In the last forty years or so attempts are being made to create an economy with emphasis on industry. This transformation began with the introduction of the Industrial Corporation Act. Industrialization further progressed with the establishment of large state sector industries. By 1980 there were nearly 5000 registered industries and 15,000 unregistered industries. Some of these have since closed down with the adoption of the open policy on imports in 1977 but with establishment on the Investment Promotion Zone (IPZ) in 1980 newer and larger scale industries have been established.

Most of the industries are concentrated in and around Colombo on the western coast. A few industries are located on the eastern coast and in the north. The main sources of industrial effluents entering the western coastal belt are Katunayake Investment Promotion Zone (over 60 manufacturing establishments). Industries located along the Kelani River and Biyagama Investment Zone (over 30), industries discharging into the Colombo Lake (over 20), industries located along the Wellawatte Canal (over 10) and the industries located in the Ratmalana and Ekala industrial areas.

Two cement factories are located in Puttalam and Galle in the western coast and one in the north at Kankesanthurai.

The chief industries on the east coast are the two paper and pulp factories at Valachchennai which have caused the heavy pollution of Baticoloa lagoon, producing black bottom sediment with lignite residues. Waste loads of 18,000 kg suspended solids and 9000 kg BOD_5 reach this lagoon daily.

Industrial wastes in Sri Lanka are subjected to little or no treatment before discharged.

A survey carried out by the Central Environment Authority of Sri Lanka (1989) estimated that 60% of the industries are likely to cause pollution.

Sri Lanka has relatively few problems from chemical industries. These industries are still confined to manufacture of paints, cosmetics, varnishes and others which have a low potential for pollution.

Pollution by Cottage and Small Industries

Coconut based industries. These are mainly dispersed on the western coast and consist of industries producing coir, dessicated coconut and coconut based alcoholic drinks.

Annual production of coconut based products is given in Table I.

In the production of white fibre, retting is carried out mostly in lagoons and in ponds in the sides of rivers and leads to the production of sediments and wastes with high contents of organic matter.

Rubber Based Industries and Production of Rubber

There is one tyre manufacturing enterprise in Sri Lanka and over a hundred rubber processing factories. The annual production of rubber processing factories is around 140,000 mt of rubber with a discharge equivalent to 330,000 tons of serum and acid. BOD_5 has been estimated to vary from 1000-1500 mg/1.

TABLE I

Annual Production of Coconut Based Products, 1992

	Production (mt)	Equivalent to
Brown Fibre (Exports)		
Mattress fibre	40,466	
Bristle fibre	9,500	717 million husks
Twisted fibre	27,866	
Yarn and twine	2,391	
(Local) Yarn & manure	11,000	
White fibre (Exports)	7,000	55 million husks
Dessicated coconut (Exports)	40,000	365 million coconuts

Batik Printing and Other Industries

Widely scattered backyard units. These discharge effluents high in dyes, oil and organic compounds (Table II).

Pollution from Solid and Liquid Wastes

Disposal of solid and liquid wastes is carried out mainly by municipal and urban councils and industrial establishments.

Coconut based industries and saw mills often dump large quantities of fibre waste, saw dust, wood shavings, etc.

Beach dwellers often use the beach as a dumping ground for litter.

The development of the tourist industry has led to the tourist hotels and tourist guest houses becoming the main sources of beach litter.

The tourist industry is presently well established in a few coastal areas such as Negombo, Mount Lavinia, Hikkaduwa, Kalkudah and Arugam Bay. Of graded hotels, 75% are located in the coastal areas (Figure 3). The tourist arrivals due to unrest decreased from 400,000 in 1982 to 230,000 in 1986. Tourism has picked up rapidly again (Anon, 1992) (Figure 4).

The tourist industry provides about 30,000 persons direct and 35,000 indirect employment.

TABLE II

Pollution Loads from Selected Industries

Sector	District	BOD_5 tons/yr	Waste Load SS tons/yr
Textile dyeing	Colombo	1,564	472
Printing	Gampaha	1,629	492
	Kurunegala	480	145
Leather	Colombo	734	510
Processing	Gampaha	767	533
Distillery	Kalutara	421	492

Source: Natural Resources of Sri Lanka, 1991.

POLLUTION OF MARINE ENVIRONMENT THROUGH RIVERS, CANALS OR OTHER WATER SOURCES INCLUDING UNDERGROUND WATER COURSES

Sri Lanka with over 20 major rivers can be divided into 103 component natural river basins on which 83 are in the Dry Zone and 20 in Wet Zone. (Arumugam, 1969). The surface water resources in terms of total annual escape to the sea and lagoons have been estimated to be of the order of 3.33x10^6 hectare meters (Ranatunga, 1985) (Table III).

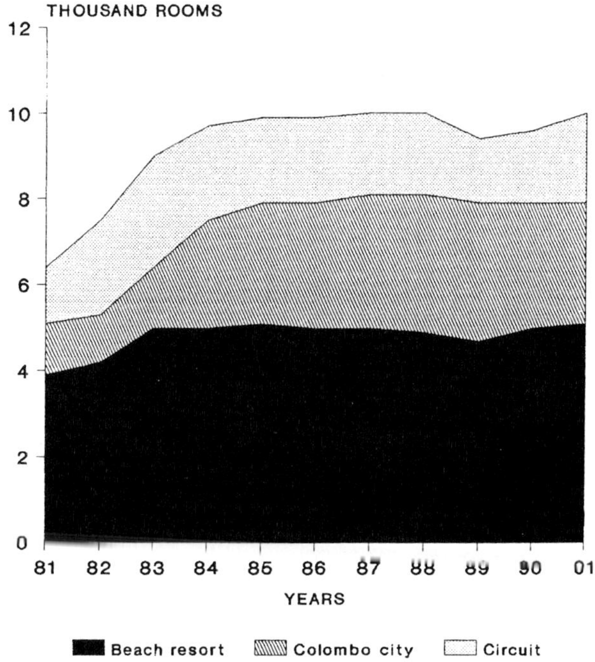

FIGURE 3. Growth of accommodation capacity (rooms) 1981-1991.

TABLE III

Surface Water Resources of Sri Lanka

	Wet Zone	Dry Zone	Total
Mean Annual Rainfall (mm)	2424	1450	1937
Mean Annual Runoff (Hectare Meters)	2.58x10^6	2.55x10^6	5.13x10^6
Escape to the sea (Hectare Meters)	2.04x10^6	1.30x10^6	3.33x10^6
Escape as a % of total runoff	78.83	51.11	64.91

Source: Ranatunga, 1985

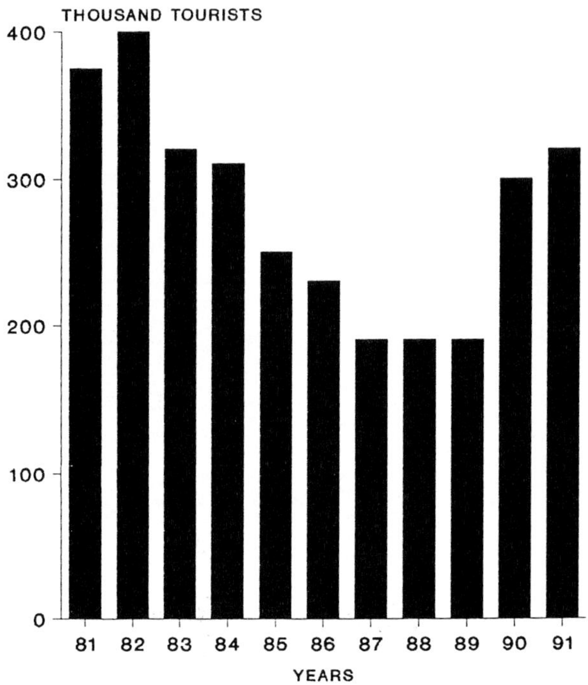

FIGURE 4. Tourist arrivals in Sri Lanka, 1981-1991.

Mining Wastes

The sand and soil supplies from river discharges have contributed to the sedimentation of certain areas. Increased human activity upstream has often aggravated these problems (Anon, 1985) (Table IV).

Table IV

Estimated volume of sand mining in Sri Lanka, 1984

Name of river	Number of cubes removed in 1984
Kelani Ganga	222,771
Maha Oya	111,720
Gin Oya	79,445
Kalu Ganga	46,667
Deduru Oya	22,896
Gin Ganga	21,563
Nilwala Ganga	507,866

Source: CCD Internal Report No. 1: A census of the exploitation of sand seashell resources in the Coastal Zone of Sri Lanka (1984).

Agricultural Wastes

Sri Lanka being predominantly an agricultural country, there has been a marked increase in the use of fertilizers and pesticides. These pollutants can get leached into the ground water or they may reach inland water ways as surface run off.

Fertilizers

A greater part of the fertilizers used in agriculture end up in ground and surface waters and ultimately in the sea if there is no degradation.

About 500,000 mt of fertilizer is used annually in agriculture (Anon, 1993) (Table V). Most fertilizers used are imported. At the fertilizer mixing/storage complex at Hunupitiya (a suburb of Colombo), due to ground water pollution, wells at a distance of 1 km or more are polluted and the water in these are not suitable for drinking or even bathing. Due to surface run off garden plants and trees have been killed in the immediate vicinity.

Pesticides

A steady increase in the use of pesticides both for control of agricultural pests and vectors of diseases is seen from 1980 onwards. The present supply of pesticides in Sri Lanka (1992) amounts to 2705 mt in Sri Lanka (Anon, 1992) (Table VI). A large quantity of the pesticides used is leached out, only about 0.1% reaching the target organisms. Pesticides are mainly utilized in river basins where cultivation takes place. Utilization differs according to crops per year. Most farmers use at least twice the recommended dose.

Aquaculture is an emerging user of pesticides and veterinary drugs. Both may affect the marine enviroment. The aquaculture farms of Sri Lanka are located in the coastal belt and cover an area of over 1500 ha. The effluents of these farms discharge directly into water ways.

TABLE V

Annual Fertilizer Imports 1991 and 1992

	1991		1992	
	Met. ton	%	Met. ton	%
Urea	183,917	36	181,352	47
Muriate of Potash	113,024	27	33,002	09
Triple Super Phosphate	41,347	08	36,817	10
Sulphate of Ammonia	93,796	18	94,343	25
Others	75,645	11	39,905	09
Total Imports	507,731	100	385,019	100
Local Production				
Ground Rock Phosphate	32,559		19,289	
	540,290		404,708	

Source: The review of Fertilizer Year 1992. Fertilizer secretatiat, Sri Lanka.

Industrial and Municipal Wastes

Industrialization has led to the gradual pollution of the major rivers such as Mahaweli and Kelani, which pass through densely populated towns. A comparison of the heavy metal pollution of these two rivers is given in Table VII. The river waters appear to be highly enriched in Copper, Lead, Cadmium and Zinc. (Dissanayake et al., 1985, Dissanayake and Weerasooriya, 1986).

River Kelani which runs through Colombo has a concentration of large water using industries (Table VIII) which in turn discharge sometimes treated and mostly untreated waters back into the river. The major waste sources that affect the waters of Kelani River and the pattern of sewage are given in Figure 5. It is a common feature to observe periodic fish kills with symptoms of ammonia poisoning (de Silva et al., 1986)

A greater part of Colombo City catchment covering 487 km^2 drains into the Kelani River through different canal openings (Figure 6). There is also some drainage through sea outfalls at Wellawatte and Galle Face. The major sources polluting the canals are the slums, oil and heavy metal discharges from the industries as well as outflows from the city's sewers (Figure 7). The BOD_5 ranges from 100-400 mg/1. During rainy weather these outflows pollute the surface water, sediment and ground water.

TABLE VI

Supply of Pesticides in Sri Lanka, 1988-91

	Quantity in mt/kl			
	1988	1989	1990	1991
Fungicides	687.60	834.91	631.60	522.24
Herbicides	1084.27	871.88	1243.73	1002.65
Insecticides	525.92	569.10	754.00	458.61
Rodenticides	13.04	6.00	1,00	1.20
Others	—	73.13	59.72	25.72
Total Supply	2310.83	2355.02	2690.02	2010.42

Source: Department of Agriculture, Economics and Product Division.

TABLE VII

Comparison of the Metal Contents of the Waters in Mahaweli and Kelani Rivers. (Dissanayake et al., 1985, 1986)

	Mahaweli River (ppb)	Kelani River (ppb)
Copper	4-37	2-22
Zinc	2-270	23-104
Lead	1-4	1-20
Cadmium	1-2	1-4.76

TABLE VIII

Estimated Waste Loads from Certain Factories, (Source: C.E.A., 1986)

Zone/Factory	Estimated waste load (kg BOD_5/day)
Wellawatte/	
Ceylon Silks Ltd.	64
Narahenpita/	
Lanka Canneries Ltd.	50
National Milk Board	318
Kolonnawa/	
Ceylon Petroleum Corporation	95 (excluding storm-water discharge)
Goverment Factories	19
Hulftsdorf/	
British Ceylon Corp. Ltd.	1000 kg COD/day
Grandpass/	
Lever Brothers Corp. Ltd.	1850
Sedawatte Exports Ltd.	2
Jafferjee Brothers	54
Macky Group of Industries	16
Kuppiyawatte	
Lankem Ceylon Ltd.	3
St. Anthony's Industries Group	15
Cyntex	84
Wilfred & Co. Ltd.	3
Pettah/	
Colombo Gas & Water Co.	9.5
Mattakuliya/	
Fertilizing Mixing Works	9.5 (excluding storm and rising water)
Fern Tea Company	4
Ceylon Leather Corporation	530
Harbour/	
Ceylon Grain Elevators	46
Ceylon Fisheries Corporation	50
Slave Island/	
Ceylon Cold Stores Ltd.	50

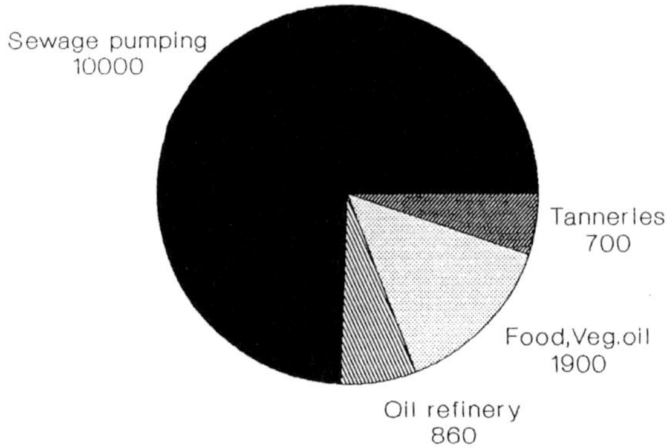

FIGURE 5. Waste sources affecting the Kelani River before outfalls (BOD_5/day). Source: CEA, 1985.

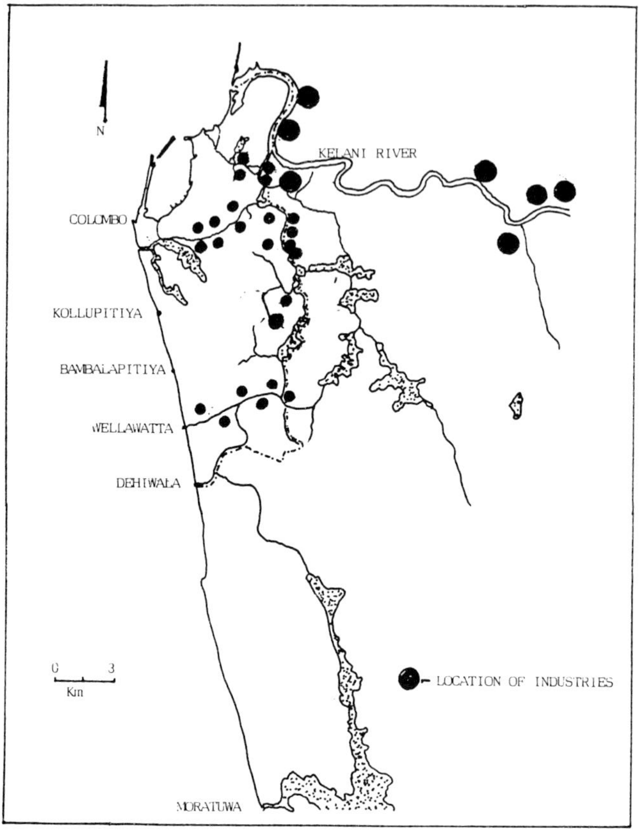

FIGURE 6. Colombo north and south drainage canals.

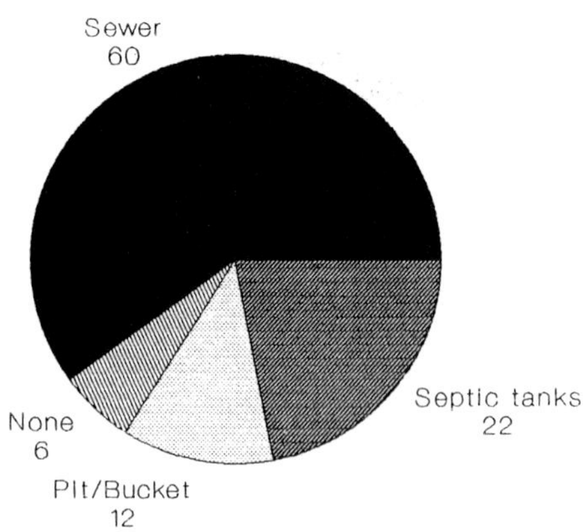

FIGURE 7. Sewage disposal in Colombo City. Source: CEA, 1985.

Pollution of Marine Environment Via the Atmosphere

The sources of atmospheric pollution are mainly industry, transport and domestic usage. The oil consumption in 1982 was 1,256,000 mt. Many factories and other installations produce smoke particulates and acid smuts with adsorbed SO_2, CO and unburned hydrocarbons all due to incomplete combustion in the furnaces.

The three cement factories situated at Kankesanthurai, Puttalam and Galle manufacture about 600,000 mt of portland cement per year using the dry process. These produce considerable amounts of dust pollution. Emissions of Kiln dust from Puttalam cement factory alone amounts to 2 tons/day, the dust plume being visible even in satellite pictures.

POLLUTION OF MARINE ENVIRONMENT FROM ACTIVITIES CONDUCTED OFFSHORE FIXED OR MOBILE FACILITIES WITHIN THE LIMITS OF NATIONAL JURISDICTION

Sri Lanka is not presently engaged in sea bed exploration or exploitation for oil or minerals.

Exploration for offshore oil and gas has been given priority in the development strategy of the country and thirteen offshore concessionary blocks have been designated for exploratory activities on a production sharing basis of which eleven are on the continental shelf and the other two are in the deeper waters. Four offshore wells have been drilled in Palk Bay and Gulf of Mannar.

These wells have not so far struck oil or gas.

MEASURES AGAINST POLLUTION FROM LAND BASED SOURCES

Although by legislation it is necessary that all industrial establishments should adopt measures to abate pollution, yet most registered and almost all unregistered industries in Sri Lanka lack facilities for the proper separation of sediments and waste material from effluents. Thus waste water from most manufacturing establishments and processing plants have significant amounts of contaminated sediments and waste material which often pollute canals, rivers and marine waters.

Many state sector industries, however, are now taking steps to reduce pollution of the environment.

Pollution of atmosphere by cement kiln dust is already being controlled at one cement factory in Kankesanthurai. Treatment plants have been planned for paper, textile and milk effluents.

Control of certain solid and liquid wastes has been brought about recently by residue utilization and waste recovery. Uses are being found for coir dust, saw dust and agrochemical residues and there is increasing use of biogas from farm and cellular wastes.

The government has recently adopted a method of zoning of industries into industrial estates and industrial promotion zones. Although effluent treatment has been neglected in the industrial estates, effluent management has been well planned in the Industrial Promotion Zones (IPZ) of which two are functioning and others are on the planning board. An example is Katunayake IPZ where there are over 50 factories (with provision for 125 more factories) and workforce of 30,000. Each factory is required to treat the effluent to a minimum standard including removal of toxic substances before waste water enters the central treatment plant. The final effluent meets standards for discharge into surface waters (De Alwis et al. 1988).

PUBLIC AWARENESS TOWARDS LAND-BASED MARINE POLLUTION

The Central Environment Authority (CEA) which is the statutory body in charge of the protection and management has initiated a nation wide public awareness program considering the fact that the country is a coastal state.

This has been created through the Institution of the public awareness program for environmental management and protection (PAPEMAP).

The project objectives of the PAPEMAP are:

a) The evolving of a strategy for a national and local level public awareness program and a plan for the implementation of the strategy.
b) The preparation of a training program based on the problem solving process at local level.
c) Designing and creating public awareness materials.

Media materials are being prepared by PAPEMAP such as videotapes, slides, 16 mm films, posters, charts, brochures and exhibits to generate public awareness in four environmental sectors identified as priority areas. They are:

1. Deforestation; includes wood cutting.
2. Soil erosion; includes upland erosion leading to siltation downstream and coastal areas.
3. Water pollution; includes contamination of water supply and coastal areas.
4. Inadequate biomass management; includes solid waste disposal and littering.

Exhibitions, seminars and workshops are being regularly conducted by Central Environment Authority, Sri Lanka Association for Advancement of Science and Non-Governmental Organizations as part of a campaign to bring about environmental awareness.

The mass media in Sri Lanka through weekly features are highlighting the environmental degradation with a view to draw the public attention to the constructive role of individuals and groups in bringing about sound environmental well being.

Coastal programs and marine pollution have almost always been a thematic area in most of these public awareness campaigns.

NATIONAL LAWS AND PROCEDURES

The 1978 constitution of Sri Lanka recognized that the state shall protect, preserve and improve the environment for the benefit of the community as state policy.

In 1980, the parliament enacted into law the "National Environmental Act No. 47" which vests authority for environmental policy formulation and coordination of various agencies in long range planning on environmental protection and management in the Central Environmental Authority (CEA).

The coast Conservation Act No. 51 came into being in 1981. It calls for a survey of the coastal zone, preparation of a coastal zone management plan, regulation and control of activities within the coastal zone and formulation and execution of schemes of work for coast conservation.

The marine Pollution Prevention Act No. 59 of 1981 gave authority for the prevention, reduction and control of pollution in Sri Lankan waters in a marine Pollution Prevention Authority (MPPA). It provided for preventive measures against pollution by criminal liabilities for enumerated acts which damage interests of Sri Lankan waters or the offshore. The latter included marine coastal or estuarine activities including fisheries activities and activities connected with the development of tourism.

The Government strategy for accelerated development brought about the establishment of several new government agencies for coastal and marine protection and management such as National Aquatic Resources Development Agency (NARA), Natural Resources, Energy and Science Authority of Sri Lanka (NARESA), Urban Development Authority (UDA), etc. The creation of several agencies has given rise to several overlaps in efforts and conflicts in jurisdiction, legal matters, functions and responsibilities.

Although there are nearly 80 different laws in the statute books, implementation is very often inadequate.

COOPERATION ON A GLOBAL, REGIONAL OR BILATERAL BASIS

A meeting of ministers in charge of environment of South Asian Countries (SAARC) was held early in 1981 to initiate the South Asia Cooperation Environment Program (SACEP). The aims of SACEP among others is to make judicious use of resources of the environment of the South Asian Countries and to make fullest use of organizational arrangements and facilities for cooperation. The protection and development of marine and coastal environments is one of the areas identified by the cooperative program.

The South Asian Regional Seas Program whose action plan is ready for adoption was drawn up by the five states (Bangladesh, India, Maldives, Pakistan and Sri Lanka) with the help of UNEP's Ocean and Coastal Areas Program Activity Centre (OCA/PAC). The South Asian countries have recently identified six priority areas for protection and management of marine environment. A survey of land-based sources of marine pollution and the formulation of guidelines and proposals for environmentally sound waste management technologies and policies is one such prioity activity area identified for immediate implementation. UNEP, UNIDO, WHO and FAO have been invited to extend their cooperation in implementing the overall aims of the action plan.

CONCLUSIONS

The following conclusions may be drawn from the meager information that is available on land-based marine pollutants.

- There is very little available data on land-based marine pollutants. This is especially so in regards to hazardous pollutants such as heavy metals and air pollution.
- The discharge of land-based pollutants to rivers, estuaries and the coastal and marine waters is not currently creating a significant problem.
- The major sources of land-based pollution result mainly from human waste and not so much from industrial sources.
- Due to the rapid growth of pollution, accelerated agricultural development and increased investment in industry and rapid urbanization significant pollution problems in the coastal and marine waters could develop in a relatively short time span.
- Treatment of industrial wastes before discharge into the environment is rare. Although a few of the newer industrial zones have central waste treatment facilities, waste is generally directly dumped or discharged into rivers, streams, lagoons or coastal waters. This is also true for sewage discharge.
- There is adequate national environmental legislation but practical measures for implementation of environmental laws, due to many reasons, are almost not existent.

ACKNOWLEDGEMENT

I am thankful to Mr. G.A. Weerasundara for typing the manuscript.

REFERENCES

1. UNEP. 1985. Marine pollution from land-based sources, Environmental Law Guidelines and Principles. No. 7, UNEP, p. 21.
2. Costa, H.H. 1980. The physical, chemical and biological characteristics of freshwater bodies in the lowlands of Sri Lanka.
3. Central Environment Authority. 1989. Survey of stationary pollution sources Volumes I & II. Colombo.
4. Ceylon Tourist Board. Monthly Bulletin of the performance of the Tourist sector January 1983.
5. Peoples Bank Research Department. 1985. Internal Report No. 7, A census of the Exploitation of Sand and Sea-shell Resources in the Coastal Zone of Sri Lanka, p. 87
6. The Review of the Fertilizer Year. 1992. Fertilizer Secretariat, Ministry of Plan Implementation.
7. Personal communication. Pesticide Registration office, Department of Agriculture, Peradeniya.
8. De Silva, M.S.K.W., C. Nandani and V. Jayaweera. 1984. Mass mortality of fish in Kelani River. Mimeographed NARA report (unpublished), p. 11.

9. Dissanayake, C.B. and S.V.R. Weerasooriya. 1986. The Environmental Chemistry of Mahaweli River, Sri Lanka. Int. J. Environmental Studies 28:207-223.
10. Dissanayake, C.B., S.V.R. Weerasooriya and A. Senaratne. 1985. The heavy metal pollution of the Kelani River in Sri Lanka. Aqua. 2:79-88.
11. De Alwis, P., Yatapana, H., Dassanayake, W.H. and R. Perera. 1988. Monitoring of industrial effluents with a view to pollution control. Paper presented at the 43rd annual sessions of the Sri Lanka Association for the Advancement of Science, Colombo, Sri Lanka. (Unpublished).
12. Ranatunge, D.G.L. 1985. Some aspects of surface water resources in Sri Lanka with special reference to hydrology. Report by the National Committee of Sri Lanka for the International Hydrological Program, pp. 95-130.
13. Central Environment Authority. 1986. Water pollution survey with an ultra light aircraft, Phase I. Colombo, p. 77.

The Oceans: Physical-Chemical Dynamics and Human Impact. Edited by S.K. Majumdar, E.W. Miller, G.S. Forbes, R.F. Schmalz and Assad A. Panah. © 1994, The Pennsylvania Academy of Science.

Chapter Twenty-Seven

HEAVY METALS IN SHELLFISH AND FINFISH OF DELTAIC SUNDERBANS, WEST BENGAL, NORTHEAST INDIA

B. BHATTACHARYA and **S.K. SARKAR***

Department of Metallurgical Engineering
Jadavpur University
Calcutta 700032, West Bengal, India
and
*Department of Marine Science
Calcutta University
Calcutta 700019

INTRODUCTION

The coastal zone is "that space in which terrestrial environments influences marine (or lacustrine) environment and vice versa (Carter, 1989)". India has a coastline of 7000 km; rivers discharge 1645 km^3 of freshwater annually, of which 75% enters the Bay of Bengal (east and northeast coast) and 25% of the Arabian Sea (west coast) (Sanziry et al. 1988). In the entire southern region of Bengal delta, Sunderban forest of the Indian part (Lat. 21°30′N-22°30′N and Long. 88°10′E-89°59′E; total area - 4266.6 km^2) is covering the south-eastern tidal areas of coastal West Bengal (210 km coastline; area 0.82 million ha). Hooghly-Matlah estuary together with the unique mangrove ecosystem makes the delta a unique biosphere characterized by luxuriant flora and fauna of high economic importance. However, these systems suffer from a variety of anthropogenic degradation and the environment is under stress disturbing its natural regeneration and purifying capacity. A total of 396x10^{6}m^3 of wastes including 122x10^{6}m^3 industrial effluents are added to this stretch every year (Ghosh et al. 1973), posing a potential pollution source to this system. Large scale accumulation of heavy metals and pesticides has been observed in the biotic and abiotic matrices (Jhingran and Yadav, 1989). Monitoring the levels of heavy metals especially those which contaminate the ecosystem is of vital importance for better management of this biosphere. The present study aims to determine the metal levels of some selective shellfish and finfish to enable future environmental impact assessment of this area which has recently been included under the International Geosphere-Biosphere Program (IGBP).

PHYSIOGRAPHY AND CLIMATE

The geographical location of the present study areas lies in the eastern and western flank of Sagar Island, embracing the western sector of deltaic Sunderbans nearly 100 km south of Calcutta (Lat. 21°37′ and 21°52′N and 88°03′ and 88°11′E) (Figure 1). The island has a triangular outline with a length of 30 km along

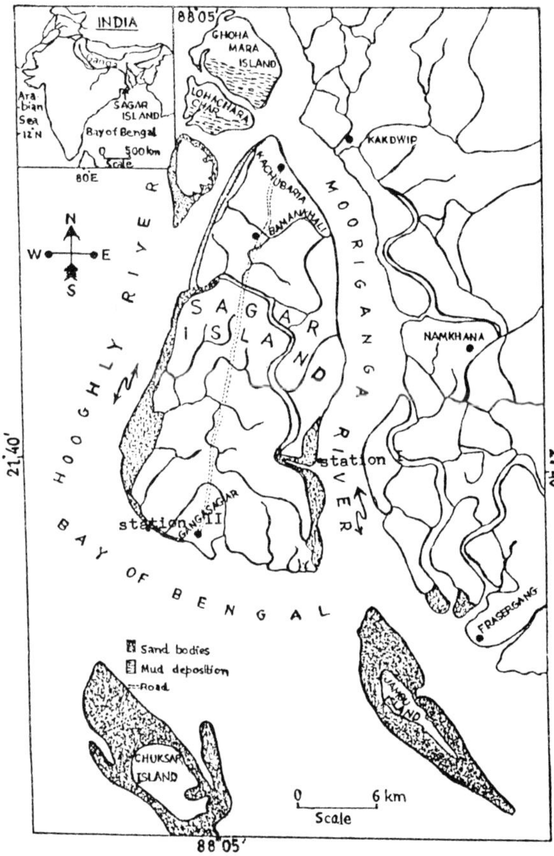

FIGURE 1. Map of Sagar Island showing the location of stations. Station I - Chemagari mud flat. Station II - light house.

north-south and a maximum width of 10 km towards south. The island is bounded on the east and west by the Mooriganga and Hooghly rivers, respectively. The southern margin directly faces the Bay of Bengal. The island is constantly receiving industrial effluents from a petrochemical refinery complex situated in the western bank of Hooghly estuary.

The climate is humid (up to 96%), tropical and the seasons are pronounced with four-month duration. Premonsoon (March-June) is the dry season with occasionally higher temperature. Monsoon season (south-west) is accompanied by heavy rainfall (annual average precipitation 1600 mm) and postmonsoon (November-February) comprised partly the winter season, with lower temperature and lesser precipitation.

PHYSIOGRAPHY OF FIELD STATIONS

Station I: Chemagri Mud-flat

Morphologically the area is characterized by a tidal-flat with admixture of fine sand, silt and clay. Towards the lower intertidal zone, the sand-mud ratio increases and ultimately the flat becomes sandy in the lowest intertidal zone. The slope is gradual to feebly terraced. This zone is covered with luxuriant but gradually declining mangrove patches (due to biotic interference) comprising of *Avicennia* sp., *Exoecaria agallocha, Rhizophora* sp.

Station II: Light House

This study area belongs to the beach face of Southern Sagar Island and is marked by a tidal-flat related beach facing Bay of Bengal. The mud flats generally occur at a depth of 0.5-1 m below the sandy beach surfaces having backshore to foreshore sedimentary facies. The existance of some ruins of an old light house has made the colonization of some invertebrates inhabiting the hard concrete substratum.

Tides

The estuary belongs to a macrotidal (>5 m) set up with semidiurnal tides. This estuarine zone is further characterized by a relatively weak tide in the winter months (3-4 m), strong tides from February-May, followed by a small period of relatively weaker tide and then again a strong tide (5-7 m) during the pre-winter season. The tidal propagation affects the estuary up to a distance of 280 km.

Sampling and Analytical Procedure

Finfish and shellfish were collected from the aforesaid eco-compartments. Gastropod molluscs (*Cerethedia cingulata, Telescopium telescopium*) and a mussel (*Macoma barmanica*) were sampled manually from intertidal mud-flat (station I). The bivalve was collected at a depth of 40-45 cm. The fishes were caught using gill net or baited hooks in the vicinity of the Chemagari creek waters. Rest gastropods (*Littorina littorea* and *Nerita articulata*) and a rock oyster (*Crassostrea cucullata*) were collected from the ruins of the sea wall (station II).

The fish samples were descaled and a portion of the dorsal muscle was removed using stainless steel knives and forceps. Different soft tissues of two bivalves were dissected out. For each species mean concentrations and standard deviations of 10-15 individuals were analyzed. In some cases, tissues from several small organisms of the same species in particular location were pooled for analysis. Immediately after the catch, specimens were collected in clean, acid-washed polythene bags and were transferred in the laboratory in an ice-box (-5° to -10°C) for further analysis.

Simultaneously, water samples were collected in previously acid-cleaned polythene bottles after filtering through 0.45 μ millipore membrane filter paper. Sediment samples (top 2.0 cm) were collected using a vanveen grab and shell and other visible organisms were removed manually.

On reaching the shore laboratory, the samples were thawed, macerated, weighed and analyzed in triplicate. The samples (1-2 gm) were refluxed in conc. HNO_3 followed by $HC10_4$ (3:1 ratio by volume) until a clean solution was obtained. Samples were diluted to volume with deionized water. Concentration of heavy metals (Fe, Zn, Cu, Mn, Co, Ni, Cd & Pb) was determined by aspiring the samples into a Perkin-Elmer Atomic Absorption Spectrophotometer (Model 2380, with background correction and acetylene as fuel) under recommended operating conditions (Perkin Elmer, 1982) and expressed as μgm metal g^{-1} dry weight. Full details of all these procedures are given in Harper et al. (1989). Estimation of Hg and As of the soft body parts of the mussel were carried out by cold-vapor and hydride technique after being oxidized with potassium permanganate solution. 3% solution of sodium borohydride E. Merck in 1% sodium hydroxide (E. Merck) has been used as a reductant. High purity argon gas was used as purge gas (Perkin-Elmer, 1978). In case of Hg, losses by volatization during sample pretreatment are a potential problem. For this reason the whole digestion procedure was investigated using spikes of both inorganic and organic Hg (methylmercury). No such losses were obtained.

The water sample was chelated with diethyldithiocarbamate at pH 3.6 $\pm$ 0.1 using phthalate buffer and extracted into methyl isobutyl ketone (Nix and Gordwin, 1970). The extracted phase was analyzed for Fe, Mn, Cu, Co, Pb, Ni and Zn in air-acetylene flame. The water sample was concentrated tenfold with a drop of nitric acid and analyzed for cadmium in air-acetylene flame. Digestion of sediment sample was carried out by Lithnor's (1975) method. Reagent blanks were also prepared in the same sequence of treatments.

All chemicals used in preparing the samples for analysis were checked for possible contamination. No evidence of contamination was found, the analyses of chemical blanks providing readings at or below the detection limits (0.02 μg g^{-1}) of the apparatus. Analysis of reference material (Kodak TEC-50-B) gave results with 5% of the certified value for all the metals, showing the analytical method to be suitably accurate. Detection limit of metals (DL) and the relative standard deviation (RSD) is shown in Table 3.

Hydrological parameters (water temperature, salinity, dissolved oxygen and pH) were estimated by standard methods as outlined by Strickland and Parsons, 1968.

Statistical correlations were computed for the concentrations of metals versus metals in each tissue and for tissue versus tissue for each metal using a statistical package.

Historical Account

Investigations on heavy metals in the Indian waters especially their distribution in water, sediment and aquatic organisms include those of Paropkari et al. (1980) in Gulf of Kutch; Murthy et al. (1985) in western continental shelf; Zingde and Desai (1981) and Matkar et al. (1981) in Bombay waters (west coast); Sankaranarayan and Reddy (1973), Zingde et al. (1976), Kamat and Sankaranarayan (1975a) and Singbal et al. (1978) in estuarine and coastal areas of Goa (west coast). Venugopal et al. (1982), Nair and Govindan (1986), Nair and Nair (1986), and Mallik and Suchidan (1984) in Cochin backwaters (south western coast). Kumaraguru (1980), Magendran (1985), Senthilnathan (1985) in Vellar estuary and its adjoining coastal waters (southern India). Ganapati and Raman (1973) and Pragatheeswaran et al. (1986) in Madras and Visakhapatnam coastal areas. Information on the metal levels in the Hooghly estuarine complex are available only to a limited extent. Kundu and Choudhury (1993) and Mitra et al. (1987, 1990) evaluated the metal levels of zooplankton and molluscs respectively. Bhattacharya et al. (1994) estimated a preliminary account of heavy metals in considerate number of flora and fauna of this area. In view of the relative paucity of information, the present investigation is undertaken with the primary objective of obtaining the base-line data and to establish the distribution of heavy metals in different shore animals which could serve as indices of pollution.

RESULTS AND DISCUSSION

Hydrology

Hydrology of the Hooghly estuary is highly dynamic and presents a cyclic pattern, characterized by large amount of precipitation and tidal interplay (Sarkar et al. 1985). No pronounced variation of the hydrological parameters was observed between the two studied stations and hence the mean values (as shown in Table 1) would be applicable for both the stations. Variations in surface water temperature

TABLE 1.

Variations of hydrological parameters of the coastal water at Sagar Island during the study period.

Month	Water temperature °C	Salinity ‰	D.O. ml/l	pH
March '92	32.0	28.50	4.70	8.3
April	29.8	30.85	4.35	8.3
July	31.0	27.55	5.35	8.0
August	31.5	11.45	5.15	7.8
November	23.5	20.52	4.88	7.56
December	20.5	24.81	4.15	8.0

were normal for a tropical estuary with maximum value in August (31.5°C) and minimum during December (20.5°C). Variations in dissolved oxygen (range 4.15-5.35 ml/L) were not as conspicuous as salinity (11.45-30.85‰). pH did not show major fluctuations.

Metal Levels in Shellfish

The capacity of bivalve molluscs to accumulate potentially toxic heavy metals in their tissues, far in excess of environmental levels, is well documented (Bryan and Gibbs, 1983) and proved to be excellent bioindicators of trace metal concentrations (Phillips, 1977). The present study also revealed the same potentiality in the bivalves and gastropods and showed either organ-specific or species-specific pattern for accumulating certain metals (Tables 3, 4 & 6) as endorsed by several workers (Bryan, 1973; Rajan and Kalyani, 1990; Mitra et al. 1990). Fe, Zn and Cu were preferentially accumulated in bivalves and gastropods as shown by several workers (Rajendran et al. 1988; Pillai et al. 1986; Mitra et al. 1987; 1990). Mn showed almost an uniform pattern of distribution throughout the study period which might be accumulated from the food source (Eisler, 1981). Excepting few cases, Co, Ni, Cd and Pb were below the spectrographic detection limit both in shellfish and finfish which was previously established by other workers in this estuary (Kundu and Choudhury, 1993; Mitra et al. 1987). This depletion might be attributed to the respective concentration of these metals in seawater. Hg and As showed higher

TABLE 2

Heavy metal concentrations in water (μgm/ml) and sediment (μgm gm⁻¹ dry wt.) during the study period

	Fe	Zn	Cu	Mn	Co	Ni	Cd	Pb
March '92								
Water	110	1.9	0.2	7.5	20	12	NT	1.0
Sediment	18,000	650	30	475	130	130	NT	NT
April								
Water	240	3.8	0.4	15	10	20	0.5	0.20
Sediment	15,590	580	35	190	155	40	NT	1.0
July								
Water	300	4.8	0.2	46	15	23	0.5	0.050
Sediment	21,000	700	20	NT	500	30	NT	NT
August								
Water	30.0	4.8	NT	4.6	1.5	2.3	0.5	0.050
Sediment	15,000	140	40	NT	20	NT	NT	NT
November								
Water	153.0	2.5	0.3	1.5	NT	0.9	NT	NT
Sediment	20,300	60	140	440	20	40	NT	1.0
December								
Water	52.0	2.1	0.6	9.1	NT	NT	0.2	0.2
Sediment	29,000	140	30	600	NT	NT	NT	1.0

NT = Not traceable

TABLE 3

Mean heavy metal concentrations (in μg g^{-1} dry wt.) in various soft tissues from the mussel, *M. barmanica* (size range 6.40 ± .5 cm) from Chemaguri mud flat (Station I).

		Fe	Zn	Cu	Mn	Co	Ni	Cd	Pb	Hg	As
RSD		1.5%	1.5%	1.2%	1.8%	1.5%	1.5%	1.5%	4.8%	4.5%	4.5%
DL (μgm/ml)		0.10	0.018 (1000 μg g^{-1}**)	0.077 (70.0 μg g^{-1}**)	0.052	0.12	0.14	0.028 (2.0 μg g^{-1}**)	0.19 (2.5 μg g^{-1}**)	0.00047	0.000095
Tissue	n	Fe	Zn	Cu	Mn	Co	Ni	Cd	Pb	Hg	As
March '92	8										
Visceral mass*		3600	300	50	90	NT	NT	NT	NT	3.67	0.53
Podium		300	290	30	50	NT	NT	NT	NT	0.86	NT
Siphon		2900	440	30	90	NT	NT	NT	NT	1.65	NT
Mantle		4200	250	110	100	NT	NT	NT	NT	2.02	0.26
Gill		2400	920	60	80	NT	NT	NT	NT	0.81	NT
Ad. muscle		1100	190	30	50	NT	NT	NT	NT	1.64	0.51
April '92	8										
Visceral mass*		2140	510	70	60	NT	NT	NT	NT	0.26	1.66
Podium		210	390	NT	30	NT	NT	NT	20	0.59	0.54
Siphon		3190	520	230	120	NT	NT	NT	NT	0.76	0.38
Mantle		5300	550	230	120	NT	NT	NT	NT	1.02	1.09
Gill		1630	1740	100	70	NT	NT	NT	NT	4.53	0.14
Ad. muscle		730	620	40	60	NT	NT	NT	NT	2.24	0.86
July -'92	11										
Visceral mass*		2690	560	10	70	30	NT	NT	NT	1.39	0.75
Podium		160	200	NT	NT	10	NT	NT	NT	0.68	NT
Siphon		2840	380	NT	60	20	NT	NT	NT	0.16	NT
Mantle		3470	250	40	100	20	NT	NT	NT	1.88	0.38
Gill		1470	500	10	20	20	NT	NT	NT	2.42	0.08
Ad. muscle		800	1600	20	10	20	NT	NT	NT	0.63	0.16

Table 3 continued on next page

Table 3 Continued

Tissue	n	Fe	Zn	Cu	Mn	Co	Ni	Cd	Pb	Hg	As
August '92	10										
Visceral mass*		2100	400	90	40	NT	20	NT	NT	0.18	1.05
Podium		150	270	20	40	NT	NT	NT	NT	0.08	NT
Siphon		900	260	40	30	NT	NT	NT	NT	0.66	NT
Mantle		3220	2700	530	70	NT	NT	NT	NT	0.89	0.54
Gill		480	1800	40	NT	NT	NT	NT	NT	1.33	NT
Ad. muscle		450	400	40	NT	NT	NT	NT	NT	0.70	NT
November '92	10										
Visceral mass*		2470	360	150	90	30	20	NT	10	0.16	0.15
Podium		390	380	130	40	20	30	10	NT	0.12	0.51
Sipohon		2040	400	120	80	30	20	NT	10	0.14	0.67
Mantle		4500	2000	230	150	20	20	NT	10	0.13	0.11
Gill		2300	760	150	100	30	30	NT	NT	0.24	0.79
Ad. muscle		740	480	140	50	40	30	NT	NT	0.27	0.15
December '92	9										
Visceral mass*		3000	460	30	120	NT	NT	NT	NT	1.06	0.43
Podium		220	910	20	NT	NT	NT	NT	NT	0.22	0.32
Siphon		3600	5500	190	110	NT	NT	NT	NT	0.06	0.21
Mantle		4000	8900	210	110	NT	NT	NT	NT	0.28	0.27
Gill		2060	7900	30	70	NT	NT	NT	NT	0.54	0.39
Ad. muscle		950	550	20	NT	NT	NT	NT	NT	0.15	NT

RSD = Relative standard deviation for 10 determinations.
DT = Detection limit of metals.
** = National Health and Medical Research Council recommended standard (Anon, 1981) for metal in molluscs.
* = Includes the gut content.
n = Number of samples analyzed.
NT = Not traceable

concentration especially in visceral mass and gill of *M. barmanica* (Table 3) with maximum concentration of 4.53 and 1.66 μg g^{-1}, respectively. This might be attributed to their bio-availability and feeding habits which results in biomagnification. These values are higher than the maximum permissible limits (0.5 ppm for Hg and 1.0 ppm for As) in seafoods for human consumption (FAO, 1983).

Considering the relative levels of each metal present, a concentration series for each tissue can be determined (Table 5). High enrichment of Zn especially in mantle (4700 μg g^{-1}) and gill (8200 μg g^{-1}) of the oyster and in mantle (8900 μg g^{-1}) of the mussel was registered (Tables 3 and 4) which might enter either by ingestion of particulate material (living and nonliving) suspended in the water or by uptake of

TABLE 4

Mean concentration of heavy metal contents in various tissues of a rock oyster, ***C. cucullata*** *(4.12 ± 1.10 cm) (in μgm g^{-1} dry wt.)*

	n	Fe	Zn	Cu	Mn	Co	Ni	Cd	Pb
March '92	12								
Visceral mass*		1000	700	270	20	NT	NT	NT	NT
Mantle		500	700	200	NT	NT	NT	NT	NT
Gill		1000	2000	800	NT	NT	NT	NT	NT
Adductor muscle		600	260	70	NT	NT	NT	NT	NT
April '92	14								
Visceral mass*		1870	4700	480	50	10	NT	30	NT
Mantle		1040	4400	540	30	20	10	NT	NT
Gill		780	1000	340	20	10	NT	20	NT
Adductor muscle		710	1900	30	30	10	NT	NT	NT
July '92	16								
Visceral mass*		1500	2300	510	80	NT	NT	60	20
Mantle		750	2300	490	30	NT	NT	40	20
Gill		1300	2200	570	20	NT	NT	40	NT
Adductor muscle		180	530	60	NT	NT	NT	20	20
August '92	14								
Visceral mass*		430	2500	610	40	NT	NT	30	NT
Mantle		250	900	230	NT	NT	NT	NT	NT
Gill		320	4500	770	40	NT	NT	30	NT
Adductor muscle		270	2800	100	40	NT	NT	NT	NT
November '92	18								
Visceral mass*		440	2500	610	10	20	20	20	10
Mantle		750	3000	600	50	20	20	20	10
Gill		480	8200	560	20	30	20	20	20
Adductor muscle		370	1800	420	40	50	30	20	NT
December '92	14								
Visceral mass*		330	3200	180	NT	NT	NT	NT	NT
Mantle		220	2200	160	NT	NT	NT	NT	NT
Gill		450	2600	540	NT	NT	NT	NT	NT
Adductor muscle		210	200	240	NT	NT	NT	NT	NT

n = number of samples taken, * = includes the gut contents, NT = not traceable dissolved metal from the water (Ayling, 1978; Luoma, 1983). Barnes (1968) studied the passage of sediments through the gills of lamellibranchiate molluscs and found that the sediment particles get trapped by secretions of the hypobranchial glands. The gills, being the vital organ in the filter-feeding, are continuously exposed to the ambient medium than the other body parts (Romeril, 1971). The mucous covering of the gill and mantle might also have increased the adsorption of metals (Kumarguru and Ramamurthy, 1978; Rajan and Kalyani, 1990). Pringle et al., (1968) reported that mucous is a complex sulphated mucopolysaccharide with exchange properties which facilitates a rapid exchange of metal ions across the cell membrane.

In general, a fair enrichment of Fe, Zn, Cu and to some extent Mn was evident in the visceral mass for both the bivalves. The digestive diverticula present in the mass are actively involved in the digestion and absorption of food and excretion. Hence, it takes an active part in heavy metal accumulation. Lysosomes present in the cells also help in increasing metal concentration in the tissues. Adductor muscle in the bivalve showed a low concentration of metals in comparison with other organs which might be attributed to their least activity and non-secretory function. The mussel may react to changes in metal levels by synthesizing metal binding protein (Noel - Lambodt, 1976) or by forming metal granules (Coombs and George, 1978), thereby increasing the storage compartment. Accumulation of trace metals by mussel is influenced by salinity, temperature and concentration of metals in water (Boyden and Romeril, 1974).

The data was transformed logarithmically and an analysis of variance (ANOVA) was performed for each metal using the tissues as the sources of variation. The result of this analysis (Table 7) showed that Fe, Cu and Mn in the mussel and only Cu in the oyster revealed significant differences between tissues. Correlation matrix of 4 metals (Fe, Zn, Cu, Mn) in different organs of the mussel and oyster established

TABLE 5

Concentration series for various tissues from (a) M. barmanica and (b) C. cucullata. Metals are ranked in order of decreasing concentration.

(a)

Tissue (*M. barmanica*)	Concentration Series
Visceral mass	Fe > Zn > Cu > Mn > Co > Ni > Hg > As > Pb > Cd
Podium	Zn > Fe > Cu > Mn > Co > Ni > Cd > Hg > As > Cd
Siphon	Fe > Zn > Cu > Mn > Co > Ni > Pb > Hg > As > Cd
Mantle	Fe > Zn > Cu > Mn > Co > Ni > Pb > Hg > As > Cd
Gill	Zn > Fe > Cu > Mn > Co > Ni > Hg > As > Cd > Pb
Adductor muscle	Fe > Zn > Cu > Mn > Co > Ni > Hg > As > Cd > Pb

(b)

Tissue (*C. cucullata*)	Concentration Series
Visceral mass	Zn > Fe > Cu > Mn > Pb > Co > Ni > Cd
Mantle	Zn > Fe > Cu > Mn > Pb > Co > Ni > Cd
Gill	Zn > Cu > Fe > Mn > Pb > Co > Ni > Cd
Adductor muscle	Zn > Fe > Cu > Mn > Co > Ni > Cd > Pb

some common features. High correlation values between Fe and Mn in the visceral mass, and Fe-Mn and Zn-Cu in the mantle for both the bivalves were observed. In the gill and adductor muscle of the oyster, Fe-Zn and Zn-Mn were significantly correlated while Fe-Mn and Cu-Mn showed the same result in the same organs of the mussel.

Among essential trace metals, Zn and Cu seemed to be preferentially concentrated in *C. cucullata* (Table 4) which might be attributed to the capability of this filter-feeding bivalve to biomagnify these elements from seawater (Eisler, 1981). The gills and mantle showed a high uptake of these metals which is consistent with the findings of Korringa (1952) that positive polyvalent ions such as Al^{+++}, Cu^{+++}, Fe^{+++}, Zn^{+++}, Hg^{+++} and Mn^{+++} are concentrated by mucous sheets of oysters. Similar

TABLE 6

*Mean concentration of heavy metals in 4 Gastropods - **C. cingulata** (2.3 ± 0 cm), **T. telescopium** (6.8 ± 0.4 cm), **L. littorea** (2.75 ± 0.34 cm) and **N. articulata** (2.24 ± 0.36 cm) (in µgm gm⁻¹ dry wt.)*

	n	Fe	Zn	Cu	Mn	Co	Ni	Cd	Pb
March '92									
C. cingulata	30	1100	180	100	840	NT	NT	NT	NT
T. telescopium	3	600	90	80	100	NT	NT	NT	NT
L. littorea	10	755	1475	2050	125	20	NT	NT	NT
N. articulata	8	485	700	20	20	NT	20	NT	NT
April '92									
C. cingulata	25	1810	80	240	660	30	NT	NT	NT
T. telescopium	2	3420	280	30	190	NT	NT	20	10
L. littorea	11	680	640	4300	60	10	NT	40	NT
N. articulata	9	820	1800	1975	90	NT	NT	20	NT
July '92									
C. cingulata	25	1460	1200	20	220	NT	10	NT	NT
T. telescopium	4	460	560	40	220	NT	20	NT	NT
L. littorea	12	890	1200	2230	70	NT	NT	50	60
N. articulata	8	230	120	10	20	NT	20	NT	NT
August '92									
C. cingulata	28	830	2500	30	160	NT	NT	NT	NT
T. telescopium	2	370	210	50	380	NT	NT	NT	NT
L. littorea	9	560	1600	2730	70	NT	NT	50	NT
N. articulata	8	520	2600	40	NT	NT	NT	NT	NT
November '92									
C. cingulata	26	1950	590	200	300	NT	NT	NT	NT
T. telescopium	4	3800	510	130	190	NT	10	NT	NT
L. littorea	10	1130	2300	1200	70	NT	NT	50	NT
N. articulata	9	2700	2200	20	100	NT	NT	NT	NT
December '92									
C. cingulata	28	4800	570	130	390	NT	NT	50	20
T. telescopium	2	540	800	80	110	NT	NT	NT	NT
L. littorea	10	630	660	520	20	NT	NT	20	NT
N. articulata	8	2700	500	150	140	NT	NT	20	10

n = number of samples taken; NT = Not traceable

observation was also reported by Arul et al. 1990; Brooks and Rumsby, 1965 and Segar et al. 1971. The present data demonstrates the potentiality of oysters as sentinel accumulators of heavy metals and can be successfully used as indicator species for pollution research and monitoring studies as emphasized by several workers (Hartwell et al. 1991).

The concentration of metals in the bivalves depends both on the level of the metal in the ambient medium as well as a variety of other ecological and biological factors. Forstner and Wittman (1981) pointed out that heavy metal concentration in oyster is a function of water quality, seasonal factors, temperature, salinity, diet, spawing and individual variation. Ayling (1974) postulated that uptake of trace metal from sediment by oysters is related to both metal availability in the sediments and the animals intake ability.

In gastropods, Fe, Zn, Cu and Mn were accumulated with marked variations and the values showed a species-specific pattern (Table 6). The herbivorous winkle, *L. littorea* registered a special affinity for accumulating Cu throughout the study period (maximum value 4300 $\mu g\ g^{-1}$) whereas *T. telescopium* and *C. cingulata* showed potentiality for Fe and Zn & Mn respectively. Substantial variation of these metals among the gastropods might be linked with the nature of diet (Young, 1975; Eisler, 1981) and pollution (Phillips, 1978). Elevated levels of Cu were attributable primarily to proximity to anthropogenic point sources as well as the copper storage granules (Martjoa et al. 1980). High values of Mn might be linked to the composition of the muddy substrates (Graham, 1972). Boyden (1977) inferred that concentration of metals in gastropods is governed to some extent by the size of the organisms and mostly values decrease with increase in size or are dependent of size.

Results of ANOVA revealed that distribution of Cu and Mn was highly significant among the four gastropods (Table 7). Correlation of 4 metals (Fe, Zn, Cu and Mn) showed significant correlation between Zn-Mn and Cu-Mn both in *C. cingulata* and *L. littorea* while Mn and Fe was significantly correlated in *N. articulata*. A high

TABLE 7

Results of analysis of variance on the logarithmically transformed data of the two bivalves, using their tissues as the source of variation and of four gastropods.

(a) M. barmanica

Metal	Fe	Zn	Cu	Mn	Hg	As
F	42.98	1.599	2.791	2.087	1.186	0.635
p	0.01	ns	0.05	0.05	ns	n

(b) *C. cucullata*

F	1.808	2.555	8.149	0.487	—	—
p	ns	ns	0.01	ns	—	—

(c) Gastropods

F	1.207	2.075	11.559	7.010	—	—
p	ns	ns	0.01	0.01	—	—

ns = not significant

correlation value was also found between Fe and Zn in *L. littorea* and *N. articulata.* In *C. cingulata,* Cu was highly correlated with Fe and Zn while such relationship was found in *N. articulata* between Fe and Mn. No significant correlation was found in *T. telescopium* in any case.

Metal Levels in Finfish

The observed level of concentration of heavy metals in the dorsal muscle of fishes were quite low excepting a few cases (Table 8) and maintained a steady state as endorsed by Bryan (1976b). In general, metal levels in fishes showed the following descending order: Fe > Zn > Cu > Mn > Ni > Co > Cd > Pb. Concentration of two essential trace elements, Fe and Zn showed high levels of accumulation which are essential for normal fish metabolism (Sakamato and Yone, 1978). Many workers also revealed an elevated concentration of Fe in fishes (Singbal et al. 1982; Namalwar, 1987). Reasons for low concentration of Cu are unkown, but suggests a discrimination against Cu by the highest trophic levels (Eisler, 1981). In marine teleosts there is evidence for the regulation of bioessential trace elements, Cu and Zn (Pentreath, 1976a) and Mn and Fe (Cross et al. 1973) and the non-essential Cd (Benoit et al. 1976) in the muscle tissues, Again, the heavy metals are probably lost from fish in urine, via the liver and over the body surface (Bryan, 1979) accounting for the low concentrations in comparison with the shellfish.

CONCLUSION

1. The metal levels reported in this paper may be taken as a convenient base-line against which any future pollution trends might be measured. Bio-monitoring of pollutants using bivalves (especially oyster) seems to be an effective way of assessing the level of contamination in this estuarine environment. A more detailed study might lead to the recognition of 'sensitive' and 'opportunistic' species of metallic pollution in this estuary.
2. The available data indicate that the estuary is getting polluted and it seems logical for immediate establishment for a scientific data base through continuous monitoring. This will not only help in understanding and identifying the environmental changes and degradation, if any, but also help in the computation of mathematical models to predict the dynamics of estuarine and coastal pollution. These environmental models would be useful for proper coastal zone management and for rational exploitation of coastal resources.
3. Further analyses of organochlorines in finfish and shellfish have also been undertaken which will be evaluated and published in near future.

ACKNOWLEDGEMENT

The authors are thankful to Dr. A. Bhattacharya, Head, Department of Marine Science, Calcutta University for critically going though the manuscript and also for suggestions for its improvement.

TABLE 8

Heavy metal concentrations in the muscle of finfish (in μgm^{-1} dry wt.)

	Fe	Zn	Cu	Mn	Co	Ni	Cd	Pb
March '92								
1. Mud-skipper (*Boleopthalmus* sp.)	140	120	NT	20	NT	NT	NT	NT
2. Parse (*Liza parsia*)	180	190	20	NT	NT	NT	NT	NT
3. Gurjali (*Eleutheronema tetradactylus*)	180	500	20	NT	NT	NT	NT	NT
4. Pomfret (*Stromateus* sp.)	320	3100	20	NT	NT	NT	NT	NT
5. Tapsee (*Polynemus paradiseus*)	100	150	20	NT	NT	NT	NT	NT
6. Bhola (*Daysciaena albida*)	120	2900	20	NT	NT	NT	NT	NT
7. *Lagocephalus* sp.	80	100	NT	NT	NT	NT	NT	NT
8. *Anguilla* sp.	280	150	NT	50	NT	NT	NT	NT
August '92								
1. Bombay duck (*Harpodon nehereus*)	1040	740	120	20	NT	30	NT	NT
2. Taugra (*Mystus* sp.)	660	570	NT	NT	NT	60	NT	NT
3. Ribbon fish (*Trichirus* sp.)	900	4900	NT	20	NT	50	NT	NT
4. Flat fish (*Cynoglosus* sp.)	950	520	NT	30	NT	40	NT	NT
5. Bhola (*Daysciaena albida*)	560	130	NT	NT	20	NT	NT	
6. Amedi (*Coila reynaldi*)	390	130	NT	NT	NT	NT	NT	NT
December '92								
1. Bombay duck (*Harpodon nehereus*)	660	650	30	30	NT	NT	NT	NT
2. Tangra (*Mystus* sp.)	530	720	40	60	NT	NT	NT	NT
3. Ribbon fish (*Trichirus* sp.)	400	330	30	30	NT	NT	NT	NT
4. Bhola (*Daysciaena albida*)	300	570	30	20	NT	NT	NT	NT
5. Amedi (*Coila reynaldi*)	390	600	30	30	NT	NT	NT	NT
6. Gurjali (*Eleutheronema tetradactylus*)	600	430	80	40	NT	NT	NT	NT
7. Phasa (*Setipinna phasa*)	1870	1100	50	110	NT	NT	NT	NT

NT = Not traceable

REFERENCES

Anon. 1973. Report on the 90th Session of the National Health and Medical Council, Australian Government Publication Service, Canberra.

Arul, V., A. Rajan and M. Kalyani. 1990. Dynamics of zinc concentration in *Katelyasia opima* (Gmelin) (Mollusca:Bivalvia) collected from Vellar estuary. *Porc. Int. Symp. Mar. Pollution.*:277-286.

Ayling, G.M. 1974. Uptake of cadmium, zinc, copper lead and chromium in the Pacific Oyster, *Crassostrea gigas. Water Res.* 8:729-738.

Barnes, R.D. 1968. Invertebrate Zoology. W.B. Sanders Co. London. pp. 311.

Benoit, D.A., E.N. Leonard, G.M. Christensen and J.T. Fiandt. 1976. Toxic effects of cadmium on three generations of brook trout (*Salvelinus fontinalis*) *Trans. Am. Fish. Soc.* 105:550-560.

Bhattacharya, B., S.K. Sarkar and P.K. Maji. 1994. Bioaccumulation of heavy metals in flora and fauna of Hooghly estuary, east coast of India. *Toxicol. and Environ. Chem.,* 42(nos. 1 & 2).

Boyden, C.R. 1977. Effect of size upon metal content of shellfish. *J. Mar. Biol. Ass. U.K.* 57:675-714.

Boyden, C.R. and M.G. Romerill. 1974. A trace metal problem in pond oyster culture *Mar. Poll. Bull.* 5:74-78.

Brooks, R.R. and D. Ramsby. 1974. Heavy metals in some New Zealand commercial sea fishes. *N.Z.J. Mar. Freshwat. Res.* 8:155-166.

Bryan, G.W. and R. Gibbs. 1983. Heavy metals in the Fal Estuary, Cornwall: A study of long-term contamination by mining waste and its effect on estuarine organisms. Occa. Pub. No. 2, *Mar. Biol. Ass. U.K.* pp. 111.

Bryan, G.W. 1976b. Heavy metal contamination in the sea. In Marine Pollution (Ed. R. Johnston). Academic Press:185-302.

Carter, R.W.G. 1989. Coastal environments. An introduction to the physical, ecological and cultural systems of coastlines. Academic Press.

Coombs, T.L. and S.G. George. 1978. Mechanisms of immobilization and detoxication of metals in marine organisms. In: Physiology and behaviour of marine organisms. D.S. McLusky and A.J. Berry (Eds.). Pergamon Press, Oxford:179-187.

Cross, F.A., L.H. Hardy, N.Y. Jones and R. Barber. 1973. Relation between total body weight and concentration of manganese, iron, copper, zinc and mercury in the white muscle of bluefish (*Pomatomus saltatrix*) and the bathyal-demersal fish *Antimora rostrata. J. Fish. Res. Bd. Canada.* 30:1287-1291.

Eisler, R. 1981. Trace metal concentration in marine organisms. Pergamon Press, N.Y.

FAO (1983). Compilation of legal limits for hazardous substances in fish and fishery products. FAO Fisheries Circular no. 764. pp. 102.

Forstner, J. and G.T.M. Wittman. 1981. Metal pollution in the aquatic environment. Springer Verlag, Berlin.:pp. 486.

Ganapati, P.N. and A.V. Raman. 1973. Pollution in Vishakapatnam Harbour. *Curr. Sci.* 42(14):490-492.

Ghosh, B.B., P. Ray and V. Gopalkrishnan. 1973. Survey and characterization of waste water into the Hooghly estuary. *J. Inland Fish. Soc. India.* 5:82-101.

Graham, D.L. 1972. Trace metal levels in intertidal molluscs of California. *Veliger.* 14(4):365-372.

Harper, D.J., C.F. Fileman, P.V. May and J.E. Portmann. 1989. The analysis of trace metals in marine and other samples. *Aquat. Environ. Port.:Analyt. Math., MAFF Direct. Fish. Res.* Lewestoft (3).

Hartwell, S.I., D.A. Wright, R. Takacs and C.H. Hocutt. 1991. Relative respiration and feeding rates of oyster and brackish water clam in variously contaminated water. *Mar. Poll. Bull.* 22(4):191-197.

Jhingran, A.G. and Y.S. Yadav. 1989. Coastal ecosystem management in West Bengal Fisheries - A perspective. *Proceed. Natn. Workshop on coastal zone management of West Bengal,* Dec. 14-16, 1989:A15-A24.

Kamat, S.B. and V.N. Sankaranarayanan. 1975a. Distribution of iron in estuarine and nearshore waters of Goa. *Indian J. Mar. Sci.* 4:30-33.

Korringa, P. 1932. Recent advances in oyster biology. *Quart. Res. Biol.* 27:266-308.

Kumaraguru, A.K. 1980. Studies on the chemical and biological transport of heavy metal pollutants copper, zinc and mercury in Vellar estuary and the toxicity of three pollutants to some estuarine fish and shellfish. Ph.D. thesis, Annamalai University, India. pp. 307.

Kumaraguru, A.K. and Ramamoorthi. 1978. Toxicity of copper to three estuarine bivalves. *Mar. Environ. Res.* 1(1):43-48.

Kundu, S.K. and A. Choudhury. 1993. Metal levels in zooplankton from Hooghly estuary, India. *Curr. Sci.* 64(2):110-112.

Lithnor, G. 1975. Manual of methodology in aquatic environment research Pt. Method for detecting measure and monitoring of water pollution. FAO Rome. pp. 41.

Magendran, A. 1985. Cycling of zinc in Vellar estuary. M. Phil., thesis, Annamalai University. pp. 93.

Mallik, T.K. and G.K. Suchidan. 1984. Some sedimentological aspects of Vembanad Lake, Kerala, Westcoast of India. *Inidan J. Mar. Sci.* 13:159-163.

Matkar, V.M., Ganapathy, S. and K.C. Pillai. 1981. Distribution of zinc, copper, manganese and iron in Bombay Harbour Bay. *Indian J. Mar. Sci.* 10(1):35-40.

Mitra, A., P.B. Ghosh and A. Choudhury. 1987. A marine bivalve *Crassostrea cucullata* can be used as an indicator species of marine pollution. *Proc. Natn. Sem. Est. Management*:177-180.

Mitra, A., B. Mukhopadhya, A.K. Ghosh and A. Choudhury. 1990. Preliminary studies on metallic pollution along the lower stretch of the Hooghly estuary, West Bengal, India. *Proc. Int. Symp. Mar. Pollution*:241-249.

Murthy, P.S.N., Paropkari, A.L. and C.H.M. Rao. 1985. Geochemistry of Zn in the sediment of the western continental shelf and slope of India. *Mahasagar Bull. Natn. Inst. Oceanogr.* 18(2):187-197.

Nair, N.U. and N. Balakrishnan Nair. 1986. Seasonality of trace metals in *Crassostrea madrasensis* (Preston) inhabiting the cochin backwaters. *Proc. Natn. Sem. Mussel Watch.* 1:6-14.

Nair, R.V. and K. Govindan. 1986. Biological characteristics as part of pollution monitoring studies. *Proc. Natn. Sem. on Mussel Watch* 1:86-92.

Nammalwar, P. 1987. Pollution impact and management of the coastal estuaries around Madras, India. *Proc. Natn. Est. Management*:190-193.

Nix, J. and T. Gordwin. 1970. At. Absorption Newsletter. 9:119.

Noel-Lambot, F. 1976. Distribution of Cd, Zn and Cu in the mussel *Mytilus edulis.* Existance of Cd-binding proteins similar to metallothioneins. *Experientia.* 32:324-325.

Paropkari, A.L., R.S. Topgi, Ch. M. Rao and P.S.N. Murthy. 1980. Distribution of Fe, Mn, Ni, Co, Cu and Zn in nonlithogenous fractions of sediment of Gulf of Kutch.

Indian J. Mar. Sci. 9:54-56.

Perkin-Elmer. 1982. Analytical Methods for Atomic Absorption Spectrophotometry. Perkin-Elmer Corporation, Norwalk.

Phillips, D.J.H. 1977. The use of biological indicator organisms to monitor trace metal pollution in marine and estuarine environments. A review. *Environ. Pollut.* 13:281-317.

Phillips, D.J.H. 1978. The common mussel, *Mytilus edulis* as an indicator of trace metals in Scandinavian water II. Lead, iron and manganese. *Mar. Biol.* 46:147-156.

Pillai, V.K., A.G. Ponniah, K.K. Valsala and P.V.R. Nair. 1986. Heavy metal load in a few bivalves from the coastal waters of South India. *Proc. Natn. Sem. on Mussel Watch.* 1:93-104.

Pragatheeswaran, V., B. Loganathan, A. Ramesh and V.K. Venugopalan. 1986. Distribution of heavy metals and organic carbon in sediments off Madras and Visakhapatnam. *Mahasagar-Bull. Natn. Inst. Oceanogr.* 19(1):39-44.

Pringle, B.H., D.E. Hissong, E.L. Katz and S.T. Mulawaka. 1968. Trace metal accumulation by estuarine molluscs. *J. Sanit. Engg. Div. A.-Soc. Civ. Engrs.* 94:435-475.

Rajendran, N., J. Tagore and R. Kasinathan. 1988. Heavy metal concentrations in oyster *Crassostrea madrasensis* (Preston) of cuddalore back waters, Southeast coast of India. *Indian J. Mar. Sci.,* 17:174-175.

Romeri, M.G. 1971. The uptake and distribution of zinc 65 in oysters. *Mar. Biol.* 9:347-354.

Sakamato, S. and Y. Yone. 1988. Requirement of red sea bream for dietary iron-II. *Bull. Japan Soc. Sci. Fish.* 44:223-225.

Sankaranarayanan, V.N. and C.V.A. Reddy. 1973. Copper content in the inshore and estuarine waters along the central west coast of India. *Curr. Sci.,* 42(7): 223-224.

Sanziry, S., A. Mesquita and T.W. Kureishy. 1988. Total mercury in water, sediments, and animals along the Indian coast. *Mar. Pollut. Bull.* 19(7):339-343.

Sarkar, S.K., A.K. Choudhury and B.N. Singh. 1986. Composition and variations in the abundance of zooplankton in the Hooghly estuary, West Bengal, India. *Proc. Ind. Acad. Sci.* (Anim. Sci.) 95(2):125-134.

Segar, D.A., J.D. Collins and J.P. Reiley. 1971. The distribution of the major and some minor elements in marine animals. Part II. Molluscs. *J. Mar. Biol. Ass. U.K.* 51:131-136.

Senthilnathan, S. 1985. Cycling of iron in vellar estuary. M. Phil., thesis, Annamalai University, pp. III.

Singhbal, S.Y.S., M.D. George, Rastogi and R. Noronha. 1982. The levels of certain heavy metals in marine organisms from Aquada Bay (Goa). *Mahasagar-Bull. Natn. Inst. Oceanogr.* 15(2):121-124.

Singbal, S.Y.S., S. Sauzgiri and R. Sen Gupta. 1978. Total mercury concentration in the Arabian sea water off the Indian coast. *Indian J. Mar. Sci.* 7(2):124.

Strickland, J.D.H. and T.R. Parsons. 1968. A practical handbook of sea water analysis. *J. Fish. Res. Bd. Can.* Bull. No. 167.

Venugopal, P., K. Sarala Devi, K.N. Remani and R.V. Unnithan. 1982. Trace metal levels in the sediments of the Cochin backwaters. *Mahasagar-Bull. Natn. Inst. Oceanogr.,* 15(4):205-214.

Young, M.L. 1975. The transfer of 65 zinc and 59 copper along a *Fucus senatus* (L) *Littorina obtusata* (L). food chain. *J. Mar. Biol. Ass. U.K.,* 55:583-610.

Zingdae, M.D. and B.W. Desai. 1981. Mercury in Thana Creek, Bombay harbor. *Mar. Pollut. Bull.* 12(7):230-231.

Zingdae, M.D., S.Y.S. Singbal, C.F. Moraes and C.V.G. Reddy. 1976. Arsenic, copper, zinc and manganese in the marine flora and fauna of coastal and estuarine waters around Goa. *Indian J. Mar. Sci.* 5:212-217.

The Oceans: Physical-Chemical Dynamics and Human Impact. Edited by S.K. Majumdar, E.W. Miller, G.S. Forbes, R.F. Schmalz and Assad A. Panah.

Chapter Twenty-Eight

OCCURRENCE AND DISTRIBUTION OF PERSISTENT CHLORINATED HYDROCARBONS IN THE SEAS AROUND INDIA

A. SARKAR
Chemical Oceanography Division
National Institute of Oceanography
Dona Paula, Goa - 403004, India

INTRODUCTION

Organochlorine compounds are well known for their high persistence, toxicity, low water solubility and lipophilic characteristics. Because of such lipophilic nature they are being accumulated in the biological organisms as well as human tissues. It has thus become a global concern due to widespread occurrence of several organochlorine pesticides such as DDTs, HCHs, aldrin, dieldrin, heptachlor, hepox, etc. and various congeners of PCBs in the environment.[1] Potential sources from which they are likely to reach the marine environment are rivers, the atmosphere, direct dumping into the sea and shipping activities. Upon entering the sea, a variety of interactions takes place with the compound. It involves many transition between different compartments such as water, suspended matter, sediment and organisms. Several groups of anthropogenic organochlorine compounds have been detected and reported frequently in the marine environment. These are DDT, cyclodiene, HCH insecticides, industrial compounds such as polychlorinated biphenyls (PCBs) and hexachlorobenzene (HCB), etc. Although most of the industrialized countries have banned or restricted the use of persistent chlorinated hydrocarbons because of their accumulative characteristics and harmful biological effects, they are in use extensively in the developing countries especially in the Southern hemisphere and tropical regions.

The use of chlorinated hydrocarbon pesticides such as DDT and HCH (BHC) is considered to be unavoidable in the developing countries for agriculture and public health in order to sustain food supply and to prevent the propagation of malaria.[2] In India around 75000 metric tons (MT) of technical grade pesticides are being used to cover 182.5 million hectares of cultivated lands.[3] The present demand for technical grade pesticide may go up to 144000 MT by the end of this century. Approximately 10000 MT of DDT and 7000 MT of BHC are being used for control of vectors of diseases as part of its vital public health programs for malaria, filarias, Kalazars, etc.[3]

As far as the occurrence and distribution of PCB in the environment are concerned it has now become an ubiquitous substance. PCBs is a generic name of many isomers and congeners with different numbers of chlorine atoms substituted in biphenyl rings. This group of chemicals was first synthesized by Schmidt & Schulz[4] (1881) before the turn of the century and has been produced commercially since 1929. Because of their large production and worldwide use, PCBs have already extended their boundaries of distribution over the global environment, as is evidenced by their detection in the Arctic and Antarctic atmosphere, hydrosphere and biosphere.[5,6] The persistent and bioaccumulative nature of PCBs have been recognized particularly in aquatic ecosystems, where the stepwise accumulation in higher-ranking predators of the foodchain is rather common.[7-9] PCBs currently present in the global environment were estimated to be about 370 x 10^3 tons. Of this, almost the entire amount, approximately 360 x 10^3 tons are retained in coastal sediments and open ocean water. Particularly the oceanic water contains the major portion of PCB residues, over 60% of the world environmental PCB load, clearly indicating that the open ocean water serves as a vast reservoir and final sink of PCBs. It has been estimated that about 780 x 10^3 tons of PCBs are still in use in the older electrical equipment and other products and deposited in landfills and dumps or in storage. The concern for PCB pollution in aquatic organisms has also been emphasized as the primary route of exposure to humans and domestic animals due to consumption of fish, etc., from contaminated waters.[10-13]

The cumulative effects of the residues of various organochlorine pesticides and PCBs over time on our coastal environment can be expected to be considerable in view of the fact that 25% of India's population lives in the coastal areas and marine fish is one of the major foods consumed by them. Uptake and accumulation of highly persistent organochlorine compounds by sediments, micro-organisms and fish lead to build up of such compounds in the foodchain thereby posing a serious threat to human health. This chapter deals with an exclusive evaluation of occurrence and distribution of various organochlorine compounds in different compartments of the seas around India extending up to the Antarctic Ocean.

MATERIALS AND METHODS

Marine environmental samples were collected from different compartments of the seas around India. The regions of investigation in different parts of the Indian Ocean are depicted in Figure 1. Water samples were collected by an *in-situ* sampler. It was deployed in different locations with the help of a self lubricating hydrographic winch PVC coated wire on board, *ORV-SAGAR KANYA*. The Sea-Star® *in-situ* sampler is an instrument designed to pump large volumes of water slowly (50 - 200 ml/min) through an extraction column which concentrates chlorinated hydrocarbons from water. It provides an accurate measurement of the volume of water extracted. The extraction column is a teflon cylinder (37 x 2.5 cm o.d) containing Amberlite XAD-2 resin. Prior to deployment of the instrument in water, the retention capacity of each of the columns for different organochlorine compounds and their recovery were determined.[14] It was observed that the retention capacity of the XAD-2 column for chlorinated hydrocarbon averaged 95 $\pm$ 12% indicating average column efficency of 94 $\pm$14%. Organochlorine compounds were eluted from the columns by 250 ml distilled methanol followed by 300 ml dichloromethane. Volumes

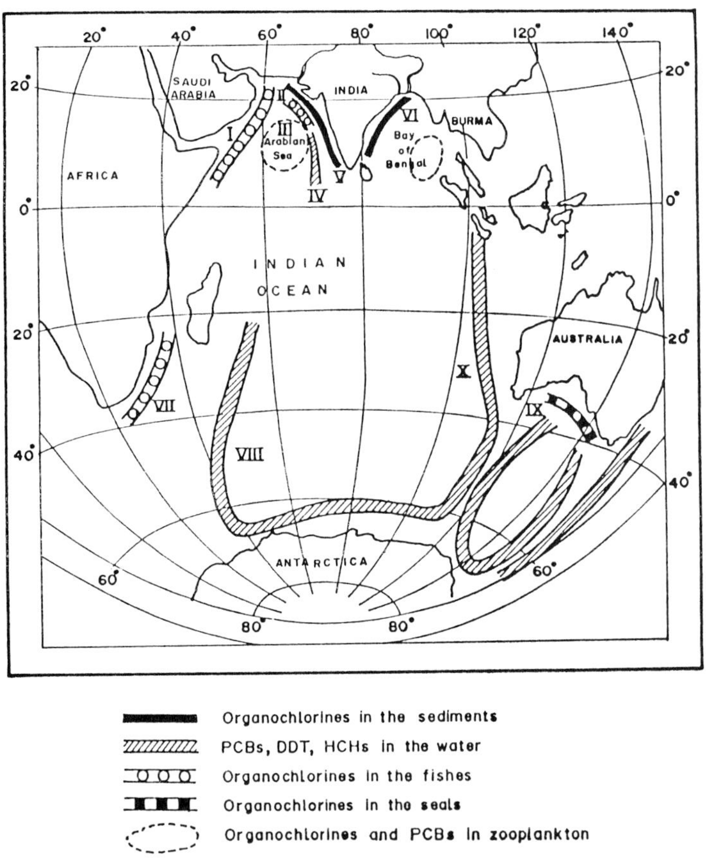

FIGURE 1. Map showing the regions of investigation of the residues of various organochlorine compounds in the Indian Ocean.

of eluates were reduced to 1 ml by evaporation on a Kuderna Danish (KD) evaporator and the aliquots were then treated with activated copper powder to remove sulfer. It was then passed through deactivated (10%) alumina followed by deactivated (5%) silica following the micro-column technique.[15]

Sediment samples were collected in aluminum foil (precleaned with distilled hexane) from different locations of the seas around India (Figure 1) with the help of Van veen grab on board *ORV-SAGAR KANYA* and *RV GAVESHANI.* Immediately after collection, sediments were preserved in deep freezer to avoid bio-degradation.

After thawing 100-150 g (wet wt.) sediments were first treated with 5 ml 1M solution of ammonium chloride and shaken thoroughly for 4 hours to facilitate the desorption of the pollutants adsorbed on the sedimentary particles. In order to determine the water content of each of the sediments a small quantity of sediment from each of the samples was dried in the oven at 120°C. Chlorinated hydrocarbons were extracted from the sediment by shaking the moistened sediment first with 150 ml acetone (nanograde) for about 8 hours followed by post extraction with 150 ml hexane (nanograde) and acetone (nanograde) (1:1) for 1 hour. The combined extracts were then transfered to hexane and the volume is reduced to 1 ml by evaporation in a KD evaporator. Just before evaporation, 1 ml iso-octane (nanograde) was added to each of the containers having the extract to prevent them from complete dryness. The aliquots were purified following the steps as described above.

Zooplankton samples were collected in aluminum foil (precleaned with distilled hexane) using Neuston net (200 u mesh size) on board *ORV-SAGAR KANYA* and *RV GAVESHANI* from different locations of the seas around India (Figure 1). Immediately after collection, the samples were preserved in the deep freezer. Organochlorine compounds were extracted from the zooplanktons by soxhlet extraction with distilled hexane (nanograde) for about 8 hours. The volume of the extracts was then reduced to 1 ml by evaporation in a KD evaporator followed by purification as described above.

Various species of fishes were collected with the help of fishing nets on board from different places of the seas around India (Figure 1) and preserved in the deep freeze. Organochlorine compounds were extracted from the fishes following the solvent extraction method and cleaned up according to the procedure described above.

CHEMICAL ANALYSIS

The determination of chlorinated hydrocarbons was performed on a gas chromatograph (Perkin Elmer 8700 and 3920B) coupled with electron capture detector (Ni^{63}). A capillary column (SE-54) (30 m in length, 0.25 mm i.d.) was used for the analysis. The temperature of the oven was programmed as follows: initial temp. 110°C for 2 min., raised to 180°C at the rate of 10°C per min. and held for 8 min., raised to 220°C at the rate of 4°C per min. and held for 5 min., finally raised to 270°C at the rate of 4°C per min. and held for 15 min. The equilibration time was 2 min. The temperature of the detector and the injection port was fixed at 340°C and 200°C respectively. In order to determine the percentage of recovery of various organochlorine compounds from the marine environmental samples, 100 μL of an internal standard (chlorobiphenyl congener, IUPAC-88) was spiked into each of the sample and was found to be 99.9%. The reproducibility of the chromatogram was checked by comparing the response of the internal standard spiked (100 μL of 0.1 ppm IU 88) into 0.9 ml standard solution of PCBs mixtures containing various congeners. For accurate determination of organochlorine compounds in each of the samples, PCBs fractions were pre-separated from organochlorine pesticides by silica (5% deactivated) column chromatography following the micro-column technique.[15] Organochlorine pesticides and various congeners of PCBs were identified and quantified with reference to the standard solution of pesticides and individual congeners of PCBs obtained from Supelco, Switzerland and Promochem, Germany, respectively.

RESULTS AND DISCUSSION

Chlorinated hydrocarbons identified and quantified in the seas around India are the compounds of the DDT family, cyclodienes, hexachlorocyclohexane and industrial compounds such as polchlorinated biphenyls (Table 1). The residue levels of various organochlorine pesticides and PCBs were determined in different compartments of the Indian Ocean. Figure 1 clearly shows the types of organochlorine compounds detected in different regions of the Indian Ocean. The compounds of the DDT family were detected in all the compartments of the ocean whereas HCH isomers in water, sediment and seals only. Aldrin and Dieldrin were detected in water and sediment from the Arabian Sea and the Bay of Bengal. Endrin is sparingly detected in various species of fishes[16] and sediments[17-19] only. PCBs were found to be present in considerable amounts in the fishes, water, seals, sediments and zooplankton. Figure 2 clearly shows the variation of the mean concentration levels

TABLE 1

Organochlorine Compounds Detected in the Seas Around India

India	
-pp′-DDT	2,2-bis-(p-chlorophenyl)-1,1,1-trichloroethane
op′-DDT	2-(o-chlorophenyl)-2(p-chlorophenyl)-1,1,1-trichloroethane
pp′-DDE	2-(o-chlorophenyl)-2-(p-chlorophenyl)-1,1-dichloroethylene
op′-DDE	2-(o-chlorophenyl)-2-(p-chlrophenyl)-1,1-dichloroethylene
pp′-DDD	2,2-bis-(p-chlrophenyl)-1,1-dichloroethane
op′-DDD	2-(o-chlorophenyl)-2-(p-chlorophenyl)-1,1-dichloroethane
HCH	A, B, Y,-isomers of 1,2,3,4,5,6-hexachlorocyclohexane.
Aldrin	1,2,3,4,10,10,-hexachloro-1,4,4a,5,8,8a-hexahydro-1,4-endo-exo-5,8-dimethanonaphthalene
Dieldrin	1,2,3,4,10,10-hexachloro-6,7-epoxy-1,4,4a,5,6,7,8a-octahydro-1,4-endo-exo-5,8-dimethanonaphthalene
Endrin	1,2,3,4,10,10-hexachloro-6,7-epoxy-1,4,4a,5,6,7,8-8a-octahydro-1,4-endo-endo-5,8-dimethanonaphthalene
PCBs	Congeners of polychlorinated biphenyls

of various chlorinated hydrocarbons in the Indian Ocean. Figure 3 exhibits a clear picture of the variation in the mean concentration levels of the isomers and metabolites of DDT in the Indian Ocean. A detailed discussion on the distribution of chlorinated hydrocarbons in different compartment of the ocean is given below.

SEDIMENT

The distribution of various organochlorine pesticides in the sediment along the east and the west coasts of India is clearly delineated in the map (Figure 4). The most interesting finding is the trends in distribution pattern of t-DDT, aldrin, dieldrin and HCHs in the sediments.[17-19] The sediment of the east coast of India was found to contain relatively higher levels of organochlorine pesticides than those in the west coast (Figure 4). Such a significant variation in the distribution pattern of pesticides may be attributed to the influx of huge amounts of pesticides in contaminated water through the runoff from the rivers on the east coast of India into the ocean. There are several other factors for which the coastal sediments are being

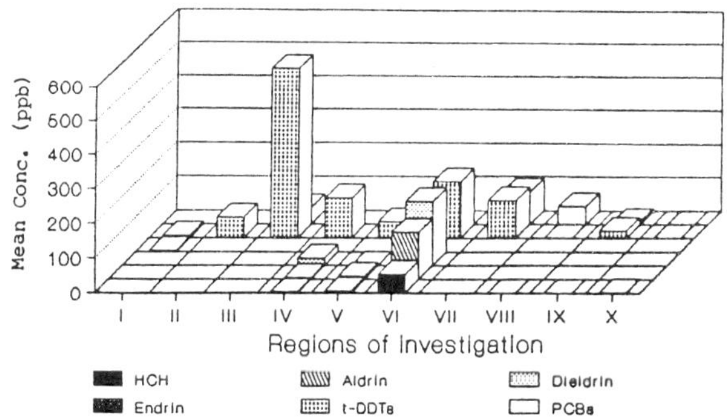

FIGURE 2. Mean concentration levels of Organochlorines in the Indian Ocean.

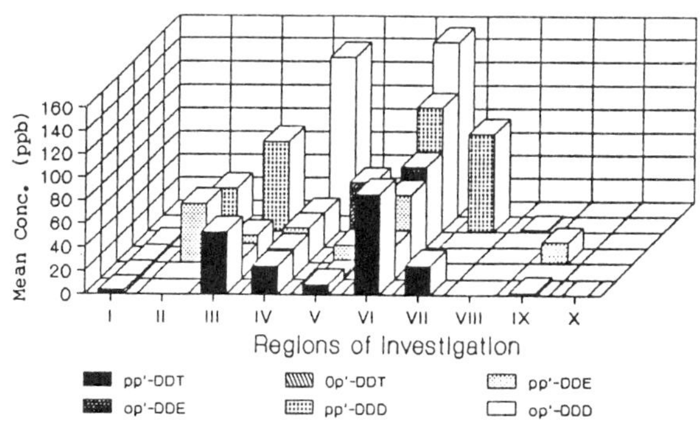

FIGURE 3. Mean concentration levels of metabolites of DDT in the Indian Ocean.

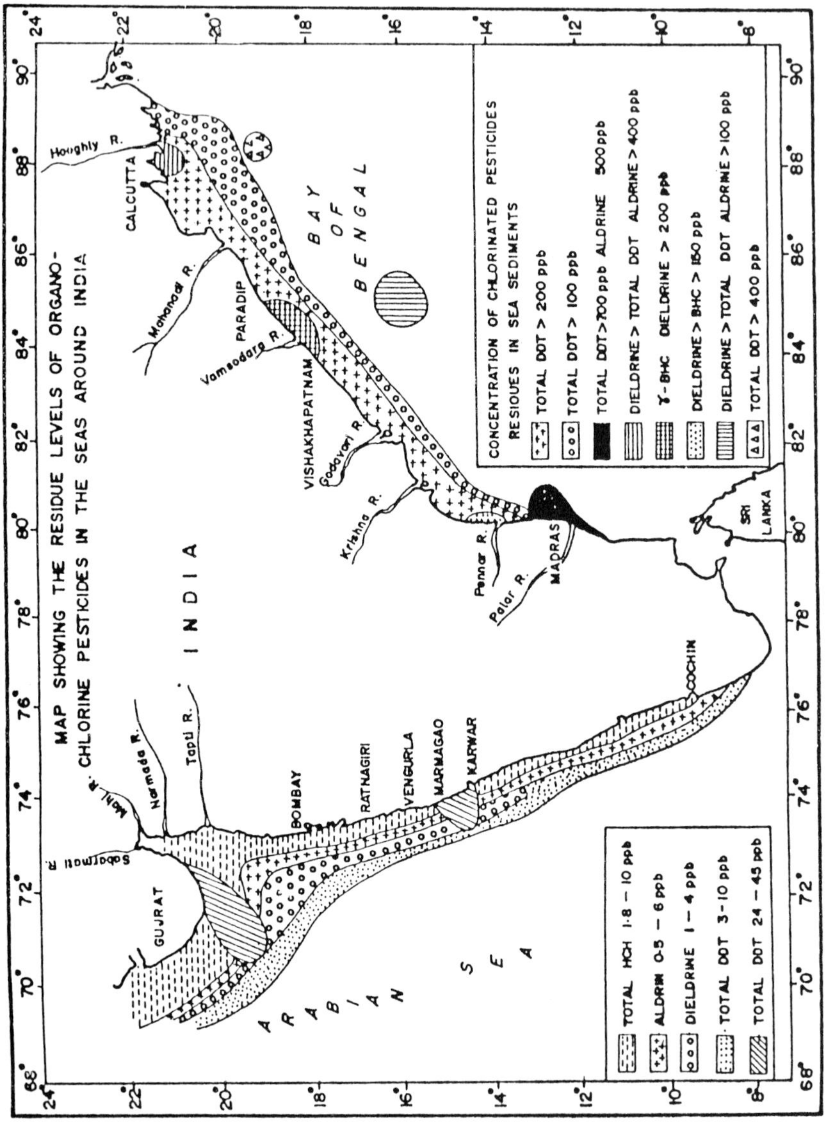

FIGURE 4. Distribution of organochlorine pesticides in the sediments from the seas around India.

contaminated with pesticides. The most striking point of observation is that there are a number of major rivers, namely, Hooghly, Mahandi, Vamsodhara, Godavari, Krishna, Pennar, Palar, etc. carrying a large volume of water into the Bay of Bengal. On the other hand, there are a few rivers, namely, Sabarmati, Mahi, Narmada and Tapti along the Gujarat coast of India. Although there are a few more small rivers falling on the west coast of India, they do not contribute as much into the seas as far as the contamination by pesticides is concerned. It was observed that the concentration levels of various organochlorine pesticides were quite high near the mouth of the rivers along the coasts and relatively less off the coasts. The most contaminated site is identified near the mouth of the Palar river where the concentration of t-DDT and aldrin were around 700 ppb and 500 ppb respectively. Dieldrin was identified discretely in a few samples collected from the east coast of India as indicated in the map (Figure 4). It has been reported that on an average 1800 MT of DDT, 270 MT aldrin, and 12000 MT of BHC were consumed every year in India.[3] Such a hugh amount of consumption of organochlorine pesticides is most likely to contribute a considerable amount of their residues in the marine environment. Among the isomers and metabolites of DDT in the sediment, pp′-DDE and op′-DDE were found to be highly dominant over the other. In fact, there is a good correlation (0.83) between the percentage composition of DDE with respect to t-DDT in the sediment along the coasts of India.

However, a very recent study on the occurrence and distribution of the residue levels of various organochlorine pesticides in a few sediment samples along the east coast of India indicated that the concentration of these compounds were quite low (HCH, 0.008-0.02 ppb, aldrin, 0.008-0.13 ppb, dieldrin, 0.017-0.15 ppb, endrin, 0.046-0.490 ppb, and t-DDT, 0.10-0.97 ppb). Such remarkably low levels of the residue of organochlorine pesticides in the sediment may be attributed to characteristics of the sediments (i.e. the texture of the sediment), clay mineral composition, retaining capacity of various organochlorine compounds, diffusivity (rate of diffusion of contaminants into the underlying layers), bioturbation, bioirrigation, population and growth of benthic organisms, etc. The texture of the sediment collected recently were mostly composed of sands and silts with a very low percentage of clay minerals whereas sediments collected during the period 1986-87 were found to be composed of more than 50% clay minerals. Thus it can be clearly understood that the retaining capacity of the sediment for organochlorine pesticides was enhanced due to higher percentage composition of clay minerals in the sediment. Above all, an awareness of environmental pollution among the people might have drastically reduced the load of input of contaminants in the coastal water. Besides, the fate and persistence of various pesticides are greatly influenced by various physio-chemical characteristics (pH, salinity, total organic carbo, metal ion concentration, clay mineral concentration, etc.) of the sediments.[20]

A few congeners of PCBs namely, CB-14, CB-16, CB-29, CB-31, CB-28, CB-66, CB-99, CB-101, CB-119, CB-153 and CB-138 were detected in the sediment along the east coast of India.[21] The variation of their concentration levels were shown in Figures 5 & 6. Of all these congeners, CB-14, CB-28, CB-138, CB-153, and CB-119 were found to be highly prominent. The presence of PCBs in the sediment along the east coast of India can be attributed to industrial discharge along the east coast of India.

WATER

Isomers of HCH, aldrin, dieldrin, DDT and PCBs were detected in water of different regions of the Indian Ocean[16] (Figure 1). Of all these organochlorine pesticides, total-DDT was found to be present in considerable amounts (15.81 - 444 ng/L). The most striking point of observation is that the concentration of t-DDT is very low in the open ocean water of the Indian Ocean. The variation in the concentration levels of various organochlorine compounds in water of different regions is shown in Figure 2. Such a remarkable variation in the concentration levels of t-DDT in the marine environment can be envisaged from the fact that huge amounts of pesticides in contaminated water are being discharged into the coastal regions regularly whereas the presence of pesticides in the open ocean water is mainly due to atmospheric transport of pollutants.[22,23] The concentration of HCHs was found to be quite low in water compared to those in the sediment which may be due to the hydrophobic characteristics of organochlorine compounds as well as dispersion of the pollutants in the open ocean.

In regards to the concentration levels of PCBs, they were present relatively in higher

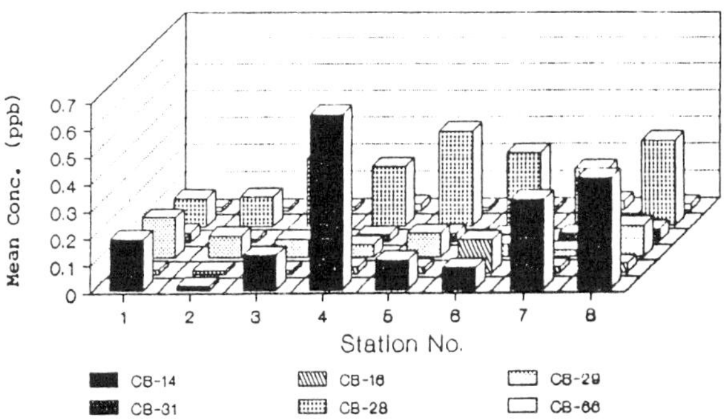

FIGURE 5. Concentration levels of PCBs in sediments along the east coast of India.

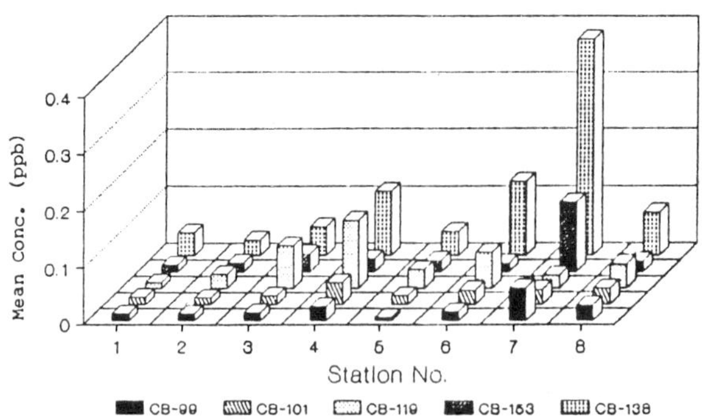

FIGURE 6. Concentration levels of PCBs in sediments along the east coast of India.

concentrations in the south west Indian Ocean than the eastern Indian Ocean.[22-25] It may be due to larger input of these compounds from the African coast.

ZOOPLANKTON

Among the organochlorine pesticides, compounds of DDT family were found to be abundant in the zooplankton from the north eastern Arabian Sea[26-30] The concentration of t-DDT in zooplankton was significantly high (mean conc. 500 ppb) in this region. This can be attributed to the input of huge amounts of DDT into the coastal water through agricultural drainage from the adjoining areas and river runoff. It is apparent that the biomass of zooplankton are quite high in the estuarine and coastal water and various depth strata will each be subjected to a different level of organochlorine contamination. The availability of organochlorine compounds in zooplankton is dependent on quality, quantity and size of competing living and inert particles because organochlorines are very hydrophobic and adsorb readily to other surfaces.[31-33] Moreover, particulate matter high in organic contents accumulates and retains a greater concentration of organochlorines than inorganic particles.[32-36] Organochlorines are also incoporated into collidal micelles believed to be formed by humic substances in the presence of ionic species[37-40] Humic substances may comprise as much as 90% of the dissolved organic matter in the seawater.[41-42] It clearly indicated that most organochlorines are probably bound to particulates or associated with the micellar structures present in seawater. Thus, the high concentration levels of DDT found in the zooplankton might be due to its associations with various other organisms being collected along with the zooplankton.

Concentration levels of various congeners of PCBs (17.62 - 105.7 ppb). were determined in the zooplankton from the Arabian Sea[29] Altogether, 31 congeners of PCBs were detected. However, only 17 congeners of PCBs, namely, 2 isomers of trichlorobiphenyl [CB-26, CB-28 (1.57-9.54 ppb)], 5 isomers of tetrachlorobiphenyl [CB-42, CB-44, CB-49, and CB-60 and CB-70 (0.55-6.85 ppb)], 6 isomers of pentachlorobiphenyl [CB-97, CB-99, CB-101, CB-105, CB-116 and CB-118 (0.55-11.28 ppb)] and 4 isomers of hexachlorobiphenyl [CB-138, CB-141, CB-149, and CB-153 (0-6.83 ppb)] were found to be present in almost all the samples. The other congeners of PCBs, namely, CB-4, CB-11, CB-14, CB-15, CB-16, CB-26, CB-31, CB-61, CB-65, CB-128, CB-151, CB-181, CB-183 and CB-198 were sparingly detected in a few samples. The distribution pattern in the concentration levels of the individual congeners of PCBs in the zooplankton showed that they were in abundance in the coastal region and steadily decreasing off the coast.[29] Such a variation in their concentration levels may be due to industrial discharge into the ocean from the adjoining areas. The decreasing trend in the distribution pattern may be attributed to the dispersion of the pollutants due to the dynamic situation of the ocean. The concentration levels of various congeners of PCBs were in the order, CB-101 > CB-99 > CB-28 > CB-52 > CB-149 > CB-42 > CB-118 > CB-60 > CB-44 > CB-138 > CB-105 > CB-116 > CB-49 > CB-97 > CB-141.

FISH

Isomers of HCHs and DDTs were detected in various species of fish collected from different regions of the seas around India[16,30] (Figure 1). However, the concentrations of the isomers of HCHs were too low to be quantified. The concentration of total-DDT was found to be in the range of 8.1-204 ng/g (dry wt.) and the mean concentration level was 59.17. It was found to be fairly high in the fish collected from the South African coast[43] probably due to extensive use of DDT in the coastal regions. Only 23 species of fish were analyzed from the Arabian Sea (Shailaja and Sen Gupta, 1989) and it was observed that the fish from the estuarine region contained the highest amount of t-DDT residues burden (344.7 ppb) followed by coastal variety. Among the metabolites of DDT, DDE constitutes a major proportion. Apart from these organochlorine pesticides, PCBs were also detected in the fish from the South African coast[43] (Figure 2). The mean concentration level of PCBs was found to be 95 ppb. This clearly shows the extent of contamination of the South African coastal water.

SEALS

Organochlorine pesticides and PCBs were also detected in the seals, *Arctocephalus pusillus doriferus* from the Australian coastal water[44] (Figure 1). Organochlorine compounds were determined in different organs of the seals. All of the seals contained PCBs, DDT, DDE and some DDD; some of them contained trace amounts of Hexachlorobenzene (Y-HCB) and HCH particularly in the blubber. In all samples the pp′-DDT were present in high proportions compared to the op′-DDT. An extensive study on the evaluation of the distribution of PCBs and organochlorine pesticides within animals clearly indicated that PCB and DDT levels in the blubber of male seals increased with age[45] but reaches a plateau value in later life. The mean concentration levels of various compounds in seal were, HCH, 0.26 ng/g; t-DDT 21.1 ng/g and PCBs, 9.02 ng/g (Figure 2).

CONCLUSION

The residue levels of various persistent chlorinated hydrocarbon pollutants such as DDTs, HCHs, aldrin, dieldrin, endrin, and PCBs were determined in different compartments (water, sediment and marine organisms) of the seas around India. The trends in distribution patterns of various organochlorine pesticides and PCBs in the seas around India indicated that the east coast of India is much more contaminated than the west coast because of the input of huge amount of pesticides and other organochlorine compounds into the ocean through the river runoff, industrial discharge, as well as agricultural drainage. The compounds of the DDT family were detected in almost all the compartments of the Indian Ocean. Concentrations levels of t-DDT were significantly high in zooplankton and various species of fish. Among the isomers and metabolites of DDT, DDE constitutes a major portion of total-DDT. The most striking point of observation is that the concentration levels of various organochlorine pollutants were quite high in the sediments near the mouth of the rivers on the east coast of India. HCH, aldrin and dieldrin were not found to be present in the marine organisms. However, endrin was sparingly detected in a few samples of fish. PCBs were detected in sediments and zooplanktons from the Indian coastal water and in various species of fish, seals and water from the southern Indian Ocean.

ACKNOWLEGEMENT

The author is thankful to Dr. B.N. Desai, Director, N.I.O., Dr. R. Sen Gupta, Emeritus scientist, N.I.O., and Dr. S.Y.S. Singbal, Head, Chemical Oceanography Division, N.I.O. for their encouragement and keen interest to the work.

REFERENCES

1. Risebrough, R.W., D.B. Menzel, D.T. Martin and H.S. Olcott. 1967. DDT residues in Pacific Sea birds: a persistent insecticide in marine food chain. *Nature*, London. 216:589-590.
2. Snelson, J.T. 1977. The importance of chlorinated hydrocarbons in the world agriculture. *Ecotoxicol. Environ. safety.* 1:17-30.
3. Bami, H.L. 1992. Pesticide's risk - an overview of Indian scene. *Pesticides Information*. vol. xviii, 3, 4-13.
4. Schmidt, H. and G. Schutz. 1981. Uber benzidin (A-di-amida-diphenyl). *Ann. Chem. Liebigs*, 207:320.
5. Atlas, E., T. Bidleman and C.S. Giam. 1986. Atmospheric transport of PCBs to the oceans. pp. 79-100. In: J.S. Waid (ed.) *PCBs and the environment.* Vol. 1, Boca Raton, Florida, CRC Press.
6. Tanabe, S. and R. Tatsukawa. 1986. Distribution, behavior and load of PCBs in the oceans, pp. 143-161. In: J.S. Waid (ed.) *PCBs and the environment,* vol. 1, Boca Raton, Florida, CRC Press.
7. Phillips, D.J.H. 1980. *Quantitative aquatic biological indicators: their use to monitor trace metal and organochlorine pollution.* London, Applied Science Publishers ltd.
8. Reijnders, P.J.H. 1980. Organochlorine and heavy metal residues in harbour seals from the Wadden Sea and their possible effects on reproduction. *Neth. J. Sea Res.* 14:30-65. 9.
9. Tanabe, S., H. Tanaka and R. Tatsukawa. 1984. Polychlorinated biphenyls, t-DDT and hexachlorocyclohexane isomers in the western North Pacific ecosystem. *Arch. Environ. Contam. Toxicol.* 13:731-738.
10. Highland, J. 1976. PCBs in food. *Environment* 18:12-16.

11. Watanabe, I., T. Yakushiji, K. Kuwabara, S. Yoshida, K. Maeda, T. Kashimoto, K. Kayama and N. Kunita. 1979. Surveillance of daily PCB intake from diet of Japanese Women from 1972-1976. *Arch Environ. Contam. Toxicol,* 8:76-75.
12. Schwartz, P.M., S.W. Jacobson, G.G. Rien, J.L. Jacobson and H.A. Price. 1983. Lake Michigan fish consumption as a source of polychlorinated biphenyls in human cord serum, maternal serum and milk. *Am. J. Public Health,* 73:293-6.
13. Sawhney, B.L. and L. Hankin. 1985. Polychlorinated biphenyls in food: A review. *J. Food Production,* 48:442-8.
14. Sarkar, A. and R. Sen Gupta. 1989. Determination of organochlorine pesticides in Indian coastal water using a moored in-situ sampler. *Wat. Res.* 23:8,975-978.
15. Duinker, J.C. and M.T.J. Hillebrand. 1983. Determination of selected organochlorines in seawater, pp. 290-309. In K. Grasshoff, M. Ehrhardt and K. Kremling, (eds.), *Methods of seawater analysis,* Verlag Chemie.
16. Shailaja, M.S. and R. Sen Gupta. 1989. DDT residues in fish from the Eastern Arabian Sea. *Mar. Pollut. Bull.* 20:12,629-630.
17. Sarkar, A. and R. Sen Gupta. 1987. Chlorinated pesticide residues in sediments from the Arabian Sea along the central west coast of India. *Bull. Environ. Contam. Toxicol.* 39:1049-1054.
18. Sarkar, A. and R. Sen Gupta. 1988a. Chlorinated pesticides in the marine sediments. *Mar. Pollut. Bull.* 19:35-37.
19. Sarkar, A. and R. Sen Gupta. 1988b. DDT residues in sediments from the Bay of Bengal. *Bull. Environ. Contam. Toxicol.* 41:664-669.
20. Sarkar, A. 1991. Evaluation of the major physico-chemical parameters influencing the degradation of DDt to DDE in the marine sediments, pp. 780-789. In: B. Nath and J.P. Robinson (ed.) *Environmental Pollution,* 1:ICEP-1, Inderscience Enterprises ltd.
21. Sarkar, A. 1993. State of pollution of the seas around India due to persistent chlorinated hydrocarbon. *Proc. First setac world congress, ecotoxicology and environmental chemistry - a global perspective.* Lisbon, March, 28-31. 1993, p. 10.
22. Tanabe, S., M. Kawano and R. Tatsukawa. 1982a. Chlorinated hydrocarbons in the Antarctic, western Pacific and eastern Indian Ocean. *Trans. Tokyo. Univ. Fisher.* 5:97-109.
23. Tanabe, S, R. Tatsukawa, M. Kawano and H. Hidaka. 1982b. Global distribution and atmospheric transport of chlorinated hydrocarbons:HCH (BHC) Isomers and DDT compounds in the western Pacific, eastern Indian and Antarctic Oceans. *J. Oceangr. Soc. Japan.* 38:137-148.
24. Tanabe, S., H. Hidaka and R. Tatsukawa. 1983. PCBs and chlorinated hydrocarbon pesticides in Antarctic atmosphere and hydrosphere. *Chemosphere.* 12:277-288.
25. Tanabe, S. and R. Tatsukawa. 1980. Chlorinated hydrocarbon in the north Pacific and Indian Ocean. J. Oceanogr. Soc. Japan. 36:217-226.
26. Kannan, S.T. and R. Sen Gupta. 1987. Organochlorine residues in Zooplankton off Saurastra coast. India. *Mar. Pollut. Bull.* 18:92-94.

27. Kureishy, T.W., M.D. George and R. Sen Gupta. 1978. DDT contamination in zooplankton from the Arabian Sea, India. *J. Mar. Sci.* 7,54-55.
28. Shailaja, M.S. and R. Sen Gupta. 1990. Residues of dichlorophenyl trichloroethane and metabolites in zooplankton from the Arabian Sea. *Current Science* 59:929-931.
29. Sarkar, A., T. Balachandran and P.N. Arvindaction. 1991. Chlorinated hydrocarbons in the zooplankton from the Arabian Sea. *Int. J. Toxicol. Occupational & Environ. Health,* 1:1,300.
30. Shailaja, M.S. and M. Nair. 1993. Observation on organochlorine pesticide residues in plankton and fish from the northern Arabian Sea, In: *Gulf oil spill and India - a report* (In press).
31. Haque, R. and D. Schmedding. 1976. Studies on the adsorption of selected polychlorianted biphenyl isomers on several surfaces. *J. Environ. Sci. Health B.* 11:129-137.
32. Picer, N., M. Picer and P. Strohal. 1977. The interaction of DDT with suspended particles in seawater. *Water Air Soil Pollut.* 8:429-440.
33. Hiraizumi, Y., M. Takahashi and H. Nishimura. 1979. Adsorption of polychlorinated biphenyl onto sea bed sediment, marine plankton and other adsorbing agents. *Environ. Sci. Tech.* 13:580-584.
34. Choi, W.W. and K.Y. Chen. 1976. Associations of chlorinated hydrocarbons with fine particles and humic substances in nearshore surficial sediments. *Environ. Sci. Technol.* 10:782-786.
35. Nau-Ritter, G.M., C.F. Wurster and R.G. Rowland. 1982. Partitioning of [^{14}ClPCB between water and particulates with various organic contents. *Wat. Res.* 16:1615-1618.
36. O'Connors, H.B. Jr., C.D. Powers, C.F. Wurster, K.D. Wyman, G.M. Nau-Ritter and R.G. Rowland. 1982. Fates and biological effects on plankton of particle-sorbed PCB's in coastal waters, pp 423-449, In: G.F. Mayer, (ed.) *Ecological stress and the New York Bight: Science and management. Estuarine research Foundation.* Charleston, South Carolina.
37. Wershaw, R.L., P.J. Burcar and M.C. Goldberg. 1969. Interaction of pesticides with natural organic material. *Environ. Sci. Technol.* 3:271-273.
38. Porrier, M.A., B.R. Bordelon and J.L. Laseter. 1972. Adsorption and concentration of dissolved carbon-14 DDT by coloring colloids in surface waters. *Environ. Sci. Technol.* 6:1033-1035.
39. Carter, C.W. and L.H. Suffet. 1982. Binding of DDT to dissolved humic materials. *Environ. Sci. Technol.* 16:735-740.
40. Hasset, J.P. and E. Milicic. 1985. Determination of equilibrium and rate constants for binding of polychlorinated biphenyl congers by dissolved humic substances. *Environ. Sci. Technol.* 19:638-643.
41. Boehm, P.D., J.G. Quinn. 1973. Solubilization of hydrocarbons by the dissolved organic matter in seawater. *Geochim. Cosmochim. Acta* 37:2459-2477.
42. Skopintsev, B.A. 1981. New data on the composition of organic matter from ocean water and the energy bound in it. *Oceanology* 21:582-588.

43. De Kock, A.C. and D.A. Lord. 1986. Chlorinated hydrocarbon residues in African Fish Eagle (Halieectus vocifer) eggs. *Institute for coastal Research Report No. 10.*
44. Smillie, R.H. and J.S. Waid. 1987. Polychlorinated biphenyls and organochlorine pesticides in the Australian Fur Seals, Arctocephalus pusillus doriferus. *Bull. Environ. Contam. Toxicol.* 39:358-364.
45. Addision, R.F. and T.G. Smith. 1974. Organochlorine residue levels in Arctic-ringed seals: Variation with age and sex. *Oikos*:335-337.

The Oceans: Physical-Chemical Dynamics and Human Impact. Edited by S.K. Majumdar, E.W. Miller, G.S. Forbes, R.F. Schmalz and Assad A. Panah. © 1994, The Pennsylvania Academy of Science.

Chapter Twenty-Nine

WASTE DUMPING IN THE OCEANS

ASSAD A. PANAH
Professor of Geology
Department of Geology and Environmental Sciences
University of Pittsburgh at Bradford
Bradford, PA 16701

INTRODUCTION

Pollutants entering the marine environment originate from a *point source* (an intentional disposal source) or from a *nonpoint and reverine source* (an accidental release or unintentional source).[1] Intentional disposals are comprised of direct dumping of municipal sewage sludge, dredged sediments, and various waste products like acid iron wastes or seafood processing water—culled from municipal sewage pipelines of industrial plants. Nonpoint sources—such as acid rain, lead, and mercury—enter the marine environment through run-offs or aquifer systems from hazardous landfills in the form of leachate. Approximately 75 percent of the waste sites along the coastal areas of the United States produce an effective source of leachate, which contaminates the marine environment.[1] Urban areas discharge lead, chromium, oil, and grease. Agricultural land produces oxygen-demanding pollutants, nutrients, pesticides, herbicides, and chlorinated hydrocarbons.[1,2,3] Ships dispose directly into the ocean. Everywhere in the United States, except in the Northeast, nonpoint sources of ocean pollution are more disastrous than point sources,[1] because seventy-eight percent of the coastal states consider water pollution by nonpoint sources to be greater than point sources.[4]

At the beginning of the industrial revolution the "out of sight, out of mind" mentality inspired replacing land disposal. As a result of the expansion of urbanization and industrialization, coastal areas endure a high concentration of pollutants. Today, organizations like the NOAA's National Estuarine Inventory (NEI), National Coastal Pollutant Discharge Inventory (NCPDI), National Status and Trend (NS&T)

program, and U.S. Environmental Protection Agency (EPA), U.S. Congress Office of Technology Assessment (OTA), Department of Natural Resources (DNR), and Center for Marine Conservation (CMC), among others, serve the will of concerned citizens to protect the environment while politicians reflect the wishes of their constituents, but every person in the entire world is morally obligated to exercise common sense to stop careless disturbance of our fragile ecosystem. Everyone must teach by example and assume responsibility for controlling pollution.

Most wastes generated in the oceans come from transportation at sea, mostly in the form of oil spills from tanker accidents, plastic disposal from ships and boats, residue after cleaning the tankers, and tailings from offshore mineral exploration. Since the onset of the Industrial Revolution, industrial, municipal, and agricultural wastes have been dumped into the ocean. In 1975, the U.S. National Academy of Science indicated that an average of 6 million tons of petroleum is released annually into the marine environment, only 10 percent of which comes from natural sources.[5]

Ocean dumping in the marine environment in the United States, as well as in the rest of the world, has caused physical, chemical, and biological contamination. The number of industrial discharge points varies considerably, depending on geographic location. In 1968 alone, more than 43 million metric tons of waste were dumped at 246 sites off the coast of the United States.[1] In 1972, federal legislation and EPA regulation reduced the number of sites to 188.[1] In spite of the control, marine organisms and their habitats are still disturbed while the marine ecosystem and food chain are particularly threatened in estuaries, bays, and coastal areas.

To assess the effects of the human impact on the quality of marine waters, scientists from the National Oceanic and Atmospheric Administration (NOAA) have developed the National Status and Trends (NS&T) Program to evaluate the relationship between chemical contamination and biological effects. Since 1984, the NS&T has studied 300 sites in the United States.[6] In 1990, the National Research Council (NRC) indicated that the United States spends $130 million on coastal environmental monitoring.[7] The special 20th anniversary report of NOAA (1990), the results of the monitoring of the NS&T, shows that on a national scale, the highest concentration of biological pollution initially corresponds to highly urbanized estuaries. The NS&T Program has also pointed out that the level of ongoing contamination has been relatively reduced in the coastal regions.

Ocean Environments

Jurisdictional boundaries and environmental laws affect waste dumping in coastal regions, estuaries, bays, and the open sea (Figure 1). All waters of the contiguous zone and beyond (located 3 to 12 nautical miles from the outer limit of the coastline) and waters of the outer continental shelf are classified as the *open sea*.Pollutants are dumped into the coastal regions and open sea by sewage pipelines, industrial waste pipelines, and dredged sediments (Figure 1). The United States coastal waters and open sea are protected by both the Federal Water Pollution Control Act, also referred to as the Clean Water Act (CWA), and the Marine Protection, Research, and Sanctuaries Act (MPRSA).[1]

Land-enclosed seawaters, which may include lagoons and tidal marshes, are classified as *estuaries*. No matter what forms they take, they provide highly valuable natural resources, an important component of the food chain. Because of their impact on the economy of an area in terms of the fish and shellfish harvesting industry, wildlife, and recreation, estuaries are protected by the Clean Water Act (CWA).

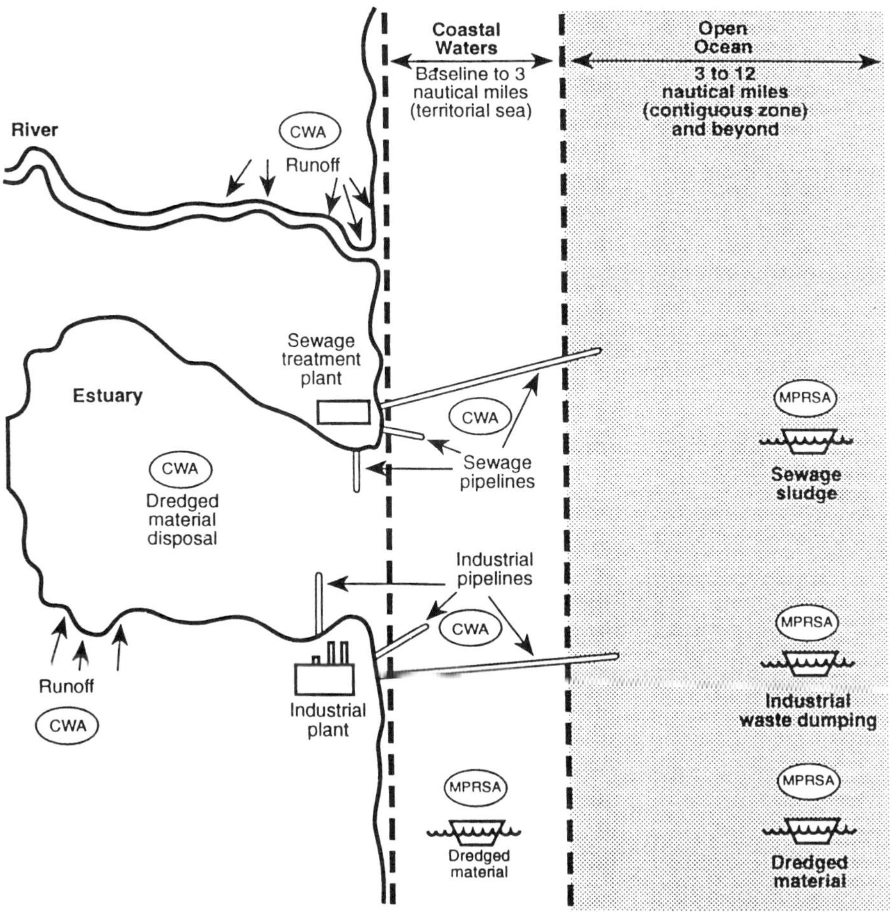

FIGURE 1. Jurisdictional Boundaries of Environmental Laws Affecting Marine Disposal.

CWA - Clean Water Act, formally known as the Federal Water Pollution Control Act.

MPRSA - Marine Protection, Research, and Sanctuaries Act.

Dumping beyond the inner boundary of the territorial seas is covered by MPRSA (CWA covers dumping within the territorial sea in principle, but is preempted by MPRSA. Estuarine dumping falls under CWA. Pipelines (wherever they are located) are covered by CWA.

Source: U.S. Congress, Office of Technology Assessment, *Waste in Marine Environments,* OTA-0-334 (Springfield, VA: National Technical Information Service, April 1987).

There are 102 estuaries listed in the U.S. National Estuarine Inventory.[8] Two-thirds of all waste dumping occurs in these estuaries, making them not only environmentally unsafe, but the most polluted. Most dredging activities occur in estuaries and produce significant sources of pollutants. Dumping in estuaries is less costly than disposal activities in upland and deep ocean environments because of their easy access and lower cost of transportation. According to the Corps of Engineers (COE), federal regulation has made it easier to obtain permits for waste dumping in estuaries than in coastal areas or the open sea. The deep ocean environment is regulated by more strict rules than are estuaries or coastal areas. Open sea dumping may be indirectly discouraged because of the high cost of transportation of waste.[1] In 1987, the Office of Technology Assessment of the U.S. Congress (OTA) estimated a total of 1,910 United States point discharges from municipal and major industrial pipelines into coastal and estuarine environments.[1]

Of these 1,910 major pipelines, 96 percent are located in estuaries. More than 69 percent of these are industrial.[1]

There are approximately 15,500 public waste treatment centers in the United States. Approximately 2.3 trillion gallons of waste are dumped into marine environments; 2 trillion gallons, into estuaries; and 0.3 trillion gallons, into coastal environments.[12] In 1982, a total of 6,645 MGD (million gallons per day) from waste treatment plants were dumped in the following coastal areas of the United States:

Area		
Northern Atlantic	....	4,150 MGD
Southern Atlantic	....	308 MGD
Gulf of Mexico	....	522 MGD
California and Hawaii	....	1,282 MGD
Northern Pacific	....	383 MGD

Environmental stress on the nation's coastal regions is ever increasing. NOAA has projected that by the year 2010 the population in all 30 coastal states will increase from 80 million to 127 million.[13] To protect the coastal waters and estuaries, the best policy is to minimize the dumping of pollutants into these environments. We must shift dumping from coastal environments and estuaries to open ocean environments by extending pipelines or by transporting waste material to deeper waters of the ocean. Today the one sewage pipeline that extends as far as 7 miles into the open sea is constructed along the California's coastline. A similar sewage outfall that will extend 9.5 miles out from the Boston Harbor is now under construction beneath the Harbor as part of a $4 billion "Boston Harbor clean up project."[9,10] The 24-feet-diameter tunnel is one of the largest civil engineering projects in the United States, and will be completed in 1999.[11] Currently, sewage from municipal and industrial treatment plants pumped into 3 meter deep harbor water which dilutes sewage discharge only by 10 to 1. Extending outfall into deeper marine environment will get 200 to 1 dilution, due to higher energy of the open sea.[10] Because of being inaccessible to the land-derived pollutants, the open ocean environments appear to be relatively unpolluted. The very small volume of dredged material, sewage sludge, and industrial wastes which is dumped into open ocean environments seems to be extensively diluted.

The ability of an estuary to flush and/or dilute disposed pollutants is measured as its relative susceptibility index. The index is estimated by measuring the nitrogen and phosphorus content in the estuaries.[6] The susceptibility index predicts nutrient enrichment problems of estuaries. The pollutants consist of both solid particles and dissolved materials. NOAA and EPA have cooperated to conduct a comprehensive susceptibility index survey of all estuaries in the United States[6] which can be used in the future to maintain or restore them.

Environmental Effects of Population Growth in Coastal Regions

Many coastal waters and estuaries exhibit a most serious degradation.[1] According to the media (e.g., CNN) in June, 1993, the population of counties adjacent to United States coastlines has risen by 43 percent during the past 30 years. Today, approximately 110 million or almost 50 percent of the nation's total population lives in coastal regions.[13] The steady increase in population along coastal areas ushers a rapid disturbance of the marine ecosystem. By the year 2010 the projected population increase of 200 percent in Florida will create a devastating environmental impact. During the past several decades, a variety of wastes, (e.g., hospital trash along the east coast), heavy metals, toxic synthetic organic chemicals, plastics, and even picnic trash like styrofoam containers have been dumped into coastal and estuarine environments.

Types of Pollutants

Environmental scientists distinguish between pollution and contamination.[6] Direct or indirect introduction of wastes into the marine environment in such a quantity that exceeds the natural ability of the recovery of the biota by removing or utilizing the harmful substances is called *pollution.* Toxic heavy metals, metals from natural ore deposits, total suspended solid (TSS) particles, oil seepage from oil deposits, organic chemicals (natural as well as synthetic), dredging wastes, sewage sludge, and radioative wastes fall are classified as pollutants. The pristine quality of marine waters is also affected by nutrients, algal blooms, oxygen deficiency, animal and plant resources, and sedimentation.[6] Pollution becomes *contamination* when the concentration of materials exceeds the natural level in any marine environment.[6]

Nutrient Enrichment

Nutrient compounds, such as phosphorus and nitrogen, may be released into the marine environment from sewage treatment plants, as well as other land discharges. An abundance of nutrients enhances or depresses biological activities. For example, blue-green algae grow excessively in nutrient rich waters, covering the entire local surface of the sea. The decomposition of blue-green algae produces an oxygen deficient marine environment. Some species of blue-green algae even generate toxic waters with a noisome smell. In decomposing organic compounds into carbon dioxide and water, bacteria also consume a considerable amount of the dissolved oxygen. The presence of excessive amounts of nutrients in seawater (eutrophication) likewise affects pH (acidity) and Eh (redox potential) of estuaries, coastal regions, and the open ocean.

Plastic Wastes

Plastics are non-degradable in the ocean environment. Although some plastics are labeled degradable, they remain not decomposed in the sea for decades,[14] perhaps forever. The plastic pollutants are hazardous to marine animals, causing suffocation and/or death by entanglement. They are also one of the main sources of the aesthetic problems causing millions of dollars of damage to the tourist industry.

Today approximately 50 billion tons of plastic wastes are produced every year. In the United States alone it has been estimated that each individual generates an average amount of 1,000 pounds of plastic garbage annually.[14] Most of the plastic in the sea, however, comes from direct disposal from ships and boats, and dumping along the shorelines. Plastics disposed into the surface waters on the land eventually reach into the sea. Having lower density than seawater, most plastic wastes remain afloat and eventually are washed onto the beach.

Oxygen Deficiency

Where pipelines discharge treated sewage into the marine environment the concentration of dissolved oxygen is so low that fish populations suffer from oxygen dificiencies. During the summer, dissolved oxygen concentrations in seawater may reach a minimum (e.g., 3 mg/liter and lower). During the winter, the concentration of the dissolved oxygen may return to normal.

Sedimentation

Land discharges originating from natural, agricultural, mining, and urban activities dump solid particles into the marine environment. Eventually, the accumulation may fill ports, marine channels, and harbors. Dredging re-releases into the seawater some of the sediments of the toxic compounds trapped in the seabed. Approximately 65 percent of dredged material in the United States is dumped within estuaries and the rest is dumped equally within the 3-mile territorial zone and the open ocean.

Depending on the geographic location, geology of the coastal region, and proximity of shorelines to the industrialized community, the dredged material shows a variable composition, generally composed of natural terrigenous sediment such as sand, gravel, silt, and clay. Such types of sediment generally contain inorganic chemicals and organic matter. They are sometimes also contaminated with heavy metals (e.g., lead, copper, chromium, cadmium, mercury, iron, nickel, and silver) and other toxic organic matter including chlorinated hydrocarbons (PCBs, DDT), polycyclic aromatic hydrocarbons (PAHs) and some petroleum products.[1] These pollutants are absorbed or closely bounded to clay particles by the process of cation capture and cation exchange in dredged material. Sand-sized particles (0.0625-2.0 mm) show less contamination. Heavy metals and organic compounds are attached to the surface of detrital particles. The finer the particles, the greater is the surface/volume ratio. The adsorption of the pollutants removes the toxic material from the environment, reducing its immediate availability to marine organisms living in the column of seawater; however, the toxic material accumulated in the sediment can affect the infauna and the epibenthonic organisms living there.[1] It is difficult to determine how much of the dredged materials is contaminated and how harmful it is to living things. According to COE's report, three percent of all dredged material in the United States coastal areas appears highly contaminated.[1] This contaminated dredged material requires special waste management. The majority of the contaminated sediment is distributed at immediate areas of direct pipeline discharges of both municipal and industrial sewage sludge in coastal regions and estuaries.

The wastes dumped into the marine environment undergo complicated physical, chemical, and biological modifications. The physical changes include dispersion and sedimentation. The chemical changes involve precipitation, solution, volatilization, buffering, and pH and redox potential modification. The biological changes include reaction of living things to the wastes by accumulation and/or dissolution of toxic wastes in the tissue of organisms.[15]

The settlement rate of dredged material dumped into the marine environment depends on the grain-size distribution of the sediment. Terrigenous material greater than silt (more than 0.0625 mm in diameter) generally descends in a shallow column of water at a rate of a few feet per second.[1] Very fine silt and clay particles remain in suspension for a longer period of time producing a water column of higher density. The clay particles, because of their surface charge, often flocculate and deposit at a somewhat more rapid rate than silt particles. Argillaceous material may spread radially from point of dump impact as a form of higly water saturated mud flow. Coarser constituents (e.g., detrital grains greater than 0.0625 mm in diameter) generally constitute more than 90 percent of the dredged material. They develop mounds of terrigenous sediment at the disposal site.[1]

The physical characteristics of dredged material depend on such factors as erosion, transportation, subaqueous currents, ocean floor topography, and the nature and the composition of the dredged material. Dredged material, especially the very fine-grained detrital grains, when dumped into a deep ocean environment (e.g., hundreds of meters deep), remains in the water column longer and spreads across a larger area without forming a noticeable mound or hill on the sea floor. In this case, there is often enough volume of dredged material in the deep ocean environment to bury the epibenthic organisms physically. In the marine environment the dredged material affects the chemistry of the water column, pelagic organisms, benthic organisms, and the seabeds. Fortunately, the infrequent dumping of dredged material, especially in small amounts, causes only short-term contamination because the damage is reversible.

Toxic Inorganic Pollutants

The concentration of toxic chemicals has been increasing in the marine environment since the onset of the Industrial Revolution. In coastal sediments, however, except for copper and mercury, such pollution has actually declined since the 1970s.[16] While mercury and copper have increased along the Hudson/Raritan Estuary, cadmium and chlordane have decreased in Long Island Sound.[6] Chemical pollutants such as cadmium, nickel, lead, methyl mercury, and chlorinated hydrocarbons are generally trapped in interstitial waters in sediments. It is difficult to discern how much bioaccumulation of the chemicals affects benthic organisms in the marine ecosystem. Metals and some organic chemicals that are tightly adsorbed in clay structures and organic matters are generally taken out of circulation in the marine environment and are not available to the living organisms. Because toxic inorganic and organic chemicals are easily taken up by plants, the food chain becomes contaminated.[17,18,19]

Metals do not generally degrade with time. Only by chemical reaction with detrital clay particles and biochemical activities can they be removed from or released into the marine water. Some metals (e.g., cadmium, nickel, copper, and zinc) are released into the seawater by the process of oxidation from the top layers or sublayers of sediments.[20] Heavy metals can also be released from the seabed into the seawater by chemical and biochemical processes associated with changes in salinity in estuaries. Heavy metals also become mobilized by the action of microorganisms. For example, in the presence of some bacteria slightly toxic inorganic mercury changes to a highly toxic organic metal in compound (e.g., methyl mercury).[21] Toxic heavy metals are also released into the marine water column by bioturbation activities of burrowing organisms.[22]

Cadmium, arsenic, lead, and mercury are four toxic metals considered to be of primary concern to human health. In both estuaries and coastal waters, their toxicity occurs in relatively high concentrations. Metals of secondary importance include chromium, copper, and selenium. Other remaining toxic metals with low concentrations in seawater have less impact on human health.[23]

Along sea floor spreading zones (e.g., Read Sea), hydrothermal vents at a depth below 4,000 meters introduce natural thermal pollution (water temperature of about 350°C.) and at the same time deploy heavy metal cations, carbon dioxide, oxygen, chlorine, and hydrogen sulfide gases, into the marine environment. The presence of these toxic metals and heat in the sea floor spreading zones makes these regions significantly dangerous environments for marine organisms.

The NS&T, while monitoring the concentration of seven trace metals from United States marine waters (Table 1), has identified abnormally high concentrations of these metals related to human activities. Most mercury pollution contributed by humans comes from burning coal. Coal contains minuscule amounts of mercury, but considering the significant volume of consumption of coal, it is estimated that more than 1,000 tons of mercury is introduced into the atmosphere annually. Natural sources, however, such as volcanoes and hot springs, produce approximately 10 times more mercury than does man.

One of the well-known paths of human exposure to toxic heavy metals in marine environments is through contaminated seafood. Swordfish take up mercury from seawater and, for unknown reasons, accumulate 10 to 100 times of that present in ocean water. Because of low concentrations of toxic metals in most seawater, direct exposure of humans to the marine environment does not cause considerable harm.[23] Some of the toxic heavy metals, however, like mercury, lead, and arsenic, have methylation capability by microorganisms that leads to high concentrations of these toxic metals in seawater. In this case, direct exposure to metal-contaminated marine water will be harmful to human health.[1] Polluted seawaters are also considered hazardous environments for divers. The contaminants, which are mostly present near the sewage outfalls, consist of microbes, toxic chemicals, radioactive wastes, and even cold or heat.[24] Toxic metals are removed from seawater by bioaccumulation and biomagnification in organisms.[25,26,27] Arsenic is mainly transformed to non-toxic forms by sea organisms.[28] Arsenic and lead generally do not accumulate in marine organisms.[23] Heavy metals adsorbed onto very fine clay particles can suffocate fish by clogging their gills.

Around the world, the toxicity of mercury and cadmium to human health ranges from lethal to sublethal, acute to chronic. Cadmium is present in a variety of wastes (e.g., wastes originating from metal electroplating industries).[23,29] In Niigata, Japan, in 1965, industrial cadmium accidentally applied to agricultural lands contaminated rice that killed 60 people.[30,31,32,33] In Japan in 1976, methyl mercury-contaminated shellfish from Minamata Bay killed 100 people and caused permanent neurological damage to 700 people.[34] Mercury can also enter human bodies and other organisms by means of inhalation, ingestion, and skin absorption.

Toxic Organic Pollutants

During 1987, 80,000 different chemicals were marketed throughout the world.[35] It is anticipated that some of these chemicals will eventually find their way into the marine environment. The bioaccumulation of these persistent chemicals in marine organisms transfers these toxic chemicals into the food chain. Each year, about 1,000 new chemicals are marketed, while 10,000 chemicals are used regularly by various industries.[36] The toxicity of organic compounds depends on the kind and nature of the host organisms, concentration of organic compounds in seawater, their bioaccumulation capacity, and the presence or absence of other chemicals in the organic system.

TABLE 1

Heavy Metals Measured in NOAA's National Status and Trends Program

Trace Metals		High Concentration μg/g (dry)
Cadmium	Cd	1.3
Chromium	Cr	230
Copper	Cu	87
Lead	Pb	87
Mercury	Hg	0.51
Silver	Ag	1.2
Zinc	Zn	280

Source: A Special NOAA 20th Anniversary Report, 1990

Toxic organic chemicals that are persistent in the environment become concentrated to a dangerous dose level. Toxic organic chemicals, like fumes and compounds from heavy metals, can be transferred into the human body directly from the water or the air, as well as through a contaminated food chain (e.g., plants and animals).[1] Coastal urbanization and industrialization, however, seem responsible for the high concentration of four synthetic organic compounds[6] (Table 2).

Since the early 1970s in the United States, the use of the most toxic organic compounds has been phased out. Total DDT (tDDT) was banned in 1970. During 1987, more than 29.4 million pounds of 35 conventionally employed agricultural pesticides have been applied in U.S. coastal watersheds.[8] These newly developed pesticides have lower environmental persistence and bioaccumulation than the older ones (e.g., DDT), but when they are disposed of into the marine environment, their impact on the marine organisms is as hazardous as the older pesticides.[8] The use of total chlordane (tCdane) first restricted in 1974 became severely restricted in 1988. Total polychlorinated biphenyls (tPCB) introduced in 1929 were gradually exploited for a number of industrial purposes (e.g., in electrical transformers and capacitors) because of their high heat capacity and low dielectric constant. By 1971, polychlorinated biphenyls (PCBs) were gradually phased out and by 1976, PCBs were banned. In spite of the controls, PCBs and PAHs are still found in land, coastal, and deep ocean sediments.[37,38,39] Because of such laws there has been a decline in such pollution along the West Coast.[40]

Since all of the banned toxic organic compounds (e.g., tDDT, tCdane, and tPCB) continue to be used without controls in other countries, they still exist in the environment[5] (Table 2) because they are highly resistant to degradation. The pesticide DDT is generally metabolized to dichloro-diphenyl-trichloroethylene (DDE) and dichloro-diphenyl-dichloroethylene (DDD) in the environment, but tDDT remains unchanged.[6]

Polycyclic aromatic hydrocarbons (PAHs) are innate organic compounds found in coal and oil and occur in both lower molecular weight and higher molecular weight.[6] The lower molecular weight compounds occur entirely naturally in petroleum, but the higher molecular weight compounds are products of human activities and are released during combustion.[6] PAHs are universal contaminants,[41] and because of their potential carcinogenic effects, their transport in the environment has been carefully studied.[42]

TABLE 2

Organic Compounds Measured in NOAA's National Status and Trend Program

	Organic Compounds	High Concentration μg/g (dry)
Total dichloro-diphenyl-trichloroethane	tDDT	40
Total chlordane	tCdane	5.5
Total polychlorinated biphenyls	tPCB	200
Total polycyclic aromatic hydrocarbons	tPAH	3900

Source: A Special NOAA 20th Anniversary Report, 1990

Biological Impact of Ocean Dumping

The presence of tumors in fish indicates pollution.[6,43] Although there is a direct relationship between contamination of environment and tumors in fish, for unknown reasons tumors have also been reported from uncontaminated regions. Older fish seem to develop a higher frequency of tumors than younger ones.[6] Different species of fish react differently to various contaminants in the marine environment, probably because of genetic make-up and inherited susceptibility.[44]

NOAA (1990) has also noticed that the biological effect of toxic contaminants in coastal marine environments is restricted to highly contaminated areas at a localized scale. Fishes dwelling at the bottom of the sea leave the dump site at the onset of contamination and may return to the site after the environment recovers to the original stable condition.[1] Some colonies of organisms are more sensitive to sediment pollution than others. Coral reef build-ups, oyster beds, and sea grasses are highly sensitive.[1]

Effects of Ocean Dumping on Human Health

The variety of health problems etiologically related to hazardous material dumped into the ocean depends on the biological characteristics and concentrations of the wastes. It is most expedient for every individual to curb such deleterious exposure to pollutants by a scrutiny of his environment, disposal techniques, and the hazardous material in food. Understanding the short term effect of toxic metals and organic compounds is somewhat easier than assessing the long term chronic effect on metabolic and biologic disorders, such as bone diseases. On May 30, 1993, the media reported health hazards related to lead sinkers used in recreational fishing. Once ingested, such contaminated waters often damage the central nervous system and may do harm to the brain and bone marrow, especially in children. Toxic metals and some synthetic organic compounds (i.e., PCBs, DDT) are considered "persistent" in the environment, while some of the volatile organic compounds, which are easily biodegradable chemicals readily removed, are considered "nonpersistent".[1]

Human pathogens enter the marine environment from a variety of sources.[1] Clay particles that adsorb viruses and bacteria protect them and increase their survival capability.[45] Some human pathogens survive better in estuaries and coastal environments than in the open sea.[46] Certain micro-organisms thrive in the sulfur-rich water found in sewage-sewer systems and produce noisome hydrogen sulfide (H_2S). Such an energy releasing reaction uses sulfur as an acceptor of hydrogen ($H_2 + S \rightarrow H_2S$ + energy). In deep marine environments, colder temperatures retard the rate of reproduction of human pathogens, but at the same time preserve viruses and bacteria that are the more important waste-borne agents of human disease in the world.

Marine waters and sediments seem to yield no human pathogens. They may be absent or only non-culturable with traditional testing methods. Pathogens become revitalized only in the presence of a host organism.[47,48,49] Moreover, most marine-generated diseases are related to exposure of marine organisms to viruses rather than bacteria.[1]

Dumping Sites in the Coastal States of the United States

As of January, 1987, OTA reports about 126 disposal sites present in the United States coastal waters. Open marine environments were almost evenly distributed along the Atlantic, Gulf of Mexico, and the Pacific coasts, with a few sites located in the South Pacific and the Caribbean Sea. Approximately 95 percent of coastal and open ocean dumping in the United States occurs at two dozen sites. 75 percent of these have a depth less than 60 feet, while only 18 sites extend over 300 feet.[1]

In 1989, OTA grouped the marine environments of the United States coastal states into five regions: Northern Atlantic, Southern Atlantic, Gulf of Mexico, West Coast, and Northern Pacific. Selected examples from these geographic areas have been chosen to demonstrate the impact of the point and nonpoint waste disposal activities on marine resources and environments.

Northern Atlantic

The *New York Bight* site is located 100 miles offshore of the coast of New York and New Jersey, one of the world's most densely populated urban and industrial centers. Open beaches and barrier islands with marshes and large bays dot the coastline of Long Island and New Jersey.[1] In this area, the Hudson River enters the Raritan Bay.

Consequently, the New York Bight receives a significant volume of wastes from both point and nonpoint sources, including industrial and municipal discharges, agricultural run-off, dredged materials, sewage sludges, and some untreated discharges. This ecosystem has been affected by a variety of pollutants. The water column, sediments, and the biota of the Bight are polluted with a variety of pathogens, heavy metals, and synthetic organic chemicals. Eutrophication and hypoxia (extremely low level of dissolved oxygen in the water) are present in several regions.[50] Sludge dumping in the Bight area was shifted to the Deepwater Municipal Sludge Site, 106 miles offshore.[1] State and local EPA and the COE monitor the site. Several public groups, such as the Coalition for the Bight and the New York Academy of Sciences, monitor their progress. Populations of organisms and bacterial contamination and accumulation of PCBs in the Bight area have been reduced because of restrictions on fishing and other marine activities.

Chesapeake Bay, the nation's largest estuary, receives waste generated by an urban area populated by approximately 13 million.[1] Nearly 200 miles long, it extends from Norfolk, Virginia, to 30 miles north of Baltimore, Maryland and holds 15 trillion gallons of water, with an average depth of slightly more than 20 feet.[51] The 64,000 square mile watershed of the Bay consists of more than 40 rivers and thousands of streams, with an areal extent of 25 times the size of the Bay.[51] Approximately one-half of the fresh water discharged into the Chesapeake Bay comes from the Susquehanna River.

The overall health of the Bay is affected by pollutants released from the James River.[52,53,54] PCBs have already reduced the dissolved oxygen and caused the population of striped bass to decline and numerous shellfish beds to close. During some summers, 15 percent of the total volume of the water in the Bay has been affected by hypoxia.[51] A large input of nutrients and suspended solid particles has decreased the submerged aquatic vegetation and increased eutrofication. These pollutants come from the municipal sewage treatment plants and from agricultural lands. Heavy metals and chemical wastes come from municipal and industrial pipelines.[1] The sediments and biota are highly contaminated with heavy metals and organic chemicals at the discharge areas of the Patapsco River (near Baltimore) and the James River.

Southern Atlantic

The *Florida Keys* coral reef is a biologically magnificent underwater world of diverse life, covering 28,000 square nautical miles south of Miami, Florida. This beautiful natural environment is now polluted by pesticides (PCBs), fertilizers, nutrients, and other wastes discharged from sewer lines at a rate of seven million gallons of sewage per day. The symptoms of environmental stress are evidenced by the drastic change in water chemistry, abundance of nutrients, and overgrowth of algae, all of which are taking over corals.

The Florida Keys is now a "Marine Sanctuary." Divers are not allowed to touch corals, collect shells, or conduct shrimp landings. Scientists believe that in order to restore the Florida Keys area to ts natural condition, we must remove the excess nutrients from the seawater and the sewage from the area.

Gulf of Mexico

The *Mississippi Sound,* located at the southern end of Mississippi, extends to the coast of Alabama. This long, shallow embayment is characterized by small bays, shallow marshes, river channels close to the land, and a series of islands. Nearby petroleum refineries, paper mills, and petrochemical plants disperse a variety of pollutants into the open coastal water. Since 1985, the Sound has been locally polluted by heavy metals and synthetic organic chemicals, a considerable amount of which is concentrated in the sediments at the point of discharge.[55] Certain parts of the Sound are contaminated with pathogens, eutrophication, and hypoxia.[56] Of all the contaminants present, chlorinated hydrocarbons are the most hazardous. Both the Bureau of Pollution and the Bureau of Marine Resources monitor such activities. The Gulf Coast Research Laboratory has conducted research in the Sound area since 1947.[1]

The *Southern Louisiana Coastline,* which has many marshes at the inward boundary and a low sloping continental shelf toward the open coastal waters, has suffered from hypoxia since 1985. The fishing industry in Louisiana has been adversely affected by extensive and cyclic hypoxia,[57,58] the exact cause of which is not known.[1] The large volume of nutrients and pollutants that flow from the Mississippi and Atchafalaya Rivers into southern Louisiana coastal waters do help to feed the organisms, but the high level of nutrients also reduces the amount of dissolved oxygen required by marine life. Even though attempts have been made by the Federal and State Pollutant Control Agency to monitor and reduce the pollutants released from municipal and industrial sources into the coastal waters of southern Louisiana, the problem of hypoxia remains.

West Coast

The *San Francisco Bay* is a vast, shallow estuary circumscribed by highly populated, densely urbanized areas such as San Francisco, Oakland, and San Jose. Industrial centers with prominent refineries take their toll on the environment. Nearby parks and wildlife reserves generate much of the nonpoint pollutants. The human impact disturbs the environment with discharges of municipal sewer sludge systems, industrial wastes, wetland filling, and diversion of freshwater inputs.[1] This environmental deterioration of the Bay area began in the early part of the twentieth century and continued into the 1960s. Since then the San Francisco Bay Regional Water Quality Control Board has monitored the pipeline discharges and the improvement of the quality of the water has reduced fish kills.[59]

Northern Pacific

Puget Sound is located in the northwest corner of the state of Washington. A large and relatively deep body of water with numerous embayments, the Sound is polluted by contaminated discharges from the highly populated urban and industrial centers, farmlands, and national parks. Commencement Bay in the Sound has been declared one of the most polluted sites in the United States.

Preventive Measures in Ocean Dumping

To control the volume of pollutants increasingly produced by residents of the coastal states all over the world, highly toxic portions of wastes should be eliminated through a treatment process prior to dumping into the ocean.

In 1987, OTA defined three municipal treatment levels required by the Clean Water Act (CWA). *Primary treatment* is characterized by the removal of approximately 30 percent of biochemical oxygen demand (BOD) factors and 60 percent of total suspended solids (TSS). The *secondary treatment* level removes both BOD and TSS to levels up to 85 percent removal. In this stage of treatment, pH of the environment ranges from 6.0 to 9.0. The *tertiary treatment* level removes 95 percent of both BOD and TSS. Tertiary treatment also requires some additional removal of the nutrients.

Untreated wastes are often released into the marine environment during a storm surge because of the limited capacity of the municipal treatment plants. These effluents, the combined sewer overflow, consist of both domestic and industrial waste waters and storm water run-off.[1]

In addition to preventive measures to protect the marine environment, it is essential to conduct research into toxic pollutants relating to marine biota, assimilative capacity and recovery potential of seawater, chemical and biological degradation, pathogenic pathways in organisms that affect human health, technology for recycling wastes, waste management and monitoring, strict national and international policing, and organizations of local volunteers interested in keeping the shoreline clean.

Monitoring Ocean Dumping

The world's oceans cover approximately 336 million square kilometers of the earth's surface. In a vast and highly dynamic marine environment, conventional monitoring by using small boats is not possible. To study solid waste particle distribution and transport within the continental shelf, a remote multi-frequency acoustic back-scattering monitoring system has been used for both sewage sludge and dredged materials.[60,61,62,63]

Satellite synoptic observations afford useful assessments of the dispersion and drift of large scale plumes of industrial wastes dumped in shelf regions.[64] The mixture of industrial wastes and seawater exhibits a reflectance peak quite different from normal unpolluted seawater. This technology facilitates tracing and studying waste plumes discharged into the sea.[59] Earth Resources Technology Satellites (ERTS-1) have been used to study sewage sludge, acid waste, suspended solid particles, and major water mass boundary features observable in the New York Bight.[65]

Considering the fact that most of ocean dumping occurs in shallow coastal and estuarine environments, most scientists believe that sediment shifting and dispersion of waste material in the marine environment can be effectively tracked. High resolution satellite and shuttle data can help to map the waste plumes in large areas. Photogrammetric and aerial photo interpretation techniques have been used successfully to monitor the dispersion patterns, diffusion coefficients, waste concentration, and nearshore currents. Air photos measure the amount of reflected light from the sea surface. A densitometer then measures the film density of images directly related to a waste dispersion parameter and other water quality parameters.[66]

The National Oceanic and Atmospheric Administration (NOAA) has supported the development of a fiber-optic chemical sensor for measuring pH, dissolved oxygen, and dissolved carbon dioxide in seawater.[67] Dissolved oxygen, very critical to the support of marine life, is monitored to evaluate the BOD status. Carbon dioxide, oxygen, and other trace gases are exchanged at the air-sea interface. The study of the gaseous constituents of seawater reveals the variation of BOD, DO, and pH in various marine environments. As technology becomes more and more sophisticated, dissolved oxygen and carbon dioxide in the marine environment will be more readily and accurately measured.

Waste Assimilation Capacity and Recovery Potential in Marine Environment

Toxic wastes and biodegradable organic wastes in seawater and seabeds can be naturally assimilated or recovered. The assimilative capacity is the reduction in the amount of waste materials in any body of water to a level that has no adverse biological effects.[68] On the other hand, recovery has been referred to as the degree of restoration of the environmental condition equal to that existing prior to an impact.[1] Assimilative capacity and recovery depend on the physical and chemical characteristics of seawater (e.g., the energy of waves, tidal currents, longshore currents, sedimentation, and adsorption rates), the recolonization and the rebound of biota that inhabit certain areas of the marine environment, and the capability of other organisms to move into and survive in that environment.

There is hope for marine environments to recover from adverse environmental impacts. When the municipal sewage treatment plants in the upper Delaware Bay estuary and in Newark Bay improved their waste treatment, the amount of dissolved oxygen increased and anoxic condition decreased.[69,70] Some organisms, however, can tolerate the severe physical onslaught of tides, waves, currents, or storms and still be able to repopulate in a short time.[71]

On the other hand, many synthetic organic chemicals cannot be assimilated by seawater, and these waters cannot recover from synthetic organic impact. Even though DDT has been banned since 1972, it still exists in some coastal sediments and fish species. Synthetic organic chemicals are highly persistent in marine environments. Their assimilative and recovery capacity exceeds or falls short of the normal limit.[1]

Future Need of Ocean Dumping

The practice of dumping wastes into the oceans may increase. As commercial traffic increases, more ports and harbors will be needed. When this increase will demand deeper and larger channels, a larger volume of material will be dredged and dumped into deeper marine waters. Construction of a large number of new ports, especially for deep-draft ships, appears to be unlikely in the near future because of the high costs and current economic constraints. Deepening of present channels seems to be the more probable solution.[1]

Ocean Dumping Laws and Regulations

The New York Harbor area has endured pollution of the sea longer than any other urban area of the United States. As early as December 30, 1675, Edmund Andros, the second British Governor of the Colony of New York, appears to have been the first administrator to issue a waste dumping regulation for New York.[72] Any violator was fined forty shillings. In spite of the 300 year existence of this regulation, both harbor water and sediments are still highly polluted.[73]

More recently, President Richard M. Nixon made the nation more aware of the dangers of pollution. In his message on April 15, 1970 to the United States Congress, he commissioned the Council on Environmental Quality (CEQ) to prepare legislative and administrative recommendations. As a result, both the Ocean Dumping Act (Public Law 92-532) and the Clean Water Act (Public Law 92-500) were enacted in 1972.

Since 1972, several national and international conventions were held to solve the problem of "ocean dumping." Forty-three nations participated in the London Convention, one of the most comprehensive. On December 19, 1972, environmental regulations were adopted. On August 30, 1975, international law banned the ocean dumping of highly toxic industrial wastes. In the United States, it is the U.S. Resource Conservation and Recovery Act that determines the toxicity of wastes disposed of in a marine environment. Some corrosive wastes, like acid or alkaline liquids, are allowed to be dumped into seawater because they become neutralized by the natural buffering capacity of marine water.

Since 1989, the MARPOL Annex V Treaty was enacted in the United States making all kinds of plastic dumping in seawater illegal. According to the MARPOL, the law enforces severe penalties to those boaters who are dumping plastics into the marine environment. When violators are caught by the U.S. Coast Guard, they will face fines as high as $50,000 and/or five years of imprisonment.[14] The U.S. Coast Guard has established a toll free hotline for people to report the violations regarding garbage dumping and oil spills in marine environments.[74]

In the 1970s other nations also enacted anti-pollution laws. Between 1976 and 1982 the London Dumping Convention (LDC) reported that an average of 300 to 400 million tons of wastes was dumped into the oceans annually by members of the LDC. Pollutant amounts by nations other than the members of LDC are not well known. The Oslo Convention covers the North Sea and the Northeastern Atlantic Ocean. The Helsinki Convention concentrated on the Baltic Sea. The Barcelona Convention discussed the Mediterranean Sea.[73]

In spite of all these laws, many nations still dump both treated and untreated sewage sludge into seawater or into rivers. Eventually much of the wastes disposed into the rivers enters the marine environment.

CONCLUSIONS

NOAA's National Estuarine Inventory (NEI), National Coastal Pollutant Discharge Inventory (NCPDI), and National Status and Trend (NS&T) Program, and U.S. Environmental Protection Agency (EPA), U.S. Congress Office of Technology Assessment (OTA), Department of Natural Resources (DNR), and Center for Marine Conservation (CMC), among other agencies and individual researchers, have made a tremendous effort to identify and analyze the quality of the marine environment and have published a huge volume of literature on this subject during the past few decades.

The published data from the U.S. and other countries around the world provides significant evidence that the world's shallow marine environment (e.g., estuaries, bays, and coastal areas) is locally polluted by municipal, agricultural, and industrial wastes since the onset of the Industrial Revolution. These pollutants originate from both point (intentional disposal source) and nonpoint (accidental release or unintentional source).

Waste dumping in the marine environment has caused physical, chemical, and biological contamination. Even though the marine environment has been carefully monitored by highly sophisticated techniques during the past few decades, the vast majority of the marine life and their habitats have been disturbed, and both the marine ecosystem and food chain are threatened in many areas around the world. Biodegradable organic wastes can become assimilated or recovered in the seawater and the seabed. On the other hand, the nondegradable organic compounds and toxic inorganic heavy metals tend to become concentrated in the tissues of the organisms and eventually transferred into the food chain.

Since 1972, several national conventions on the subject of "Ocean Dumping" came into existence, each leading to new sets of laws and regulations. Today, a voluminous body of the published environmental laws and regulations exist, but to administer these laws effectively we must educate the individual citizens in the society. For example, the laws and regulations of waste dumping into the ocean in the New York Harbor date back to December 30, 1675, but, the New York Harbor area stands as an example of disregard by the citizens.

Overpopulation, urbanization, and industrialization are considered to be the main causes of increased volume of waste disposal in the marine environment. Primary, secondary, and tertiary sewage sludge treatments have been administered to remove the toxic nondegradable and nonrecoverable contaminants from the municipal and the industrial wastes prior to their disposal into the sea.

To protect the marine environment, it is essential to conduct research into toxic pollutants and to expand and support the present recycling technology. By improving the waste treatment techniques, and by close monitoring of ocean dumping, along with the natural ability of the marine environment to assimilate and recover, there is hope that the marine environment will recover from the present adverse impact of the waste dumping in the ocean.

REFERENCES

1. U.S. Congress, Office of Technology Assessment. 1987. Waste in Marine Environments, OTA-0-334 (Springfield, VA: National Technical Information Service).
2. Resources For the Future. 1986. Pollutant Discharges to Surface Waters in Coastal Regions. Contract Prepared for U.S. Congress, Office of Technology Assessment, Washington, DC.
3. U.S. Department of Commerce, National Oceanic and Atmospheric Administration, Ocean Assessments Division. 1985-1986. Tables From National Coastal Pollutant Discharge Inventory. Tables Prepared for U.S. Congress, Office of Technology Assessment, Rockville, MD.
4. U.S. Environmental Protection Agency, Office of Water Regulations and Standards. 1985. National Water Quality Inventory, 1984 Report to Congress. EPA 440/4-85-029, Washington, DC.
5. National Academy of Sciences. 1975. Petroleum in the marine environment. National Academy of Sciences, Washington, DC.
6. O'Connor, T.P. and C.N. Ehler. 1990. Coastal Environmental Quality in the United States. Chemical Contamination in Sediments and Tissues. A Special NOAA 20th Anniversary Report, U.S. Department of Commerce, National Oceanic and Atmospheric Administration, National Ocean Service.
7. National Research Council (NRC). 1990. Managing Troubled Waters: The Role of Marine Environmental Monitoring. National Academy Press, Washington, DC.
8. National Oceanic and Atmospheric Administration (NOAA). 1992. Agricultural Pesticide Use in Coastal Areas: A National Summary. U.S. Department of Commerce.

9. Dolin, E.J. 1992. Boston Harbor's Political Waters. *Environment* 34:7-33.
10. Engineering News Record. 1993. Boston Pipe Ok for Whales. *Engineering News Record* 230:20.
11. Hecht, J. 1992. Raising a Stink in Boston. *New Scientist,* No. 1850, 136:32-36.
12. Science Applications International Corp. 1986. Overview of Sewage Sludge and Effluent Management. Contract Prepared for U.S. Congress, Office of Technology Assessment, McLean, VA.
13. Culliton, T.J., M.A. Warren, T.R. Goodspeed, D.G. Remer, C.M. Blackwell and J.J. McDonough, III. 1990. 50 Years of Population Change Along the Nation's Coasts, 1960-2010. Coastal Trends Series, Report No. 2, National Oceanic and Atmospheric Administration, Strategic Assessment Branch, Rockville, MD.
14. MIT Sea Grant. 1992. MARPOL Annex V: No Dumping Plastics in the Ocean—Plastic in the Ocean is a Killer, MIT-G-92-002.
15. Duedall, I.W., H. Ketchum, P.K. Park and D.R. Kester (Ed.). 1983. Industrial and Sewage Wastes in the Ocean, pp. 5-45. In: *Wastes in the Ocean.*
16. Long, E.R. and L.G. Morgan. 1990. The Potential for Biological Effects of Sediment-Sorbed Contaminants Tested in the National Status and Trends Program. NOAA Technical Memorandum NOS OMA 52, NOAA Office of Oceanography and Marine Assessment, Ocean Assessments Division, Seattle, WA.
17. Gambrell, R.P., R.A. Khalid and W.J. Patrick, Jr. 1978. Disposal Alternatives of Contaminated Dredged Material as a Management Tool to Minimize Adverse Environmental Effects. Technical Report DS-78-8, U.S. Army Engineer Waterways Experiment Station, Vicksburg, MS.
18. Khalid, R.A., et al. 1984. Toxic Contaminant Uptake in Dredged Sediment Marshes, pp. 500-508. In: *Dredging and Dredged Material Disposal* (Conference Proceedings). American Society of Civil Engineers, New York.
19. Rhan, W.R., Jr. 1973. The Role of Spartina Alterniflora in the Transfer of Mercury in a Salt Marsh. M.S. Thesis, Georgia Institute of Technology, Atlanta, GA.
20. Davis, W.R. 1983. Sediment-Copper Reservoir Formation by the Burrowing Polychaete *Nephtys Incisa,* pp. 173-184. In: D.R. Kester, et al. (Ed.) *Wastes in the Ocean*, Vol. 2. John Wiley & Sons, New York, NY.
21. Beier, K. and A. Jerelov. 1979. Methylation of Mercury in Aquatic Environments, pp. 203-210. In: J. Nriaqu (Ed.) *The Biochemistry of Mercury in the Environment.* Elsevier, New York.
22. Young, R.A., et al. 1985. Dispersal Pathways for Particle-Associated Pollutants. *Science* 229:431-435.
23. O'Connor, J.M. and T.J. Kneip. 1986. Human Health Impacts of Waste Constituents. New York University Medical Center, New York Institute of Environmental Medicine. Sponsor: U.S. Congress, Office of Technology Assessment, Washington, DC.
24. Maryland Sea Grant College. 1992. The Hazards of Diving in Polluted Waters. Proceedings of an International Symposium, NOAA National Undersea Research Program, Undersea and Hyperbaric Medical Society.
25. Gusman, S. and V. Huser. 1984. Mediation in the Estuary. *Coastal Zone Management Journal* 11:272-295.

26. Hirsch, N.D., L.H. DiSalvo and R. Peddicord. 1978. Effects of Dredging and Disposal on Aquatic Organisms. Technical Report DS-78-5, U.S. Army Engineer Waterways Experiment Station, Vicksburg, MS.
27. Kay, S.H. 1984. Potential for Biomagnification of Contaminants Within Marine and Freshwater Food Webs. Technical Report D-84-7, U.S. Army Engineer Waterways Experiment Station, Vicksburg, MS.
28. Doull, J., C. Klaassen and M. Amdur. 1980. *Toxicology—The Basic Science of Poisons.* MacMillan, New York.
29. Klein, L., et al. 1974. Sources of Metals in New York City's Wastewater. Water Pollution Control Association, New York, NY.
30. Nriagu, J. (Ed.). 1979. *The Biogeochemistry of Mercury in the Environment.* Elsevier, New York.
31. Reddy, C., C. Dorn, et al. 1985. Municipal Sewage Sludge Application on Ohio Farms: Tissue Metal Residues and Infections. *Environmental Research* 38:360-376.
32. Reddy, C. and C. Dorn. 1985. Municipal Sewage Sludge Application on Ohio Farms: Estimation of Cadmium Intake. *Environmental Research* 38:377-388.
33. Takizawa, Y. 1979. Epidemiology of Mercury Poisoning, pp. 325-366. In: J. Nriagu (Ed.) *The Biogeochemistry of Mercury in the Environment.* Elsevier, New York.
34. D'Itri, P. and F. D'Itri. 1977. *Mercury Contamination: A Human Tragedy.* John Wiley & Sons, New York.
35. Geyer, H.J., I. Scheunert, and F. Korte. 1987. Correlation Between the Bioconcentration Potential of Organic Environmental Chemicals in Humans and Their N-Octanol Water Partition Coefficients. *Chemosphere* 16:239.
36. Hermens, J., P. Leeuwangh and A. Husch. 1984. Quantitative Structure-Activity Relationships and Mixture Toxicity Studies of Chloro- and Alkyl Anilines at an Acute Lethal Toxicity Level to the Guppy (*Poecilia reticulata*). *Ecotoxicology and Environmental Safety* 8:388-394.
37. Burks, S.A. and R.M. Engler. 1978. Water Quality Impacts of Aquatic Dredged Material Disposal (Laboratory Investigations). Technical Report DS-78-4, U.S. Army Engineer Waterways Experiment Station, Vicksburg, MS.
38. MacLeod, W., et al. 1981. Analysis of Residual Chlorinated Hydrocarbons and Aromatic Hydrocarbons and Related Compounds in Selected Sources, Sinks, and Biota of New York Bight. NOAA Technical Memorandum, OMPA-6, National Oceanic and Atmospheric Administration, Boulder, CO.
39. Risebrough, R.W., P. Reichle, et al. 1968. Polychlorinated Biphenyls in the Global Ecosystem. *Nature* 220:1098-1102.
40. Matta, M.B., A.J. Mearns and M.F. Buchman. 1986. Trends in DDT and PCBs in U.S. West Coast Fish and Invertebrates. National Oceanic and Atmospheric Administration, National Ocean Service, Seattle, WA.
41. Mangani, R., G. Cresentini, E. Sisti, F. Bruner and S. Cannarsa. 1991. PAHs, PCBs, and Chlorinated Pesticides in Mediterranean Coastal Sediments. *Intern. J. Environ. Anal. Chem.* 45:89-100.
42. Sportsol, S., N. Gjos, R.G. Lichthentharel, K.O. Gustavsen, K. Urdal, F. Oreld and J. Skei. 1983. Source Identification of Aromatic Hydrocarbons in Sediments Using GC/MS. *Environ. Sci. Technol.* 17:282-286.

43. Susani, L. 1986. Liver Lesions in Feral Fish: A Discussion of Their Relationship to Environmental Pollutants. NOAA Technical Memorandum NOS OMA 27, NOAA Office of Oceanography and Marine Assessment, Rockville, MD.
44. Varanasi, U., J.E. Stein, M. Nishimoto, W.L. Reichert and T.K. Collier. 1987. Chemical Carcinogenesis in Feral Fish: Uptake, Activation, and Detoxification of Organic Xenobiotics. *Environ. Health Perspect.* 71:155-170.
45. Gerba, C.P., E.M. Smith and J.L. Melnick. 1977. Development of a Quantitative Method for Detecting Enteroviruses in Estuarine Sediments. *Applied Environmental Microbiology* 34:158-163.
46. Jones, G.E. 1971. The Fate of Freshwater Bacteria in the Sea. *Developments in Industrial Microbiology* 12:141-151.
47. Colwell, R.R., et al. 1985. Viable, But Non-Culturable *Vibrio cholerae* and Related Pathogens in the Environment: Implications for Release of Genetically Engineered Microorganisms. *Biotechnology* 3:817-823.
48. Grimes, D.J 1986. Human Health Impacts of Waste Constituents, Pathogens and Antibiotic- and Heavy Metal-Resistant Bacteria. Contract Prepared for U.S. Congress, Office of Technology Assessment, College Park, MD.
49. Grimes, D.J., R.W. Attwell, et al. 1986. Fate of Enteric Pathogenic Bacteria in Estuarine and Marine Environments. *Microbiological Sciences:* 324-329.
50. Boesch, D.F. and N.N. Rabalais. 1991. Effects of Hypoxia on Continental Shelf Benthos: Comparison Between the New York Bight and the Northern Gulf of Mexico. pp. 27-34. In: R.V. Tyson, and T.H. Pearson (Ed.) *Center for Environmental and Estuarine Studies,* University of Maryland System, Camabridge, MA. Geological Society, London, Special Publication 58.
51. Horton, T. 1993. Hanging in the Balance Chesapeake Bay. *National Geographic* 183:4-35.
52. Martin, S.O. 1986. Last Chance for Chesapeake Bay. *Planning Magazine* 52:12-19.
53. Sun, M. 1986. The Chesapeake Bay's Difficult Comeback. *Science* 233:715-717.
54. U.S. Environmental Protection Agency, Region III, Chesapeake Bay Program. 1983. Chesapeake Bay: A Profile of Environmental Change, PB 84119197, Philadelphia, PA.
55. Lytle, T.F. and J.S. Lytle. 1985. Pollutant Transport in Mississippi Sound. Mississippi-Alabama Sea Grant Consortium, Ocean Springs, MS.
56. Rabalais, N.N., R.E. Turner, W.J. Wiseman Jr. and D.F. Boesch. 1991. A Brief Summary of Hypoxia on the Northern Gulf of Mexico Continental Shelf: 1985-1988, pp. 35-47. In: R.V. Tyson, and T.H. Pearson (Ed.) Modern and Ancient Continental Shelf Anoxia. Louisiana Universities Marine Consortium, Chauvin, LA. Geological Society, London, Special Publication 58.
57. Rabalais, N.N., M.J. Dagg and D.F. Boesch. 1985. Oxygen Depletion on the Inner Continental Shelf of the Northern Gulf of Mexico. Biennial International Estuarine Research Conference, Durhan, NH. Louisiana University Marine Consortium, Chauvin, LA. *Estuaries* 8:44A.
58. Renaud, M.L. 1986. Hypoxia in Louisiana Coastal Waters During 1983: Implications for Fisheries. *Fishery Bulletin* 84:19-26.
59. California State Water Resources Control Board. 1984. Plan for Assessing the

Effects of Pollutants in the San Francisco Bay-Delta Estuary. Aquatic Habitat Program Publication No. 82-7SP, Sacramento, CA, 1982, reprinted October 1984.

60. Proni, J.R., F.C. Newman, D.C. Rona, D.E. Drake, G.A. Berberian, C.A. Lauter, Jr. and R.L. Sellers. 1976. On the Use of Acoustics for Studying Suspended Oceanic Sediment and for Determining the Onset of the Shallow Thermocline. *Deep-Sea Res.* 23:831-837.
61. Proni, J.R., F.C. Newman, R.L. Sellers and C. Parker. 1976. Acoustic Tracking of Ocean-Dumped Sewage Sludge. *Science* (AAAS) 193:1005-1007.
62. Orr, M.H. and F.R. Hess. 1978. Acoustic Monitoring of Industrial Chemical Waste Release at Deep Water Dump Site Log, Woods Hole Oceanographic Institution, MA, Department of Ocean Engineering. *Journal of Geophysical Research* 83:6145-6154.
63. Tsai, J.J. 1984. Acoustic Remote Sensing of Waste Disposal in the Ocean. Atlantic Oceanographic and Meteorological Laboratory (AOML-59), Miami, FL.
64. Klemas, V., G.R. Davis and R.D. Henry. 1977. Satellite and Current Drougue Studies of Ocean-Disposed Waste Drift. *Journal of the Water Pollution Control Federation* 49:757-763.
65. Wezernak, C.T. and F.J. Thomson. 1973. Monitoring of Dumping by Means of Satellite Remote Sensing. Environmental Research Institute of Michigan, Ann Arbor. *AMBIO* 2:84-86.
66. Burgess, F.J. and J.P. Wesley. 1971. Airphoto Analysis of Ocean Outfall Dispersion. Water Pollution Control Research Series. EPA-16070-ENS, Oregon State University, Corrallis, School of Engineering, Report No. W72-10146.
67. Bell, F.W. and V.R. Leeworthy. 1986. An Economic Analysis of the Importance of Saltwater Beaches in Florida. FLSGP-T-86-003, University of Florida Sea Grant Program, Gainesville, FL.
68. Goldberg, E.D. (Ed.). 1979. *Proceedings of a Workshop on Scientific Problems Relating to Ocean Pollution, Held at Estes Park, Colorado, On July 10-14, 1978.* National Oceanic and Atmospheric Administration, Environmental Research Laboratory, Boulder, CO. Sponsor: Scripps Institution of Oceanography, La Jolla, CA.
69. McCormick, J.M. et al. 1983. Partial Recovery of Newark Bay, NJ, Following Pollution Abatement. *Marine Pollution Bulletin* 14:188-197.
70. Walsh, J. 1984. Delaware Bay on the Rebound. *Science* 223:1375.
71. Florida Institute of Technology. 1986. Definition of the Oceanic, Atmospheric, and Terrestrial Media, An Overview of Resources and Processes in These Media, and A Discussion of the Concepts of Assimilitative Capacity and Resiliency. Contract Prepared for U.S. Congress, Office of Technology Assessment, Melbourne, FL.
72. Forsythe, E.B. 1977. Ocean Dumping Authorization Hearing Statement, March 9, 1977. Ninety-Fifth Congress, U.S. Government Printing Office, 94-4960.
73. Park, P.K. and T.P. O'Connor. 1978. Ocean Dumping Research: Historical and International Development, pp. 3-24. In: B.H. Ketchum, D.R. Kester, and P.K. Park (Ed.) *Ocean Dumping of Industrial Wastes.*
74. North Hills News Record. 1993. A Gannett Newspaper, Pittsburgh, Pennsylvania, May 12.

The Oceans: Physical-Chemical Dynamics and Human Impact. Edited by S.K. Majumdar, E.W. Miller, G.S. Forbes, R.F. Schmalz and Assad A. Panah.

Chapter Thirty

THE ROLE OF THE OCEAN IN RADIOACTIVE WASTE MANAGEMENT

ROBERT F. SCHMALZ
The Pennsylvania State University
University Park, PA 16802

INTRODUCTION

At the present time, world power consumption is at a level slightly greater than 10^{13} watts (Table 1), and is rising at the rate of 2-4% annually as the world population grows and as the peoples of underdeveloped nations strive to achieve living standards more nearly comparable to those enjoyed in the nations of the West. At this rate, the world wide demand for power will double in the next twenty years, and double again by 2035 A.D. The fossil fuels which supply nearly ninety percent of our energy today consume atmospheric oxygen and yield large volumes of ozone, greenhouse gases, acid precipitation and solid waste. The environmental consequences

TABLE 1
World Energy Consumption, 1988

	Total demand	Power level	Percent of total
Petroleum	127,222 petajoules*	4.03×10^{12} watts	37.6%
Natural gas	68,290 petajoules*	2.16×10^{12} watts	20.2%
Coal	101,660 petajoules*	3.22×10^{12} watts	30.0%
Nuclear	18,373 petajoules*	5.83×10^{11} watts	5.4%
Hydropower	22,493 petajoules*	7.13×10^{11} watts	6.6%
Other	—	2.14×10^{10} watts	0.2%
TOTAL	338,037 petajoules*	1.07×10^{13} watts	

*A petajoule is 10^{15} joules; a watt is one joule/second.

Source: Hammond, A.L., et al., eds., 1990.

of doubling or quadrupling their use to satisfy the anticipated future demand are probably unacceptable. Most of the environmentally benign renewable energy sources appear inadequate to satisfy more than a fraction of the projected demand (Table 2). Solar energy technology, which affords the greatest potential is far from the state of development where its contribution can be relied upon to fill our needs.

Faced with a choice between diminished world wide average *per capita* power consumption (with correspondingly reduced living standards) and a substantial increase in power generation, it is essential that we carefully weigh the costs and benefits of alternative sources of power. For many renewable power sources such as wind or hydropower, the critical factor is the practical limit of potential power. For others, such as solar conversion, the limitation imposed by current technological development is critical. For others, appropriate technologies exist and power potential is sufficient to meet our needs, but the critical factors governing their use are the human and environmental hazards they represent. Among existing technologies, it appears that nuclear fission is most likely to offer the proven efficiency and potential capacity to deliver energy at the level necessary to meet world demand in the next two decades, without unacceptable environmental damage and public risk (Lewis, 1990).

COMMERCIAL NUCLEAR POWER

The Shippingport, Pennsylvania, nuclear reactor began commercial service in 1957 amid optimistic predictions of almost unlimited, cheap electric power free of the technological hazards and environmental problems associated with burning coal or oil. In the years that followed, nuclear power enjoyed relatively rapid growth in the United States, the U.S.S.R., Europe and Japan. Despite growing public concern about radiation hazards, increasingly burdensome government regulation and escalating costs, the contribution of nuclear power rose steadily through the 1960's and 1970's. Although public response to reactor accidents at Three Mile Island and Chernobyl effectively halted all new construction in the United States, the commercial nuclear power industry continued to grow through the 1980's as reactors already contracted for were completed and began operation. Today, nuclear fission provides nearly 7% of total world power needs. In the United States, nuclear reactors supply more than 22% of our electricity and almost 7% of our total power

TABLE 2

*Estimated **Recoverable** Power from Principal Alternative Energy Sources*

Total solar energy at earth surface	10^{16} watts
Photosynthetic biomass conversion	10^{13+} watts
Available ocean heat	10^{13} watts
Available wind	10^{12} watts
Total hydropower (including tides)	10^{12} watts

Adapted fron von Arx, 1974

requirements (Tables 1, 3). Compared to any other power source, nuclear reactors enjoy a remarkable safety record. Moreover, reactors produce no greenhouse gases, ozone or acid precipitation and generate trivial volumes of solid waste compared to coal-fired plants. Realistically, however, in the absence of a power shortage of crisis proportions, additional nuclear reactors are unlikely to be welcome in any of the western nations until the public is convinced that improved reactor designs are safe and that suitable means have been devised to dispose of both high- and low-level radioactive wastes.

RADIOACTIVE WASTE

Radioactive wastes are divided arbitrarily into two categories (Dornsife, 1985): *High level radioactive waste* (HLRW) which is intensely radioactive and may have a hazardous life in excess of 500 years, and much less intensely radioactive *low-level radioactive waste* (LLRW) with a hazardous life of one hundred years or less.

High level radioactive waste produced by commercial reactors consists almost exclusively of spent fuel with its cladding and support assembly. Uranium mine and mill tailings generated in the production of uranium metal for reactor fuel are also classed as "high level" waste, as are the wastes generated by fuel reprocessing (in those countries where reprocessing is permitted). In the United States, management of all high level waste is the responsibility of the federal government. Under current federal law, spent fuel may not be reprocessed to recover unused enriched uranium and fissionable transuranic radionuclides (TRU's), but the HLRW must be disposed of in a manner which will allow recovery in the event that reprocessing is permitted at some future date. In the United States several alternative plans for high level waste management have been investigated (Bennett & McAneny, 1985; Schmalz, 1989; Majumdar & Miller, 1985; de Marsily et al., 1977), but at this time [1993] no high level disposal facility is available for commercial users, and there seems to be little probability that a disposal facility for commercially generated HLRW will be in operation before the end of the century. In the interval, commercial generators must store high level waste on site. (All military radioactive waste is managed by the government, and little information is available about prevailing management practices or the quantities of wastes involved.)

Low level reactor wastes include "housekeeping" trash (protective clothing,

TABLE 3

Sources of Electrical Power in the United States, 1991 (gigawatt-hours).

Coal	1,548,373 gw-hrs	54.8%
Nuclear	612,596	21.7
Oil	111,417	3.9
Natural gas	263,548	9.3
Hydroelectric	275,475	9.7
Other (wind, solar, geothermal, etc.)	10,163	0.3

clean-up materials, etc.), filters and exchange resins used to purify cooling water, and some irradiated components and tools. The hazardous life of most low level radioactive wastes is thirty to one hundred years, and such materials will decay to the background level of surface soil and rock within three hundred years. The safe disposal of LLRW is primarily a political rather than a technical problem. The Low-Level Waste Policy Act of 1980 (as amended in 1984) made each state responsible for disposing of non-military low level radioactive waste generated within its borders. Public opposition stemming from environmental concerns and a general fear of radioactivity (the "NIMBY" phenomenon) has slowed progress in developing the necessary facilities, and since enactment of the law, not a single new facility has been placed in operation. Of the three commercial facilities which accepted low level waste prior to 1980, only one, that at Barnwell, S.C., continues to receive waste. Pennsylvania, which has probably made as rapid progress as any state toward establishing a facility to serve the Appalachian States Compact, does not expect to have a LLRW disposal facility ready for use before 1996.

It is evident that management of reactor waste is fast becoming a critical problem. The day-to-day operation of existing commercial reactors (as well as medical treatment and diagnosis centers, industries and research laboratories which depend upon radionuclides) generates thousands of cubic feet of low level radioactive waste annually. At the present time, all of this LLRW must be packaged and shipped to South Carolina for disposal. It would appear that the risks involved in transcontinental shipment of LLRW might exceed those resulting from disposal of the waste at a number of well-designed and carefully operated regional disposal facilities. High level waste, mainly in the form of spent fuel assemblies, is held in "temporary" storage at more than one hundred reactor sites around the United States. The fuel assemblies in these storage facilities, many of which are located in populated areas, are highly radioactive and contain chemical toxins as well (Choppin & Rydberg, 1980). Until a solution to the problem of radioactive waste management is in sight, it would be unreasonable to consider adding to the radioactive waste stream by building additional power reactors. In view of the urgency of our need for additional power sources and the difficulty of locating suitable HLRW disposal facilities on the continents, however, it is logical to examine the possibility of disposing of radioactive waste in the deep sea.

DEEP SEA DISPOSAL OF RADIOACTIVE WASTE

The sea has served as a receptacle for radioactive waste for forty years or more. Prior to 1970 the United States and several European nations routinely disposed of radioactive materials in the ocean. The practice continued in Europe for several years after it was stopped in the United States. Although the volume of radioactive waste dumped at sea declined after 1967, the total radioactivity of waste dumped in the ocean increased (Council on Environmental Quality, 1970). Radioactive liquid wastes from fuel reprocessing plants in Britain ("Windscale") and France ("Centre de la Manche") were routinely discharged into the sea for at least twenty years.

Radioactive decay coupled with oceanic mixing and dilution were thought sufficient to ensure public safety. Little attention was accorded the effect of the waste materials on the bathyl environment, on marine organisms or to the possibility of bioaccumulation of radionuclides at higher trophic levels in the food chain. Solid wastes and slurries, including irradiated reactor components and fuel assemblies, were also dumped at sea at designated dump sites far from land and far from recognized fishing grounds. Where it was convenient to do so, the wastes were packed in steel drums, and simply dropped from ships or barges. Sinking was often hastened by puncturing the containers with a few blows from a fire axe or several rounds of small arms fire. Large reactor components might be dumped without any sort of packaging (Deese, 1977; Schmalz, 1984-b). Since 1970, routine ocean disposal of high level radioactive waste has been prohibited by Annex I of the International Ocean Dumping Convention (the "London Convention") and for nations subscribing to that Convention dumping low- and intermediate-level radioactive waste is strictly regulated (Lettow, 1974; USEPA, 1979).

Despite efforts to curtail the practice of dumping at sea, substantial quantities of radioactive materials are already lying on the sea bed. Recently released miliary records of the former Soviet Union indicate that at least seventeen "spent" reactors were dumped in shallow Arctic Ocean bays off the coast of Novaya Zemlya, together with other radioactive material comprising at least 250 million Curies (Marshall, 1992; Anderson, 1993). Because of the secrecy which surrounds the management of all U.S. military waste, reliable data are not available. However, two fully fueled U.S. Navy reactors are believed to have lain on the floor of the Atlantic Ocean since the sinking of nuclear submarines *Thresher* and *Scorpion* in 1963 and 1968 respectively. It is not improbable that the U.S. Navy (perhaps the army as well), like their Soviet counterparts, may have disposed of additional radioactive waste at sea. Certainly in Operation "Desert Storm" the use of artillery projectiles and small arms ammunition made from depleted uranium (which civilian regulations would designate HLRW ("mill tailings") or Class "A" LLRW) suggests a rather casual miliary approach to radioactive waste management.

We may never be sure just how much radioactive waste has already been dumped in the world oceans, although at this time it does not appear that concentrations of radionuclides have reached significant levels. Abnormal concentrations of ^{137}Cs have been observed in the North Sea, and studies of both Russian and Western European disposal sites are under way (Anderson, 1993). The results of these investigation may determine whether further radioactive waste disposal at sea is practicable. We are justified, however, in weighing the relative advantages and disadvantages of the marine environment as a waste management resource (Schmalz, 1984-b).

THE MARINE REPOSITORY

Like any large body of water, the ocean has always been treated as a convenient receptacle for waste material of all sorts. Heavy debris sinks out of sight (and out of mind) almost immediately, organic wastes are quickly consumed by the marine community, water-soluble solids dissolve and may soon be diluted to trace concentrations. Even floating debris is carried away by waves and currents to reappear as flotsam on some distant shore. Until the mid-twentieth century, recovery of cast-off material from the sea bed was a rare occurrence, and most of the material which was returned to the surface by deep sea trawls or sample collection was relatively innocuous. The occasional bottle, ballast stone or fragment from a wrecked ship was likely to be viewed as a curiosity rather than as evidence of a burgeoning problem of waste management. In the years following World War II, photographic and, later, video techniques revealed that there was coming to be a substantial quantity of trash on the sea bed - a conclusion supported by direct observation from early manned submersibles like Piccard's *Trieste* and Cousteau's diving saucers. That the disposal of wastes in the ocean had become more than a matter of casual interest was made abundantly clear by a series of incidents in the 1960's and 1970's. In Japan, Minamata disease, which killed dozens of people and crippled scores, was found to result from bioaccumulation of mercury dumped into the sea from a chemical plant, whose operators were confident that oceanic dilution would eliminate any potential hazard (Kurland, 1960; Smith & Smith, 1975). Raw sewage from coastal outfalls and floating hospital wastes from disposal sites far out in the New York bight made New Jersey and Long Island beaches unattractive and dangerous for weeks at a time. In two extreme cases, solid waste accumulating at dump sites in the Atlantic off New York and in Lake Ontario near Toronto grew to become recognized hazards to navigation (NOAA, 1979)!

Today, the characteristics which made the ocean an attractive waste repository in the past retain their appeal. Waste on the sea floor beneath 5,000-10,000 feet of water is out of sight and easily forgotten. With an areal extent in excess of 136 billion square miles, the capacity of the ocean to absorb waste appears almost boundless. The bottom fauna of the deep sea is sparse, and the bathyl ecosystem would be little disturbed by the addition of insoluble, non-toxic wastes. In shallow waters, in fact, packaged domestic wastes, crushed auto bodies, demolition debris and other forms of innocuous solid waste have been used to provide solid substrates to foster the growth of sessile "artificial reef" communities (Loder, 1975). Remote from land, and inaccessible to all but the most determined and technologically sophisticated probing, the sea floor appears to provide almost complete isolation of hazardous waste materials. Unfortunately, the advances in oceanology and marine technology which might facilitate our use of the sea bed as a waste repository serve equally to increase the likelihood of inadvertent or deliberate intrusion into the waste materials. Satellite navigation techniques which permit the precise location of a dump site serve equally well to guide an unauthorized recovery vessel. Transponder "beacons" which mark the limits of a hazardous waste dumping area would guide deliberate intruders to the site just as effectively as they might warn inadvertent intruders away. Such considerations assume significance when the hazardous waste under consideration may include, for example, spent reactor fuel elements rich in fissionable plutonium.

Scientific and commercial interest in the sea and the sea floor which grew rapidly in the years following World War II increases the probability of inadvertent intrusion upon a hazardous waste disposal site on the sea floor. Research vessels intent upon collecting sediment or benthic faunal samples, deep sea trawlers, cable laying and recovery vessels, and ships exploring for or attempting to exploit sea bed resources (manganese, petroleum, etc.) are operating today out of virtually every maritime nation. Submersibles capable of exploring all but the very deepest parts of the ocean are common. Ideally, of course, such vessels are all equipped with sophisticated navigational and SONAR instruments and would heed posted warnings and avoid charted hazards. But even the most elaborate equipment occasionally breaks down, and we have yet to eliminate all human errors. In 1871 the probability of a *Challenger* core sampler or trawl bringing up debris of human origin was negligible. Today, increased maritime activity of all kinds has raised that probability to a near certainty in many parts of the ocean. Clearly, while the sea floor remains inaccessible and remote for the general public, it has become anything but a secure repository for hazardous materials.

Finally, it is essential to consider the peculiar characteristics of the radioactive waste as distinct from ordinary biohazardous or chemo-toxic waste materials. First,

TABLE 4

Comparative Merits of Radioactive Waste Management Loci

Seabed	*vs. Continental*
Advantages:	
Extensive area available	Limited available area makes commitment to long-term exclusionary use inadvisable.
Remote, uninhabited, stable environment.	Expanding population makes isolation from human activity (direct or indirect) increasingly difficult.
Low temperature (~0°C) slows reaction rates.	Elevated temperatures, chemically active environment.
Anoxic reducing environments retard corrosion, stabilize uranium and other nuclides.	Abundant oxygen, CO_2 and organics favor chemical weathering, release of nuclides.
Inaccessible in course of ordinary human activities	Artificial (and possibly fallible) barriers required to prevent intrusion.
Groundwater contamination extremely unlikely.	Leakage jeopardizes groundwater and other resources.
Reduced economic and political cost.	High economic cost; strong political opposition.
Disadvantages:	
Long-term site maintenance virtually impossible, but need is minimal.	Long-term maintenance essential; difficult and costly, but feasible.
Waste retrieval very difficult and costly; may be impossible.	Waste retrievable in suitably designed facility.
Inaccessibility makes site design, operation, maintenance and remediation difficult.	Accessible site makes construction, operation maintenance and remedial action (if necessary) relatively simple.
Remote location makes supervision and control difficult.	Continental site can be regulated and policed.

radioactive waste, especially HLRW, generates heat. In the case of spent fuel, the heat generated may be sufficient to raise the temperature of its surroundings 100 to 300 degrees (C). A concentration of spent fuel assemblies on the sea floor would generate sufficient heat to initiate thermal convection in the overlying water column. The effects of such convective circulation on biota, transport and mixing of dissolved materials, and bottom currents, for example, would have to be weighed carefully before embarking on any large scale waste disposal program. Second, radioactive waste is inherently radioactive. Alpha and beta emissions from the waste could be effectively shielded by appropriate containers, and any escaping emission would be quickly absorbed in the seawater. Gamma radiation, however, would be more difficult to shield, and its effects on bottom fauna would have to be considered. Finally, radioactive and chemotoxic constituents in HLRW (and to a lesser extent, in LLRW) have hazardous lives measured in decades, centuries or millenia. Obviously, therefore, effective isolation of the waste requires a waste package of exceptional stability, even in the chemically aggressive marine environment. At the present time, none of these issues has been addressed in sufficient detail to make seabed disposal of high- or low-level radioactive wastes practicable even on an experimental scale.

GEOGRAPHIC AND TECTONIC CONSIDERATIONS

Widespread acceptance of the hypothesis of sea-floor spreading and plate tectonics has stimulated new concepts of hazardous waste disposal in the ocean (MacLeish, 1977; Kelly & Shea, 1982). Two proposals in particular are worthy of discussion here.

The first suggests that waste with extended hazardous life—specifically HLRW—after appropriate packaging, be emplaced on the ocean side of deep oceanic trenches which mark active zones of crustal subduction. The Peru-Chile Trench, Kurile Trench or possibly the Puerto Rico Trench might serve. The water depth in the trenches would reduce accessibility to the emplaced waste and thus increase security, but more importantly, the slow subductive movement of the oceanic plate should carry the hazardous materials downward into the upper mantle with the subducting plate. Ultimately, the waste would (presumably) undergo reaction with rocks and fluids of the lower crust or upper mantle, and be incorporated in metamorphic or magmatic rocks of the Benioff zone. It is possible, of course, that some or all of the waste might be returned to the surface in volcanic eruptions, but proponents argue that most hazardous compounds would be reduced to elements at the temperatures of metamorphism or magmatic melting. Toxic elements in the waste would be reduced to non-hazardous concentrations by magmatic dilution. The time interval involved in subduction and subsequent melting and eruption would be sufficiently long to ensure that all but the longest-lived radio nuclides would have decayed to acceptable levels prior to eruption on the surface.

Appealing as it may be, this waste subduction model appears to have a serious flaw. Studies of active subduction zones associated with continental margins suggest that most of the sedimentary blanket is not carried downward with the subducting crustal plate, but is scraped off the downward moving slab by the overriding plate. The resulting complexly folded, crushed and metamorphosed rock sequence, termed a "mélange" by Dickinson (Dickinson, 1971) is exposed along the convergence. The Franciscan series exposed along the coast of California is just such a subduction mélange. Where it is not actively forced upward by compressive stresses, the mélange sequence will be buoyed up isostatically by its low density. In either case, the oceanic sediments, together with any waste deposited on or emplaced within them, would be returned to the surface, perhaps long before the close of the hazardous life. For this reason, many geologists believe that oceanic trenches may be inappropriate sites for the disposal of hazardous wastes.

A second proposal advocates emplacement of hazardous wastes in mid-ocean, mid-plate locations. The remote mid-ocean site implies inaccessibility, and a central plate position should minimize any risk of tectonic or volcanic disturbance of the emplaced waste. Proponents suggest that wastes could be sealed into chemically inert and thermally stable cylinders. The packages would be streamlined to achieve terminal velocity in water after a free fall of 100-200 feet. The packages would then be lowered to within 100 feet of the sea bed at a selected dump site and released. The site would be chosen for a thick sediment blanket, and the waste packages should penetrate thirty feet (or more) into the sediment, effectively removing the waste from contact with the biosphere, even in the event that the package should ultimately fail. Buried in the ocean sediments, the waste package would be exposed to a stable chemical and thermal environment, altered only by the heat generated within the package itself. That heat, however, could present a serious problem if interstitial water in the sediment adjacent to the waste package should be heated above the (pressure adjusted) boiling point. Explosive eruption of HLRW packages on the sea bed could result from falling sea-level or passing internal waves if a substantial safety margin were not required. Current estimates of the rate of plate movement (1-5 cm/year) suggest that waste emplaced in a mid-plate site would not be carried into a zone of active tectonism (such as a subduction zone) within the hazardous life of any but the longest-lived alpha-emitters in HLRW, and these would be adquately shielded by overlying sediments.

CONCLUSION

Nuclear fission appears to offer the most satisfactory and environmentally acceptable means of generating power needed to meet anticipated world demands for the immediate future. Before the nuclear power industry can expand, however, it is essential that public confidence in reactor safety be established and that politically and economically realistic means be devised to dispose of high- and low-level radioactive wastes. The deep sea promises to provide the badly needed repository for nuclear waste with almost unlimited capacity. Mid-ocean, mid-plate emplacement of specially packaged waste should ensure long-term isolation with minimal risk to biosphere and hydrosphere. Packaging materials and package designs must be developed before large-scale waste disposal can be initiated, and the stability of packages and simulated waste materials must be assessed in both laboratory and field conditions. Particular attention must be directed toward the problem of dissipating heat generated in encapsulated HLRW in water-saturated sediments under confining pressures of 100-300 psi. Finally, if simulations confirm the stability of waste and waste packages in the seafloor environment, the terms of Annex I of the "London Treaty" must be modified to allow sea bed disposal of radioactive waste, and new international agreements must be negotiated to regulate disposal packages and techniques, to identify disposal sites, to establish standards for identifying disposal areas, charting them and providing interactive navigational aids to reduce the possibility of inadvertent intrusion by vessels engaged in fishing, research or sea bed resource exploration or exploitation. The process of preparing for utilization of the marine waste repository will be long and complex, and it will doubtless be opposed by some environmental organizations, but whatever its drawbacks, the marine repository may offer the most practical and secure alternative to waste disposal on the continents. The need for an effective plan for radioactive waste management is urgent, however, and it is timely to initiate serious evaluation of the marine waste repository today.

REFERENCES

Anderson, C., Ed. 1993. Random Samples: Russians Admit Reactor Dumping. *Science,* vol. 259, p. 1119.

Bennett, R.D. and C.C. McAneny. *Inspection of Low- and Intermediate-Level Radioactive Waste Disposal Facilities and Research and Development Projects in France, Sweden and West Germany, 30 April-16 May, 1984.* Geotechnical Laboratory, USAE Waterways Experiment Station, Vicksburg, MS, July, 1985. 143 pp. + 4 appendices.

Bord, R.J. 1987. Siting low-level radioactive waste in Pennsylvania: Socio-political problems and the search for solutions. Chapter 34 *in* Majumdar, S.K., F.J. Brenner and E.W. Miller, Eds., 1987, pp. 474-486.

Choppin, G.R. and J. Rydberg. *Nuclear Chemistry: Theory and Applications.* New York, N.Y., Pergamon Press, c1980. 574 pp + 14 appendices, indices.

deMarsily, G., E. Ledoux, A. Barbreau and J. Margat. 1977. Nuclear waste disposal: Can the geologist guarantee isolation? *Science,* vol. 197, pp. 519-527.

Deese, D.A. 1977. Seabed emplacement and political reality. *Oceanus,* vol. 20, pp. 47-63.

Dickinson, W.R. 1971. Plate tectonics in geologic history. *Science,* vol. 174, pp. 107-113.

Hollister, C.D. 1977. The seabed option. *Oceanus.* vol. 20, pp. 18-25.

Kates, R.W. and A.M. Weinberg. *Hazards: Technology and Fairness.* Washington, D.C., National Academy Press, c1986. 225 pp.

Kelly, J.E. and C.E. Shea. 1982. The seabed disposal program for highlevel radioactive waste: public response. *Oceanus.* vol. 25, pp. 42-53.

Kurland, L.T., S.N. Faro and H. Seidler. 1960. Minamata disease. *World Neurology,* vol. 1, pp. 370-395.

Lettow, C.F. The control of marine pollution. *in* Dolgin, E.L. and T.G.P. Guilbert, Eds. *Federal Environmental Law,* St. Paul, MN, West Publishing Co., c1974. pp. 662-667.

Lewis, H.W., *Technological Risk.* New York, Norton & Co., c1990. 338 pp. + index.

Lipschutz, R.O. *Radioactive Waste: Politics, Technology and Risk.* Cambridge, MA, Ballinger Press, c1980. 247 pp.

Loder, T.C. Effects of baled solid waste disposal in the marine environment - a descriptive model. *in* Church, T.M. (Ed.) *Marine Chemistry in the Coastal Environment.* Amer. Chem. Soc. Symposium Series, #18. Washington, D.C., American Chemical Society. c1975. 710 pp.

MacLeish, W.H. (Ed.) *Burying Faust* (Special issue dedicated to radioactive waste disposal.) *Oceanus.* vol. 20, pp. 1-67, Winter, 1977.

Majumdar, S.K. and E.W. Miller, eds. *Hazardous & Toxic Wastes: Technology, Management & Health Effects.* Phillipsburg, NJ. Typehouse of Easton, c1984-a. 405 pp. + appendix, index.

Majumdar, S.K. and E.W. Miller, eds. *Solid & Liquid Wastes: Management, Methods and Socioeconomic Considerations.* Phillipsburg, NJ. Typehouse of Easton, c1984-b. 376 pp. + 3 appendices, index.

Majumdar, S.K. and E.W. Miller, eds. *Management of Radioactive Materials & Wastes.* Phillipsburg, NJ. Typehouse of Easton, c1985. 355 pp. +2 appendices, index.

Majumdar, S.K., F.J. Brenner and E.W. Miller, eds. *Environmental Consequences of Energy Production: Problems and Prospects.* Phillipsburg, NJ. Typehouse of Easton, c1987. 522 pp. + index.

Majumdar, S.K., E.W. Miller and R.F. Schmalz, eds. *Management of Hazardous Materials & Wastes: Treatment, Minimization and Environmental Impacts.* Phillipsburg, NJ. Typehouse of Easton, c1989. 454 pp. + appendix, index.

Majumdar, S.K., G.S. Forbes, E.W. Miller and R.F. Schmalz. *Natural and Technolgical Disasters: Causes, Effects and Preventive Measures.* Phillipsburg, NJ. Typehouse of Easton, c1992. 552 pp. + index.

Marshall, E. 1992. A Scramble for Data on Arctic Radioactive Dumping. *Science.* vol. 257, pp. 608-609.

Miller, E.W. The Nuclear Energy Industry: Catastrophe or Salvation? Chapter 28 *in* Majumdar, Forbes, Miller & Schmalz, Eds., 1992. pp. 366-376.

National Oceanographic and Atmospheric Administration, (NOAA). *Report to Congress on Ocean Dumping, Monitoring and Research, January through December, 1978.* U.S. Department of Commerce, Washington, D.C. 1979.

Rasmussen, N.C. 1975. *Reactor Safety Study: An Assessment of Accident Risks in U.S. Commercial Nuclear Power Plants.* Washington, D.C., U.S. Nuclear Regulatory Commission, WASH-1400 (NUREG 75/014).

Schmalz, R.F. The Marine Repository - A Vital Waste Management Resource. Chapter 20 *in* Majumdar and Miller, Eds., 1984-b, pp. 254-272.

Smith, W.E. and A.M. Smith. *Minamata.* New York. Holt, Rinehardt & Winston, c1975.

vonArx, W.S. 1974. Energy: Natural limits and abundances. *Oceanus,* vol. XVII, pp. 2-13.

Witzig, W., K. Sahadewan and J.K. Shillen. Environmental Impacts of Nuclear Power Production. Chapter 25 *in* Majumdar, Brenner and Miller, 1987. pp. 369-385.

Subject Index

A

B

C

H

I

J

K

L

M

N

O

P

Q

R

S

T

U

V

W

Z